The Princeton Review®

D1211626

AP® PHYSICS C PREP

2023 Edition

The Staff of The Princeton Review

PrincetonReview.com

Penguin
Random
House

The Princeton Review, Inc.
110 East 42nd St, 7th Floor
New York, NY 10017

Published in the United States by Penguin Random House LLC, New York.

ISBN: 978-0-593-45086-4
ISSN: 2690-6511

The material in this book is up-to-date at the time of publication. However, changes may have been instituted by the testing body in the test after this book was published.

If there are any important late-breaking developments, changes, or corrections to the materials in this book, we will post that information online in the Student Tools. Register your book and check your Student Tools to see if there are any updates posted there.

Editor: Laura Rose
Production Editors: Emily Epstein White and Becky Radway
Production Artist: Deborah Weber
Content Contributor: Dave Mackenzie

Printed in the United States of America.

10 9 8 7 6 5 4 3 2 1

2023 Edition

The Princeton Review Publishing Team
Rob Franek, Editor-in-Chief
David Soto, Senior Director, Data Operations
Stephen Koch, Senior Manager, Data Operations
Deborah Weber, Director of Production
Jason Ullmeyer, Production Design Manager
Jennifer Chapman, Senior Production Artist
Selena Coppock, Director of Editorial
Aaron Riccio, Senior Editor
Meave Shelton, Senior Editor
Chris Chimera, Editor
Orion McBean, Editor
Patricia Murphy, Editor
Laura Rose, Editor
Alexa Schmitt Bugler, Editorial Assistant

Penguin Random House Publishing Team
Tom Russell, VP, Publisher
Alison Stoltzfus, Senior Director, Publishing
Brett Wright, Senior Editor
Emily Hoffman, Assistant Managing Editor
Ellen Reed, Production Manager
Suzanne Lee, Designer
Eugenia Lo, Publishing Assistant

For customer service, please contact **editorialsupport@review.com**, and be sure to include:

- full title of the book
- ISBN
- page number

Acknowledgments

The Princeton Review would like to thank Dave Mackenzie for his valuable contribution to the 2023 edition of this book.

Our gratitude also goes out to Deborah Weber, Emily Epstein White, and Becky Radway for their careful attention to each page.

Contents

Get More (Free) Content
at **PrincetonReview.com/prep**

As easy as 1·2·3

1 Go to PrincetonReview.com/prep or scan the **QR code** and enter the following ISBN for your book: **9780593450864**

2 Answer a few simple questions to set up an exclusive Princeton Review account. *(If you already have one, you can just log in.)*

3 Enjoy access to your **FREE** content!

Once you've registered, you can...

- Get our take on any recent or pending updates to the AP Physics C Exams

- Access comprehensive study guides and a variety of printable resources, including bubble sheets for the practice tests in the book as well as important equations and formulas

- Take a full-length practice SAT and ACT

- Get valuable advice about the college application process, including tips for writing a great essay and where to apply for financial aid

- If you're still choosing between colleges, use our searchable rankings of *The Best 388 Colleges* to find out more information about your dream school

- Check to see if there have been any corrections or updates to this edition

Need to report a potential **content** issue?

Contact **EditorialSupport@review.com** and include:

- full title of the book
- ISBN
- page number

Need to report a **technical** issue?

Contact **TPRStudentTech@review.com** and provide:

- your full name
- email address used to register the book
- full book title and ISBN
- Operating system (Mac/PC) and browser (Chrome, Firefox, Safari, etc.)

Look For These Icons Throughout The Book

 ONLINE ARTICLES

 PROVEN TECHNIQUES

 STUDY BREAK

 OTHER REFERENCES

 ASK YOURSELF

Part I
Using This Book to Improve Your AP Score

- Preview: Your Knowledge, Your Expectations
- Your Guide to Using This Book
- How to Begin

PREVIEW: YOUR KNOWLEDGE, YOUR EXPECTATIONS

Your route to a high score on the AP Physics C Exams will depend on how you plan to use this book. Please respond to the following questions.

1. Rate your level of confidence about your knowledge of the content tested by the AP Physics C Exams.

    A. Very confident—I know it all
    B. I'm pretty confident, but there are topics for which I could use help
    C. Not confident—I need quite a bit of support
    D. I'm not sure

2. If you have a goal score in mind, circle your goal score for the AP Physics C Exams.

    5   4   3   2   1   I'm not sure yet

3. What do you expect to learn from this book? Circle all that apply to you.

    A. A general overview of the test and what to expect
    B. Strategies for how to approach the test
    C. The content tested by these exams
    D. I'm not sure yet

Course Changes

For information on the College Board updates to the AP Physics C courses, visit your Student Tools (see Get More (Free) Content on the previous pages for details).

YOUR GUIDE TO USING THIS BOOK

This book is organized to provide as much—or as little—support as you need, so you can use this book in whatever way will be most helpful for improving your score on the AP Physics C Exams.

*   The remainder of **Part I** will provide guidance on how to use this book and help you determine your strengths and weaknesses.

*   **Part II** of this book contains your first set of practice tests for Mechanics and Electricity & Magnetism, the Diagnostic Answer Keys, and answers and explanations for each question. (Bubble sheets can be found in the very back of the book for easy tear-out.) This is where you should begin your test preparation in order to realistically determine:
    o   your starting point right now
    o   which question types you're ready for and which you might need to practice
    o   which content topics you are familiar with and which you will want to carefully review

Once you have nailed down your strengths and weaknesses with regard to this exam, you can focus your test preparation, build a study plan, and be efficient with your time. Our Diagnostic Answer Key will assist you with this process.

- **Part III** of this book will:
  o inform you about the structure, scoring, and content of the AP Physics C Exams
  o help you make a study plan
  o point you toward additional resources

- **Part IV** of this book will explore various strategies including:
  o how to attack multiple-choice questions
  o how to write high-scoring free-response answers
  o how to manage your time to maximize the number of points available to you

For more guidance on managing your time, access your online Student Tools to find some sample study guides.

- **Part V** of this book covers the content you need to know for your exams.

- **Part VI** of this book contains your second set of practice tests for Mechanics and for Electricity & Magnetism, another diagnostic answer key, and explanations. (Bubble sheets can be found in the very back of the book for easy tear-out.) If you skipped Practice Test 1, we recommend that you take both tests (with at least a day or two between them) so that you can compare your progress between them. Additionally, this will help to identify any external issues: If you get a certain type of question wrong both times, you probably need to review it. If you only got it wrong once, you may have run out of time or been distracted by something. In either case, this will allow you to focus on the factors that caused the discrepancy and to be as prepared as possible on the day of the tests.

You may choose to use some parts of this book over others, or you may work through the entire book. This will depend on your needs and how much time you have. Let's now look at how to make this determination.

HOW TO BEGIN

1. **Take Practice Test 1**

   Before you can decide how to use this book, you need to take a practice test. Doing so will give you insight into your strengths and weaknesses, and the test will also help you make an effective study plan. If you're feeling test-phobic, remind yourself that a practice test is a tool for diagnosing yourself—it's not how well you do that matters but how you use information gleaned from your performance to guide your preparation.

   So, before you read further, take Practice Test 1 starting on page 7 of this book. If you plan to take both the AP Physics C: Mechanics and AP Physics C: Electricity & Magnetism exams, then take both tests included in Practice Test 1. If you're planning to just take one of those exams, then only take that test in Practice Test 1. Be sure to do so in one sitting, following the instructions that appear before the test. If you're taking both exams, make sure to use a separate bubble sheet for each.

2. **Check Your Answers**

Using the Diagnostic Answer Key on pages 40 and 61, follow our three-step process to determine how you did on each section of the test.

3. **Reflect on the Test**

After you take your first test(s), respond to the following questions:

- How much time did you spend on the multiple-choice questions?

- How much time did you spend on each free-response question?

- How many multiple-choice questions did you miss?

- Do you feel you had the knowledge to address the subject matter of the free-response questions?

- Do you feel you wrote well-organized, thoughtful free responses?

4. **Read Part III of this Book and Complete the Self-Evaluation**

Part III will provide information on how the test is structured and scored. It will also set out areas of content that are tested.

As you read Part III, re-evaluate your answers to the questions above. At the end of Part III, you will revisit the questions on the previous page and refine your answers to them. Use the diagnostic answer keys to identify the content chapters of this book in which you missed the most questions, as that may present you with a good place to begin your review. Make a study plan, based on your needs and time available, that will help you to use this book most effectively.

5. **Engage with Parts IV and V as Needed**

For a great study tool, go online to your Student Tools and download the Reflection worksheet! You can fill it out as you finish strategy and content chapters to help you assess how to best move forward in your study.

Notice the word *engage*. You'll get more out of this book if you use it intentionally than if you read it passively, hoping for an improved score through osmosis.

Strategy chapters will help you think about your approach to the question types on this exam. Content chapters are designed to provide a review of the content tested on the AP Physics C Exams, including the level of detail you need to know and how the content is tested. You will have the opportunity to assess your mastery of the content of each chapter through test-appropriate questions.

6.   Take Practice Test 2 and Assess Your Performance

Once you feel you have developed the strategies you need and gained the knowledge you lacked, you should take Practice Test 2. You should do so in one sitting, following the instructions at the beginning of the test.

When you are done, check your answers to the multiple-choice sections. See if a teacher will read your essays and provide feedback.

Once you have taken the test, reflect on what areas you still need to work on, and revisit the chapters in this book that address those deficiencies. Through this type of reflection and engagement, you will continue to improve.

7.   Keep Working

After you have revisited certain chapters in this book, continue the process of testing, reflecting, and engaging with the next practice test in this book. Consider what additional work you need to do and how you will change your strategic approach to different parts of the test. You can continue to explore areas that can stand to be improved and engage in those areas right up to the day of the test. As we will discuss in Part III, there are other resources available to you, including a wealth of information for AP students.

The Princeton Review writes tons of books to guide you through test preparation and college admissions. If you're thinking about college, check out our wildly popular book *The Best 388 Colleges* and visit our website PrincetonReview.com for tons of college rankings and ratings.

Part II
Practice Test 1

Practice Test 1

- Mechanics
- Electricity & Magnetism

AP® Physics C: Mechanics Exam

SECTION I: Multiple-Choice Questions

DO NOT OPEN THIS BOOKLET UNTIL YOU ARE TOLD TO DO SO.

At a Glance

Total Time
45 minutes
Number of Questions
35
Percent of Total Grade
50%
Writing Instrument
Pen required

Instructions

Section I of this examination contains 35 multiple-choice questions. Fill in only the ovals for numbers 1 through 35 on your answer sheet.

CALCULATORS MAY BE USED IN BOTH SECTIONS OF THE EXAMINATION.

Indicate all of your answers to the multiple-choice questions on the answer sheet. No credit will be given for anything written in this exam booklet, but you may use the booklet for notes or scratch work. After you have decided which of the suggested answers is best, completely fill in the corresponding oval on the answer sheet. Give only one answer to each question. If you change an answer, be sure that the previous mark is erased completely. Here is a sample question and answer.

Sample Question   Sample Answer

Chicago is a   Ⓐ ● Ⓒ Ⓓ Ⓔ
(A) state
(B) city
(C) country
(D) continent
(E) planet

Use your time effectively, working as quickly as you can without losing accuracy. Do not spend too much time on any one question. Go on to other questions and come back to the ones you have not answered if you have time. It is not expected that everyone will know the answers to all the multiple-choice questions.

About Guessing

Many candidates wonder whether or not to guess the answers to questions about which they are not certain. Multiple-choice scores are based on the number of questions answered correctly. Points are not deducted for incorrect answers, and no points are awarded for unanswered questions. Because points are not deducted for incorrect answers, you are encouraged to answer all multiple-choice questions. On any questions you do not know the answer to, you should eliminate as many choices as you can, and then select the best answer among the remaining choices.

GO ON TO THE NEXT PAGE.

ADVANCED PLACEMENT PHYSICS C TABLE OF INFORMATION

CONSTANTS AND CONVERSION FACTORS

Proton mass,	$m_p = 1.67 \times 10^{-27}$ kg	Electron charge magnitude,	$e = 1.60 \times 10^{-19}$ C
Neutron mass,	$m_n = 1.67 \times 10^{-27}$ kg	1 electron volt,	1 eV $= 1.60 \times 10^{-19}$ J
Electron mass,	$m_e = 9.11 \times 10^{-31}$ kg	Speed of light,	$c = 3.00 \times 10^8$ m/s
Avogadro's number,	$N_A = 6.02 \times 10^{23}$ mol^{-1}	Universal gravitational constant,	$G = 6.67 \times 10^{-11}$ (N•m^2)/kg^2
Universal gas constant,	$R = 8.31$ J/(mol K)	Acceleration due to gravity	$g = 9.8$ m/s^2
Boltzmann's constant,	$k_B = 1.38 \times 10^{-23}$ J/K	at Earth's surface,	

1 unified atomic mass unit,	1 u $= 1.66 \times 10^{-27}$ kg $= 931$ MeV/c^2
Planck's constant,	$h = 6.63 \times 10^{-34}$ J•s $= 4.14 \times 10^{-15}$ eV•s
	$hc = 1.99 \times 10^{-25}$ J•m $= 1.24 \times 10^3$ eV•nm
Vacuum permittivity,	$\varepsilon_0 = 8.85 \times 10^{-12}$ C^2/(N•m^2)
Coulomb's law constant,	$k = 1/(4\pi\varepsilon_0) = 9.0 \times 10^9$ (N•m^2)/C^2
Vacuum permeability,	$\mu_0 = 4\pi \times 10^{-7}$ (T•m)/A
Magnetic constant,	$k' = \mu_0/(4\pi) = 1 \times 10^{-7}$ (T•m)/A
1 atmosphere pressure,	1 atm $= 1.0 \times 10^5$ N/m$^2 = 1.0 \times 10^5$ Pa

UNIT SYMBOLS							
meter,	m	mole,	mol	watt,	W	farad,	F
kilogram,	kg	hertz,	Hz	coulomb,	C	tesla,	T
second,	s	newton,	N	volt,	V	degree Celsius,	°C
ampere,	A	pascal,	Pa	ohm,	Ω	electron volt,	eV
kelvin,	K	joule,	J	henry,	H		

VALUES OF TRIGONOMETRIC FUNCTIONS FOR COMMON ANGLES

θ	0°	30°	37°	45°	53°	60°	90°
$\sin\theta$	0	1/2	3/5	$\sqrt{2}/2$	4/5	$\sqrt{3}/2$	1
$\cos\theta$	1	$\sqrt{3}/2$	4/5	$\sqrt{2}/2$	3/5	1/2	0
$\tan\theta$	0	$\sqrt{3}/3$	3/4	1	4/3	$\sqrt{3}$	∞

PREFIXES		
Factor	Prefix	Symbol
10^9	giga	G
10^6	mega	M
10^3	kilo	k
10^{-2}	centi	c
10^{-3}	milli	m
10^{-6}	micro	μ
10^{-9}	nano	n
10^{-12}	pico	p

The following assumptions are used in this exam.

I. The frame of reference of any problem is inertial unless otherwise stated.

II. The direction of current is the direction in which positive charges would drift.

III. The electric potential is zero at an infinite distance from an isolated point charge.

IV. All batteries and meters are ideal unless otherwise stated.

V. Edge effects for the electric field of a parallel-plate capacitor are negligible unless otherwise stated.

GO ON TO THE NEXT PAGE.

MECHANICS			ELECTRICITY & MAGNETISM	

MECHANICS

$v_x = v_{x0} + a_x t$

$x = x_0 + v_{x0}t + \frac{1}{2}a_x t^2$

$v_x^2 = v_{x0}^2 + 2a_x(x - x_0)$

$\vec{a} = \frac{\Sigma \vec{F}}{m} = \frac{\vec{F}_{net}}{m}$

$\vec{F} = \frac{d\vec{p}}{dt}$

$\vec{J} = \int \vec{F}\,dt = \Delta \vec{p}$

$\vec{p} = m\vec{v}$

$\left|\vec{F}_f\right| \le \mu \left|\vec{F}_N\right|$

$\Delta E = W = \int \vec{F}\cdot d\vec{r}$

$K = \frac{1}{2}mv^2$

$P = \frac{dE}{dt}$

$P = \vec{F}\cdot\vec{v}$

$\Delta U_g = mg\Delta h$

$a_c = \frac{v^2}{r} = \omega^2 r$

$\vec{\tau} = \vec{r}\times\vec{F}$

$\vec{\alpha} = \frac{\Sigma\vec{\tau}}{I} = \frac{\vec{\tau}_{net}}{I}$

$I = \int r^2\,dm = \Sigma mr^2$

$x_{cm} = \frac{\Sigma m_i x_i}{\Sigma m_i}$

$v = r\omega$

$\vec{L} = \vec{r}\times\vec{p} = I\vec{\omega}$

$K = \frac{1}{2}I\omega^2$

$\omega = \omega_0 + \alpha t$

$\theta = \theta_0 + \omega_0 t + \frac{1}{2}\alpha t^2$

$\vec{F}_S = -k\Delta\vec{x}$

$U_S = \frac{1}{2}k(\Delta x)^2$

$x = x_{max}\cos(\omega t + \phi)$

$T = \frac{2\pi}{\omega} = \frac{1}{f}$

$T_S = 2\pi\sqrt{\frac{m}{k}}$

$T_P = 2\pi\sqrt{\frac{\ell}{g}}$

$\left|\vec{F}_G\right| = \frac{Gm_1 m_2}{r^2}$

$U_G = \frac{Gm_1 m_2}{r}$

a = acceleration
E = energy
F = force
f = frequency
h = height
I = rotational inertia
J = impulse
K = kinetic energy
k = spring constant
ℓ = length
L = angular momentum
m = mass
P = power
p = momentum
r = radius or distance
T = period
t = time
U = potential energy
v = velocity or speed
W = work done on a system
x = position
μ = coefficient of friction
θ = angle
τ = torque
ω = angular speed
α = angular acceleration
ϕ = phase angle

ELECTRICITY & MAGNETISM

$\left|\vec{F}_E\right| = \frac{1}{4\pi\varepsilon_0}\left|\frac{q_1 q_2}{r^2}\right|$

$\vec{E} = \frac{\vec{F}_E}{q}$

$\oint\vec{E}\cdot d\vec{A} = \frac{Q}{\varepsilon_0}$

$E_x = -\frac{dV}{dx}$

$\Delta V = -\int\vec{E}\cdot d\vec{r}$

$V = \frac{1}{4\pi\varepsilon_0}\sum_i \frac{q_i}{r_i}$

$U_E = qV = \frac{1}{4\pi\varepsilon_0}\frac{q_1 q_2}{r}$

$\Delta V = \frac{Q}{C}$

$C = \frac{\kappa\varepsilon_0 A}{d}$

$C_p = \sum_i C_i$

$\frac{1}{C_s} = \sum_i \frac{1}{C_i}$

$I = \frac{dQ}{dt}$

$U_C = \frac{1}{2}Q\Delta V = \frac{1}{2}C(\Delta V)^2$

$R = \frac{\rho\ell}{A}$

$\vec{E} = \rho\vec{J}$

$I = Nev_d A$

$I = \frac{\Delta V}{R}$

$R_s = \sum_i R_i$

$\frac{1}{R_p} = \sum_i \frac{1}{R_i}$

$P = I\Delta V$

A = area
B = magnetic field
C = capacitance
d = distance
E = electric field
ε = emf
F = force
I = current
J = current density
L = inductance
ℓ = length
n = number of loops of wire per unit length
N = number of charge carriers per unit volume
P = power
Q = charge
q = point charge
R = resistance
r = radius or distance
t = time
U = potential or stored energy
V = electric potential
v = velocity or speed
ρ = resistivity
Φ = flux
κ = dielectric constant

$\vec{F}_M = q\vec{v}\times\vec{B}$

$\oint\vec{B}\cdot d\vec{\ell} = \mu_0 I$

$d\vec{B} = \frac{\mu_0}{4\pi}\frac{I d\vec{\ell}\times\vec{r}}{r^2}$

$\vec{F} = \int I d\vec{\ell}\times\vec{B}$

$B_S = \mu_0 n I$

$\Phi_B = \int\vec{B}\cdot d\vec{A}$

$\varepsilon = \oint\vec{E}\cdot d\vec{\ell} = -\frac{d\Phi_B}{dt}$

$\varepsilon = -L\frac{dI}{dt}$

$U_L = \frac{1}{2}LI^2$

GO ON TO THE NEXT PAGE.

GEOMETRY AND TRIGONOMETRY		CALCULUS

GEOMETRY AND TRIGONOMETRY

Rectangle

$A = bh$

Triangle

$A = \frac{1}{2}bh$

Circle

$A = \pi r^2$

$C = 2\pi r$

$s = r\theta$

Rectangular Solid

$V = \ell wh$

Cylinder

$V = \pi r^2 \ell$

$S = 2\pi r\ell + 2\pi r^2$

Sphere

$V = \frac{4}{3}\pi r^3$

$S = 4\pi r^2$

Right Triangle

$a^2 + b^2 = c^2$

$\sin\theta = \dfrac{a}{c}$

$\cos\theta = \dfrac{b}{c}$

$\tan\theta = \dfrac{a}{b}$

A = area
C = circumference
V = volume
S = surface area
b = base
h = height
ℓ = length
w = width
r = radius
s = arc length
θ = angle

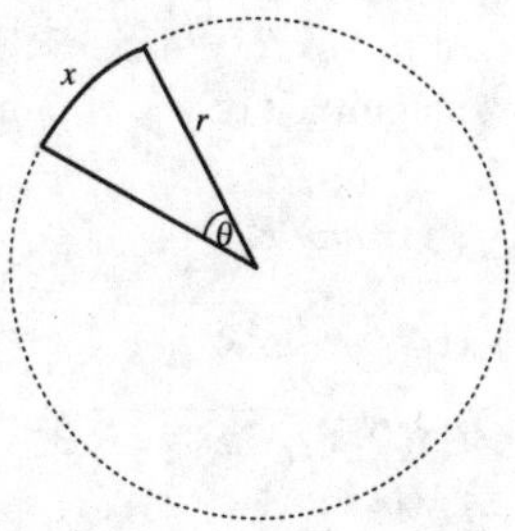

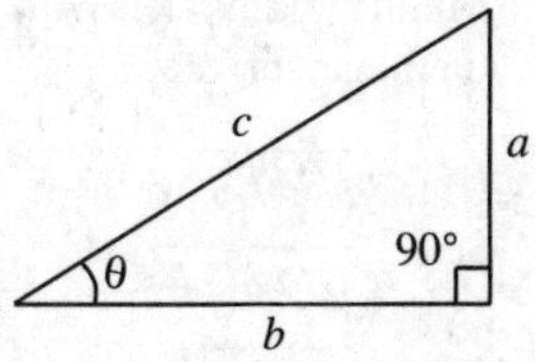

CALCULUS

$$\frac{df}{dx} = \frac{df}{du}\frac{du}{dx}$$

$$\frac{d}{dx}(x^n) = nx^{n-1}$$

$$\frac{d}{dx}(e^{ax}) = ae^{ax}$$

$$\frac{d}{dx}(\ln ax) = \frac{1}{x}$$

$$\frac{d}{dx}[\sin(ax)] = a\cos(ax)$$

$$\frac{d}{dx}[\cos(ax)] = -a\sin(ax)$$

$$\int x^n\, dx = \frac{1}{n+1}x^{n+1}, n \neq -1$$

$$\int e^{ax}\, dx = \frac{1}{a}e^{ax}$$

$$\int \frac{dx}{x+a} = \ln|x+a|$$

$$\int \cos(ax)\, dx = \frac{1}{a}\sin(ax)$$

$$\int \sin(ax)\, dx = -\frac{1}{a}\cos(ax)$$

VECTOR PRODUCTS

$$\vec{A} \cdot \vec{B} = AB\cos\theta$$

$$|\vec{A} \times \vec{B}| = AB\sin\theta$$

GO ON TO THE NEXT PAGE.

PHYSICS C: MECHANICS
SECTION I
Time—45 minutes
35 Questions

Directions: Each of the questions or incomplete statements below is followed by five suggested answers or completions. Select the one that is best in each case and then mark it on your answer sheet.

1. A rock is dropped off a cliff and falls the first half of the distance to the ground in t_1 seconds. If it falls the second half of the distance in t_2 seconds, what is the value of t_2/t_1? (Ignore air resistance.)

   (A) $1/(2\sqrt{2})$

   (B) $1/\sqrt{2}$

   (C) $1/2$

   (D) $1-(1/\sqrt{2})$

   (E) $\sqrt{2}-1$

2. A bubble starting at the bottom of a soda bottle experiences constant acceleration, a, as it rises to the top of the bottle in some time, t. How much farther does it travel in the last second of its journey than in the first second? Assume that the journey takes longer than 2 seconds.

   (A) $a(t+1\text{ s})^2$
   (B) $a(t-1\text{ s})^2$
   (C) at^2
   (D) $a(t+1\text{ s})(1\text{ s})$
   (E) $a(t-1\text{ s})(1\text{ s})$

3. An object initially at rest experiences a time-varying acceleration given by $a = (2\text{ m/s}^3)t$ for $t \geq 0$. How far does the object travel in the first 3 seconds?

   (A) 9 m
   (B) 12 m
   (C) 18 m
   (D) 24 m
   (E) 27 m

4. What is the benefit of raising an object using an inclined plane instead of simply lifting the object? Assume ideal conditions.

   (A) The amount of force needed to move the object is reduced.
   (B) The amount of time needed to move the object is reduced.
   (C) The distance the object must be moved is reduced.
   (D) The amount of work needed to move the object is reduced.
   (E) The power that must be exerted will be reduced.

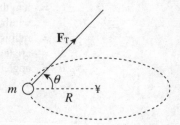

5. In the figure shown, a tension force $\mathbf{F}_T$ causes a particle of mass m to move with constant angular speed ω in a horizontal circular path (in a plane perpendicular to the page) of radius R. Which of the following expressions gives the magnitude of $\mathbf{F}_T$? (Ignore air resistance.)

   (A) $m\omega^2 R$

   (B) $m\sqrt{\omega^4 R^2 - g^2}$

   (C) $m\sqrt{\omega^4 R^2 + g^2}$

   (D) $m(\omega^2 R - g)$

   (E) $m(\omega^2 R + g)$

6. An object (mass = m) above the surface of the Moon (mass = M) is dropped from an altitude h equal to the Moon's radius (R). With what speed will the object strike the lunar surface?

   (A) $\sqrt{GM/R}$

   (B) $\sqrt{GM/(2R)}$

   (C) $\sqrt{2GM/R}$

   (D) $\sqrt{2GMm/R}$

   (E) $\sqrt{GMm/(2R)}$

GO ON TO THE NEXT PAGE.

7. Two pendulums are constructed in such a way that they are identical except that one has a hanging mass of m and the other has a hanging mass of $2m$. Both hanging masses are set into motion such that each system has the same total mechanical energy. Assume that the motion approximates simple harmonic motion. Comparatively, the system with the lesser mass will have greater

   (A) maximum speed and maximum angle of displacement
   (B) frequency and maximum angle of displacement
   (C) period and maximum speed
   (D) frequency only
   (E) maximum angle of displacement only

8. A uniform cylinder of mass m and radius r unrolls without slipping from two strings tied to a vertical support. If the rotational inertia of the cylinder is $\frac{1}{2}mr^2$, find the acceleration of its center of mass.

   (A) $\frac{1}{4}g$

   (B) $\frac{1}{2}g$

   (C) $\frac{1}{3}g$

   (D) $\frac{2}{3}g$

   (E) $\frac{3}{4}g$

9. A space shuttle is launched from Earth. As it travels up, it moves at a constant velocity of 150 m/s straight up. If its engines provide 1.5×10^8 W of power, what is the shuttle's mass? You may assume that the shuttle's mass and the acceleration due to gravity are constant.

   (A) 6.7×10^2 kg
   (B) 1.0×10^5 kg
   (C) 6.7×10^5 kg
   (D) 1.0×10^6 kg
   (E) 2.3×10^6 kg

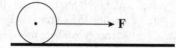

10. A uniform cylinder, initially at rest on a frictionless, horizontal surface, is pulled by a constant force **F** from time $t = 0$ to time $t = T$. From time $t = T$ on, this force is removed. Which of the following graphs best illustrates the speed, v, of the cylinder's center of mass from $t = 0$ to $t = 2T$?

(A)
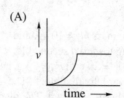

(B)
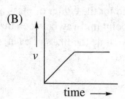

(C)
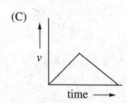

(D)
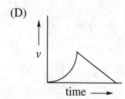

(E)
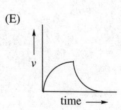

GO ON TO THE NEXT PAGE.

11. A satellite is in circular orbit around Earth. If the work required to lift the satellite to its orbital height is equal to the satellite's kinetic energy while in this orbit, how high above the surface of Earth (radius = R) is the satellite?

    (A) $\frac{1}{2}R$
    (B) $\frac{2}{3}R$
    (C) R
    (D) $\frac{3}{2}R$
    (E) $2R$

12. A block of length $\ell_1 = 10$ cm, mass $m_1 = 10$ kg, and $v_1 = 2$ m/s is currently moving toward a second block. The second block has length $\ell_2 = 20$ cm, $m_2 = 2$ kg, and $v_2 = 8$ m/s, and is moving toward the first block. The blocks' closest edges are currently 1 m apart. The blocks will experience a perfectly inelastic collision. How long after the collision does it take for the center of mass of the new object to cross the point midway between the blocks' starting positions? Assume the mass of each block is uniformly distributed.

    (A) 0.225 s
    (B) 0.325 s
    (C) 0.525 s
    (D) 0.900 s
    (E) 0.975 s

13. A rubber ball (mass = 0.08 kg) is dropped from a height of 3.2 m and, after bouncing off the floor, rises almost to its original height. If the impact time with the floor is measured to be 0.04 s, what average force did the floor exert on the ball?

    (A) 0.16 N
    (B) 16 N
    (C) 32 N
    (D) 36 N
    (E) 64 N

14. A disk of radius 0.1 m initially at rest undergoes an angular acceleration of 2.0 rad/s². If the disk only rotates, find the total distance traveled by a point on the rim of the disk in 4.0 s.

    (A) 0.4 m
    (B) 0.8 m
    (C) 1.2 m
    (D) 1.6 m
    (E) 2.0 m

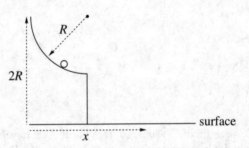

15. In the figure above, a small object slides down a frictionless quarter-circular slide of radius R. If the object starts from rest at a height equal to $2R$ above a horizontal surface, find its horizontal displacement, x, at the moment it strikes the surface.

    (A) $2R$
    (B) $\frac{5}{2}R$
    (C) $3R$
    (D) $\frac{7}{2}R$
    (E) $4R$

GO ON TO THE NEXT PAGE.

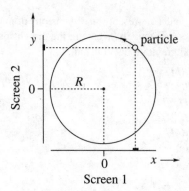

16. The figure above shows a particle executing uniform circular motion in a circle of radius R. Light sources (not shown) cause shadows of the particle to be projected onto two mutually perpendicular screens. The positive directions for x and y along the screens are denoted by the arrows. When the shadow on Screen 1 is at position $x = -(0.5)R$ and moving in the $+x$ direction, what is true about the position and velocity of the shadow on Screen 2 at that same instant?

(A) $y = -(0.866)R$; velocity in $-y$ direction
(B) $y = -(0.866)R$; velocity in $+y$ direction
(C) $y = -(0.5)R$; velocity in $-y$ direction
(D) $y = +(0.866)R$; velocity in $-y$ direction
(E) $y = +(0.866)R$; velocity in $+y$ direction

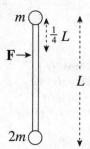

17. The figure above shows a view from above of two objects attached to the end of a rigid massless rod at rest on a frictionless table. When a force $\mathbf{F}$ is applied as shown, the resulting rotational acceleration of the rod about its center of mass is $kF/(mL)$. What is k?

(A) $\frac{3}{8}$

(B) $\frac{1}{2}$

(C) $\frac{5}{8}$

(D) $\frac{3}{4}$

(E) $\frac{5}{6}$

18. Old cars were made with rigid frames that could retain their shape in a collision. Modern cars are made with frames that crumple in a collision. If each type of vehicle were to crash into a wall and come to a complete stop, which of the following would NOT be true regarding those collisions? Assume the vehicles were of equal mass and traveling at the same speed before the collision.

(A) The modern vehicle would experience less force.
(B) The modern vehicle would experience less impulse.
(C) The modern vehicle's magnitude of acceleration would be less.
(D) The modern vehicle's collision would take less time.
(E) More than one of the above would not be true.

19. A homogeneous bar is lying on a flat table. Besides the gravitational and normal forces (which cancel), the bar is acted upon by exactly two other external forces, $\mathbf{F}_1$ and $\mathbf{F}_2$, which are parallel to the surface of the table. If the net force on the rod is zero, which one of the following is also true?

(A) The net torque on the bar must also be zero.
(B) The bar cannot accelerate translationally or rotationally.
(C) The bar can accelerate translationally if $\mathbf{F}_1$ and $\mathbf{F}_2$ are not applied at the same point.
(D) The net torque will be zero if $\mathbf{F}_1$ and $\mathbf{F}_2$ are applied at the same point.
(E) None of the above

20. An astronaut lands on a planet whose mass and radius are each twice that of Earth. If the astronaut weighs 800 N on Earth, how much will he weigh on this planet?

(A) 200 N
(B) 400 N
(C) 800 N
(D) 1,600 N
(E) 3,200 N

GO ON TO THE NEXT PAGE.

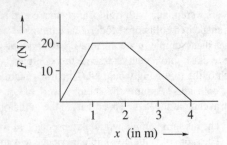

21. A particle of mass $m = 1.0$ kg is acted upon by a variable force, $F(x)$, whose strength is given by the graph above. If the particle's speed was zero at $x = 0$, what is its speed at $x = 4$ m?

    (A) 5.0 m/s
    (B) 8.7 m/s
    (C) 10 m/s
    (D) 14 m/s
    (E) 20 m/s

22. The radius of a collapsing spinning star (assumed to be a uniform sphere with a constant mass) decreases to $\frac{1}{16}$ of its initial value. What is the ratio of the final rotational kinetic energy to the initial rotational kinetic energy?

    (A) 4
    (B) 16
    (C) 16^2
    (D) 16^3
    (E) 16^4

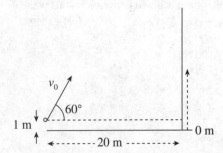

23. A ball is projected with an initial velocity of magnitude $v_0 = 40$ m/s toward a vertical wall as shown in the figure above. How long does the ball take to reach the wall?

    (A) 0.25 s
    (B) 0.6 s
    (C) 1.0 s
    (D) 2.0 s
    (E) 3.0 s

24. If M, L, and T represent the dimensions of mass, length, and time, respectively, what are the dimensions of the universal gravitational constant, G?

    (A) $L^2/(MT^2)$
    (B) $L^3/(MT^2)$
    (C) $(MT^2)/L^3$
    (D) $(ML^3)/T^2$
    (E) $L^3/(MT)$

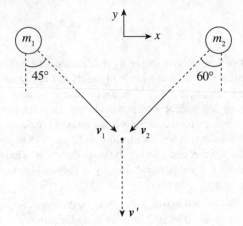

25. The figure shown above is a view from above of two clay balls moving toward each other on a frictionless surface. They collide perfectly inelastically at the indicated point and are observed to then move in the direction indicated by the post-collision velocity vector, v'. If $m_1 = 2m_2$, and v' is parallel to the negative y-axis, what is v_2?

    (A) $v_1(\sin 45°)/(2\sin 60°)$
    (B) $v_1(\cos 45°)/(2\cos 60°)$
    (C) $v_1(2\cos 45°)/(\cos 60°)$
    (D) $v_1(2\sin 45°)/(\sin 60°)$
    (E) $v_1(\cos 45°)/(2\sin 60°)$

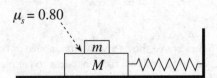

26. In the figure above, the coefficient of static friction between the two blocks is 0.80. If the blocks oscillate with a frequency of 2.0 Hz, what is the maximum amplitude of the oscillations if the small block is not to slip on the large block?

    (A) 3.1 cm
    (B) 5.0 cm
    (C) 6.2 cm
    (D) 7.5 cm
    (E) 9.4 cm

GO ON TO THE NEXT PAGE.

27. When two objects collide, the ratio of the relative speed after the collision to the relative speed before the collision is called the *coefficient of restitution*, *e*. If a ball is dropped from height H_1 onto a stationary floor, and the ball rebounds to height H_2, what is the coefficient of restitution of the collision?

    (A) H_1/H_2

    (B) H_2/H_1

    (C) $\sqrt{H_1/H_2}$

    (D) $\sqrt{H_2/H_1}$

    (E) $(H_1/H_2)^2$

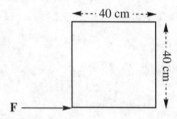

28. The figure above shows a square metal plate of side length 40 cm and uniform density, lying flat on a table. A force **F** of magnitude 10 N is applied at one of the corners, as shown. Determine the torque produced by **F** relative to the center of rotation.

    (A) 0 N·m
    (B) 1.0 N·m
    (C) 1.4 N·m
    (D) 2.0 N·m
    (E) 4.0 N·m

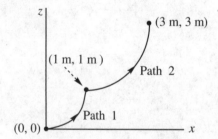

29. A small block of mass $m = 2.0$ kg is pushed from the initial point $(x_i, z_i) = (0$ m, 0 m) upward to the final point $(x_f, z_f) = (3$ m, 3 m) along the path indicated. Path 1 is a portion of the parabola $z = x^2$, and Path 2 is a quarter circle whose equation is $(x - 1)^2 + (z - 3)^2 = 4$. How much work is done by gravity during this displacement?

    (A) −60 J
    (B) −80 J
    (C) −90 J
    (D) −100 J
    (E) −120 J

30. A horizontal spring of spring constant k is experiencing simple harmonic motion between points $x = -A$ and $x = A$ with a block of mass M attached to the end. A bullet of mass m is fired from a gun with speed v so that it collides perfectly inelastically with the block when it is at position $x = -A$. How far beyond $x = -A$ will the spring be compressed as a result of this collision?

    (A) $mv\sqrt{kM}$

    (B) $mv\sqrt{k(M+m)}$

    (C) $mv\sqrt{\dfrac{1}{k(M+m)}}$

    (D) $\sqrt{\dfrac{mv}{k(M+m)}}$

    (E) $mv\sqrt{\dfrac{1}{kM}}$

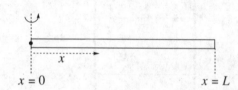

31. The rod shown above can pivot about the point $x = 0$ and rotates in a plane perpendicular to the page. Its linear density, λ, increases with x such that $\lambda(x) = kx$, where k is a positive constant. Determine the rod's moment of inertia in terms of its length, L, and its total mass, M.

    (A) $\dfrac{1}{6} ML^2$

    (B) $\dfrac{1}{4} ML^2$

    (C) $\dfrac{1}{3} ML^2$

    (D) $\dfrac{1}{2} ML^2$

    (E) $2ML^2$

GO ON TO THE NEXT PAGE.

32. A particle is subjected to a conservative force whose potential energy function is

$$U(x) = (x - 2)^3 - 12x$$

where U is given in joules when x is measured in meters. Which of the following represents a position of stable equilibrium?

(A) $x = -4$
(B) $x = -2$
(C) $x = 0$
(D) $x = 2$
(E) $x = 4$

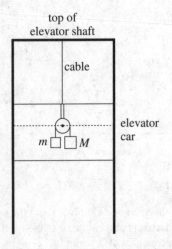

33. A light, frictionless pulley is suspended from a rigid rod attached to the roof of an elevator car. Two masses, m and M (with $M > m$), are suspended on either side of the pulley by a light, inextensible cord. The elevator car is descending at a constant velocity. Determine the acceleration of the masses.

(A) $(M - m)g$

(B) $(M + m)g$

(C) $\dfrac{M + m}{M - m}g$

(D) $\dfrac{M - m}{M + m}g$

(E) $(M - m)(M + m)g$

34. A block of mass $m = 2$ kg starts at rest at the top of a ramp of angle θ_r such that $\sin \theta_r = 0.45$ and $\cos \theta_r = 0.9$. The length of the ramp consists of alternating stretches of frictionless portions followed by portions with a coefficient of friction $\mu = \dfrac{2}{3}$. The first meter of the ramp's length is frictionless, followed by 1 m of friction. The next meter is frictionless again, followed by 2 m of friction. This alternating pattern continues with the frictionless stretch always being 1 m and the friction stretch increasing in length by 1 m each time. How far does the block travel before coming to a stop? Assume the ramp has infinite length.

(A) 4 m
(B) 8 m
(C) 12 m
(D) 19 m
(E) 24 m

35. An object of mass 2 kg is acted upon by three external forces, each of magnitude 4 N. Which of the following could NOT be the resulting acceleration of the object?

(A) 0 m/s²
(B) 2 m/s²
(C) 4 m/s²
(D) 6 m/s²
(E) 8 m/s²

STOP

END OF SECTION I, MECHANICS

PHYSICS C: MECHANICS
SECTION II
Time—45 minutes
3 Questions

Directions: Answer all three questions. The suggested time is about 15 minutes per question for answering each of the questions, which are worth 15 points each. The parts within a question may not have equal weight.

1. An ideal projectile is launched from the ground at an angle θ to the horizontal, with an initial speed of v_0. The ground is flat and level everywhere. Write all answers in terms of v_0, θ, and fundamental constants.

   (a) Calculate the time the object is in the air.

   (b) Calculate the maximum height the object reaches.

   (c) What is the *net* vertical displacement of the object?

   (d) Calculate the range (horizontal displacement) of the object.

   (e) What should θ be so that the projectile's range is equal to its maximum vertical displacement?

GO ON TO THE NEXT PAGE.

2. A narrow tunnel is drilled through Earth (mass = M, radius = R), connecting points P and Q, as shown in the diagram on the left below. The perpendicular distance from Earth's center, C, to the tunnel is x. A package (mass = m) is dropped from point P into the tunnel; its distance from P is denoted y and its distance from C is denoted r. See the diagram on the right.

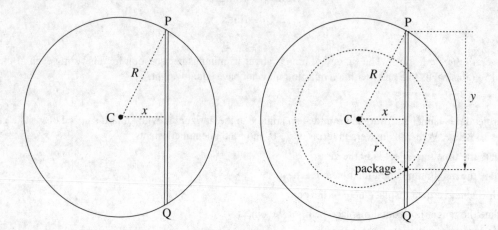

(a) Assuming that Earth is a homogeneous sphere, the gravitational force F on the package is due to m and the mass contained within the sphere of radius $r < R$. Use this fact to show that

$$F = -\frac{GMm}{R^3} r$$

(b) Use the equation $F(r) = -dU/dr$ to find an expression for the change in gravitational potential energy of the package as it moves from point P to a point where its distance from Earth's center is r. Write your answer in terms of G, M, m, R, and r.

(c) Apply Conservation of Energy to determine the speed of the package in terms of G, M, R, x, and y. (Ignore friction.)

(d)

   i. At what point in the tunnel—that is, for what value of y—will the speed of the package be maximized?

   ii. What is this maximum speed? (Write your answer in terms of G, M, R, and x.)

GO ON TO THE NEXT PAGE.

3. The diagram below is a view from above of three sticky hockey pucks on a frictionless horizontal surface. The pucks with masses m and $2m$ are connected by a massless, rigid rod of length L and are initially at rest. The puck of mass $3m$ is moving with velocity v directly toward puck m. When puck $3m$ strikes puck m, the collision is perfectly inelastic.

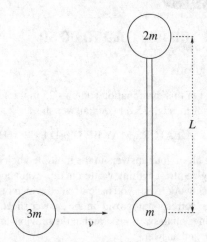

(a) Immediately after the collision,

   i.   where is the center of mass of the system?

   ii. what is the speed of the center of mass? (Write your answer in terms of v.)

   iii. what is the angular speed of the system? (Write your answer in terms of v and L.)

(b) What fraction of the system's initial kinetic energy is lost as a result of the collision?

STOP

END OF SECTION II, MECHANICS

AP® Physics C: Electricity & Magnetism Exam

SECTION I: Multiple-Choice Questions

DO NOT OPEN THIS BOOKLET UNTIL YOU ARE TOLD TO DO SO.

At a Glance

Total Time
45 minutes
Number of Questions
35
Percent of Total Grade
50%
Writing Instrument
Pen required

Instructions

Section I of this examination contains 35 multiple-choice questions. Fill in only the ovals for numbers 1 through 35 on your answer sheet.

CALCULATORS MAY BE USED IN BOTH SECTIONS OF THE EXAMINATION.

Indicate all of your answers to the multiple-choice questions on the answer sheet. No credit will be given for anything written in this exam booklet, but you may use the booklet for notes or scratch work. After you have decided which of the suggested answers is best, completely fill in the corresponding oval on the answer sheet. Give only one answer to each question. If you change an answer, be sure that the previous mark is erased completely. Here is a sample question and answer.

Sample Question   Sample Answer

Chicago is a   Ⓐ ● Ⓒ Ⓓ Ⓔ
(A) state
(B) city
(C) country
(D) continent
(E) planet

Use your time effectively, working as quickly as you can without losing accuracy. Do not spend too much time on any one question. Go on to other questions and come back to the ones you have not answered if you have time. It is not expected that everyone will know the answers to all the multiple-choice questions.

About Guessing

Many candidates wonder whether or not to guess the answers to questions about which they are not certain. Multiple-choice scores are based on the number of questions answered correctly. Points are not deducted for incorrect answers, and no points are awarded for unanswered questions. Because points are not deducted for incorrect answers, you are encouraged to answer all multiple-choice questions. On any questions you do not know the answer to, you should eliminate as many choices as you can, and then select the best answer among the remaining choices.

GO ON TO THE NEXT PAGE.

This page intentionally left blank.

GO ON TO THE NEXT PAGE.

PHYSICS C: ELECTRICITY & MAGNETISM
SECTION I
Time—45 minutes
35 Questions

Directions: Each of the questions or incomplete statements below is followed by five suggested answers or completions. Select the one that is best in each case and mark it on your answer sheet.

1. A nonconducting sphere is given a nonzero net electric charge, $+Q$, and then brought close to a neutral conducting sphere of the same radius. Which of the following will be true?

   (A) An electric field will be induced within the conducting sphere.
   (B) The conducting sphere will develop a net electric charge of $-Q$.
   (C) The spheres will experience an electrostatic attraction.
   (D) The spheres will experience an electrostatic repulsion.
   (E) The spheres will experience no electrostatic interaction.

2. If the total resistance of a circuit with fixed voltage, V, were doubled, the power dissipated by that same circuit would

   (A) increase by a factor of 4
   (B) increase by a factor of 2
   (C) decrease by a factor of 2
   (D) decrease by a factor of 4
   (E) Cannot be determined

3. Each of the following ionized isotopes is projected with the same speed into a uniform magnetic field **B** such that the isotope's initial velocity is perpendicular to **B**. Which combination of mass and charge would result in a circular path with the largest radius?

   (A) $m = 16$ u, $q = -5\,e$
   (B) $m = 17$ u, $q = -4\,e$
   (C) $m = 18$ u, $q = -3\,e$
   (D) $m = 19$ u, $q = -2\,e$
   (E) $m = 20$ u, $q = -1\,e$

GO ON TO THE NEXT PAGE.

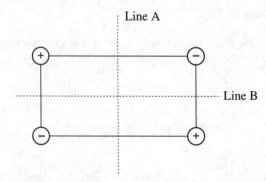

Line A

Line B

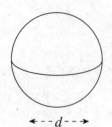

A ———

B ————————

C **————**

D **——————**

4. The picture above shows 4 charges fixed in position at the corners of a rectangle measuring 2 cm by 4 cm. Assuming the charges are all of equal magnitude, how many locations on either Line A or Line B would be places with 0 net electric field?

(A) 1
(B) 5
(C) All of Line A
(D) All of Line B
(E) All of both Line A and Line B

5. The four wires shown above are each made of aluminum. Which wire will have the greatest resistance?

(A) Wire A
(B) Wire B
(C) Wire C
(D) Wire D
(E) All the wires have the same resistance because they're all composed of the same material.

6. Which of the following is NOT equal to one tesla?

(A) $1 \ J/(A \cdot m^2)$
(B) $1 \ kg/(C \cdot s)$
(C) $1 \ N/(A \cdot m)$
(D) $1 \ V \cdot s/m^2$
(E) $1 \ A \cdot N/V$

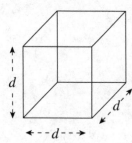

7. The figure above shows two Gaussian surfaces: a cube with side length d and a sphere with diameter d. The net electric charge enclosed within each surface is the same, $+Q$. If Φ_C denotes the total electric flux through the cubical surface, and Φ_S denotes the total electric flux through the spherical surface, then which of the following is true?

(A) $\Phi_C = (\pi/6)\Phi_S$
(B) $\Phi_C = (\pi/3)\Phi_S$
(C) $\Phi_C = \Phi_S$
(D) $\Phi_C = (3/\pi)\Phi_S$
(E) $\Phi_C = (6/\pi)\Phi_S$

GO ON TO THE NEXT PAGE.

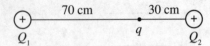

8. The picture above shows two positive charges, Q_1 and Q_2, of equal magnitude that are fixed in place. If a third positive charge, q, were released from rest at the position shown, at what position would it first return to rest? Assume ideal conditions and that Q_1 occupies the position of $x = 0$ cm and Q_2 occupies the position of $x = 1$ cm.

(A) $x = 30$ cm
(B) $x = 50$ cm
(C) $x = 70$ cm
(D) Cannot be determined without knowing the magnitude of q
(E) It never returns to rest.

9. An object carries a charge of –1 C. How many excess electrons does it contain?

(A) 6.25×10^{18}
(B) 8.00×10^{18}
(C) 1.60×10^{19}
(D) 3.20×10^{19}
(E) 6.25×10^{19}

Questions 10–11

Each of the resistors shown in the circuit below has a resistance of 200 Ω. The emf of the ideal battery is 24 V.

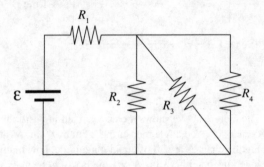

10. How much current is supported by the voltage source?

(A) 30 mA
(B) 48 mA
(C) 64 mA
(D) 72 mA
(E) 90 mA

11. What is the ratio of the power dissipated by R_1 to the power dissipated by R_4?

(A) 1/9
(B) 1/4
(C) 1
(D) 4
(E) 9

GO ON TO THE NEXT PAGE.

12. What is the value of the following product?

$$20 \ \mu F \times 500 \ \Omega$$

(A) 0.01 henry
(B) 0.01 ampere per coulomb
(C) 0.01 weber
(D) 0.01 second
(E) 0.01 volt per ampere

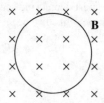

13. A copper wire in the shape of a circle of radius 1 m, lying in the plane of the page, is immersed in a magnetic field, **B**, that points into the plane of the page. The strength of **B** varies with time, t, according to the equation

$$B(t) = 2t(1 - t)$$

where B is given in teslas when t is measured in seconds. What is the magnitude of the induced electric field in the wire at time $t = 1$ s?

(A) $(1/\pi)$ N/C
(B) 1 N/C
(C) 2 N/C
(D) π N/C
(E) 2π N/C

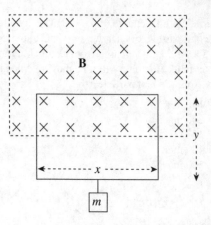

14. In the figure above, the top half of a rectangular loop of wire, x meters by y meters, hangs vertically in a uniform magnetic field, **B**. Describe the magnitude and direction of the current in the loop necessary for the magnetic force to balance the weight of the mass m supported by the loop.

(A) $I = mg/xB$, clockwise

(B) $I = mg/xB$, counterclockwise

(C) $I = mg \Big/ \left(x + \dfrac{1}{2} y \right) B$, clockwise

(D) $I = mg \Big/ \left(x + \dfrac{1}{2} y \right) B$, counterclockwise

(E) $I = mg/(x + y)B$, clockwise

GO ON TO THE NEXT PAGE.

15. A particle with a charge of $q_1 = 10$ nC is held in place. A second particle of charge $q_2 = -5$ nC and mass $m = 5 \times 10^{-10}$ kg is released from rest 2 cm away from the first particle. How fast will it be moving when it is 1 cm away from the first particle?

    (A) 100 m/s
    (B) 200 m/s
    (C) 300 m/s
    (D) 400 m/s
    (E) 500 m/s

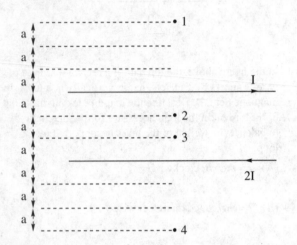

16. The figure above shows a pair of long, straight current-carrying wires and four marked points. At which of these points is the net magnetic field zero?

    (A) Point 1 only
    (B) Points 1 and 2 only
    (C) Point 2 only
    (D) Points 3 and 4 only
    (E) Point 3 only

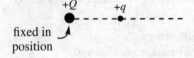

fixed in position

17. The figure above shows two positively charged particles. The $+Q$ charge is fixed in position, and the $+q$ charge is brought close to $+Q$ and released from rest. Which of the following graphs best depicts the acceleration of the $+q$ charge as a function of its distance r from $+Q$?

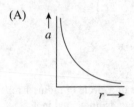

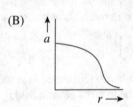

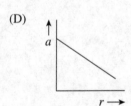

GO ON TO THE NEXT PAGE.

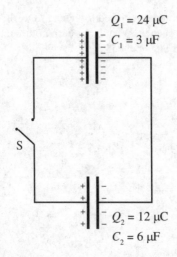

18. Once the switch S in the figure above is closed and electrostatic equilibrium is regained, how much charge will be stored on the positive plate of the 6 μF capacitor?

    (A) 9 μC
    (B) 18 μC
    (C) 24 μC
    (D) 27 μC
    (E) 36 μC

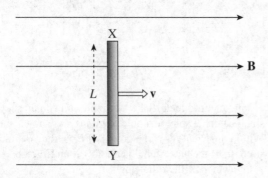

19. A metal bar of length L is pulled with velocity $\mathbf{v}$ through a uniform magnetic field, $\mathbf{B}$, as shown above. What is the voltage produced between the ends of the bar?

    (A) vB, with point X at a higher potential than point Y
    (B) vB, with point Y at a higher potential than point X
    (C) vBL, with point X at a higher potential than point Y
    (D) vBL, with point Y at a higher potential than point X
    (E) None of the above

20. An electric dipole consists of a pair of equal but opposite point charges of magnitude 4.0 nC separated by a distance of 2.0 cm. What is the electric field strength at the point midway between the charges?

    (A) 0
    (B) 9.0×10^4 V/m
    (C) 1.8×10^5 V/m
    (D) 3.6×10^5 V/m
    (E) 7.2×10^5 V/m

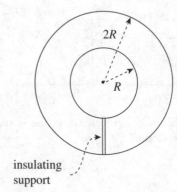

insulating support

21. The figure above shows a cross section of two concentric spherical metal shells of radii R and $2R$, respectively. Find the capacitance.

    (A) $1/(8\pi\varepsilon_0 R)$
    (B) $1/(4\pi\varepsilon_0 R)$
    (C) $2\pi\varepsilon_0 R$
    (D) $4\pi\varepsilon_0 R$
    (E) $8\pi\varepsilon_0 R$

22. Traveling at an initial speed of 1.5×10^6 m/s, a proton enters a region of constant magnetic field, $\mathbf{B}$, of magnitude 1.0 T. If the proton's initial velocity vector makes an angle of 30° with the direction of $\mathbf{B}$, compute the proton's speed 4 s after entering the magnetic field.

    (A) 5.0×10^5 m/s
    (B) 7.5×10^5 m/s
    (C) 1.5×10^6 m/s
    (D) 3.0×10^6 m/s
    (E) 6.0×10^6 m/s

GO ON TO THE NEXT PAGE.

Questions 23–25

There is initially no current through any circuit element in the following diagram.

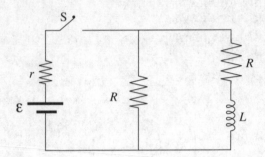

23. What is the current through r immediately after the switch S is closed?

   (A) 0

   (B) $\dfrac{\varepsilon}{r+R}$

   (C) $\dfrac{\varepsilon}{r+2R}$

   (D) $\dfrac{\varepsilon(r+R)}{rR}$

   (E) $\dfrac{\varepsilon(2R)}{2Rr+2R}$

24. After the switch has been kept closed for a long time, how much energy is stored in the inductor?

   (A) $\dfrac{L\varepsilon^2}{2(r+R)^2}$

   (B) $\dfrac{L\varepsilon^2}{2(2r+R)^2}$

   (C) $\dfrac{L\varepsilon^2}{4(2r+R)^2}$

   (D) $\dfrac{L(\varepsilon R)^2}{8(2r+R)^2}$

   (E) $\dfrac{L\varepsilon^2}{8(2r+R)^2}$

25. After having been closed for a long time, the switch is suddenly opened. What is the current through r immediately after S is opened?

   (A) 0

   (B) $\dfrac{\varepsilon}{r+R}$

   (C) $\dfrac{\varepsilon}{r+2R}$

   (D) $\dfrac{\varepsilon(r+R)}{rR}$

   (E) $\dfrac{\varepsilon(2R)}{r(2R)+2R}$

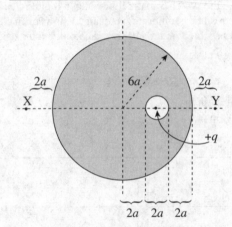

26. A solid, neutral metal sphere of radius $6a$ contains a small cavity, a spherical hole of radius a as shown above. Within this cavity is a charge, $+q$. If E_X and E_Y denote the strength of the electric field at points X and Y, respectively, which of the following is true?

   (A) $E_Y = 4E_X$
   (B) $E_Y = 16E_X$
   (C) $E_Y = E_X$
   (D) $E_Y = (11/5)E_X$
   (E) $E_Y = (11/5)^2 E_X$

GO ON TO THE NEXT PAGE.

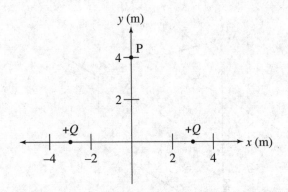

27. Two particles of charge $+Q$ are located on the x-axis, as shown above. Determine the work done by the electric field to move a particle of charge $-Q$ from very far away to point P.

(A) $\dfrac{2kQ}{5}$

(B) $\dfrac{2kQ^2}{5}$

(C) $-\dfrac{2kQ^2}{5}$

(D) $\dfrac{kQ^2}{5}$

(E) $-\dfrac{3kQ^2}{5}$

28. A battery is connected in a series with a switch, a resistor of resistance R, and an inductor of inductance L. Initially, there is no current in the circuit. Once the switch is closed and the circuit is completed, how long will it take for the current to reach 99% of its maximum value?

(A) $(\ln \dfrac{99}{100})RL$

(B) $(\ln 99)RL$

(C) $(\ln \dfrac{1}{100})\dfrac{L}{R}$

(D) $\dfrac{L}{R}(\ln \dfrac{100}{99})$

(E) $(\ln 100)\dfrac{L}{R}$

29. What is the maximum number of 40 W light bulbs that could be connected in parallel with a 120 V source? The total current cannot exceed 5 A or the circuit will blow a fuse.

(A) 3
(B) 6
(C) 9
(D) 12
(E) 15

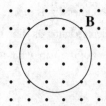

30. The metal loop of wire shown above is situated in a magnetic field **B** pointing out of the plane of the page. If **B** decreases uniformly in strength, the induced electric current within the loop is

(A) clockwise and decreasing
(B) clockwise and increasing
(C) counterclockwise and decreasing
(D) counterclockwise and constant
(E) counterclockwise and increasing

GO ON TO THE NEXT PAGE.

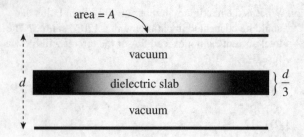

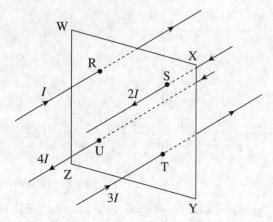

31. A dielectric of thickness $\dfrac{d}{3}$ is placed between the plates of a parallel-plate capacitor, as shown above. If K is the dielectric constant of the slab, what is the capacitance?

(A) $\dfrac{\varepsilon_0 A(2+3K)}{d}$

(B) $\dfrac{d}{\varepsilon_0 A(2+3K)}$

(C) $\dfrac{3\varepsilon_0 A}{d(2K+1)}$

(D) $\dfrac{3K\varepsilon_0 A}{d(2K+1)}$

(E) $\dfrac{3K\varepsilon_0 A}{d}$

33. The figure above shows four current-carrying wires passing perpendicularly through the interior of a square whose vertices are W, X, Y, and Z. The points where the wires pierce the plane of the square (namely, R, S, T, and U) themselves form the vertices of a square each side of which has half the length of each side of WXYZ. If the currents are as labeled in the figure, what is the absolute value of

$$\oint \mathbf{B} \cdot d\ell$$

where the integral is taken around WXYZ?

(A) $\dfrac{1}{2}\mu_0 I$

(B) $\mu_0 I$

(C) $\sqrt{2}\,\mu_0 I$

(D) $2\mu_0 I$

(E) $5\mu_0 I$

 $+Q$ $-Q$

32. Consider the two source charges shown above. At how many points in the plane of the page, in a region around these charges, are both the electric field and the electric potential equal to zero?

(A) 0
(B) 1
(C) 2
(D) 3
(E) 4

GO ON TO THE NEXT PAGE.

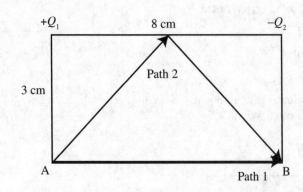

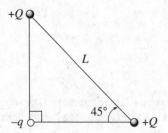

35. Two point charges, each $+Q$, are fixed a distance L apart. A particle of charge $-q$ and mass m is placed as shown in the figure above. What is this particle's initial acceleration when released from rest?

34. A particle with charge $+q$ is moved from point A to point B along Path 1 in the picture above and experiences a change in potential energy given by ΔU. What would the change in potential energy be if the charge had the opposite sign and moved along Path 2 from point A to point B instead of along Path 1?

(A) $-\left(\dfrac{10}{8}\right)\Delta U$

(B) $-\Delta U$

(C) ΔU

(D) $\left(\dfrac{10}{8}\right)\Delta U$

(E) Cannot be determined without knowing the values of Q_1 and Q_2

(A) $\dfrac{\sqrt{2}Qq}{2\pi\varepsilon_0 L^2 m}$

(B) $\dfrac{\sqrt{2}Qq}{\pi\varepsilon_0 L^2 m}$

(C) $\dfrac{2Qq}{\pi\varepsilon_0 L^2 m}$

(D) $\dfrac{2\sqrt{2}Qq}{\pi\varepsilon_0 L^2 m}$

(E) $\dfrac{4Qq}{\pi\varepsilon_0 L^2 m}$

STOP

END OF SECTION I, ELECTRICITY & MAGNETISM

PHYSICS C: ELECTRICITY & MAGNETISM
SECTION II
Time—45 minutes
3 Questions

Directions: Answer all three questions. The suggested time is about 15 minutes per question for answering each of the questions, which are worth 15 points each. The parts within a question may not have equal weight.

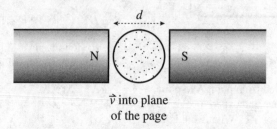

$\vec{v}$ into plane
of the page

1. A stream of equal but oppositely charged particles is moving through a cylinder that is between two magnets as shown above. The magnets create a magnetic field **B** that points from the north pole to the south pole. The particles move with speed v, and the cylinder has a diameter d.

   (a) The magnetic field will cause the charged particles to move in a way that creates an electric field. Indicate this electric field's direction and qualitatively describe the magnitude of the resulting electric forces on the particles in relation to the magnetic forces the particles experience.

   (b) Assume enough time has passed that a constant potential difference V is established in the cylinder. What is the value of the particles' speed v in terms of the other known values?

   (c) Given that one of the positive particles has a mass of m, how much work is done on that particle by the magnetic force in 1 second?

   (d) An overall neutral particle is structured in such a way that it has a slight positive charge on its right side and a slight negative charge on its left side. If this particle were moved through the cylinder, describe the effect the magnetic field would have on such a particle. Ignore any electric forces for this question.

GO ON TO THE NEXT PAGE.

2. In the circuit shown below, the capacitor is initially uncharged and there is no current in any circuit element.

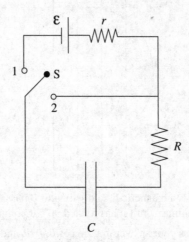

In each of the following, k is a number greater than 1; write each of your answers in terms of ε, r, R, C, k, and fundamental constants.

(a) At $t = 0$, the switch S is moved to position 1.

    i. At what time t is the current through R equal to $\dfrac{1}{k}$ of its initial value?

    ii. At what time t is the charge on the capacitor equal to $\dfrac{1}{k}$ of its maximum value?

    iii. At what time t is the energy stored in the capacitor equal to $\dfrac{1}{k}$ of its maximum value?

(b) After the switch has been at position 1 for a very long time, it is then moved to position 2. Let this redefine $t = 0$ for purposes of the following questions.

    i. How long will it take for the current through R to equal $\dfrac{1}{k}$ of its initial value?

    ii. At what time t is the charge on the capacitor equal to $\dfrac{1}{k}$ of its initial value?

GO ON TO THE NEXT PAGE.

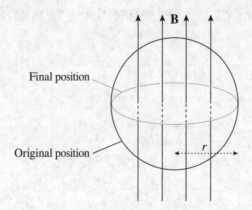

Final position

Original position

r

3. The metal ring of radius r shown above has a magnetic field of strength B passing through it as shown. The ring begins to rotate at a steady angular velocity and continues until it has rotated 90°. It completes the rotation in T seconds.

   (a) Find the average induced emf. Express your answer in terms of B, r, T, and fundamental constants.

   (b) During the rotation, what is the direction of the induced current in the ring?

   (c) If the ring has a resistance of R Ω, what is the magnitude of the average induced current? Express your answer in terms of B, r, R, T, and fundamental constants.

   (d) If the time at which the maximum instantaneous magnitude of induced emf occurs is given as t_{max}, what is the magnitude of that emf? Express your answer in terms of B, r, T, t_{max}, and fundamental constants.

   (e) In terms of known values, what would be the value of t_{max} in (d)?

STOP

END OF EXAM

Practice Test 1: Diagnostic Answer Keys and Explanations

PRACTICE TEST 1: MECHANICS DIAGNOSTIC ANSWER KEY

Let's take a look at how you did on Practice Test 1. Follow the three-step process in the diagnostic answer key below and go read the explanations for any questions you got wrong or you struggled with but got correct. Once you finish working through the answer key and the explanations, go to the next chapter to make your study plan.

 **Check your answers and mark any correct answers with a ✔ in the appropriate column.**

Section I—Multiple Choice								
Q #	Ans.	✔	Chapter #, Section Title	Q #	Ans.	✔	Chapter #, Section Title	
1	E		**5,** Free Fall	19	D		**6,** The Second Law **9,** Torque	
2	E		**5,** Uniformly Accelerated Motion and the Big Five	20	B		**11,** Acceleration of Gravity Due to Large Bodies	
3	A		**5,** Acceleration	21	C		**7,** Work, The Work–Energy Theorem	
4	A		**7,** Work	22	C		**9,** Kinetic Energy of Rotation, Conservation of Angular Momentum	
5	C		**6,** Uniform Circular Motion **9,** Rotational Kinematics	23	C		**5,** Projectile Motion	
6	A		**11,** Gravitational Potential Energy	24	B		**11,** Newton's Law of Gravitation	
7	A		**10,** Pendula	25	D		**8,** Collisions	
8	D		**9,** Rotational Inertia	26	B		**6,** Friction **10,** The Sinusoidal Description of Simple Harmonic Motion	
9	B		**7,** Power	27	D		**7,** Conservation of Mechanical Energy	
10	B		**5,** Kinematics with Graphs **6,** The Second Law	28	D		**9,** Torque	
11	A		**11,** Gravitational Potential Energy	29	A		**7,** Potential Energy	
12	E		**5,** Speed and Velocity **8,** Collisions, Center of Mass	30	C		**8,** Collisions **10,** Simple Harmonic Motion in Terms of Energy	
13	C		**7,** Conservation of Mechanical Energy **8,** Impulse	31	D		**9,** Rotational Inertia	
14	D		**9,** Rotational Kinematics	32	E		**7,** Potential Energy Curves	
15	C		**5,** Projectile Motion **7,** Conservation of Mechanical Energy	33	D		**6,** Pulleys	
16	A		**6,** Uniform Circular Motion	34	E		**7,** Conservation of Mechanical Energy	
17	C		**8,** Center of Mass **9,** Rotational Inertia	35	E		**6,** The Second Law	
18	E		**8,** Impulse					

Section II—Free Response			
Q #	Ans.	✔	Chapter #, Section Title
1a	See Explanation		**5,** Projectile Motion
1b	See Explanation		**5,** Projectile Motion
1c	See Explanation		**5,** Projectile Motion
1d	See Explanation		**5,** Projectile Motion
1e	See Explanation		**5,** Projectile Motion
2a	See Explanation		**11,** The Gravitational Attraction Due to an Extended Body
2b	See Explanation		**11,** Gravitational Potential Energy
2c	See Explanation		**7,** Conservation of Mechanical Energy
2d(i)	See Explanation		**7,** Conservation of Mechanical Energy
2d(ii)	See Explanation		**7,** Conservation of Mechanical Energy
3a(i)	See Explanation		**8,** Center of Mass
3a(ii)	See Explanation		**8,** Collisions
3a(iii)	See Explanation		**9,** Conservation of Angular Momentum
3b	See Explanation		**9,** Kinetic Energy of Rotation

 **STEP 2 »** Tally your correct answers from Step 1 by chapter. For each chapter, write the number of correct answers in the appropriate box. Then, divide your correct answers by the number of total questions (which we've provided) to get your percent correct.

CHAPTER 5 TEST SELF-EVALUATION

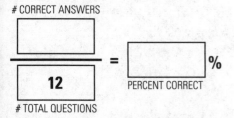

CHAPTER 7 TEST SELF-EVALUATION

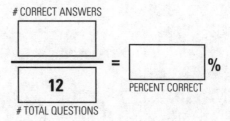

CHAPTER 6 TEST SELF-EVALUATION

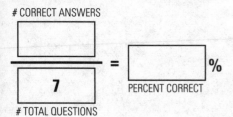

CHAPTER 8 TEST SELF-EVALUATION

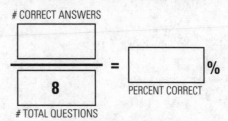

CHAPTER 9 TEST SELF-EVALUATION

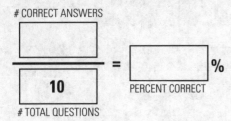

CORRECT ANSWERS

10

TOTAL QUESTIONS

=   %

PERCENT CORRECT

CHAPTER 11 TEST SELF-EVALUATION

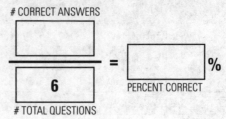

CORRECT ANSWERS

6

TOTAL QUESTIONS

=   %

PERCENT CORRECT

CHAPTER 10 TEST SELF-EVALUATION

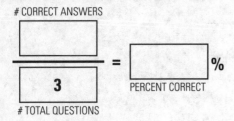

CORRECT ANSWERS

3

TOTAL QUESTIONS

=   %

PERCENT CORRECT

STEP 3 ❯❯ **Use the results above to customize your study plan. You may want to start with, or give more attention to, the chapters with the lowest percents correct.**

PRACTICE TEST 1: MECHANICS ANSWERS AND EXPLANATIONS

Section I—Multiple-Choice

1. **E**   Let y denote the total distance that the rock falls and let T denote the total time of the fall. Then

$$\frac{1}{2}y = \frac{1}{2}gt_1^2 \text{ and } y = \frac{1}{2}gT^2, \text{ so}$$

$$\frac{t_2}{t_1} = \frac{T - t_1}{t_1} = \frac{T}{t_1} - 1 = \frac{\sqrt{2y/g}}{\sqrt{y/g}} - 1 = \sqrt{2} - 1$$

2. **E**   A trick for solving problems of this nature is to make up numbers for a and t. Pretend that total time = 10 seconds, and acceleration = 2 m/s². Then make a table:

Time, s	Speed, m/s
0	0
1	2
9	18
10	20

So in the first second, v_{avg} = 1 m/s, and d = 1 m. In the last (tenth) second, v_{avg} = 19 m/s, and d = 19 m. The answer to the question is 19 m − 1 m = 18 m. Which answer choice gives us 18 m?

(A) $a(t + 1 \text{ s})^2 = (2 \text{ m/s}^2)(10 \text{ s} + 1 \text{ s})^2 = (2 \text{ m/s}^2) (11 \text{ s})^2 = (2 \text{ m/s}^2)(121 \text{ s}^2) = 242 \text{ m}$

(B) $a(t − 1 \text{ s})^2 = (2 \text{ m/s}^2)(10 \text{ s} − 1 \text{ s})^2 = (2 \text{ m/s}^2)(9 \text{ s})^2 = (2 \text{ m/s}^2)(81 \text{ s}^2) = 162 \text{ m}$

(C) $at^2 = (2 \text{ m/s}^2)(10 \text{ s})^2 = (2 \text{ m/s}^2)(100 \text{ s}^2) = 200 \text{ m}$

(D) $a(t + 1 \text{ s})(1 \text{ s}) = (2 \text{ m/s}^2)(10 \text{ s} + 1 \text{ s})(1 \text{ s}) = (2 \text{ m/s}^2)(11 \text{ s})(1 \text{ s}) = 22 \text{ m}$

(E) $a(t − 1 \text{ s})(1 \text{ s}) = (2 \text{ m/s}^2)(10 \text{ s} − 1 \text{ s})(1 \text{ s}) = (2 \text{ m/s}^2)(9 \text{ s})(1 \text{ s}) = 18 \text{ m}$

Therefore (E) is the correct answer.

If you really want to do it algebraically, here's one way. Solve for the distance traveled in the first second.

$$d = v_0 \Delta t + \frac{1}{2}a\Delta t^2$$

$$= (0 \text{ m/s})(1 \text{ s}) + (1/2)(a)(1 \text{ s})^2$$

$$= (1/2)(1 \text{ s}^2)a$$

Next, solve for the distance in the final second.

$$d = v_0 \Delta t + \frac{1}{2} a \Delta t^2$$

$$= a(t - 1 \text{ s})(1 \text{ s}) + \frac{1}{2}(a)(1 \text{ s})^2$$

$$= a(t - 1 \text{ s})(1 \text{ s}) + \frac{1}{2}(1 \text{ s}^2)(a)$$

The difference between the two is $a(t - 1 \text{ s})(1 \text{ s})$ meters.

3. **A** Integrating $a(t)$ gives $v(t)$:

$$v(t) = \int a(t)\,dt = \int 2t\,dt = t^2 + v_0$$

Since $v_0 = 0$, we have $v(t) = t^2$, which is never negative. This means the object never moves backward, so displacement is equal to the distance traveled. Now integrating $v(t)$ from $t = 0$ to $t = 3$ s gives the object's displacement during this time interval:

$$s = \int_0^3 v(t)\,dt = \int_0^3 t^2\,dt = \frac{1}{3}t^3 \Big|_0^3 = 9 \text{ m}$$

4. **A** Raising an object in ideal conditions involves overcoming gravity, which is a conservative force. Therefore, the work required will always be the same. If we think of work as $W = Fd\cos\theta$ and treat W as a constant, the purpose of an inclined ramp is to increase the distance of the movement so that the minimum force can be reduced.

5. **C** The figure below shows that $F_T \sin\theta = mg$ and $F_T \cos\theta = mv^2/R = m\omega^2 R$:

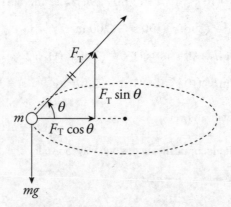

We can eliminate θ from these equations by remembering that $\sin^2\theta + \cos^2\theta$ is always equal to 1:

$$\left(\frac{mg}{F_T}\right)^2 + \left(\frac{m\omega^2 R}{F_T}\right)^2 = 1 \Rightarrow \frac{m^2 g^2 + m^2 \omega^4 R^2}{F_T^2} = 1 \Rightarrow F_T = m\sqrt{\omega^4 R^2 + g^2}$$

That's enough information to get you to the correct answer, (C), but here's another way to think about this problem.

$$y$$
$$\longrightarrow x$$

If you look at the image of $F_T \sin \theta$ and $F_T \cos \theta$, you might recognize that this is a right triangle, to which you can apply the Pythagorean Theorem. $F_{T,x} = \dfrac{mv^2}{R} = m\omega^2 R$ and $F_{T,y} = mg$, which means that $F_T^2 = (mg)^2 + (m\omega^2 R)^2 \rightarrow F_T^2 = m^2(g^2 + \omega^4 R^2) \rightarrow m\sqrt{g^2 + \omega^4 R^2}$.

6. **A** You can eliminate some choices right off the bat if you do some basic dimensional analysis. Choices (A), (B), and (C) all have dimensions of $\sqrt{GM/R}$. Looking up the units of G, you'll see that GM/R has units of N·m/kg = m²/s². Choices (D) and (E) have an extra factor of mass in the numerator ($\sqrt{\text{kg}}$), which doesn't belong in an expression of speed, so they can be ruled out. Because the initial height of the object is comparable to the radius of the Moon, we cannot simply use mgh for its initial potential energy. Instead, we must use the more general expression $U = -GMm/r$, where r is the distance from the center of the Moon. Notice that since $h = R$, the object's initial distance from the Moon's center is $h + R = R + R = 2R$. Conservation of Energy then gives

$$K_i + U_i = K_f + U_f$$
$$0 - \frac{GMm}{2R} = \frac{1}{2}mv^2 - \frac{GMm}{R}$$
$$\frac{GMm}{2R} = \frac{1}{2}mv^2$$
$$v = \sqrt{\frac{GM}{R}}$$

7. **A** The formulas for frequency and period of a pendulum show that these values depend only on the length of the string and the acceleration of gravity. Because these are the same for both pendulums, (B), (C), and (D) can be eliminated. If the total mechanical energies of the systems are the same, that means the maximum kinetic energy, which is found at the bottom of the pendulum's movement, will be the same for both systems. Therefore,

$$KE_1 = KE_2$$
$$\frac{1}{2}mv_1^2 = \frac{1}{2}(2m)v_2^2$$
$$v_1 = \sqrt{2}\,v_2$$

Thus, the system with the smaller mass will have the greater maximum speed. Through Process of Elimination, you can arrive at (A) without doing any further work; however, it's also helpful to know that the maximum angle of displacement must also be greater for the smaller mass. That is the only way to ensure the pendulum with the smaller mass will also have the same maximum potential energy.

8. **D** Let F_T be the tension in each string. Then $F_{net} = ma$ becomes $mg - 2F_T = ma$. Also, the total torque exerted by the tension forces on the cylinder is $2rF_T$, so $\tau_{net} = I\alpha$ becomes $2rF_T = (\frac{1}{2}mr^2)\alpha$. Because the cylinder doesn't slip, $\alpha = a/r$, so $2F_T = \frac{1}{2}ma$. These equations can be combined to give $mg - \frac{1}{2}ma = ma$, which implies $mg = \frac{3}{2}ma$. Therefore, $a = \frac{2}{3}g$.

9. **B**

$$P = Fv = mgv$$

$$m = P/(gv)$$

$$= (1.5 \times 10^8 \text{ W})/[(10 \text{ m/s}^2)(150 \text{ m/s})] = 1.0 \times 10^5 \text{ kg}$$

10. **B** The cylinder slides across the surface with acceleration $a = F/m$ until time $t = T$, when a drops to zero (because **F** becomes zero). Therefore, from time $t = 0$ to $t = T$, the velocity is steadily increasing (because the acceleration is a positive constant), but, at $t = T$, the velocity remains constant. This is illustrated in graph (B).

11. **A** Let m and M denote the mass of the satellite and Earth, respectively. The work required to lift the satellite to height h above the surface of Earth is

$$W = \Delta U = -\frac{GMm}{R+h} - \left(-\frac{GMm}{R}\right) = \frac{GMmh}{R(R+h)}$$

Earth's gravitational pull provides the necessary centripetal force on the satellite, so $F_G = F_c$.

$$\frac{GMm}{(R+h)^2} = \frac{mv^2}{R+h} \Rightarrow mv^2 = \frac{GMm}{R+h} \Rightarrow K = \frac{GMm}{2(R+h)}$$

Because we're told that $W = K$, we find that

$$\frac{GMmh}{R(R+h)} = \frac{GMm}{2(R+h)} \Rightarrow \frac{h}{R} = \frac{1}{2} \Rightarrow h = \frac{1}{2}R$$

12. **E** First, calculate the location of the collision. Given that the objects start 1 m apart and they move toward one another at a net rate of 10 m/s, 0.1 s will pass before the collision occurs. Therefore, the collision point will be 20 cm from the first object's starting point.

Next, calculate the velocity of the combined blocks post-collision using conservation of linear momentum for a perfectly inelastic collision:

$$m_1\mathbf{v}_1 + m_2\mathbf{v}_2 = (m_1 + m_2)\mathbf{v}_f \rightarrow \mathbf{v}_f$$

$$= \frac{m_1\mathbf{v}_1 + m_2\mathbf{v}_2}{m_1 + m_2}$$

$$= \frac{(10 \text{ kg})(2 \text{ m/s}) + (2 \text{ kg})(-8 \text{ m/s})}{10 \text{ kg} + 2 \text{ kg}}$$

$$= \frac{1}{3} \text{ m/s}$$

Next, find the center of mass of the newly combined objects.

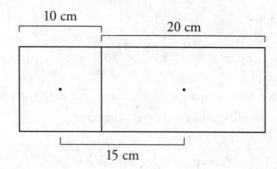

Treating the center of the first object as a 0 point, we get

$$x_{\text{CoM}} = (m_1 x_1 + m_2 x_2)/(m_1 + m_2) = (10 \text{ kg})(0 \text{ m}) + (2 \text{ kg})(0.15 \text{ m})/(10 \text{ kg} + 2 \text{ kg}) = 0.025 \text{ m}$$

This means that, in addition to the 0.3 m needed to get from the collision point back to the midpoint, there is an additional 0.025 m that must be traveled because the center of mass location is not precisely the same as the collision location. Therefore

$$v = d/t \rightarrow t = d/v = (0.325 \text{ m})/(1/3 \text{ m/s}) = 0.975 \text{ s}$$

13.   **C**   Because the ball rebounded to the same height from which it was dropped, its takeoff speed from the floor must be the same as the impact speed. Calling *up* the positive direction, the change in linear momentum of the ball is $p_f - p_i = mv - (-mv) = 2mv$, where $v = \sqrt{2gh}$ (this last equation comes from the equation $\frac{1}{2}mv^2 = mgh$). The impulse–momentum theorem, $J = \Delta p$, now gives

$$F_{\text{avg}}\,\Delta t = 2m\sqrt{2g\,h} \Rightarrow F_{\text{avg}} = \frac{2m\sqrt{2g\,h}}{\Delta t} = \frac{2(0.08\,\text{kg})\sqrt{2(10\text{ m/s}^2)(3.2\,\text{m})}}{0.04\,\text{s}} = 32\,\text{N}$$

14.   **D**   Using the Big Five equation $\Delta\theta = \omega_0 t + \frac{1}{2}\alpha t^2$, we find that

$$\Delta\theta = \frac{1}{2}(2\text{ rad/s}^2)(4.0\text{ s})^2 = 16\text{ rad}$$

Therefore, $\Delta s = r\Delta\theta = (0.1\text{ m})(16\text{ rad}) = 1.6\text{ m}$. If you're worried about memorizing the Big Five equations, here's another method. The angular acceleration is 2.0 rad/s^2, which means it gets 2.0 rad/s faster every second. It starts from rest. After 1 s, $\omega = 2$ rad/s. After 2 s, $\omega = 4$ rad/s. After 3 s, $\omega = 6$ rad/s. After 4.0 s, $\omega = 8.0$ rad/s. That makes the average angular velocity during these four seconds 4.0 rad/s (ω increased linearly from 0 s to 8 s). Since the disk averages 4.0 rad/s for 4 seconds, the angular distance traveled will be 16 rad, which brings you right back to $\Delta s = r\Delta\theta$.

One final alternative is to make a graph with ω on the vertical axis and t on the horizontal axis. The slope of an ω-versus-t plot is $\alpha = 2.0$ rad/s^2. The area under the graph from 0 to 4 is the angular distance traveled by the point in the first 4.0 seconds.

15. **C** The object's initial velocity from the slide is horizontal, so $v_{0y} = 0$, which implies that $\Delta y = -\frac{1}{2}gt^2$. Since $\Delta y = -R$,

$$-R = -\frac{1}{2}g\,t^2 \Rightarrow t = \sqrt{\frac{2R}{g}}$$

The (horizontal) speed with which the object leaves the slide is found from the Conservation of Energy: $mgR = \frac{1}{2}mv^2$, which gives $v = \sqrt{2gR}$. Therefore,

$$\Delta x = v_{0x}t = \sqrt{2g\,R} \cdot \sqrt{\frac{2R}{g}} = 2R$$

Since the object travels a horizontal distance of $2R$ from the end of the slide, the total horizontal distance from the object's starting point is $R + 2R = 3R$.

16. **A** From the diagram below,

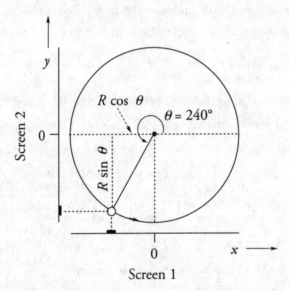

if $R \cos \theta = -(0.5)R$, then $\theta = 240°$. Therefore, $y = R \sin \theta = R \sin 240° = -(0.866)R$. Also, it is clear that the particle's subsequent motion will cause the shadow on Screen 2 to continue moving in the $-y$ direction.

17. **C** The center of mass of the system is at a distance of

$$y_{cm} = \frac{(m)(0) + (2m)(L)}{m + 2m} = \frac{2}{3}L$$

below the mass m. With respect to this point, the clockwise torque produced by the force **F** has magnitude

$$\tau = rF = \left(\frac{2}{3}L - \frac{1}{4}L\right)F = \frac{5}{12}LF$$

Since the rotational inertia of the system about its center of mass is

$$I = \Sigma m_i r_i^2 = m\left(\frac{2}{3}L\right)^2 + (2m)\left(\frac{1}{3}L\right)^2 = \frac{2}{3}mL^2$$

the equation $\tau = I\alpha$ becomes

$$\frac{5}{12}LF = \left(\frac{2}{3}mL^2\right)\alpha \Rightarrow \alpha = \frac{5}{8}\frac{F}{mL}$$

18. **E**   If the two vehicles have the same mass and speed before the collision and both come to a complete stop, then they have the same change in momentum. Impulse is just another name for change in momentum, so (B) is not true. If one vehicle is able to crumple, then it has more distance in which to come to a stop. This will require more time to stop. Therefore, (D) is also not true, which means (E) is the correct answer.

Both (A) and (C) are true statements.

19. **D**   Since $\mathbf{F}_{net} = \mathbf{F}_1 + \mathbf{F}_2 = 0$, the bar cannot accelerate translationally, so (C) is false. The net torque does *not* need to be zero, as the following diagram shows, eliminating (A) and (B).

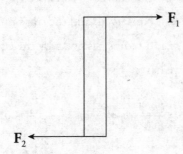

However, since $\mathbf{F}_2 = -\mathbf{F}_1$, (D) is true; one possible illustration of this is given below:

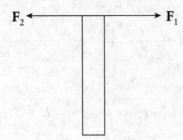

20. **B** The value of g near the surface of a planet depends on the planet's mass and radius:

$$mg = \frac{GMm}{r^2} \Rightarrow g = \frac{GM}{r^2}$$

Therefore, calling this Planet X, we find that

$$g_X = \frac{GM_X}{r_X^2} = \frac{G(2M_{Earth})}{(2r_{Earth})^2} = \frac{1}{2}\frac{GM_{Earth}}{r_{Earth}^2} = \frac{1}{2}g_{Earth}$$

Since g is half as much on the planet as it is on Earth, the astronaut's weight (mg) will be half as much, as well. The astronaut would weigh half of 800 N, which is 400 N, (B).

21. **C** The work done by the force $F(x)$ is equal to the area under the given curve. The region under the graph can be broken into two triangles and a rectangle, and it is then easy to figure out that the total area is 50 N·m. Since work is equal to change in kinetic energy (and $v_i = 0$),

$$W = \Delta K = \frac{1}{2}m(v_f^2 - v_i^2) \Rightarrow v_f = \sqrt{\frac{2W}{m}} = \sqrt{\frac{2(50\,\text{N}\cdot\text{m})}{1.0\,\text{kg}}} = 10\,\text{m/s}$$

22. **C** Since no external torques act on the star as it collapses (just like a skater when she pulls in her arms), angular momentum, $I\omega$, is conserved, and the star's rotational speed increases. The moment of inertia of a sphere of mass m and radius r is given by the equation $I = kmr^2$ (with $k = \frac{2}{5}$, but the actual value is irrelevant), so we have:

$$\omega_f = \frac{I_i}{I_f}\omega_i = \frac{kmr_i^2}{kmr_f^2}\omega_i = \frac{r_i^2}{r_f^2}\omega_i = \frac{r_i^2}{\left(\frac{1}{16}r_i\right)^2}\omega_i = 16^2\omega_i$$

Therefore,

$$\frac{K_f}{K_i} = \frac{\frac{1}{2}I_f\omega_f^2}{\frac{1}{2}I_i\omega_i^2} = \frac{\frac{1}{2}kmr_f^2\omega_f^2}{\frac{1}{2}kmr_i^2\omega_i^2} = \frac{r_f^2\omega_f^2}{r_i^2\omega_i^2} = \frac{\left(\frac{1}{16}r_i\right)^2\left(16^2\omega_i\right)^2}{r_i^2\omega_i^2} = 16^2$$

23. **C** The projectile's horizontal speed is $v_0\cos 60° = v_0 \cdot 1/2 = 20$ m/s, so it reaches the wall in 1.0 s.

24. **B** Use the units for the universal gravitational constant found on the exam's Table of Information pages.

$$[G] = (N \cdot m^2)/(kg^2)$$

Note that a newton, N, is the same as kg m/s^2 and substitute this into the expression above.

$$[G] = (kg \cdot m/s^2)(m^2)/(kg^2) = m^3/(kg \cdot s^2)$$

Finally, substitute the letters as the problem describes, and you end up with (B).

25. **D** Right off the bat, you can eliminate (A) and (E), as they are equivalent to one another. Whether you noticed that or not, the diagram given with the question shows that after the clay balls collide, they move in the $-y$ direction, which means that the horizontal components of their linear momenta canceled. In other words, $\mathbf{p}_{1x} + \mathbf{p}_{2x} = 0$:

$$m_1\mathbf{v}_{1x} + m_2\mathbf{v}_{2x} = 0$$
$$(m_1 v_1 \sin 45°) - (m_2 v_2 \sin 60°) = 0$$
$$v_2 = \frac{m_1 v_1 \sin 45°}{m_2 \sin 60°}$$
$$= \frac{2\sin 45°}{\sin 60°} v_1$$

26. **B** The horizontal position of the blocks can be given by the equation

$$x = A \cos(\omega t + \phi_0)$$

where A is the amplitude, ω is the angular frequency, and ϕ_0 is the initial phase. Differentiating this twice gives the acceleration:

$$a = \frac{d^2 x}{dt^2} = -A\omega^2 \cos(\omega t + \phi_0) \Rightarrow a_{max} = A\omega^2$$

This means that the maximum force on block m is $F_{max} = ma_{max} = mA\omega^2$. The static friction force must be able to provide this same amount of force; otherwise, the block will slip. Therefore,

$$F_{static, \, max} \geq mA\omega^2$$
$$\mu mg \geq mA\omega^2$$
$$A \leq \frac{\mu mg}{m\omega^2} = \frac{\mu g}{(2\pi f)^2} = \frac{(0.80)(10 \text{ m/s}^2)}{4\pi^2 (2.0 \text{ Hz})^2} \approx 0.05 \text{ m} = 5.0 \text{ cm}$$

27. **D** The ball strikes the floor with speed $v_1 = \sqrt{2gH_1}$, and it leaves the floor with speed $v_2 = \sqrt{2gH_2}$. Therefore,

$$e = \frac{v_2}{v_1} = \frac{\sqrt{2gH_2}}{\sqrt{2gH_1}} = \sqrt{\frac{H_2}{H_1}}$$

28. **D** The center of rotation is the center of mass of the plate, which is at the geometric center of the square because the plate is homogeneous. Since the line of action of the force coincides with one of the sides of the square, the lever arm of the force, ℓ, is simply equal to $\frac{1}{2}s$. Therefore,

$$\tau = \ell F = \frac{1}{2}sF = \frac{1}{2}(0.40 \text{ m})(10 \text{ N}) = 2.0 \text{ N·m}$$

29. **A** Since gravity is a conservative force, the actual path taken is irrelevant. If the block rises a vertical distance of $z = 3$ m, then the work done by gravity is

$$W = -mgz = -(2.0 \text{ kg})(10 \text{ N/kg})(3 \text{ m}) = -60 \text{ J}$$

30. **C** Applying conservation of linear momentum for a perfectly inelastic collision, we see that the speed of the combined objects just after the collision will be

$$m_1\mathbf{v}_1 + m_2\mathbf{v}_2 = (m_1 + m_2)\mathbf{v}_f$$

$$\rightarrow \mathbf{v}_f = \frac{m\mathbf{v}}{M + m}$$

This speed means the block has additional kinetic energy, which will be converted into potential energy as the spring compresses.

$$KE = PE$$

$$\frac{1}{2}m_{net}v_f^2 = \frac{1}{2}kx^2$$

$$\rightarrow x = v_f\sqrt{\frac{m_{net}}{k}}$$

$$= \left(\frac{mv}{M + m}\right)\sqrt{\frac{M + m}{k}}$$

$$= mv\sqrt{\frac{1}{k(M + m)}}$$

31. **D** Consider an infinitesimal slice of width dx at position x; its mass is $dm = \lambda \, dx = kx \, dx$. Then, by definition of rotational inertia,

$$I = \int r^2 \, dm = \int_0^L x^2 (kx \, dx) = \frac{1}{4}kx^4\Big|_0^L = \frac{1}{4}kL^4$$

Because the total mass of the rod is

$$M = \int dm = \int_0^L kx \, dx = \frac{1}{2}kx^2\Big|_0^L = \frac{1}{2}kL^2$$

we see that $I = \frac{1}{2}ML^2$.

32. **E** Points of equilibrium occur when dU/dx is equal to zero:

$$\frac{dU}{dx} = 0 \Rightarrow 3(x-2)^2 - 12 = 0 \Rightarrow (x-2)^2 = 4 \Rightarrow x = 0 \text{ or } x = 4$$

There's more than one way to determine which equilibrium points are stable equilibrium points. One way is to graph $U(x)$ and see which equilibrium point is a minimum. Another way is to use the first derivative. The following explanation uses the second derivative test, but feel free to use the method you're most comfortable with. Points of *stable* equilibrium occur when $dU/dx = 0$ and d^2U/dx^2 is positive (because then the equilibrium point is a relative minimum). Since

$$\frac{d^2U}{dx^2} = 6(x-2)$$

the point $x = 4$ is a point of stable equilibrium ($x = 0$ is unstable).

33. **D** Since the pulley is not accelerating, it can serve as the origin of an inertial reference frame. Applying Newton's Second Law to the masses using the following free-body diagrams,

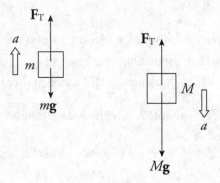

gives

$$F_T - mg = ma \quad \text{and} \quad Mg - F_T = Ma$$

Adding these equations eliminates F_T, and we find that

$$Mg - mg = (m + M)a \quad \Rightarrow \quad a = \frac{M - m}{M + m}g$$

The acceleration of block m is $(M - m)g/(M + m)$ upward, and the acceleration of block M is this value, $(M - m)g/(M + m)$, downward, which is (D).

Alternatively, you can treat the two objects as one system. Tension is an internal force and cancels. The net force on the system is $Mg - mg$. The total mass is $M + m$. If you simplify this expression, you get the same answer, (D).

34. **E** This can be solved using kinematics or work and energy. We'll use the latter solution because it is faster, but both methods are valid.

First, determine how much potential energy is converted to kinetic for each meter the block slides down the ramp.

$$\Delta PE = -\Delta KE \rightarrow mg\Delta h = -\Delta KE \rightarrow \Delta KE = -(mg(-d\sin\theta_r)) = -(2\text{ kg})(10\text{ m/s}^2)(-1\text{ m})(0.45) = 9\text{ J}$$

On a frictionless portion, this 9 J will convert to kinetic without issue. On the friction portions, however, some energy will be lost due to the friction. Each meter the block slides over the friction surface loses

$$W_f = F_f d\cos\theta = (\mu F_N)d\cos\theta = \mu(mg\cos\theta_r)d\cos\theta = (\tfrac{2}{3})(2\text{ kg})(10\text{ m/s}^2)(0.9)(1)(-1) = -12\text{ J}$$

The value of the $\cos\theta = -1$ because the angle used for θ is not θ_r, the angle of incline of the ramp. When calculating the work lost to friction, $W_f = F_f d\cos\theta$, θ is the angle that the friction force acts relative to the motion of the block. Since friction always acts against motion, $\theta = 180°$, so then $\cos\theta = -1$.

This means that each meter of sliding on a friction portion results in +9 J from the conversion of *PE* and –12 J from the friction, producing a net result of –3 J per meter when sliding over the friction portions of the ramp. The ramp's pattern of friction and non-friction will be

1 m without, 1 m with, 1 m without, 2 m with, 1 m without, 3 m with, 1 m without, 4 m with, 1 m without, 5 m with (total of 20 m)

This produces *KE* of

$$+9\text{ J} - 3\text{ J} + 9\text{ J} - 6\text{ J} + 9\text{ J} - 9\text{ J} + 9\text{ J} - 12\text{ J} + 9\text{ J} - 15\text{ J} = 0\text{ J}$$

Although the block has no kinetic energy after passing the 5 m friction portion, it will continue because it is now on another frictionless portion. It gains the 9 J from this frictionless portion. This is enough to travel through 3 m of the next friction portion. Adding these additional 4 m to the previous 20 m gives the final answer of 24 m.

35. **E** The maximum net force on the object occurs when all three forces act in the same direction, giving $F_{net} = 3F = 3(4\text{ N}) = 12\text{ N}$, and a resulting acceleration of $a = F_{net}/m = (12\text{ N})/(2\text{ kg}) = 6\text{ m/s}^2$. These three forces could not give the object an acceleration greater than 6 m/s². Because the question looks for the acceleration that is NOT possible, the answer is 8 m/s, (E).

Section II—Free Response

1.   This problem involves motion in two dimensions, so it's important to begin by drawing a large picture of the situation. Note that our coordinate system is set up in the usual way with x pointing to the right, y pointing up, and the origin at the point where the projectile is launched.

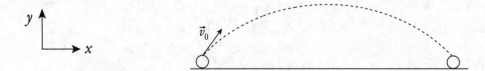

Next, draw a second diagram that splits $\vec{v}_0$ into vertical and horizontal components. Note that you can use trigonometric properties to derive the values of v_{0x} and v_{0y}.

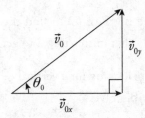

Now we're ready to start answering the questions.

(a)   For this problem, time of flight is determined by vertical quantities. This is because the flight ends when the object hits the ground, that is, when the y-coordinate is zero again. If $\Delta y = 0$, the projectile is either beginning or ending its flight.

You need to find t and you're missing $v_{f,y}$, so use Big Five #3.

$$\Delta y = v_{0y}t + \frac{1}{2}a_y t^2$$

$$0 = \left(v_0 \sin\theta\right)t + \frac{1}{2}\left(-g\right)t^2$$

$$\frac{1}{2}gt^2 - \left(v_o \sin\theta\right)t = 0$$

$$t\left(\frac{g}{2}t - v_0 \sin\theta\right) = 0$$

At the beginning of the flight, t is equal to 0. However, at the end of the flight, you would have the following:

$$\frac{g}{2}t - v_0 \sin\theta = 0$$

$$\frac{g}{2}t = v_0 \sin\theta$$

$$t = \frac{2v_0 \sin\theta}{g}$$

The total amount of time in the air, then, is $\dfrac{2v_0 \sin\theta}{g}$.

As an alternative, you could use Big Five #2 to find the amount of time in which the projectile moves upward and then multiply that number by 2 to get the total time.

(b) As with the previous step, you are looking for a vertical quantity. Since it's t that you no longer care about and Δy that you need, use Big Five #5.

$$v_{f,y}^2 = v_{0y}^2 + 2a(\Delta y)$$

At the top of the trajectory, $v_{f,y} = 0$, which means you can now solve for Δy.

$$0 = v_{0y}^2 + 2a\Delta y$$

$$-v_{0y}^2 = 2a\Delta y$$

$$\frac{-v_{0y}^2}{2a} = \Delta y$$

If you plug in the known values of v_{0y} and a, you get the following:

$$\Delta y = \frac{-\left(v_0 \sin\theta\right)^2}{2(-g)}$$

$$\Delta y = \frac{v_0^2 \sin^2\theta}{2g}$$

Because $y_0 = 0$, that value of Δy is also the value of y_{max}.

(c) The object starts and ends on the ground ($y = 0$), so its net vertical displacement is 0.

(d)   You're asked for a horizontal quantity, so let's review our horizontal components.

$$\Delta x = ?$$

$$v_x = v_0 \cos\theta$$

$$t = \frac{2v_0 \sin\theta}{g}$$

Velocity is defined as $v_x = \frac{\Delta x}{t}$, and this can be rewritten as $\Delta x = v_x t$. You've found two of these values already, so plug them in and solve.

$$\Delta x = (v_0 \cos\theta)\left(\frac{2v_0 \sin\theta}{g}\right)$$

$$\Delta x = \frac{2v_0^2 \sin\theta \cos\theta}{g}$$

(e)   We're told that the range equals the maximum vertical displacement, so $\Delta x = y_{max}$.

$$\frac{2v_0^2 \sin\theta \cos\theta}{g} = \frac{v_0^2 \sin^2\theta}{2g}$$

Canceling out from this equation, you're left with the following:

$$2\cos\theta = \frac{\sin\theta}{2}$$

$$4 = \tan\theta$$

$$\theta = \tan^{-1}(4)$$

You can check this by noting that the arctan of 4 is a fairly large angle (about 76°), which indicates that $\vec{v}_0$ points more upward than rightward.

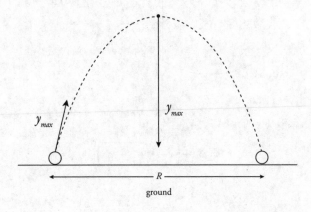

2.   (a)   The mass contained in a sphere of radius $r < R$ is

$$M_{\text{within } r} = \rho V_{\text{within } r} = \frac{M}{\frac{4}{3}\pi R^3}\left(\tfrac{4}{3}\pi r^3\right) = \frac{r^3}{R^3}M$$

so

$$F = -\frac{Gm}{r^2}\left(\frac{r^3}{R^3}M\right) = -\frac{GMm}{R^3}r$$

(b)   Since $dU = -F(r)\,dr$,

$$\Delta U\Big|_R^r = \int_R^r dU = \int_R^r -F(r)\,dr = \int_R^r \frac{GMm}{R^3}r\,dr = \frac{GMm}{R^3}\left[\tfrac{1}{2}r^2\right]_R^r = \frac{GMm}{2R^3}(r^2 - R^2)$$

(c)   Conservation of Energy, $K_i + U_i = K_f + U_f$, can be written in the form $\Delta K = -\Delta U$, which gives

$$\tfrac{1}{2}mv^2 = -\frac{GMm}{2R^3}(r^2 - R^2)$$

$$v = \sqrt{\frac{GM}{R^3}(R^2 - r^2)}$$

By applying the Pythagorean Theorem to the two right triangles in the figure,

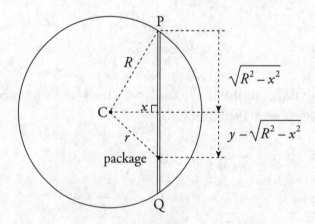

we see that

$$r^2 = x^2 + \left(y - \sqrt{R^2 - x^2}\right)^2$$

so the expression given on the previous page for v can be rewritten in the form

$$v = \sqrt{\frac{GM}{R^3} \left[R^2 - x^2 - \left(y - \sqrt{R^2 - x^2} \right)^2 \right]}$$

$$= \sqrt{\frac{GM}{R^3} \left[2y\sqrt{R^2 - x^2} - y^2 \right]}$$

(d) (i) From the expression derived for r^2 in part (c) on the previous page, we can see that the value of y which will maximize v is

$$y = \sqrt{R^2 - x^2}$$

This is the midpoint of the tunnel.

(ii) The maximum value for v is

$$v_{max} = \sqrt{\frac{GM}{R^3} \left[R^2 - x^2 \right]}$$

3. (a) (i) Once the $3m$ puck sticks to the m puck, the center of mass of the system is

$$y_{cm} = \frac{(m + 3m)(0) + (2m)(L)}{(m + 3m) + 2m} = \tfrac{1}{3}L$$

above the bottom pucks on the rod.

(ii) Applying Conservation of Linear Momentum, we find that

$$(3m)v = (3m + m + 2m)v'_{cm} \quad \Rightarrow \quad v'_{cm} = \tfrac{1}{2}v$$

(iii) To keep this as straightforward as possible, choose our origin to be the center of mass of the system. Now, applying Conservation of Angular Momentum, we find that

$$L_i = L_f$$

$$L_i = I_f \omega_f$$

$$\omega_f = \frac{L_i}{I_f}$$

$$= \frac{(\tfrac{1}{3}L)(3m)v}{\left[(m + 3m)(\tfrac{1}{3}L)^2 + (2m)(\tfrac{2}{3}L)^2 \right]}$$

$$= \frac{3v}{4L} \text{ (counterclockwise)}$$

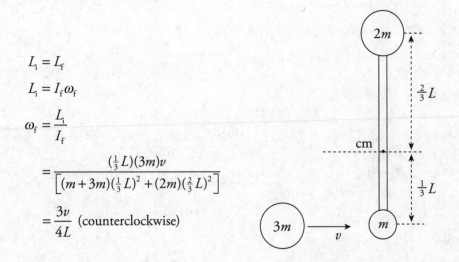

(b)   The kinetic energy before the collision was the initial translational kinetic energy of the $3m$ puck:

$$K_i = \tfrac{1}{2}(3m)v^2$$

After the collision, the total kinetic energy is the sum of the system's translational kinetic energy and its rotational kinetic energy:

$$
\begin{aligned}
K' &= K'_{\text{translational}} + K'_{\text{rotational}} \\
&= \tfrac{1}{2}Mv_{\text{cm}}^2 + \tfrac{1}{2}I\omega^2 \\
&= \tfrac{1}{2}(3m+m+2m)(\tfrac{1}{2}v)^2 + \tfrac{1}{2}\left[(3m+m)(\tfrac{1}{3}L)^2 + (2m)(\tfrac{2}{3}L)^2\right]\left(\tfrac{3v}{4L}\right)^2 \\
&= \tfrac{9}{8}mv^2
\end{aligned}
$$

Therefore, the fraction of the initial kinetic energy that is lost due to the collision is

$$\frac{K-K'}{K} = \frac{\tfrac{3}{2}mv^2 - \tfrac{9}{8}mv^2}{\tfrac{3}{2}mv^2} = \frac{1}{4}$$

PRACTICE TEST 1: ELECTRICITY & MAGNETISM DIAGNOSTIC ANSWER KEY

Let's take a look at how you did on Practice Test 1. Follow the three-step process in the diagnostic answer key below and go read the explanations for any questions you got wrong or you struggled with but got correct. Once you finish working through the answer key and the explanations, go to the next chapter to make your study plan.

STEP 1 » **Check your answers and mark any correct answers with a ✔ in the appropriate column.**

Section I—Multiple Choice							
Q #	Ans.	✔	Chapter #, Section Title	Q #	Ans.	✔	Chapter #, Section Title
1	C		12, Coulomb's Law 13, Conductors and Insulators	19	E		15, The Magnetic Force on a Moving Charge
2	C		14, Energy and Power	20	E		12, The Electric Field
3	E		15, The Magnetic Force on a Moving Charge	21	E		13, Capacitors of Other Geometries
4	A		12, The Electric Field	22	C		15, The Magnetic Force on a Moving Charge
5	B		14, Resistivity	23	B		16, RL Circuits
6	E		15, The Magnetic Force on a Moving Charge	24	B		16, RL Circuits
7	C		12, Gauss's Law	25	A		16, RL Circuits
8	A		12, Coulomb's Law	26	C		12, The Electric Field 13, Conductors and Insulators
9	A		12, Introduction	27	B		12, Electrical Potential Energy
10	E		14, Combinations of Resistors	28	E		16, RL Circuits
11	E		14, Energy and Power	29	E		14, Energy and Power
12	D		14, Charging or Discharging a Capacitor in an RC Circuit	30	D		16, Faraday's Law of Electromagnetic Induction
13	B		16, Faraday's Law of Electromagnetic Induction	31	D		13, Combinations of Capacitors, Dielectrics
14	A		15, The Magnetic Force on a Current-Carrying Wire	32	A		12, The Electric Field, Electric Potential
15	C		12, Electrical Potential Energy	33	D		15, Ampere's Law
16	A		15, Magnetic Fields Created by Current-Carrying Wires	34	B		12, Electrical Potential Energy
17	A		12, Coulomb's Law	35	A		12, Coulomb's Law
18	C		13, Capacitors				

Section II—Free Response			
Q #	Ans.	✔	Chapter #, Section Title
1a	See Explanation		**12,** The Electric Field **15,** The Magnetic Force on a Moving Charge
1b	See Explanation		**12,** The Potential in a Uniform Field
1c	See Explanation		**15,** The Magnetic Force on a Moving Charge
1d	See Explanation		**15,** The Magnetic Force on a Moving Charge
2a(i)	See Explanation		**14,** Combinations of Resistors, Charging a Capacitor
2a(ii)	See Explanation		**13,** Capacitors **14,** Charging a Capacitor
2a(iii)	See Explanation		**13,** The Energy Stored in a Capacitor **14,** Charging a Capacitor
2b(i)	See Explanation		**14,** Discharging a Capacitor
2b(ii)	See Explanation		**13,** Capacitors **14,** Discharging a Capacitor
3a	See Explanation		**16,** Faraday's Law of Electromagnetic Induction
3b	See Explanation		**16,** Faraday's Law of Electromagnetic Induction
3c	See Explanation		**14,** Resistance
3d	See Explanation		**16,** Faraday's Law of Electromagnetic Induction
3e	See Explanation		**16,** Faraday's Law of Electromagnetic Induction

 **STEP 2 »** Tally your correct answers from Step 1 by chapter. For each chapter, write the number of correct answers in the appropriate box. Then, divide your correct answers by the number of total questions (which we've provided) to get your percent correct.

CHAPTER 12 TEST SELF-EVALUATION

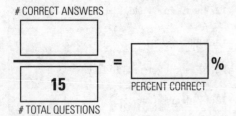

CHAPTER 13 TEST SELF-EVALUATION

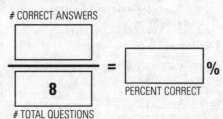

CHAPTER 14 TEST SELF-EVALUATION

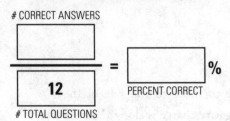

CORRECT ANSWERS

────────── = ┌──────────┐ %
12   └──────────┘
TOTAL QUESTIONS   PERCENT CORRECT

CHAPTER 16 TEST SELF-EVALUATION

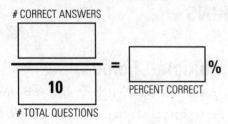

CORRECT ANSWERS

────────── = ┌──────────┐ %
10   └──────────┘
TOTAL QUESTIONS   PERCENT CORRECT

CHAPTER 15 TEST SELF-EVALUATION

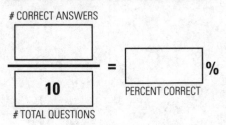

CORRECT ANSWERS

────────── = ┌──────────┐ %
10   └──────────┘
TOTAL QUESTIONS   PERCENT CORRECT

STEP 3 》 Use the results above to customize your study plan. You may want to start with, or give more attention to, the chapters with the lowest percents correct.

PRACTICE TEST 1: ELECTRICITY & MAGNETISM ANSWERS AND EXPLANATIONS

Section I—Multiple-Choice

1. **C** The proximity of the charged sphere will induce negative charge to move to the side of the uncharged sphere closer to the charged sphere. Since the induced negative charge is closer than the induced positive charge to the charged sphere, there will be a net electrostatic attraction between the spheres.

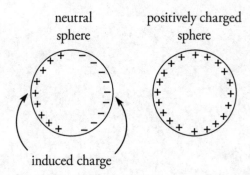

neutral sphere

positively charged sphere

induced charge

2. **C** Typically, the formula for power dissipated by a circuit is expressed as $P = IV$. However, it can also be written as $P = I^2R$ or $P = V^2/R$. Using $P = V^2/R$, if the voltage remains fixed while the resistance doubles, the power dissipated will be cut in half.

3. **E** When the particle enters the magnetic field, the magnetic force provides the centripetal force to cause the particle to execute uniform circular motion:

$$|q|vB = \frac{mv^2}{r} \Rightarrow r = \frac{mv}{|q|B}$$

Since v and B are the same for all the particles, the largest r is found by maximizing the ratio $m/|q|$. Furthermore, since the ratio of u/e (atomic mass unit/magnitude of elementary charge) is a constant, the answer depends only on the ratio of "factor of u" to "factor of charge." The ratio 20/1 is largest, so it will have the largest r.

4. **A** The only place with 0 net electric field would be in the center of the rectangle. Moving up along Line A would give a net electric field pointing to the right. This is because the top two charges, which would both force a positive particle to the right (in addition to vertical effects), are closer and therefore more impactful than the bottom two charges, which would oppose them by trying to make a positive particle move to the left. Similarly, moving down along Line A would give a net electric field to the left. Moving to the right along Line B would give a net electric field upward. Moving to the left along Line B would give a net electric field downward.

5. **B** The resistance of a wire made of a material with resistivity ρ, length L, and cross-sectional area A is given by the equation $R = \rho L/A$. Since Wire B has the greatest length and smallest cross-sectional area, it has the greatest resistance.

6. **E** From the equation $F = qvB$, we see that 1 tesla is equal to $1 \text{ N·s}/(\text{C·m})$. Since

$$\frac{1\,\text{N} \cdot \text{s}}{\text{C} \cdot \text{m}} = \frac{1\,\text{N}}{\text{A} \cdot \text{m}} = \frac{1\,\text{J}}{\text{A} \cdot \text{m}^2},$$

(A) and (C) are eliminated.

Furthermore,

$$\frac{1\,\text{N} \cdot \text{s}}{\text{C} \cdot \text{m}} = \frac{1\,\text{kg} \cdot \text{m}/\text{s}}{\text{C} \cdot \text{m}} = \frac{1\,\text{kg}}{\text{C} \cdot \text{s}}$$

eliminating (B). Finally, since

$$\frac{1\,\text{J}}{\text{A} \cdot \text{m}^2} = \frac{1\,\text{C} \cdot \text{V}}{(\text{C}/\text{s}) \cdot \text{m}^2} = \frac{1\,\text{V} \cdot \text{s}}{\text{m}^2},$$

(D) is also eliminated.

7. **C** Gauss's law states that the total electric flux through a closed surface is equal to $(1/\varepsilon_0)$ times the net charge enclosed by the surface. Both the cube and the sphere contain the same net charge, so Φ_C must be equal to Φ_S.

8. **A** The particle would experience oscillations between the positions $x = 30$ cm and $x = 70$ cm. At the position $x = 50$ cm, it would have no net force acting on it, but this would be similar to a spring having no net force as it passes through the equilibrium position. It will continue moving until an opposing force reduces its speed to 0, which would occur at $x = 30$ cm. It would then be pushed to the right until it reached $x = 70$ cm, and the cycle would repeat indefinitely.

9. **A** The Table of Information provides the following conversion factor: $1\ e = 1.6 \times 10^{-19}$ C. Multiplying the magnitude of charge, 1 C, by this conversion factor gives the following:

$$1\,\text{C} = \frac{-1\,e}{1.60 \times 10^{-19}\,\text{C}} = 6.25 \times 10^{18}\,e$$

There are 6.25×10^{18} electrons in 1 coulomb of negative charge.

10. **E** Resistors R_2, R_3, and R_4 are in parallel, so their equivalent resistance, $R_{2\text{-}3\text{-}4}$, satisfies

$$\frac{1}{R_{2\text{-}3\text{-}4}} = \frac{1}{R_2} + \frac{1}{R_3} + \frac{1}{R_4} = \frac{3}{200\,\Omega} \Rightarrow R_{2\text{-}3\text{-}4} = \frac{200}{3}\,\Omega$$

Since this is in series with R_1, the total resistance in the circuit is

$$R = R_1 + R_{2-3-4} = \frac{800}{3} \, \Omega$$

The current provided by the source must therefore be

$$I = \frac{\mathcal{E}}{R} = \frac{24 \, \text{V}}{\frac{800}{3} \, \Omega} = \frac{9}{100} \, \text{A} = \frac{90}{1,000} \, \text{A} = 90 \, \text{mA}$$

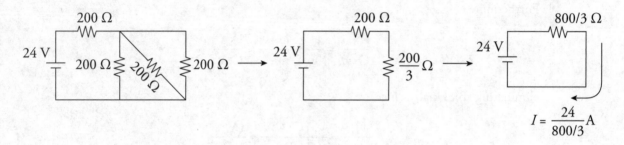

Drawing a series of equivalent circuits can make it easier to see what the resistance should be at each step.

11. **E** The current through R_4 is $\frac{1}{3}$ the current through R_1, and $R_4 = R_1$, so

$$\frac{P_1}{P_4} = \frac{I_1^2 R_1}{I_4^2 R_4} = \frac{I_1^2 R_1}{\left(\frac{1}{3} I_1\right) R_1} = 9$$

12. **D** Note that all answers have the same number, so for this problem, don't even worry about multiplying the numbers—just focus on the units. Because the time constant for an RC circuit is equal to the product of resistance and capacitance, $\tau = RC$, this product has the dimensions of time. If you don't know that off the top of your head, just be sure to carefully break the farad and ohm into their subunits, simplifying where possible:

$1 \, \text{F} \times 1 \, \Omega = 1 \, \text{C/V} \times 1 \, \text{V/A} = 1 \, \text{C/A} = 1 \, \text{s}$

13. **B** Apply Faraday's Law of Electromagnetic Induction:

$$\oint E \cdot d\ell = -\frac{d\Phi_B}{dt}$$

$$E(2\pi r) = -\frac{d}{dt}(BA)$$

$$= -A\frac{dB}{dt}$$

$$= -\pi r^2 \cdot \frac{d}{dt}(2t - 2t^2)$$

$$= -\pi r^2(2 - 4t)$$

$$E = -r(1 - 2t)$$

$$E = r(2t - 1)$$

Since $r = 1$ m, the value of E at $t = 1$ s is $E = 1$ N/C.

14. **A** The magnetic force, $\mathbf{F}_B$, on the top horizontal wire of the loop must be directed upward and have magnitude mg. (The magnetic forces on the vertical portions of the wire will cancel.) By the right-hand rule, the current in the top horizontal wire must be directed to the right—because $\mathbf{B}$ is directed into the plane of the page—in order for $\mathbf{F}_B$ to be directed upward. Therefore, the current in the loop must be clockwise, eliminating (B) and (D). Since $F_B = IxB$, we must have

$$IxB = mg \Rightarrow I = \frac{mg}{xB}$$

15. **C** First, find the difference in potential between the starting location (2 cm away) and the final location (1 cm away):

$$\Delta\Phi = \Phi_f - \Phi_0$$

$$= \frac{kQ}{r_f} - \frac{kQ}{r_0}$$

$$= \frac{(9 \times 10^9)(10 \times 10^{-9})}{1 \times 10^{-2}} - \frac{(9 \times 10^9)(10 \times 10^{-9})}{2 \times 10^{-2}}$$

$$= 4{,}500 \text{ V}$$

Next, find the change in potential energy.

$$\Delta U = q\Delta\Phi = (-5 \times 10^{-9})(4{,}500) = -2.25 \times 10^{-5} \text{ J}$$

Because total energy will be conserved, the change in kinetic energy must be 2.25×10^{-5} J. This can be used to find the final speed.

$$\Delta K = K_f - K_i \rightarrow 2.25 \times 10^{-5} = \frac{1}{2} m v_f^2$$

$$\rightarrow v_f = \sqrt{\frac{2\left(2.25 \times 10^{-5}\right)}{5 \times 10^{-10}}}$$

$$= 300 \text{ m/s}$$

16. **A** Call the top wire (the one carrying a current I to the right) Wire 1, and call the bottom wire (carrying a current $2I$ to the left) Wire 2. Then, in the region between the wires, the individual magnetic field vectors due to the wires are both directed into the plane of the page, so they could not cancel in this region. Therefore, the total magnetic field could not be zero at either Point 2 or Point 3. This eliminates (B), (C), (D), and (E), so the answer must be (A). Since the magnetic field created by a current-carrying wire is proportional to the current and inversely proportional to the distance from the wire, the fact that Point 1 is in a region where the individual magnetic field vectors created by the wires point in opposite directions and that Point 1 is twice as far from Wire 2 as from Wire 1 implies that the total magnetic field there will be zero.

17. **A** If the mass of the $+q$ charge is m, then its acceleration is

$$a = \frac{F_E}{m} = \frac{1}{4\pi\varepsilon_0} \frac{Qq}{m r^2}$$

Because a should decrease as r increases, you can eliminate (C) and (E). The graph in (A) best depicts an inverse-square relationship between a and r.

18. **C** Once the switch is closed, the capacitors are in parallel, which means the voltage across C_1 must equal the voltage across C_2. Since $V = Q/C$,

$$\frac{Q_2'}{C_2} = \frac{Q_1'}{C_1} \Rightarrow Q_2' = \frac{C_2}{C_1} Q_1' = \frac{6 \text{ μF}}{3 \text{ μF}} Q_1' = 2Q_1'$$

This makes sense qualitatively: C_2 has twice the capacitance of C_1, so it should hold twice the charge. The total charge, 24 μC + 12 μC = 36 μC, must be redistributed so that $Q_2' = 2Q_1'$. Therefore, we see that $Q_2' = 24$ μC (and $Q_1' = 12$ μC).

19. **E** Because **v** is parallel to **B**, the charges in the bar feel no magnetic force, so there will be no movement of charge in the bar and no motional emf.

20. **E** Along the line joining the two charges of an electric dipole, the individual electric field vectors point in the same direction, so they add constructively. At the point equidistant from both charges, the total electric field vector has magnitude

$$E_+ + E_- = 2E = 2\frac{kQ}{\left(\frac{1}{2}d\right)^2} = \frac{8kQ}{d^2}$$

$$= \frac{8(9\times10^9\,\text{N}\cdot\text{m}^2/\text{C}^2)(4.0\times10^{-9}\,\text{C})}{(2.0\times10^{-2}\,\text{m}^2)^2}$$

$$= 72\times10^4\ \text{N/C}$$

$$= 7.2\times10^5\ \text{N/C (or V/m)}$$

21. **E** Start with basic dimensional analysis. Choices (A) and (B) can be eliminated because those expressions do not yield units of capacitance. Imagine placing equal but opposite charges on the spheres, say, $+Q$ on the inner sphere and $-Q$ on the outer sphere. Then the electric field between the spheres is radial and depends only on $+Q$, and its strength is

$$E = \frac{1}{4\pi\varepsilon_0}\frac{Q}{r^2}$$

Therefore, the potential difference between the spheres is

$$V = \int_R^{2R} E\cdot d\mathbf{r} = \frac{Q}{4\pi\varepsilon_0}\int_R^{2R}\frac{1}{r^2}dr = \frac{Q}{4\pi\varepsilon_0}\left(-\frac{1}{r}\right)_R^{2R} = \frac{Q}{4\pi\varepsilon_0}\left(\frac{1}{R}-\frac{1}{2R}\right) = \frac{Q}{8\pi\varepsilon_0 R}$$

Now, by definition, $C = Q/V$, so

$$C = \frac{Q}{V} = Q\cdot\frac{1}{V} = Q\cdot\frac{8\pi\varepsilon_0 R}{Q} = 8\pi\varepsilon_0 R$$

22. **C** Since the magnetic force is always perpendicular to the object's velocity, it does zero work on any charged particle. Zero work means zero change in kinetic energy, so the speed remains the same. Remember, the magnetic force can only change the direction of a charged particle's velocity, not its speed.

23. **B** The presence of the inductor in the rightmost branch effectively removes that branch from the circuit at time $t = 0$ (the inductor produces a large back emf when the switch is closed and the current jumps abruptly). Initially, then, current only flows in the loop containing the resistors r and R. Since their total resistance is $r + R$, the initial current is $\mathcal{E}/(r + R)$.

24. **B** The correct answer should have dimensions of energy. Using the equation $U_L = \frac{1}{2}LI^2$ from the Table of Information, L should have units of J/A^2. Choices (A), (B), (C), and (E) all have units of $J/A^2 \times V^2/\Omega^2 = J/A^2 \times A^2 = J$, which is a unit of energy. Choice (D) has an extra factor of resistance in the numerator, and can be eliminated. After a long time, the current through the branch containing the inductor increases to its maximum value. With all three resistors in play, the total resistance is $R_{eq} = r + \frac{R}{2}$, since $\frac{R}{2}$ is the equivalent resistance of the two resistors in parallel. The current provided by the source, $\frac{\varepsilon}{R_{eq}}$, splits at the junction leading to the parallel combination. The amount which flows through the inductor is half the total.

$$I = \frac{1}{2} \cdot \frac{\varepsilon}{\left(r + \dfrac{R}{2} \right)} = \frac{\varepsilon}{2r + R}$$

so the energy stored in the inductor is

$$U_L = \frac{LI^2}{2} = \frac{L}{2}\left(\frac{\varepsilon}{(2r+R)} \right)^2 = \frac{L\varepsilon^2}{2(2r+R)^2}$$

25. **A** Once the switch is opened, the resistor r is cut off from the circuit, so no current passes through it. (Current *will* flow around the loop containing the resistors labeled R, gradually decreasing until the energy stored in the inductor is exhausted.)

26. **C** If a conducting sphere contains a charge of $+q$ within an inner cavity, a charge of $-q$ will move to the wall of the cavity to "guard" the interior of the sphere from an electrostatic field, regardless of the size, shape, or location of the cavity. As a result, a charge of $+q$ is left on the exterior of the sphere (and it will be uniform). So, at points outside the sphere, the sphere behaves as if this charge $+q$ were concentrated at its center, so the electric field outside is simply kQ/r^2. Since points X and Y are at the same distance from the center of the sphere, the electric field strength at Y will be the same as at X.

27. **B** A particle of negative charge is attracted to the two positive charges in the problem, and will gain kinetic energy as it approaches. By the work–energy theorem, the electric field must be doing positive work on the particle. Alternatively, the particle is attracted to the two charges it's moving toward, which means $\mathbf{F}_E$ and $\mathbf{d}$ are in the same direction, so the work done by $\mathbf{E}$ is positive. Eliminate (C) and (E).

To decide between the remaining choices, first determine the electric potential at point P. Realize that point P is 5 meters away from each charge because it is a 3-4-5 right triangle.

$$V = \sum \frac{kQ}{r}$$

$$V_P = \frac{kQ}{5} + \frac{kQ}{5} = \frac{2kQ}{5}$$

The work done by the electric field will be positive because a $-Q$ charge would be attracted to point P by the two positive Q charges. Also, the work done by the force is the negative of the change in potential energy.

$$\Delta V = \frac{\Delta U}{q}$$

$$V_P - V_\infty = \frac{\Delta U}{-Q}$$

$$\frac{2kQ}{5} - 0 = \frac{\Delta U}{-Q}$$

$$\Delta U = \frac{-2kQ^2}{5}$$

$$W = -\Delta U = \frac{2kQ^2}{5}$$

28. **E** The current in an RL series circuit—in which initially no current flows—increases with time t according to the equation

$$I(t) = I_{max}(1 - e^{-t/\tau})$$

where $I_{max} = \mathcal{E}/R$ and τ is the inductive time constant, L/R. The time t at which $I = (99\%)I_{max}$ is found as follows:

$$1 - e^{-t/\tau} = \frac{99}{100}$$

$$e^{-t/\tau} = \frac{1}{100}$$

$$-t/\tau = \ln \frac{1}{100}$$

$$t = (-\ln \frac{1}{100})\tau = (\ln 100)\tau = (\ln 100)L/R$$

Alternatively, the solution to this question can be found by using Process of Elimination. Choice (B) can be eliminated because it does not incorporate the 99%. Choices (A) and (C) can be eliminated because they give negative results. Choice (D) can be eliminated because ln(100/99) is close to zero. This leaves only (E), which satisfies the given conditions.

29.   **E**   First, solve for the resistance of each lightbulb as shown below:

$$P = V^2/R$$

$$R = V^2/P = (120 \text{ V})^2/(40 \text{ W}) = 360 \text{ } \Omega$$

Next, solve for the equivalent resistance, R_{eq}, of the circuit as follows:

$$R_{eq} = V/I = (120 \text{ V})/(5 \text{ A}) = 24 \text{ } \Omega$$

If you used a current less than 5 A (the maximum allowable before the fuse blows), then the R_{eq} would increase. Finally, solve for the number of 360 Ω resistors (lightbulbs) that can be placed in a parallel circuit with an equivalent resistance of 24 Ω.

$$1/R_{eq} = n/R$$

$$n = R/R_{eq} = (360 \text{ } \Omega)/(24 \text{ } \Omega) = 15$$

Thus, a maximum of 15 lightbulbs can be connected in parallel with the 120 V source.

Alternatively, note that the maximum power that can be supplied to the circuit by the battery is $P = IV = (5 \text{ A})(120 \text{ V}) = 600 \text{ W}$. Therefore, the maximum power dissipated by the resistors is also 600 W. Since each resistor dissipates 40 W of power, the maximum number of resistors is 600/40 = 15.

30.   **D**   Since the magnetic flux out of the page is decreasing, the induced current will oppose this change, acting to create more magnetic flux out of the page. According to the right-hand rule, the current must circulate in the counterclockwise direction. Eliminate (A) and (B). As B decreases *uniformly* (that is, while dB/dt is negative and *constant*), the induced emf,

$$\mathcal{E} = \frac{d\Phi_B}{dt} = -A\frac{dB}{dt},$$

is nonzero and constant, which implies that the induced current, $I = \mathcal{E}/R$, is also nonzero and constant.

31. **D** Treat the configuration as three capacitors in series, each with a distance between the plates of $\dfrac{d}{3}$. The capacitance of the two vacuum capacitors will be $\dfrac{\varepsilon_o A}{\left(\dfrac{d}{3}\right)} = \dfrac{3\varepsilon_o A}{d}$, and the capacitance of the dielectric capacitor will be that value times K. Now solve for the total capacitance for the three capacitors in series.

$$\frac{1}{C_{eq}} = \frac{1}{C_1} + \frac{1}{C_2} + \frac{1}{C_3}$$

$$\frac{1}{C_{eq}} = \frac{1}{\left(\dfrac{3\varepsilon_o A}{d}\right)} + \frac{1}{\left(\dfrac{3K\varepsilon_o A}{d}\right)} + \frac{1}{\left(\dfrac{3\varepsilon_o A}{d}\right)}$$

$$\frac{1}{C_{eq}} = \frac{d}{3\varepsilon_o A} + \frac{d}{3K\varepsilon_o A} + \frac{d}{3\varepsilon_o A}$$

$$\frac{1}{C_{eq}} = \frac{2Kd + d}{3K\varepsilon_o A}$$

$$C_{eq} = \frac{3K\varepsilon_o A}{d(2K + 1)}$$

Alternatively, this problem can be solved by Process of Elimination. First, use dimensional analysis. The Table of Information provided with the AP Exams notes that capacitance should have units consistent with $\varepsilon_0 A/d$; (B) fails this test. Second, pretend that the dielectric constant $K = 1$. This is the same as not having a dielectric, so plugging in $K = 1$ should give $C = \varepsilon_0 A/d$. Eliminate (A) and (E). Finally, as K increases, the capacitor becomes stronger, so C should increase. Once (C) fails this test, only (D) is left.

32. **A** The potential is zero at the point midway between the charges, but nowhere is the electric field equal to zero (except at infinity).

33. **D** Ampere's law states that

$$\oint B \cdot d\ell = \mu_0 I_{\text{net through loop}}$$

Because a total current of $2I + 4I = 6I$ passes through the interior of the loop in one direction, and a total current of $I + 3I = 4I$ passes through in the opposite direction, the net current passing through the loop is $6I - 4I = 2I$. Therefore, the absolute value of the integral of **B** around the loop WXYZ is equal to $\mu_0(2I)$, which is equivalent to (D).

34.　**B**　Because $\Delta U = q(\Delta V)$, our answer must be negative. ΔV will be the same in both cases, and the sign on q changed, so the sign on ΔU has to change, as well. Eliminate (C) and (D). Next, notice that the formula for ΔU does not depend on the distance traveled. That means the different path will not impact the magnitude of ΔU. Eliminate (A). Finally, knowing the values of Q_1 and Q_2 would be important if we wanted to calculate a numerical answer, but that knowledge is unnecessary for determining the relationship in the question.

35.　**A**　First, note that the distance between $-q$ and each charge $+Q$ is $\dfrac{1}{\sqrt{2}} L$. Now, refer to the following diagram, where we've invoked Coulomb's law to determine the two electrostatic forces:

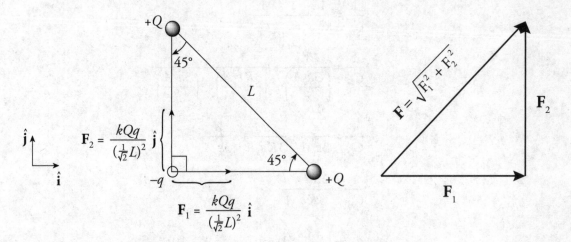

We find that the magnitude of **F** is

$$F = \frac{kQq}{\left(\dfrac{1}{\sqrt{2}}L\right)^2}\sqrt{1^2 + 1^2} = k\frac{2\sqrt{2}\cdot Qq}{L^2} = \frac{1}{4\pi\varepsilon_0}\frac{2\sqrt{2}\cdot Qq}{L^2} = \frac{\sqrt{2}\cdot Qq}{2\pi\varepsilon_0 L^2}$$

so dividing this by m gives the initial acceleration of the charge $-q$.

Section II—Free Response

1. (a) According to the right-hand rule, the magnetic force will push the positive particles to the bottom of the cylinder and the negative particles to the top. This will create an electric field pointing toward the top of the cylinder. As the negative and positive particles accumulate on their respective sides, the electric field and therefore the electric force will increase. This will continue until the electric force (pushing positive particles to the top and negative ones to the bottom) and the magnetic force (pushing positive particles to the bottom and negative ones to the top) reach an equilibrium.

   (b) This situation is similar to a capacitor, so use the equation $V = Ed$. From part (a), we know the magnetic force and electric force will eventually balance, meaning $F_B = F_E$.

   $$F_B = F_E \rightarrow qE = qvB \rightarrow E = vB$$

   Combining this with $V = Ed$ gives

   $$V = (vB)d \rightarrow v = V/(Bd)$$

   (c) Magnetic forces can never do work, so the answer is 0 J.

   (d) Applying the right-hand rule to each side of the particle yields opposite results. The positive end of the particle will experience a push downward, and the negative end will experience a push upward. This means the net force will be zero, resulting in no translational motion. However, forces applied to an object in this way will result in a nonzero net torque. Thus, the object will begin to rotate clockwise.

2. (a) (i) The current decays exponentially according to the equation

   $$I(t) = \frac{\mathcal{E}}{r + R} e^{-t/(r+R)C}$$

   where the initial current, $I(0)$, is equal to $\mathcal{E}/(r + R)$. To find the time t at which $I(t) = I(0)/k$, we solve the following equation:

   $$\frac{\mathcal{E}}{r + R} e^{-t/(r+R)C} = \frac{1}{k} \cdot \frac{\mathcal{E}}{r + R}$$

   $$e^{-t/(r+R)C} = \frac{1}{k}$$

   $$-\frac{t}{(r + R)C} = \ln \frac{1}{k}$$

   $$-\frac{t}{(r + R)C} = -\ln k$$

   $$t = (\ln k)(r + R)C$$

(ii) The charge on the capacitor increases according to the equation

$$Q(t) = C\mathcal{E}\left[1 - e^{-t/(r+R)C}\right]$$

The maximum value is $C\mathcal{E}$ (when the capacitor is fully charged). To find the time t at which $Q(t) = C\mathcal{E}/k$, we solve the following equation:

$$C\mathcal{E}\left[1 - e^{-t/(r+R)C}\right] = \frac{1}{k}C\mathcal{E}$$

$$1 - e^{-t/(r+R)C} = \frac{1}{k}$$

$$e^{-t/(r+R)C} = 1 - \frac{1}{k}$$

$$-\frac{t}{(r+R)C} = \ln\frac{k-1}{k}$$

$$\frac{t}{(r+R)C} = \ln\frac{k}{k-1}$$

$$t = \left(\ln\frac{k}{k-1}\right)(r+R)C$$

(iii) The field energy stored in the capacitor can be written in the form $U_E(t) = [Q(t)]^2/(2C)$, where $Q(t)$ is the function of time given in part (ii) above. We are asked to find the time t at which

$$U(t) = \frac{1}{k}U_{max}$$

$$\frac{[Q(t)]^2}{2C} = \frac{1}{k} \times \frac{(C\mathcal{E})^2}{2C}$$

$$\left\{C\mathcal{E}\left[1 - e^{-t/(r+R)C}\right]\right\}^2 = \frac{1}{k}(C\mathcal{E})^2$$

$$1 - e^{-t/(r+R)C} = \frac{1}{\sqrt{k}}$$

$$e^{-t/(r+R)C} = 1 - \frac{1}{\sqrt{k}}$$

$$-\frac{t}{(r+R)C} = \ln\frac{\sqrt{k}-1}{\sqrt{k}}$$

$$\frac{t}{(r+R)C} = \ln\frac{\sqrt{k}}{\sqrt{k}-1}$$

$$t = \left(\ln\frac{\sqrt{k}}{\sqrt{k}-1}\right)(r+R)C$$

(b) (i) Once the switch is turned to position 2, the capacitor discharges through resistor R (only). Since the capacitor is fully charged (to $V = \mathcal{E}$) when the switch is moved to position 2, the current will decrease exponentially according to the equation

$$I(t) = I_0 e^{-t/RC}$$

where $I_0 = I_{max} = \mathcal{E}/R$. To find the time at which I is equal to I_0/k, we solve the following equation:

$$I_0 e^{-t/RC} = \frac{1}{k} I_0$$

$$e^{-t/RC} = \frac{1}{k}$$

$$-\frac{t}{RC} = \ln \frac{1}{k}$$

$$\frac{t}{RC} = \ln k$$

$$t = (\ln k) RC$$

(ii) Since the charge on the capacitor obeys the same equation as the current—as given in (b) (i) above, simply replacing I by Q—the time t at which the charge drops to $1/k$ times its initial value is the same as the time t at which the current drops to $1/k$ times its initial value: $t = (\ln k) RC$.

3. (a) The magnitude of the average induced emf is given by the equation

$$\mathcal{E}_{avg} = \frac{-\Delta \Phi_B}{\Delta t} = \frac{-\left(BA \cos \theta_f - BA \cos \theta_0\right)}{t_f - t_0}$$

$$= -\frac{B\left(\pi r^2\right)\cos(90) - B\left(\pi r^2\right)\cos(0)}{T - 0}$$

$$= \frac{\pi B r^2}{T}$$

(b) During the rotation, the upward magnetic flux through the ring would be increasing, so the induced current would have to be in a direction that produces downward magnetic flux. Using the right-hand rule, that means the current must run clockwise.

(c) Use Ohm's Law:

$$V = IR \rightarrow \mathcal{E}_{avg} = I_{avg} R \rightarrow I_{avg} = \mathcal{E}_{avg}/R = (\pi B r^2/T)/R = \pi B r^2/RT$$

(d)   First, because the ring rotates steadily through $\pi/2$ radians in T seconds, we can say that the angle θ between **B** and **A** at any given time t will be $\pi/2 \times t/T = \pi t/(2T)$. From there,

$$|\varepsilon| = d\Phi_B/dt = d/dt\,(BA\cos\theta) = d/dt\,(BA\cos(\pi t/(2T))) = BA\,d/dt\cos(\pi t/(2T)) =$$
$$B(\pi r^2)(-\sin(\pi t/(2T)))(\pi/(2T)) = \pi^2 r^2 B(-\sin(\pi t/(2T))/(2T)$$

Inserting t_{max} for t and dropping the negative sign (since we want magnitude) then gives

$$|\varepsilon| = \pi^2 r^2 B\sin(\pi t_{max}/(2T))/(2T)$$

(e)   The maximum value of a function can be found by taking the derivative and setting it equal to 0. Therefore, building on the work from (d), we can say the equation we want to maximize is

$$|\varepsilon| = \pi^2 r^2 B(-\sin(\pi t/(2T)))/(2T)$$

Taking the derivative gives

$$d/dt\,[\pi^2 r^2 B(-\sin(\pi t/(2T)))/(2T)] = \pi^2 r^2 B\,d/dt\,[-\sin(\pi t/(2T))/(2T)] = \pi^2 r^2 B\,(-\cos(\pi t/(2T))/(2T))(\pi/(2T))$$

Finally, setting this equal to 0 will give an equation that could be solved for time.

$$0 = \pi^3 r^2 B(-\cos(\pi t/(2T)))/(4T^2) \rightarrow \cos(\pi t/(2T)) = 0 \rightarrow \pi t/(2T) = \pi/2 \rightarrow t = T$$

Part III
About the
AP Physics C
Exams

- The Structure of the AP Physics C Exams
- How the AP Physics C Exams Are Scored
- Overview of Content Topics
- How AP Exams Are Used
- Other Resources
- Designing Your Study Plan

THE STRUCTURE OF THE AP PHYSICS C EXAMS

The AP Physics C Exam is actually composed of two separate exams: one in Mechanics and one in Electricity and Magnetism (E & M). You can take just the Mechanics, just the E & M, or both. They are administered at different times, and separate scores are reported for each.

Equations and Formulas

Not only are you allowed to use a calculator on the Physics C exams, but you'll also be given the AP Physics C Table of Information, which includes important formulas and equations you'll likely need to apply on the test. We've included these in each of the Practice Tests in this book, as well as online. See "Get More (Free) Content" on page x to gain access to this resource and more!

Whether you are taking Mechanics or E & M, your exam will contain two sections: a multiple-choice section and a free-response section. Questions in the multiple-choice section are each followed by five possible responses (only one of which is correct), and your job, of course, is to choose the right answer. Each right answer is worth one point, and there is no penalty for a wrong answer. There are 35 total multiple-choice questions, and you are given 45 minutes to complete this section. (Note that the Mechanics and E & M tests have completely different sets of questions; the timing is the only thing they share in common.) Calculators can be used for the whole exam on both the multiple-choice and free-response sections. Scientific or graphing calculators are allowed, including the approved calculators listed on the College Board website at apcentral.collegeboard.org/ap-coordinators/on-exam-day/calculator-policy.

The free-response section consists of three multi-part questions that require you to write out your solutions, showing your work. The total amount of time for this section is 45 minutes, so you have an average of 15 minutes per question. Unlike the multiple-choice section, which is scored by computer, the free-response section is graded by high school and college teachers. They have guidelines for awarding partial credit, so you don't need to correctly answer every part in order to get points. Again, you are allowed to use a calculator for the free-response sections, and a table of information (mostly equations) is provided for your use. (We've included a set of AP Physics C equation tables in both Practice Tests.) The two sections—multiple choice and free response—are weighted equally, so each is worth 50 percent of your grade.

HOW THE AP PHYSICS C EXAMS ARE SCORED

Grades on the AP Physics C Exams are reported as a number: either 1, 2, 3, 4, or 5. The descriptions for each of these five numerical scores are as follows, along with the percentage of test-takers who got each grade.

Score	2021 Mechanics Percentage	2021 Electricity & Magnetism Percentage	Credit Recommendation	College Grade Equivalent
5	23.5%	32.6%	Extremely Well Qualified	A
4	28.6%	23.1%	Well Qualified	A−, B+, B
3	21.3%	13.8%	Qualified	B−, C+, C
2	14.9%	18.0%	Possibly Qualified	−
1	11.6%	12.5%	No Recommendation	−

Scores from May 2021 test administration. Data taken from the College Board website.

Colleges are generally looking for a 4 or 5, but some may grant credit for a 3. How well do you have to do to earn such a grade? Each test is curved, and specific cut-offs for each grade vary a little from year to year, but here's a rough idea of how many points you must earn—as a percentage of the maximum possible raw score—to achieve each of the grades 2 through 5:

AP Exam Grade	Percentage Needed
5	≥ 75%
4	≥ 60%
3	≥ 45%
2	≥ 35%

For late-breaking information about test dates, exam formats, and any other changes pertaining to AP Physics C, make sure to check the College Board's website at apstudents. collegeboard.org/ courses

The percentages needed are usually a little lower for the E & M section of Physics C.

OVERVIEW OF CONTENT TOPICS

So, what exactly is on the exams and how do you prepare for them? Here's a listing of the major topics covered on the AP Physics C Exams and the approximate amount of the test devoted to each subject.

Mechanics
> Kinematics (14–20%)
> Newton's Laws of Motion (17–23%)
> Work, Energy, and Power (14–17%)
> Systems of Particles and Linear Momentum (14–17%)
> Rotation (14–20%)
> Oscillations (6–14%)
> Gravitation (6–14%)

Electricity & Magnetism
> Electrostatics (26–34%)
> Conductors, Capacitors, and Dielectrics (14–17%)
> Electric Circuits (17–23%)
> Magnetic Fields (17–23%)
> Electromagnetism (14–20%)

Naturally, it's important to be familiar with the topics—to understand the basics of the theory, to know the definitions of the fundamental quantities, and to recognize and be able to use the equations. Then, you must practice applying what you've learned to answering questions like you'll see on the exam. This book is designed to review all of the content areas covered on the exam, illustrated by hundreds of examples. Also, each chapter (except the first) is followed by practice multiple-choice and free-response questions, and perhaps even more important, *answers and explanations are provided for every example and question in this book*. You'll learn as much—if not more—from actively reading the solutions as you will from reading the text and examples.

Also, two full-length Practice Tests (with solutions) are provided for each course, Mechanics and E&M, one in Part II and one at the end of the book in Part VI. The difficulty level of the examples and questions is equal to or slightly above AP level, so if you have the time and motivation to attack these questions and learn from the solutions, you should feel confident that you can do your very best on the real thing.

HOW AP EXAMS ARE USED

Different colleges use AP Exams in different ways, so it is important that you go to a particular college's website to determine how it uses AP Exams. The three items below represent the main ways in which AP Exam scores can be used:

How Will I Know?
Your dream college's website should tell you whether it accepts AP Exams, and, if so, the minimum score that it requires. You can also contact that school's admissions department, or use the College Board's Credit Policy Search: (apstudent. collegeboard.org/ creditandplacement/ search-credit-policies)

- **College Credit.** Some colleges will give you college credit if you score well on an AP Exam. These credits count towards your graduation requirements, meaning that you can take fewer courses while in college. Given the cost of college, this could be quite a benefit, indeed.

- **Satisfy Requirements.** Some colleges will allow you to "place out" of certain requirements if you do well on an AP Exam, even if they do not give you actual college credits. For example, you might not need to take an introductory-level course, or perhaps you might not need to take a class in a certain discipline at all.

- **Admissions Plus.** Even if your AP Exam will not result in college credit or even allow you to place out of certain courses, most colleges will respect your decision to push yourself by taking an AP course or even an AP Exam outside of a course. A high score on an AP Exam shows mastery of more difficult content than is taught in many high school courses, and colleges may take that into account during the admissions process.

OTHER RESOURCES

There are many resources available to help you improve your scores on the AP Physics C Exams, not the least of which are your teachers. If you are taking an AP class, you may be able to get extra attention from your teacher, such as obtaining feedback on your essays. If you are not in an AP course, reach out to a teacher who teaches AP Physics C and ask if the teacher will review your essays or otherwise help you with content.

Another wonderful resource is **AP Students,** the College Board's official site for all AP Exam materials and information for students. The home page is: http://apstudent.collegeboard.org. The scope of the information at this site is quite broad and includes the following:

- Course description, which includes details on what content is covered
- Free-response question prompts and multiple-choice questions from previous years
- Up-to-date information about changes to the format of the AP Physics C Exams

The AP Students home page address of AP Physics C: Electricity & Magnetism is: apstudent.collegeboard.org/apcourse/ap-physics-c-electricity-and-magnetism.

The home page address of AP Physics C: Mechanics is: apstudent.collegeboard.org/apcourse/ap-physics-c-mechanics.

Finally, The Princeton Review offers tutoring and small group instruction. Our expert instructors can help you refine your strategic approach and add to your content knowledge. For more information, call 1-800-2REVIEW.

AP Students
The College Board provides a bunch of helpful information about the AP Physics C Exams on its AP Students pages, including scoring details and sample free-response questions.

DESIGNING YOUR STUDY PLAN

As part of the Introduction, you identified some areas of potential improvement. Let's now delve further into your performance on Practice Test 1, with the goal of developing a study plan appropriate to your needs and time commitment.

Read the answers and explanations associated with the multiple-choice questions (starting on page 40). After you have done so, respond to the following questions:

- Review the topics list on page 81, and next to each topic, indicate your rank as follows: "1" means "I need a lot of work on this," "2" means "I need to beef up my knowledge," and "3" means "I know this topic well."

- How many days/weeks/months away is your exam?

- What time of day is your best, most-focused study time?

- How much time per day/week/month will you devote to preparing for your exam?

- When will you do this preparation? (Be as specific as possible: Mondays and Wednesdays from 3:00 to 4:00 P.M., for example.)

- Based on the answers above, will you focus on strategy (Part IV) or content (Part V) or both?

- What are your overall goals in using this book?

Part IV
Test-Taking Strategies for the AP Physics C Exams

PREVIEW

Review your Practice Test 1 results and then respond to the following questions:

- How many multiple-choice questions did you miss even though you knew the answer?

- On how many multiple-choice questions did you guess blindly?

- How many multiple-choice questions did you miss after eliminating some answers and guessing based on the remaining answers?

- Did you find any of the free-response questions easier or harder than the others—and, if so, why?

HOW TO USE THE CHAPTERS IN THIS PART

For the following Strategy chapters, think about what you are doing now before you read the chapters. As you read and engage in the directed practice, be sure to appreciate the ways you can change your approach.

Chapter 1
How to Approach
Multiple-Choice
Questions

MULTIPLE-CHOICE QUESTIONS ON THE AP PHYSICS C EXAMS

Use the Answer Sheet

For the multiple-choice section, you write the answers not in the test booklet but on a separate answer sheet (very similar to the ones we've supplied at the very end of this book). Five oval-shaped bubbles follow the question number, one for each possible answer. Don't forget to fill in all your answers on the answer sheet. Marks in the test booklet will not be graded. Also, make sure that your filled-in answers correspond to the correct question numbers! Check your answer sheet after every five answers to make sure you haven't skipped any bubbles by mistake.

Guess Away!

AP Exams do not penalize you for wrong answers, which means you should always guess on questions you do not know. Use Process of Elimination to eliminate answer choices you know are wrong. Eliminating even one wrong answer increases your chances of choosing the right choice.

Should You Guess?

Use Process of Elimination (POE) to rule out answer choices you know are wrong and increase your chances of guessing the right answer. Read all the answer choices carefully. Eliminate the ones that you know are wrong. If you only have one answer choice left, choose it, even if you're not completely sure why it's correct. Remember: Questions in the multiple-choice section are graded by a computer, so it doesn't care how you arrived at the correct answer.

Even if you can't eliminate answer choices, go ahead and guess. The AP Exams no longer include a guessing penalty of a quarter of a point for each incorrect answer. You will be assessed only on the total number of correct answers, so be sure to fill in all the bubbles even if you have no idea what the correct answers are. When you get to questions that are too time-consuming, or you don't know the answer to (and can't eliminate any options), don't just fill in any answer. Use what we call your "letter of the day" (LOTD). Selecting the same answer choice each time you guess will increase your odds of getting a few of those skipped questions right.

Use the Two-Pass System

Remember that you have about one and a quarter minutes per question on this section of the exam. Do not waste time by lingering too long over any single question. If you're having trouble, move on to the next question. After you finish all the questions, you can come back to the ones you skipped.

The best strategy is to go through the multiple-choice section twice. The first time, do all the questions that you can answer fairly quickly—the ones where you feel confident about the correct answer. On this first pass, skip the questions that seem to require more thinking or the ones you need to read two or three times before you understand them. Circle the questions that you've skipped in the question booklet so that you can find them easily in the second pass. You must be very careful with the answer sheet by making sure the filled-in answers correspond correctly to the questions.

Once you have gone through all the questions, go back to the ones that you skipped in the first pass. But don't linger too long on any one question even in the second pass. Spending too much time wrestling over a hard question can cause two things to happen: One, you may run out of time and miss out on answering easier questions in the later part of the exam. Two, your anxiety might start building up, and this could prevent you from thinking clearly, making it even more difficult to answer other questions. If you simply don't know the answer, or can't eliminate any of them, just use your LOTD and move on.

General Advice

Answering 35 multiple-choice questions in 45 minutes can be challenging. Make sure to pace yourself accordingly and remember that you do not need to answer every question correctly to do well. Exploit the multiple-choice structure of this section. There are four wrong answers and only one correct one, so even if you don't know exactly which one is the right answer, you can eliminate some that you know for sure are wrong. Then you can make an educated guess from among the answers that are left and greatly increase your odds of getting that question correct.

Problems with graphs and diagrams are usually the fastest to solve and problems with long sentences for each answer choice usually take the longest to work through. Do not spend too much time on any one problem or you may not get to easier problems further into the test.

These practice exams are written to give you an idea of the format of the test, the difficulty of the questions, and to allow you to practice pacing yourself. Take them in the same circumstances that you will encounter during the real exam: 45 minutes for each of the two multiple-choice sections. And remember, you are allowed to use a calculator!

It Takes Two
In the two-pass system, the first pass involves answering all the questions you know fairly quickly. During the second pass, return to the questions you skipped. (Be sure to circle the skipped questions during your first pass so that you can find them easily during your second pass!)

Chapter 2
How to Approach
Free-Response
Questions

FREE-RESPONSE QUESTIONS ON THE AP PHYSICS C EXAMS

Section II is worth 50 percent of your grade on each of the AP Physics C Exams. You're given a total of 45 minutes to answer three free-response questions on both the Mechanics test and the E & M test.

Pace Yourself

The time constraints of answering three multiple-part free-response questions in 45 minutes can be a challenge. Again, pace yourself so you get to some part of each problem. Make sure to skim all the questions quickly to determine which will be the easiest to solve and start with that one. Often some specific parts of a question are easy, so even if the rest of the question is very difficult, make sure to answer these easier parts. Often students spend too long to get all the parts of one question correct and get almost nothing correct on the other two problems because they are rushed. Pace yourself so that you get to answer the easiest parts of all three questions, and leave the parts that stump you for last.

In the Mechanics free-response section, there is almost always one general mechanics question that involves a variety of principles: energy, momentum, Newton's laws, and other physics concepts. This problem synthesizes a lot of concepts, but each part is usually straightforward. Another one of the questions is almost always a rotational motion problem: rolling motion, fixed-axis rotation, and/or angular momentum. Though you might encounter a second general mechanics question, one of the three problems usually comes from among the following topics: resistive forces, simple harmonic motion, potential energy functions, circular motion and orbits, and gravitational forces. One of these three questions usually involves an experimental component. This is a problem in which an experiment is explained, data or a graph is given, and you must use the information given to solve the problem. The other type of experimental problem asks you to design an experiment that would determine some value you have previously solved for. The experimental problems may make up just one part of one question, or an entire question may be centered on the experiment.

On the Electricity & Magnetism free-response section, you will almost always see one electrostatics question, one circuits question, and one magnetism question. The electrostatics question is often on Gauss's law or several point charges distributed in a plane. The circuit question almost always involves an RC circuit and occasionally includes an inductor. The circuit will feature a switch to add different electrical components at different times. The magnetism question almost always involves induced emf, Faraday's law, and occasionally includes Ampere's law to determine the magnetic field of a wire, solenoid, or toroid. This portion of the test can also include an experimental component. Infrequently there are two magnetism questions and no circuits questions on the free-response section of the test.

Clearly Explain and Justify Your Answers

Remember that your answers to the free-response questions are graded by readers and not by computers. Communication is a very important part of AP Physics C. Compose your answers in precise sentences. Just getting the correct numerical answer is not enough. You should be able to explain your reasoning behind the technique that you selected and communicate your answer in the context of the problem. Even if the question does not explicitly say so, always explain and justify every step of your answer, including the final answer. Do not expect the graders to read between the lines. Explain everything as though somebody with no knowledge of physics is going to read it. Be sure to present your solution in a systematic manner using solid logic and appropriate language. And remember: Although you won't earn points for neatness, the graders can't give you a grade if they can't read and understand your solution!

Use Only the Space You Need

Do not try to fill up the space provided for each question. The space given is usually more than enough. The people who design the tests realize that some students write in big letters and some students make mistakes and need extra space for corrections. So if you have a complete solution, don't worry about the extra space. Writing more will not earn you extra credit. In fact, many students tend to go overboard and shoot themselves in the foot by making a mistake after they've already written the right answer.

Complete the Whole Question

Some questions might have several subparts. Try to answer them all, and don't give up on the question if one part is giving you trouble. For example, if the answer to part (b) depends on the answer to part (a), but you think you got the answer to part (a) wrong, you should still go ahead and do part (b) using your answer to part (a) as required. Chances are that the grader will not mark you wrong twice, unless it is obvious from your answer that you should have discovered your mistake.

Use Common Sense

Always use your common sense in answering questions. For example, on a free-response question that asked students to compute the mean weight of newborn babies from given data, some students answered 70 pounds. It should have been immediately obvious that the answer was probably off by a decimal point. A 70-pound baby would be a giant! This is an important mistake that should be easy to fix. Some mistakes may not be so obvious from the answer. However, the grader will consider simple, easily recognizable errors to be very important.

Think Like a Grader

When answering questions, try to think about what kind of answer the grader is expecting. Look at past free-response questions and grading rubrics on the College Board website. These examples will give you some idea of how the answers should be phrased. The graders are told to keep in mind that there are two aspects to the scoring of free-response answers: showing physics knowledge and communicating that knowledge. Again, responses should be written as clearly as possible in complete sentences.

Think Before You Write

Abraham Lincoln once said that if he had eight hours to chop down a tree, he would spend six of them sharpening his axe. Be sure to spend some time thinking about what the question is, what answers are being asked for, what answers might make sense, and what your intuition is before starting to write. These questions aren't meant to trick you, so all the information you need is given. If you think you don't have the right information, you may have misunderstood the question. In some calculations, it is easy to get confused, so think about whether your answers make sense in terms of what the question is asking. If you have some idea of what the answer should look like before starting to write, then you will avoid getting sidetracked and wasting time on dead-ends.

Chapter 3
Using Time Effectively to Maximize Points

BECOMING A BETTER TEST-TAKER

Very few students stop to think about how to improve their test-taking skills. Most assume that if they study hard, they will test well, and if they do not study, they will do poorly. Most students continue to believe this even after experience teaches them otherwise. Have you ever studied really hard for an exam and then blown it on test day? Have you ever aced an exam for which you thought you weren't well prepared? Most students have had one, if not both, of these experiences. The lesson should be clear: Factors other than your level of preparation influence your final test score. This chapter will provide you with some insights that will help you perform better on the AP Physics C Exams and on other exams, as well.

PACING AND TIMING

Looking for More Help with Your APs?
We now offer specialized AP tutoring and course packages that guarantee a 4 or 5 on the AP. To see which courses are offered and available, and to learn more about the guarantee, visit PrincetonReview.com/college/ap-test-prep

A big part of scoring well on an exam is working at a consistent pace. The worst mistake made by inexperienced or unsavvy test-takers is that they come to a question that stumps them and rather than just skip it, they panic and stall. Time stands still when you're working on a question you cannot answer, and it is not unusual for students to waste five minutes on a single question (especially a question involving a graph or the word EXCEPT) because they are too stubborn to cut their losses. It is important to be aware of how much time you have spent on a given question and on the section you are working. There are several ways to improve your pacing and timing for the test:

- **Know your average pace.** While you prepare for your test, try to gauge how long you take on 5, 10, or 20 questions. Knowing how long you spend on average per question will help you identify how many questions you can answer effectively and how best to pace yourself for the test.

- **Have a watch or clock nearby.** You are permitted to have a watch or clock nearby to help you keep track of time. However, it's important to remember that constantly checking the clock is in itself a waste of time and can be distracting. Devise a plan. Try checking the clock every 10 or 20 questions to see if you are keeping the correct pace or whether you need to speed up. This will ensure that you're cognizant of the time but will not permit you to fall into the trap of dwelling on it.

- **Know when to move on.** Since all questions are scored equally, investing appreciable amounts of time on a single question is inefficient and can potentially deprive you of the chance to answer easier ones later on. You should eliminate answer choices if you are able to, but don't worry about picking a random answer and moving on if you cannot find the correct answer. Remember, tests are like marathons; you do best when you work through them at a steady pace. You can always come back to a question you don't know. When you do, very often you will find that your previous mental block is gone and you will wonder why the question perplexed you the first time around (as you gleefully move on to the next question). Even if you still don't know the answer, you will not have wasted valuable time you could have spent on easier questions.

- **Be selective.** You don't have to do any of the questions in a given section in order. If you are stumped by an essay or multiple-choice question, skip it or choose a different one. Select the questions or essays that you can answer and work on them first. This will make you more efficient and give you the greatest chance of getting the most questions correct.

- **Use Process of Elimination (POE) on multiple-choice questions.** Many times, one or more answer choices can be eliminated. Every answer choice that can be eliminated increases the odds that you will answer the question correctly.

Remember, when all the questions on a test are of equal value, no one question is that important and your overall goal for pacing is to get the most questions correct. Finally, you should set a realistic goal for your final score. In the next section, we will break down how to achieve your desired score and ways of pacing yourself to do so.

Go Online!
Check us out on YouTube for test-taking tips and techniques to help you ace your next exam at www.youtube.com/ThePrincetonReview

GETTING THE SCORE YOU WANT

Depending on the score you need, it may be in your best interest not to try to work through every question. Check with the schools to which you are applying to determine your needed score.

AP Exams in all subjects no longer include a "guessing penalty" of a quarter of a point for every incorrect answer. Instead, students are assessed only on the total number of correct answers. A lot of AP materials, even those you receive in your AP class, may not include this information. It's really important to remember that if you are running out of time, you should fill in all the bubbles before the time for the multiple-choice section is up. Even if you don't plan to spend a lot of time on every question or even if you have no idea what the correct answer is, you need to fill something in.

TEST ANXIETY

Everybody experiences anxiety before and during an exam. To a certain extent, test anxiety can be helpful. Some people find that they perform more quickly and efficiently under stress. If you've ever pulled an all-nighter to write a paper and ended up doing good work, you know the feeling.

However, too much stress is definitely a bad thing. Hyperventilating during the test, for example, almost always leads to a lower score.

If you find that you stress out during exams, here are a few preemptive actions you can take.

- **Take a reality check.** Evaluate your situation before the test begins. If you have studied hard, remind yourself that you are well prepared. Remember that many others taking the test are not as well prepared, and (in your classes, at least) you are being graded against them, so you have an advantage. If you didn't study, accept the fact that you will probably not ace the test. Make sure you get to every question you know something about. Don't stress out or fixate on how much you don't know. Your job is to score as high as you can by maximizing the benefits of what you do know. In either scenario, it's best to think of a test as if it were a game. How can you get the most points in the time allotted to you? Always answer questions you can answer easily and quickly before tackling those that will take more time.

Keep Calm

Trying to relax and de-stress isn't important only on test day; it's also necessary for your test prep! As you work your way through this book, be sure to take intermittent study breaks to help yourself unwind and then refocus.

- **Try to relax.** Slow, deep breathing works for almost everyone. Close your eyes, take a few slow, deep breaths, and concentrate on nothing but your inhalation and exhalation for a few seconds. This is a basic form of meditation that should help you to clear your mind of stress and, as a result, concentrate better on the test. If you have ever taken yoga classes, you probably know some other good relaxation techniques. Use them when you can (obviously, anything that requires leaving your seat and, say, assuming a handstand position won't be allowed by any but the most free-spirited proctors).

- **Eliminate as many surprises as you can.** Make sure you know where the test will be given, when it starts, what type of questions are going to be asked, and how long the test will take. You don't want to be worrying about any of these things on test day or, even worse, after the test has already begun.

The best way to avoid stress is to study both the test material and the test itself. Congratulations! By buying or reading this book, you are taking a major step toward stress-free AP Physics C Exams.

REFLECT

Respond to the following questions:

- How much time will you spend on multiple-choice questions?

- How will you change your approach to multiple-choice questions?

- What is your multiple-choice guessing strategy?

- How much time will you spend on the free-response questions?

- What will you do before you begin writing your free-response answers?

- Will you seek further help, outside of this book (such as a teacher, tutor, or AP Students), on how to approach the questions that you will see on the AP Physics C Mechanics and/or E & M Exams?

Part V
Content Review for the AP Physics C Exams

Chapter 4
Vectors

INTRODUCTION

Vectors will show up all over the place in our study of physics. Some physical quantities that are represented as vectors are: displacement, velocity, acceleration, force, momentum, and electric and magnetic fields. Since vectors play such a recurring role, it's important to become comfortable working with them; the purpose of this chapter is to provide you with a mastery of the fundamental vector algebra we'll use in subsequent chapters. For now, we'll restrict our study to two-dimensional vectors (that is, ones that lie flat in a plane).

DEFINITION

A **vector** is a quantity that involves both magnitude and direction and obeys the **commutative law for addition,** which we'll explain in a moment. A quantity that does not involve direction is a **scalar**. For example, the quantity *55 miles per hour* is a scalar, while the quantity *55 miles per hour to the north* is a vector. Other examples of scalars include: mass, work, energy, power, temperature, and electric charge.

Vectors can be denoted in several ways, including:

$$\mathbf{A}, \boldsymbol{A}, \overrightarrow{\mathrm{A}}$$

In textbooks, you'll usually see one of the first two, but when it's handwritten, you'll see the last one.

Displacement (which is the difference between the final and initial positions) is the prototypical example of a vector:

$$\underbrace{\mathbf{A}}_{\text{displacement}} = \underbrace{\text{4 miles}}_{\text{magnitude}} \underbrace{\text{to the north}}_{\text{direction}}$$

When we say that vectors obey the commutative law for addition, we mean that if we have two vectors of the same type, for example another displacement,

$$\mathbf{B} = \underbrace{\text{3 miles}}_{\text{magnitude}} \underbrace{\text{to the east}}_{\text{direction}}$$

then **A** + **B** must equal **B** + **A**. The vector sum **A** + **B** means *the vector A followed by B*, while the vector sum **B** + **A** means *the vector B followed by A*. That these two sums are indeed identical is shown in the following figure:

Two vectors are equal if they have the same magnitude and the same direction.

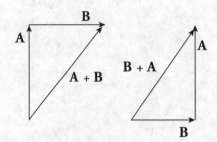

VECTOR ADDITION (GEOMETRIC)

The figure above illustrates how vectors are added to each other geometrically. Place the tail (the initial point) of one vector at the tip of the other vector, then connect the exposed tail to the exposed tip. The vector formed is the sum of the first two. This is called the "tip-to-tail" method of vector addition.

Example 1 Add **A** + **B** using the two following vectors:

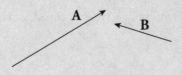

Solution. Place the tail of **B** at the tip of **A** and connect them:

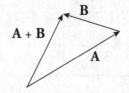

> To construct the resultant vector, connect the tail of the first vector (where you started drawing) to the tip of the final vector (where you stopped drawing).

SCALAR MULTIPLICATION

A vector can be multiplied by a scalar (that is, by a number), and the result is a vector. If the original vector is **A** and the scalar is k, then the scalar multiple $k\mathbf{A}$ is as follows:

$$\text{magnitude of } k\,\mathbf{A} = |k|\,\mathbf{A} \text{ (magnitude of } \mathbf{A})$$

$$\text{direction of } k\mathbf{A} = \begin{cases} \text{the same as } \mathbf{A} \text{ if } k \text{ is positive} \\ \text{the opposite of } \mathbf{A} \text{ if } k \text{ is negative} \end{cases}$$

Example 2 Sketch the scalar multiples $2\mathbf{A}$, $\frac{1}{2}\mathbf{A}$, $-\mathbf{A}$, and $-3\mathbf{A}$ of the vector **A**:

Solution.

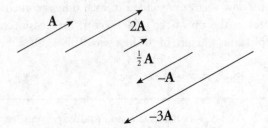

Keep in mind that when you multiply a vector times a scalar of k, the vector becomes k times longer.

VECTOR SUBTRACTION (GEOMETRIC)

To subtract one vector from another, for example, to get $\mathbf{A} - \mathbf{B}$, simply form the vector $-\mathbf{B}$, which is the scalar multiple $(-1)\mathbf{B}$, and add it to $\mathbf{A}$:

$$\mathbf{A} - \mathbf{B} = \mathbf{A} + (-\mathbf{B})$$

Example 3 For the two vectors $\mathbf{A}$ and $\mathbf{B}$, find the vector $\mathbf{A} - \mathbf{B}$.

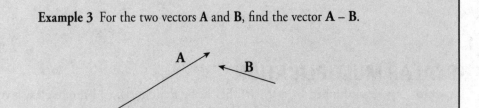

Solution. Flip $\mathbf{B}$ around—thereby forming $-\mathbf{B}$—and add that vector to $\mathbf{A}$:

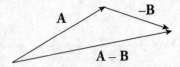

It is important to know that vector subtraction is not commutative: you must perform the subtraction in the order stated in the problem.

STANDARD BASIS VECTORS

Two-dimensional vectors, that is, vectors that lie flat in a plane, can be written as the sum of a horizontal vector and a perpendicular vertical vector. For example, in the following diagram, the vector **A** is equal to the horizontal vector **B** plus the vertical vector **C**:

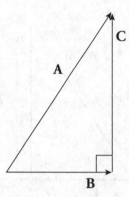

The horizontal vector is always considered a scalar multiple of what's called the **horizontal basis vector**, **i**, and the vertical vector is a scalar multiple of the **vertical basis vector**, **j**. Both of these special vectors have a magnitude of 1, and for this reason, they're called **unit vectors.**

Unit vectors are often represented by placing a hat (caret) over the vector; for example, the unit vectors **i** and **j** are sometimes denoted $\hat{\mathbf{i}}$ and $\hat{\mathbf{j}}$, or $\hat{\imath}$ and $\hat{\jmath}$.

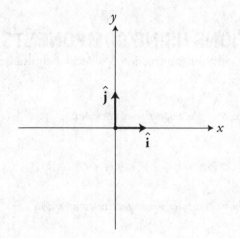

For instance, the vector **A** in the figure below is the sum of the horizontal vector **B** = $3\hat{\mathbf{i}}$ and the vertical vector **C** = $4\hat{\mathbf{j}}$.

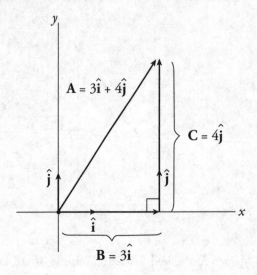

The vectors **B** and **C** are called the **vector components** of **A**, and the scalar multiples of $\hat{\mathbf{i}}$ and $\hat{\mathbf{j}}$ which give **A**—in this case, 3 and 4—are called the **scalar components** of **A**. So vector **A** can be written as the sum **B** + **C** = $A_x\hat{\mathbf{i}} + A_y\hat{\mathbf{j}}$, where A_x and A_y are the scalar components of **A**. The component A_x is called the **horizontal** scalar component of **A**, and A_y is called the **vertical** scalar component of **A**.

VECTOR OPERATIONS USING COMPONENTS

The vector operations of addition, subtraction, and scalar multiplication can now be shown using components.

Vector addition: *Add the respective components*

$$\mathbf{A} + \mathbf{B} = (A_x + B_x)\hat{\mathbf{i}} + (A_y + B_y)\hat{\mathbf{j}}$$

Vector subtraction: *Subtract the respective components*

$$\mathbf{A} - \mathbf{B} = (A_x - B_x)\hat{\mathbf{i}} + (A_y - B_y)\hat{\mathbf{j}}$$

Scalar multiplication: *Multiply each component by k*

$$k\mathbf{A} = (kA_x)\hat{\mathbf{i}} + (kA_y)\hat{\mathbf{j}}$$

Example 4 If $\mathbf{A} = 2\hat{\mathbf{i}} - 3\hat{\mathbf{j}}$ and $\mathbf{B} = -4\hat{\mathbf{i}} + 2\hat{\mathbf{j}}$, compute each of the following vectors: $\mathbf{A} + \mathbf{B}$, $\mathbf{A} - \mathbf{B}$, $2\mathbf{A}$, and $\mathbf{A} + 3\mathbf{B}$.

Solution. It's very helpful that the given vectors **A** and **B** are written explicitly in terms of the standard basis vectors $\hat{\mathbf{i}}$ and $\hat{\mathbf{j}}$:

$$\mathbf{A} + \mathbf{B} = (2 - 4)\hat{\mathbf{i}} + (-3 + 2)\hat{\mathbf{j}} = -2\hat{\mathbf{i}} - \hat{\mathbf{j}}$$

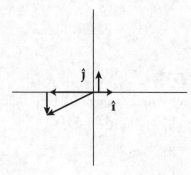

> **Brush Up On the Basics**
>
> If you are taking AP Physics C, you should be familiar with the basics of calculus. If you wish to brush up on your calculus skills, check out our other books, *AP Calculus AB Prep* and/or *AP Calculus BC Prep*.

$$\mathbf{A} - \mathbf{B} = [2 - (-4)]\hat{\mathbf{i}} + (-3 - 2)\hat{\mathbf{j}} = 6\hat{\mathbf{i}} - 5\hat{\mathbf{j}}$$

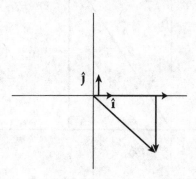

$$2\mathbf{A} = 2(2)\hat{\mathbf{i}} + 2(-3)\hat{\mathbf{j}} = 4\hat{\mathbf{i}} - 6\hat{\mathbf{j}}$$

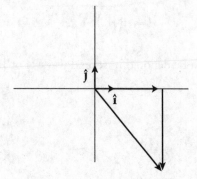

$$\mathbf{A} + 3\mathbf{B} = [2 + 3(-4)]\hat{\mathbf{i}} + [-3 + 3(2)]\hat{\mathbf{j}} = -10\hat{\mathbf{i}} + 3\hat{\mathbf{j}}$$

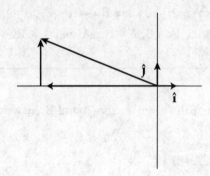

MAGNITUDE OF A VECTOR

As shown in the figure, vector $\mathbf{A}$ is equal to the sum of the perpendicular component vectors $A_x + A_y$. In terms of the components of A_x and A_y, vector $\mathbf{A}$ is equal to $A_x\hat{\mathbf{i}} + A_y\hat{\mathbf{j}}$.

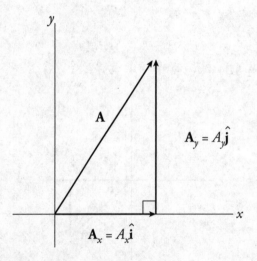

The magnitude of any vector, $\mathbf{A}$, can by computed by applying the Pythagorean Theorem to the scalar components of vectors $A_x\hat{\mathbf{i}}$ and $A_y\hat{\mathbf{j}}$, A_x and A_y.

$$A = \sqrt{\left(A_x\right)^2 + \left(A_y\right)^2}$$

The magnitude of vector $\mathbf{A}$ can be denoted in several ways: A or $|\mathbf{A}|$ or $\|\mathbf{A}\|$.

DIRECTION OF A VECTOR

The direction of a vector can be specified by the angle it makes with the positive *x*-axis. You can sketch the vector and use its components to determine the angle. For example, if θ denotes the angle that the vector $\mathbf{A} = 3\hat{\mathbf{i}} + 4\hat{\mathbf{j}}$ makes with the +*x*-axis, then $\tan\theta = A_y/A_x = 4/3$. You can solve for θ using the inverse trig function, $\theta = \tan^{-1}(4/3) = 53.1°$. Make sure your calculator is in the correct mode, either radian or degree, when using trig functions.

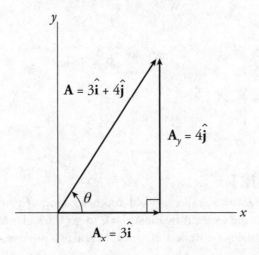

If $\mathbf{A}$ makes the angle θ with the +*x*-axis, then its *x*- and *y*-components are $A\cos\theta$ and $A\sin\theta$, respectively (where A is the magnitude of $\mathbf{A}$).

$$\mathbf{A} = \underbrace{\left(A\cos\theta\right)\hat{\mathbf{i}}}_{A_x} + \underbrace{\left(A\sin\theta\right)\hat{\mathbf{j}}}_{A_y}$$

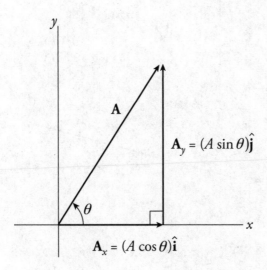

In general, any vector in the plane can be written in terms of two perpendicular component vectors. For example, vector **W** (shown below) is the sum of two component vectors whose magnitudes are $W \cos \theta$ and $W \sin \theta$:

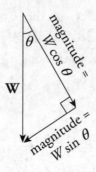

THE DOT PRODUCT

A vector can be multiplied by a scalar to yield another vector, as you saw in Example 2. For instance, we can multiply the vector **A** by the scalar 2 to get the scalar multiple 2**A**. The product is a vector that has twice the magnitude and, since the scalar is positive, the same direction as **A**.

Additionally, we can form a scalar by multiplying two vectors. The product of the vectors in this case is called the **dot product** or the **scalar product**, since the result is a scalar. Several physical concepts (work, electric and magnetic flux) require that we multiply the magnitude of one vector by the magnitude of the component of the other vector that's parallel to the first. The dot product was invented specifically for this purpose.

Consider these two vectors, **A** and **B**:

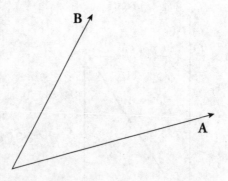

In order to find the component of **B** that's parallel to **A**, we do the following:

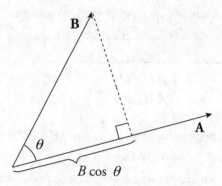

As this figure shows, the component of **B** that's parallel to **A** has the magnitude $B \cos \theta$. So if we multiply the magnitude of **A** by the magnitude of the component of **B** that's parallel to **A**, we would form the product $A(B \cos \theta)$. This is the definition of the dot product of the vectors **A** and **B**

$$\mathbf{A} \cdot \mathbf{B} = AB \cos \theta$$

where θ is the angle between **A** and **B**. Notice that the dot product of two vectors is a scalar.

The angle between the vectors is crucial to the value of the dot product. If $\theta = 0$, then the vectors are already parallel to each other, so we can simply multiply their magnitudes: $\mathbf{A} \cdot \mathbf{B} = AB$. If **A** and **B** are perpendicular, then there is no component of **B** that's parallel to **A** (or vice versa), so the dot product should be zero. And if θ is greater than 90°,

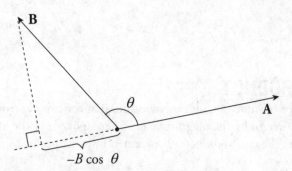

then the component of one that's *parallel* to the other is actually *antiparallel* (backwards), and this will give the dot product a negative value (because $\cos \theta < 0$ if $90° < \theta \leq 180°$).

The value of the dot product can, of course, be figured out using the definition above ($AB \cos \theta$) if θ is known. If θ is not known, the dot product can be calculated from the components of **A** and **B** in this way:

$$\mathbf{A} \cdot \mathbf{B} = (A_x \hat{\mathbf{i}} + A_y \hat{\mathbf{j}}) \cdot (B_x \hat{\mathbf{i}} + B_y \hat{\mathbf{j}}) = A_x B_x + A_y B_y$$

This means that, to form the dot product, simply add the product of the scalar x-components and the product of the scalar y-components.

> **Example 5**
> (a) What is the dot product of the unit vectors $\hat{i}$ and $\hat{j}$?
> (b) What is the dot product of the unit vectors $\hat{i}$ and $\hat{i}$?

Solution.

(a) Since $\hat{i}$ and $\hat{j}$ are perpendicular to each other ($\theta = 90°$), their dot product must be zero, because $\cos 90° = 0$. (In fact, unless **A** or **B** already has magnitude zero, it's also true that two vectors are perpendicular to each other when their dot product is zero.)

(b) Because $\hat{i}$ and $\hat{i}$ are parallel to each other, their dot product is just the product of the magnitudes, which is $1 \cdot 1 = 1$.

> **Example 6** If $\mathbf{A} = -2\hat{i} + 4\hat{j}$ and $\mathbf{B} = 6\hat{i} + B_y\hat{j}$, find the value of B_y such that the vectors **A** and **B** will be perpendicular to each other.

Solution. Two vectors are perpendicular to each other if their dot product is zero. That is, $\mathbf{A} \cdot \mathbf{B} = 0$. The scalar x-component is $(-2)(6) = -12$. The scalar y-component is $(4)(B_y)$. Since the scalar x-component + scalar y-component = 0, $-12 + 4B_y = 0$. This means that B_y must equal 3.

THE CROSS PRODUCT

Some physical concepts (torque, angular momentum, magnetic force) require that we multiply the magnitude of one vector by the magnitude of the component of the other vector that's *perpendicular* to the first. The cross product was invented for this specific purpose.

Consider these two vectors **A** and **B**:

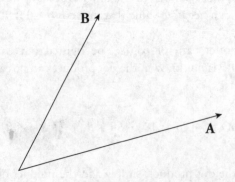

In order to find the component of one of these vectors that's perpendicular to the other one, we do the following:

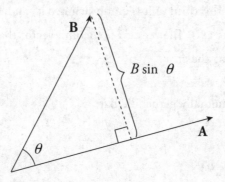

As this figure shows, the component of **B** that's perpendicular to **A** has magnitude $B \sin \theta$. Therefore, if we multiply the magnitude of **A** by the magnitude of the component of **B** that's perpendicular to **A**, we form the product $A(B \sin \theta)$. This is the magnitude of what's called the **cross product** of the vectors **A** and **B**, denoted $\mathbf{A} \times \mathbf{B}$:

$$|\mathbf{A} \times \mathbf{B}| = AB \sin \theta$$

where θ is the angle between **A** and **B** (such that $0° \leq \theta \leq 180°$).

The previous equation gives the magnitude of the cross product. The cross product of two vectors is another vector that's always perpendicular to both **A** and **B**, with its direction determined by a procedure known as the *right-hand rule*. The direction of $\mathbf{A} \times \mathbf{B}$ is perpendicular to the plane that contains **A** and **B**, but this leads to an ambiguity, since there are two directions perpendicular to a plane (one on either side; they point in opposite directions). The following description resolves this ambiguity.

Make sure you are using your right hand. Point your index finger in the direction of the first vector, **A**, then point your middle finger in the direction of the second vector, **B**, and your thumb now points in the direction of the cross product, $\mathbf{A} \times \mathbf{B}$.

> Sometimes, the unit vectors $\hat{i}$, $\hat{j}$, and $\hat{k}$ are instead denoted by $\hat{x}$, $\hat{y}$, and $\hat{z}$, respectively.

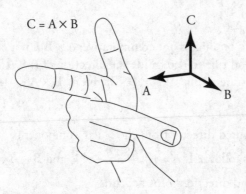

This is called the **right-hand rule**.

You know that the standard basis vectors $\hat{i}$ and $\hat{j}$ can be used to write any two-dimensional vector, but now that we have introduced the cross product, we need a third unit vector to write three-dimensional vectors. This third unit vector is denoted $\hat{k}$, and, like $\hat{i}$ and $\hat{j}$, it points along the direction of a coordinate axis. The vector $\hat{i}$ is the unit vector that points along the x-axis, $\hat{j}$ along the y, and now $\hat{k}$ along the z.

The three unit vectors are mutually perpendicular:

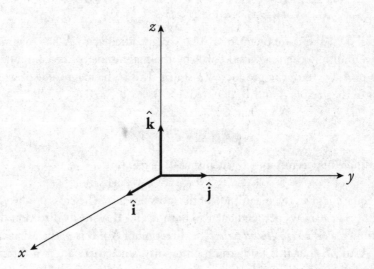

The coordinate axes shown above define a *right-handed* coordinate system, because the directions of the x-, y-, and z-axes obey the right-hand rule. That is, $\hat{i} \times \hat{j}$ points in the direction of $\hat{k}$. In fact, $\hat{i} \times \hat{j}$ actually equals $\hat{k}$, since the magnitude of $\hat{i} \times \hat{j}$ is 1. As a *right-handed* coordinate system, $\hat{j} \times \hat{k} = \hat{i}$ and $\hat{k} \times \hat{i} = \hat{j}$.

Unlike the dot product, the cross product is not commutative; $\mathbf{A} \times \mathbf{B}$ is not equal to $\mathbf{B} \times \mathbf{A}$. This is because applying the right-hand rule to determine the direction of $\mathbf{B} \times \mathbf{A}$ would give a vector that points in the direction opposite to that of $\mathbf{A} \times \mathbf{B}$. Therefore, $\mathbf{B} \times \mathbf{A} = -(\mathbf{A} \times \mathbf{B})$.

The cross product can be computed directly from the scalar components of $\mathbf{A}$ and $\mathbf{B}$ without first determining the angle θ, as follows: If $\mathbf{A} = A_x\hat{i} + A_y\hat{j} + A_z\hat{k}$ and $\mathbf{B} = B_x\hat{i} + B_y\hat{j} + B_z\hat{k}$, then the cross product (magnitude and direction) of $\mathbf{A}$ and $\mathbf{B}$ is

$$\mathbf{A} \times \mathbf{B} = (A_yB_z - A_zB_y)\hat{i} + (A_zB_x - A_xB_z)\hat{j} + (A_xB_y - A_yB_x)\hat{k}$$

This formula requires a lot of memorization. Another method for determining the cross product is to realize that the cross product is the determinant of the following 3×3 matrix.

$$\mathbf{A} \times \mathbf{B} = \begin{vmatrix} \hat{\mathbf{i}} & \hat{\mathbf{j}} & \hat{\mathbf{k}} \\ A_x & A_y & A_z \\ B_x & B_y & B_z \end{vmatrix} = \begin{vmatrix} A_y & A_z \\ B_y & B_z \end{vmatrix} \hat{\mathbf{i}} - \begin{vmatrix} A_x & A_z \\ B_x & B_z \end{vmatrix} \hat{\mathbf{j}} + \begin{vmatrix} A_x & A_y \\ B_x & B_y \end{vmatrix} \hat{\mathbf{k}}$$

By taking the determinant of each 2×2 matrix, you will realize that you get the same formula as shown above. This may appear to be equally difficult to memorize; however, you can just memorize two facts. Each 2×2 matrix is missing the column associated with the basis vector that multiplies it and the terms alternate in sign.

> **Example 7** Calculate the cross product of the vectors $\mathbf{A} = 2\hat{\mathbf{i}} + 3\hat{\mathbf{j}}$ and $\mathbf{B} = -\hat{\mathbf{i}} + \hat{\mathbf{j}} + 4\hat{\mathbf{k}}$, and verify that it's perpendicular to both $\mathbf{A}$ and $\mathbf{B}$.

Solution. One way to calculate the cross product is by using the formula method.

$$\mathbf{A} \times \mathbf{B} = [(3)(4) - (0)(1)]\hat{\mathbf{i}} - [(2)(4) - (0)(-1)]\hat{\mathbf{j}} + [(2)(1) - (3)(-1)]\hat{\mathbf{k}} = 12\hat{\mathbf{i}} - 8\hat{\mathbf{j}} + 5\hat{\mathbf{k}}$$

Using the determinant method,

$$\mathbf{A} \times \mathbf{B} = \begin{vmatrix} \hat{\mathbf{i}} & \hat{\mathbf{j}} & \hat{\mathbf{k}} \\ 2 & 3 & 0 \\ -1 & 1 & 4 \end{vmatrix} = \begin{vmatrix} 3 & 0 \\ 1 & 4 \end{vmatrix} \hat{\mathbf{i}} - \begin{vmatrix} 2 & 0 \\ -1 & 4 \end{vmatrix} \hat{\mathbf{j}} + \begin{vmatrix} 2 & 3 \\ -1 & 1 \end{vmatrix} \hat{\mathbf{k}}$$

$$\mathbf{A} \times \mathbf{B} = [(3)(4) - (0)(1)]\hat{\mathbf{i}} - [(2)(4) - (0)(1)]\hat{\mathbf{j}} + [(2)(1) - (3)(-1)]\hat{\mathbf{k}}$$

$$\mathbf{A} \times \mathbf{B} = 12\hat{\mathbf{i}} - 8\hat{\mathbf{j}} + 5\hat{\mathbf{k}}$$

Now, to verify that this vector is perpendicular to both $\mathbf{A}$ and $\mathbf{B}$, we use the property of the dot product: Two vectors are perpendicular to each other if their dot product is zero. Extending the computation of the dot product in terms of the scalar components to three dimensions, we get:

$$(\mathbf{A} \times \mathbf{B}) \cdot \mathbf{A} = (12\hat{\mathbf{i}} - 8\hat{\mathbf{j}} + 5\hat{\mathbf{k}}) \cdot (2\hat{\mathbf{i}} + 3\hat{\mathbf{j}}) = (12)(2) + (-8)(3) + (5)(0) = 0$$

$$(\mathbf{A} \times \mathbf{B}) \cdot \mathbf{B} = (12\hat{\mathbf{i}} - 8\hat{\mathbf{j}} + 5\hat{\mathbf{k}}) \cdot (-\hat{\mathbf{i}} + \hat{\mathbf{j}} + 4\hat{\mathbf{k}}) = (12)(-1) + (-8)(1) + (5)(4) = 0$$

Chapter 4 Drill

The answers and explanations can be found in Chapter 17.

Section I: Multiple Choice

1. Vector **A** has components (15, 8), and vector **B** has components (3, 3). If **C** = **B** – **A**, what is the magnitude of vector **C**?

   (A) 10.5
   (B) 13.0
   (C) 15.6
   (D) 16.5
   (E) 21.2

2. If $\mathbf{a} = 2\hat{\mathbf{i}} + y\hat{\mathbf{j}}$ and $\mathbf{b} = x\hat{\mathbf{i}} + 3\hat{\mathbf{j}}$ and $\mathbf{c} = \mathbf{a} + \mathbf{b} = 3\hat{\mathbf{i}} - 10\hat{\mathbf{j}}$, what is the value of the ordered pair (x, y)?

   (A) (−1, −13)
   (B) (−1, 13)
   (C) (1, −13)
   (D) (1, 7)
   (E) (1, 13)

3. Let $\mathbf{u} = -2\hat{\mathbf{i}} + 2\hat{\mathbf{j}}$ and $\mathbf{v} = 4\hat{\mathbf{i}} + 5\hat{\mathbf{j}}$. What is the angle formed between vectors **u** and **v**?

   (A) 83.7°
   (B) 89.7°
   (C) 90.0°
   (D) 90.3°
   (E) 96.3°

4. Which of the following is vector **B** = (2, B_y, B_z) if the cross product of $\mathbf{v} \times \mathbf{B} = 10$ directed out of the page and $\mathbf{v} = 2\hat{\mathbf{i}} + \hat{\mathbf{j}}$?

   (A) (2, −6, 0)
   (B) (2, 0, −6)
   (C) (2, 0, 6)
   (D) (2, 2, 0)
   (E) (2, 6, 0)

5. Let $\mathbf{A} = 4\hat{\mathbf{i}} + 5\hat{\mathbf{j}} + 8\hat{\mathbf{k}}$, $\mathbf{B} = 2\hat{\mathbf{i}} + \hat{\mathbf{j}} - \hat{\mathbf{k}}$, and $\mathbf{C} = \hat{\mathbf{i}} - 3\hat{\mathbf{j}} + 2\hat{\mathbf{k}}$. What is the value of $\mathbf{A} \cdot (\mathbf{B} \times \mathbf{C})$?

   (A) −101
   (B) −85
   (C) −35
   (D) 19
   (E) 85

Section II: Free Response

1. A force of 10 N acted 30° above the +x-axis is used to displace a 5 kg mass horizontally in the +x-axis for 10 m.

    (a) Considering that force, **F**, and displacement, **d**, are vector quantities,

    (i) calculate $F_{parallel}$, the component of **F** that aligns with **d**.

    (ii) calculate the work done by this force using $W = F_{parallel}d$. The S.I. unit for work is N·m = J for joules.

    (b) Let $\mathbf{F} = 5\sqrt{3} \text{ N } \hat{\mathbf{i}} + 5 \text{ N } \hat{\mathbf{j}}$ and $\mathbf{d} = 10 \text{ m } \hat{\mathbf{i}}$. Calculate $\mathbf{F} \cdot \mathbf{d}$.

    (c) Compare your answers in (a)(ii) and (b). Briefly explain why they should or shouldn't be the same.

    (d) What is the direction of $\mathbf{F} \times \mathbf{d}$? Briefly explain your reasoning without calculating the cross product.

Summary

Definitions

o A vector is a quantity that involves both magnitude and direction.

o A quantity that does not involve direction is a scalar.

Vector Addition/Subtraction (Geometric)

o Vector addition is carried out with the "tip-to-tail" method: place the tail of one vector at the tip of the other vector, then connect the exposed tail to the exposed tip.

o Changing the sign of a vector changes its direction by 180°.

o Vector subtraction is carried out with the "tip-to-tail" method, but the negative of the second vector is added: $\mathbf{A} - \mathbf{B} = \mathbf{A} + (-\mathbf{B})$.

Standard Basis Vectors

o Vectors can be expressed in terms of components using the unit basis vectors to indicate direction: $\hat{\mathbf{i}}$ indicates the +x-direction, $\hat{\mathbf{j}}$ indicates the +y-direction, and $\hat{\mathbf{k}}$ indicates the +z-direction.

o Vector addition: add the respective components

o Vector subtraction: subtract the respective components

o Scalar multiplication: multiply each component by the scalar

Magnitude and Direction of a Vector

o The magnitude of any vector, **A**, can be computed by applying the Pythagorean Theorem to the scalar components of the vector, A_x and A_y: $A = \sqrt{\left(A_x\right)^2 + \left(A_y\right)^2}$.

o The direction of any vector, **A**, can be specified by the angle, θ, it makes with the positive x-axis such that $\tan(\theta) = A_y / A_x$.

Dot Product and Cross Product

o The dot product or scalar product is a scalar that results from multiplying the magnitude of one vector with the magnitude of the component of another vector that's parallel to the first: $\mathbf{A} \cdot \mathbf{B} = AB\cos(\theta)$.

o The cross product is a vector that has a magnitude that results from multiplying the magnitude of one vector by the magnitude of the component of another vector that's perpendicular to the first: $|\mathbf{A} \times \mathbf{B}| = AB\sin(\theta)$. The direction of the cross product is perpendicular to both vectors.

o The right-hand rule can be used to determine the direction of the cross product: point your index finger in the direction of the first vector, then point your middle finger in the direction of the second vector, and your thumb now points in the direction of the cross product.

o The magnitude and direction of the cross product can be found from the determinant of a 3×3 matrix with the unit basis vectors forming the top row, the scalar components of the first vector forming the middle row, and the scalar components of the second vector forming the bottom row of the matrix.

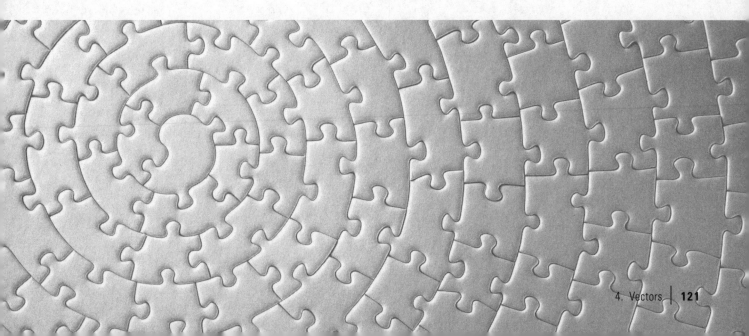

Chapter 5
Kinematics

INTRODUCTION

Kinematics is the study of an object's motion in terms of its displacement, velocity, and acceleration. Questions such as *How far does this object travel?, How fast and in what direction does it move?,* or *At what rate does its speed change?* all properly belong to kinematics. In the next chapter, we will study **dynamics**, which delves more deeply into *why* objects move the way they do.

POSITION, DISTANCE, AND DISPLACEMENT

Position is an object's relation to a coordinate axis system. **Distance** is a scalar that represents the total amount traveled by an object. **Displacement** is an object's change in position. It's the vector that points from the object's initial position to its final position, regardless of the path actually taken. Since displacement means *change in position*, it is generically denoted $\Delta\mathbf{s}$, where Δ denotes *change in* and $\mathbf{s}$ means spatial location. (The letter $\mathbf{p}$ is not used for position because it's reserved for another quantity: **momentum.**) If it's known that the displacement is horizontal, then it can be called $\Delta\mathbf{x}$; if the displacement is vertical, then it's $\Delta\mathbf{y}$. The magnitude of this vector is the *net* distance traveled and, sometimes, the word *displacement* refers just to this scalar quantity. Since a distance is being measured, the correct unit for displacement is the meter.

Example 1 Traveling along a single axis, a car starts 10 m from the origin. The car then moves 8 m directly away from the origin and then turns around and moves 12 m back toward the origin. Determine the final position of the car, the distance the car traveled, and the displacement of the car.

Solution. The car ends at the 6 m mark, so the final position is 6 m. The car moves a total of 8 m + 12 m = 20 m, so the distance traveled is 20 m. The displacement of the car only refers to the car's final position minus its initial position. Because the car ended 4 m behind where it started, its displacement would be −4 m.

Example 2 An infant crawls 5 m east, then 3 m north, then 1 m east. Find the magnitude of the infant's displacement.

Solution. Although the infant crawled a total distance of 5 m + 3 m + 1 m = 9 m, this is not the magnitude of the infant's displacement, which is merely the net distance traveled.

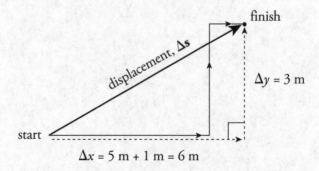

Using the Pythagorean Theorem, we can calculate that the magnitude of the displacement is

$$\Delta s = \sqrt{\left(\Delta x\right)^2 + \left(\Delta y\right)^2} = \sqrt{\left(6\,\mathrm{m}\right)^2 + \left(3\,\mathrm{m}\right)^2} = \sqrt{45\,\mathrm{m}^2} = 6.7\,\mathrm{m}$$

> **Example 3** In a track-and-field event, an athlete runs exactly once around an oval track, a total distance of 400 m. Find the runner's displacement for the race.

Solution. If the runner returns to the same position from which she left, then her displacement is zero.

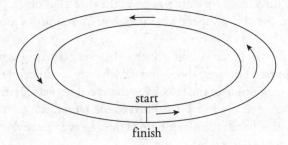

The *total distance* covered is 400 m, and the displacement is 0 m.

SPEED AND VELOCITY

When we're in a moving car, the speedometer tells us how fast we're going; it gives us our speed. But what does it mean to have a speed of say, 10 m/s? It means that we're covering a distance of 10 meters every second. By definition, **average speed** is the ratio of the total distance traveled to the time required to cover that distance:

$$\text{average speed} = \frac{\text{total distance}}{\text{time}}$$

The car's speedometer doesn't care in what direction the car is moving (as long as the wheels are moving forward). You could be driving north, south, east, west, whatever; the speedometer would make no distinction. *55 miles per hour, north* and *55 miles per hour, east* register the same on the speedometer: 55 miles per hour. Speed is a scalar.

However, we will also need to include *direction* in our descriptions of motion. We just learned about displacement, which takes both (net) distance and direction into account. The single concept that embodies both displacement and direction is called **velocity**, and the definition of average velocity is:

$$\text{average velocity} = \frac{\text{displacement}}{\text{time}}$$

$$\overline{\mathbf{v}} = \frac{\Delta \mathbf{s}}{\Delta t}$$

(The bar over the **v** means *average*.) Because $\Delta \mathbf{s}$ is a vector, $\overline{\mathbf{v}}$ is also a vector, and because Δt is a *positive* scalar, the direction of $\overline{\mathbf{v}}$ is the same as the direction of $\Delta \mathbf{s}$. The magnitude of the velocity vector is called the object's **speed**, and is expressed in units of meters per second (m/s).

Note the distinction between speed and velocity. In everyday language, they're often used interchangeably. In physics, however, speed and velocity are technical terms whose definitions are not the same. The magnitude of the velocity is the speed, so velocity is speed plus direction. Because speed is a scalar, average speed is defined as the *total* distance traveled divided by the elapsed time. On the other hand, the magnitude of the average velocity, which is a vector, is the *displacement* divided by the elapsed time.

However, the car's speedometer is not intended to tell you an average speed over a long period of time: It's supposed to tell you the current speed at the present moment! For instantaneous speed or velocity, we need calculus. Any shift from an average quantity with deltas in the definition to an instantaneous quantity, especially if the quantity is a rate of change (that is, if Δt is in the denominator), is likely to involve changing "Δ", a difference, to "d", a derivative, and leaving everything else the same, and this is no exception:

$$\mathbf{v} = \frac{d\mathbf{s}}{dt}$$

Got a Question?
For answers to test-prep questions for all your tests and additional test-taking tips, subscribe to our YouTube channel at www.youtube.com/ThePrincetonReview

Often, a derivative with respect to time is denoted by a dot above the quantity, so an alternative notation for the equation above is $\mathbf{v} = \dot{\mathbf{x}}$.

Because taking an integral is the inverse of taking a derivative, this definition also implies that displacement is given by the integral of velocity with respect to time:

$$\mathbf{s} = \int \mathbf{v}\, dt$$

Example 4 If the infant in Example 2 completes his journey in 20 seconds, find the magnitude of his average velocity.

Solution. Since the magnitude of the displacement is 6.7 m, the magnitude of his average velocity is

$$\overline{v} = \Delta s/\Delta t = (6.7 \text{ m})/(20 \text{ s}) = 0.34 \text{ m/s}$$

Example 5 Assume that the runner in Example 3 completes the race in 1 minute and 18 seconds. Find her average speed and the magnitude of her average velocity.

Solution. *Average speed is total distance divided by elapsed time, a scalar.* Since the length of the track is 400 m, the runner's average speed was (400 m)/(78 s) = 5.1 m/s. However, since her displacement was zero, the magnitude of her average velocity was zero also: $\overline{v} = \Delta s/\Delta t = (0 \text{ m})/(78 \text{ s}) = 0$ m/s.

Example 6 Is it possible to move with constant speed but not constant velocity? Is it possible to move with constant velocity but not constant speed?

Solution. The answer to the first question is *yes*. For example, if you set your car's cruise control at 55 miles per hour but turn the steering wheel to follow a curved section of road, then the direction of your velocity changes (which means your velocity is not constant), even though your speed doesn't change.

The answer to the second question is *no*. Velocity means speed and direction; if the velocity is constant, then that means both speed and direction are constant. If speed were to change, then the velocity vector's magnitude would change (by definition), which immediately implies that the vector changes.

Example 7 An object's position x, in meters, obeys the equation $x = \sin(t)$, where t is the time in seconds since the object began moving. How fast is the object moving at $t = \dfrac{\pi}{2}$ seconds?

Solution. This question asks for the velocity of the object at $t = \dfrac{\pi}{2}$ seconds. Since velocity is the time derivative of position, the velocity is given by $v = \dfrac{d}{dt}(\sin(t)) = \cos(t)$. Then plug in the value of t: $v\left(\frac{\pi}{2}\right) = \cos\left(\frac{\pi}{2}\right) = 0$ m/s .

ACCELERATION

When you step on the gas pedal in your car, the car's speed increases; step on the brake, and the car's speed decreases. Turn the wheel, and the car's direction of motion changes. In all of these cases, the velocity changes. To describe this change in velocity, we need a new term: **acceleration**. In the same way that velocity measures the rate-of-change of an object's position, acceleration measures the rate-of-change of an object's velocity. An object's average acceleration is defined as follows:

$$\text{average acceleration} = \frac{\text{change in velocity}}{\text{time}}$$

$$\bar{\mathbf{a}} = \frac{\Delta \mathbf{v}}{\Delta t}$$

The units of acceleration are meters per second, per second: $[a] = \text{m/s}^2$. Because $\Delta \mathbf{v}$ is a vector, $\bar{\mathbf{a}}$ is also a vector; and because Δt is a *positive* scalar, the direction of $\bar{\mathbf{a}}$ is the same as the direction of $\Delta \mathbf{v}$.

Furthermore, if we take an object's original direction of motion to be positive, then an increase in speed corresponds to a positive acceleration, while a decrease in speed corresponds to a negative acceleration (deceleration).

Note that an object can accelerate even if its speed doesn't change. (Again, it's a matter of not allowing the everyday usage of the word *accelerate* to interfere with its technical, physics usage.) This is because acceleration depends on $\Delta \mathbf{v}$, and the velocity vector $\mathbf{v}$ changes if (1) speed changes, (2) direction changes, or (3) both speed and direction change. For instance, a car traveling around a circular racetrack is constantly accelerating even if the car's *speed* is constant because the direction of the car's velocity vector is constantly changing.

Just as a transition from average velocity to instantaneous velocity changes "Δ" to "d" in an equation, so too does a transition from average acceleration to instantaneous acceleration. Instantaneous acceleration is given by:

$$\mathbf{a} = \frac{d\mathbf{v}}{dt}$$

Since velocity is itself the derivative of position, this implies that acceleration is the second derivative of position:

$$\mathbf{a} = \frac{d^2\mathbf{s}}{dt^2}$$

A dot over a quantity denotes a derivative with respect to time acting on that quantity.

In dot notation, $\mathbf{a} = \dot{\mathbf{v}} = \ddot{\mathbf{x}}$. The above definition of acceleration also implies that velocity is the integral of acceleration:

$$\mathbf{v} = \int \mathbf{a} \, dt$$

The relationships described here can be summarized in the following diagram:

$$\text{position} \xleftarrow{\text{differentiate}} \text{velocity} \xleftarrow{\text{differentiate}} \text{acceleration}$$
$$\Delta x(t) \xleftarrow{\text{integrate}} \Delta v(t) \xleftarrow{\text{integrate}} a(t)$$

Example 8 A car is traveling in a straight line along a highway at a constant speed of 80 miles per hour for 10 seconds. Find its acceleration.

Solution. Neither the direction nor the speed change, so the car is traveling at a constant velocity. If the velocity is constant, then the acceleration is zero. If there's no change in velocity, then there's no acceleration.

Example 9 A car is traveling along a straight highway at a speed of 20 m/s. The driver steps on the gas pedal and, 3 seconds later, the car's speed is 32 m/s. Find its average acceleration.

Solution. Here, the direction does not change but the speed does, so there is a change in velocity. To obtain the average acceleration, simply divide the change in velocity, 32 m/s – 20 m/s = 12 m/s, by the time interval during which the change occurred: $\overline{\mathbf{a}} = \mathbf{v}/t = (12 \text{ m/s})/(3 \text{ s}) = 4 \text{ m/s}^2$.

Example 10 Spotting a police car ahead, the driver of the car in the previous example slows from 32 m/s to 20 m/s in 2 seconds. Find the car's average acceleration.

Solution. Dividing the change in velocity, 20 m/s – 32 m/s = –12 m/s, by the time interval during which the change occurred, 2 s, gives us $\overline{\mathbf{a}} = \mathbf{v}/t = (-12 \text{ m/s}) / (2 \text{ s}) = -6 \text{ m/s}^2$. The negative sign here means that the direction of the acceleration is opposite the direction of the velocity, which describes slowing down.

Example 11 The position of an object (measured in meters from the origin, where $x = 0$) moving along a straight line is given as a function of time t (measured in seconds) by the equation $x(t) = 4t^2 - 6t - 40$. Find

(a) its velocity at time t

(b) its acceleration at time t

(c) the time at which the object is at the origin

(d) the object's velocity and acceleration at the time calculated in (c)

Solution.

(a) The direction does not change but the position does. The velocity as a function of t is the derivative of the position function:

$$v(t) = \dot{x}(t) = \frac{d}{dt}\left(4t^2 - 6t - 40\right) = 8t - 6 \ \left(\text{in m/s}\right)$$

(b) The acceleration as a function of t is the derivative of the velocity function:

$$a(t) = \dot{v}(t) = \frac{d}{dt}(8t - 6) = 8 \text{ m/s}^2$$

Notice that the acceleration is constant (because it doesn't depend on t).

(c) The object is at the origin when $x(t)$ is equal to 0; that is, when

$$
\begin{aligned}
4t^2 - 6t - 40 &= 0 \\
2(2t^2 - 3t - 20) &= 0 \\
2(2t + 5)(t - 4) &= 0
\end{aligned}
$$

$$t = -\frac{5}{2} \text{ or 4 seconds}$$

(d) Disregarding the negative value for t, we can say that the object passes through the origin at $t = 4$ s. At this time, the object's velocity is

$$v(4) = v(t)\big|_{t=4} = (8t - 6)\big|_{t=4} = (8)(4) - 6 = 26 \text{ m/s}$$

The object's acceleration is a constant 8 m/s² throughout its motion, so, in particular, at $t = 4$ s, the acceleration is 8 m/s².

Example 12 An object is moving along the x-axis with an acceleration given by the function, $a(t) = (4t + 7)$ m/s². At time $t_0 = 0$, the object is at $x = 6$ m, and it is moving at 2 m/s. How fast will the object be traveling at time $t = 4$ s? Where will the object be at time $t = 4$ s?

Solution. By integrating the acceleration with respect to time, we find the velocity as a function of time.

$$v(t) = \int a(t)dt = \int (4t + 7)dt = 2t^2 + 7t + c$$

We determine the value of c, the constant of integration, by using the given initial velocity, $v(0) = 2$ m/s.

$$v(0) = 2 \Rightarrow (2t^2 + 7t + c)\big|_{t=0} = 2 \Rightarrow c = 2$$

(Notice that c is the initial velocity, v_0). So the velocity function is given by: $v(t) = 2t^2 + 7t + 2$ (in m/s). Therefore, we can evaluate the function at $t = 4$ s and determine the velocity at that moment of time.

$$v(4) = (2t^2 + 7t + 2)\big|_{t=4} = (2(4)^2 + 7(4) + 2) = 62 \text{ m/s}$$

Integrating this velocity function with respect to time will give us the position function.

$$x(t) = \int v(t)dt = \int (2t^2 + 7t + 2)dt = \frac{2}{3}t^3 + \frac{7}{2}t^2 + 2t + c_1$$

Again, we determine the constant of integration, c_1, by using the given initial position, $x(0) = 6$ m.

$$x(0) = 6 \Rightarrow \left(\frac{2}{3}t^3 + \frac{7}{2}t^2 + 2t + c_1\right)\bigg|_{t=0} = 6 \Rightarrow c_1 = 6$$

(Again, notice that c_1 is just the initial position, x_0). So the position function is given by:

$$x(t) = \left(\frac{2}{3}t^3 + \frac{7}{2}t^2 + 2t + 6\right) \text{ in meters.}$$

Therefore, we can evaluate the function at $t = 4$ s and determine the position at that moment of time.

$$x(4) = \left(\frac{2}{3}t^3 + \frac{7}{2}t^2 + 2t + 6\right)\bigg|_{t=4} = \left(\frac{2}{3}(4)^3 + \frac{7}{2}(4)^2 + 2(4) + 6\right) = 112.\overline{66} \text{ m}$$

UNIFORMLY ACCELERATED MOTION AND THE BIG FIVE

The simplest type of motion to analyze is motion in which the acceleration is *constant* (possibly equal to zero). Although true uniform acceleration is rarely achieved in the real world, many common motions are governed by approximately constant acceleration. In these cases, the kinematics of uniformly accelerated motion provide an approximate description of what's happening. Notice that if the acceleration is constant, then the average acceleration is also constant, so $\bar{\mathbf{a}} = a$.

We can simplify our analysis by considering only motion that takes place along a straight line. In these cases, there are only two possible directions of motion. One is positive, and the other, which is in the opposite direction, is negative. Displacement, velocity, and acceleration are all vectors, which means that they include both a magnitude and a direction. With straight-line motion, direction can be specified simply by attaching a + or − sign to the magnitude of the quantity. Therefore, although we will often abandon the use of bold letters to denote the vector quantities of displacement, velocity, and acceleration, the fact that these quantities include direction will still be indicated by a positive or negative sign.

Let's review the quantities we've seen so far. The fundamental quantities are displacement (Δx), velocity (v), and acceleration (a). Acceleration is a change in velocity, from an initial velocity (v_i or v_0) to a final velocity (v_f or simply v—with no subscript) during some elapsed time interval, Δt. At $t_0 = 0$, $\Delta t = t - 0 = t$. We will use $\Delta t = t$ to match the AP equation sheet. Therefore, we have five kinematics quantities: Δx, v_0, v, a, and t.

These five quantities are related by a group of five equations that we call the *Big Five*. They work in cases where acceleration is uniform, which are the ones we're considering.

The Big Five

The three boxed equations to the right are the ones that will be provided to you on the AP Physics C Table of Information. (Remember, you can find this resource in each Practice Test.) Throughout this book, equations from the Table of Information will be enclosed in a box.

		Variable that is missing
Big Five #1	$\Delta x = x - x_0 = \bar{v}t$	a
Big Five #2	$v = v_0 + at$	Δx
Big Five #3	$x = x_0 + v_0 t + \frac{1}{2}at^2$	v
Big Five #4	$x = x_0 + vt - \frac{1}{2}at^2$	v_0
Big Five #5	$v^2 = v_0^2 + 2a(x - x_0)$	t

The change in the position of an object is the displacement: $\Delta x = x - x_0$. Equations #2, #3, and #5 are given on the AP Physics C Table of Information. They are written here exactly as they will appear on the sheet, which is why we are using Δx rather than Δs for the displacement. Because the acceleration is constant, the average velocity is $\bar{v} = \frac{1}{2}\left(v_o + v\right)$.

Each of the Big Five equations is missing one of the five kinematic quantities. To decide which equation to use when solving a problem, determine which of the kinematic quantities is missing from the problem—that is, which quantity is neither given nor asked for—and then use the equation that doesn't contain that variable. A good strategy is to make a list of your "knowns" and your "unknowns." For example, if the problem never mentions the final velocity—v is neither given nor asked for—the equation that will work is the one that's missing v. That's Big Five #3.

Big Five #1 and #2 are simply the definitions of $\bar{v}$ and $\bar{a}$ written in forms that don't involve fractions. The other Big Five equations can be derived from these two definitions and the equation $\bar{v} = \frac{1}{2}(v_0 + v)$ by using a bit of algebra.

> **Example 13** An object with an initial velocity of 4 m/s moves along a straight axis under constant acceleration. Three seconds later, its velocity is 14 m/s. How far did it travel during this time?

Solution. We're given v_0, t, and v, and we're asked for a change in position, x. So a is missing; it isn't given and it isn't asked for, and we use Big Five #1:

$$\Delta x = \bar{v}t$$

$$\Delta x = \frac{1}{2}(14 + 4)(3) = 27 \text{ m}$$

It's okay to leave off the units in the middle of the calculation as long as you remember to include them in your final answer. Leaving units off your final answer will cost you points on the free-response section of the AP Exam.

> **Example 14** A car that's initially traveling at 10 m/s accelerates uniformly for 4 seconds at a rate of 2 m/s² in a straight line. How far does the car travel during this time?

Solution. We're given v_0, t, and a, and we're asked for x. So, v is missing; it isn't given and it isn't asked for, and we use Big Five #3:

$$\Delta x = v_0 t + \frac{1}{2}a(t)^2 = (10 \text{ m/s})(4 \text{ s}) + \frac{1}{2}(2 \text{ m/s}^2)(4 \text{ s})^2 = 56 \text{ m}$$

> **Example 15** A rock is dropped off a cliff that's 80 m high. If it strikes the ground with an impact velocity of 40 m/s, what acceleration did it experience during its descent?

Solution. If something is dropped, then that means it has no initial velocity: $v_0 = 0$. So, we're given v_0, x, and v, and we're asked for a. Since t is missing, we use Big Five #5, choosing positive as downward:

$$v^2 = v_0^2 + 2a\Delta x \Rightarrow v^2 = 2a\Delta x \quad (\text{since } v_0 = 0)$$

$$a = \frac{v^2}{2\Delta x} = \frac{(40 \text{ m/s})^2}{2(80 \text{ m})} = 10 \text{ m/s}^2 \text{ downward}$$

Note that since a has the same sign as Δx, the acceleration vector points in the same direction as the displacement vector. This makes sense since the object moves downward and the acceleration it experiences is due to gravity, which also points downward.

KINEMATICS WITH GRAPHS

So far, we have dealt with kinematics problems algebraically, but you should also be able to handle kinematics questions in which information is given graphically. The two most popular graphs in kinematics are position-vs.-time graphs and velocity-vs.-time graphs. For example, consider an object that's moving along an axis in such a way that its position x as a function of time t is given by the following position-vs.-time graph:

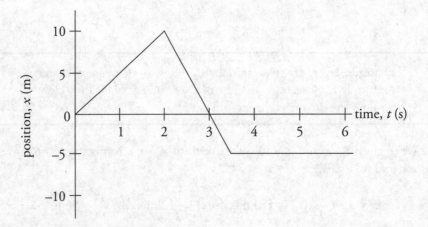

What does this graph tell us? It says that at time $t = 0$, the object was at position $x = 0$. Then, in the next two seconds, its position changed from $x = 0$ to $x = 10$ m. Then, at time $t = 2$ s, it reversed direction and headed back toward its starting point, reaching $x = 0$ at time $t = 3$ s, and continued, reaching position $x = -5$ m at time $t = 3.5$ s. Then the object remained at this position, $x = -5$ m, at least through time $t = 6$ s. Notice how economically the graph embodies all this information!

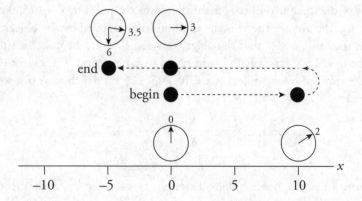

We can also determine the object's average velocity (and average speed) during particular time intervals. For example, its average velocity from time $t = 0$ to time $t = 2$ s is equal to the object's displacement, $10 - 0 = 10$ m, divided by the elapsed time, 2 s:

$$\bar{v} = \frac{\Delta x}{\Delta t} = \frac{(10-0)\text{ m}}{(2-0)\text{ s}} = 5 \text{ m/s}$$

Note, however, that the ratio that defines the average velocity, $\Delta x/\Delta t$, also defines the slope of the x-vs.-t graph. Therefore, we know the following important fact:

> The slope of a position-vs.-time graph gives the average velocity.

What was the average velocity from time $t = 2$ s to time $t = 3.5$ s? The slope of the line segment joining the point $(t, x) = (2 \text{ s}, 10 \text{ m})$ to the point $(t, x) = (3.5 \text{ s}, -5 \text{ m})$ is

$$\bar{v} = \frac{\Delta x}{\Delta t} = \frac{(-5-10)\text{ m}}{(3.5-2)\text{ s}} = -10 \text{ m/s}$$

The fact that $\bar{v}$ is negative tells us that the object's displacement was negative during this time interval; that is, it moved in the negative x direction. The fact that $\bar{v}$ is negative agrees with the observation that the slope of a line that falls to the right is negative. What is the object's average velocity from time $t = 3.5$ s to time $t = 6$ s? Since the line segment from $t = 3.5$ s to $t = 6$ s is horizontal, its slope is zero, which implies that the average velocity is zero, but we can also figure this out from looking at the graph, since the object's position did not change during that time.

Finally, let's figure out the object's average velocity and average speed for its entire journey (from $t = 0$ to $t = 6$ s). The average velocity is

$$\bar{v} = \frac{\Delta x}{\Delta t} = \frac{(-5-0)\text{ m}}{(6-0)\text{ s}} = -0.83 \text{ m/s}$$

This is the slope of the imagined line segment that joins the point $(t, x) = (0\text{ s}, 0\text{ m})$ to the point $(t, x) = (6\text{ s}, -5\text{ m})$. The average speed is the total distance traveled by the object divided by the elapsed time. In this case, notice that the object traveled 10 m in the first 2 s, then 15 m (albeit backward) in the next 1.5 s; it covered no additional distance from $t = 3.5$ s to $t = 6$ s. Therefore, the total distance traveled by the object is $x = 10 + 15 = 25$ m, which took 6 s, so

$$\text{average speed} = \frac{x}{\Delta t} = \frac{25\text{ m}}{6\text{ s}} = 4.2\text{ m/s}$$

Let's next consider an object moving along a straight axis in such a way that its velocity, v, as a function of time, t, is given by the following velocity-vs.-time graph:

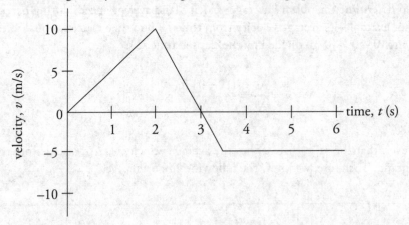

What does this graph tell us? It says that, at time $t = 0$, the object's velocity was $v = 0$. Over the first two seconds, its velocity increased steadily to 10 m/s. At time $t = 2$ s, the velocity then began to decrease (eventually becoming $v = 0$, at time $t = 3$ s). The velocity then became negative after $t = 3$ s, reaching $v = -5$ m/s at time $t = 3.5$ s. From $t = 3.5$ s on, the velocity remained a steady -5 m/s.

> The velocity changed signs from positive to negative at $t = 3$ s. Therefore, the object turned around at $t = 3$ s.

What can we ask about this motion? First, the fact that the velocity changed from $t = 0$ to $t = 2$ s tells us that the object accelerated. The acceleration during this time was

$$a = \frac{\Delta v}{\Delta t} = \frac{(10 - 0)\text{ m/s}}{(2 - 0)\text{ s}} = 5\text{ m/s}^2$$

Note, however, that the ratio that defines the acceleration, $\Delta v / \Delta t$, also defines the slope of the v-vs.-t graph. Therefore,

> The slope of a velocity-vs.-time graph gives the average acceleration.

What was the acceleration from time $t = 2$ s to time $t = 3.5$ s? The slope of the line segment joining the point $(t, v) = (2$ s, 10 m/s$)$ to the point $(t, v) = (3.5$ s, -5 m/s$)$ is

$$a = \frac{\Delta v}{\Delta t} = \frac{(-5-10)\ \text{m/s}}{(3.5-2)\ \text{s}} = -10\ \text{m/s}^2$$

The fact that a is negative tells us that the object's velocity change was also negative during this time interval; that is, the object accelerated in the negative direction, or slowed down. In fact, after time $t = 3$ s, the velocity became more negative, indicating that the direction of motion was negative with increasing speed. What is the object's acceleration from time $t = 3.5$ s to time $t = 6$ s? Since the line segment from $t = 3.5$ s to $t = 6$ s is horizontal, its slope is zero, which indicates that the acceleration is zero, but you can also see this from looking at the graph; the object's velocity did not change during this time interval.

Another question can be asked when a velocity-vs.-time graph is given: How far did the object travel during a particular time interval? For example, let's figure out the displacement of the object from time $t = 4$ s to time $t = 6$ s. During this time interval, the velocity was a constant -5 m/s, so the displacement was $\Delta x = v\Delta t = (-5$ m/s$)(2$ s$) = -10$ m.

Geometrically, we've determined the area between the graph and the horizontal axis. After all, the area of a rectangle is *base* × *height* and, for the shaded rectangle shown below, the *base* is Δt and the *height* is v. So, *base* × *height* equals $\Delta t \times v$, which is displacement.

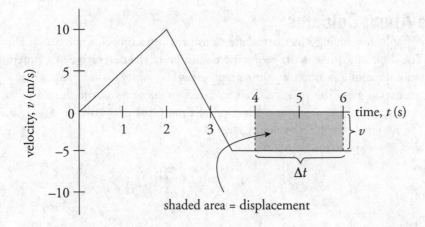

Counting areas above the horizontal axis as positive and areas below the horizontal axis as negative, we can make the following claim:

> Given a velocity-vs.-time graph, the area between the graph and the t-axis equals the object's displacement.

What is the object's displacement from time $t = 0$ to $t = 3$ s? Using the fact that displacement is the area bounded by the velocity graph, we can calculate the area of the triangle shown below:

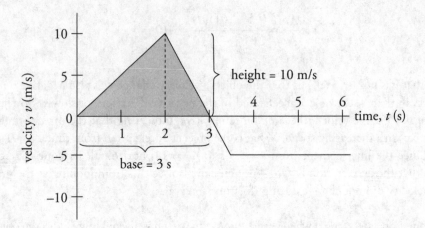

Since the area of a triangle is $\left(\dfrac{1}{2}\right) \times$ base $\times$ height, we find that

$$\Delta x = \frac{1}{2}\left(3 \text{ s}\right)\left(10 \text{ m/s}\right) = 15 \text{ m}$$

A Note About Calculus

The use of graphs for solving kinematics questions provides a link to some basic definitions in calculus. The slope of a curve is the geometric definition of the **derivative** of a function. Since the slope of a position-vs.-time graph gives the velocity, the derivative of a position function gives the velocity. That is, given an equation of the form $x = x(t)$ for the position x of an object as a function of time t, the derivative of $x(t)$—with respect to t—gives the object's velocity $v(t)$:

$$v\left(t\right) = \frac{dx}{dt}$$

Often, a derivative with respect to time is denoted by a dot above the quantity, so an alternative notation for the equation above is $v\left(t\right) = \dot{x}\left(t\right)$.

Next, since the slope of a velocity-vs.-time graph gives the acceleration, the derivative of a velocity function gives the acceleration. That is, given an equation of the form $v = v(t)$ for the velocity v of an object as a function of time t, the derivative of $v(t)$—with respect to t—gives the object's acceleration $a(t)$:

$$a\left(t\right) = \frac{dv}{dt}$$

**Go Crazy
with Calculus**

Are you also a calculus whiz who is taking the AP Calculus AB or BC Exam? We've got you covered either way! Check out our AP Calculus prep guides, *AP Calculus AB Prep* and *AP Calculus BC Prep*.

Combining this equation with the previous one, we see that the acceleration is the second derivative of the position:

$$a(t) = \frac{dv}{dt} = \frac{d}{dt}\left(\frac{dx}{dt}\right) = \frac{d^2x}{dt^2}$$

> Equations written in terms of differentials like *d/dt* are called **differential equations.** If you need to solve a differential equation on either of the AP Physics C Exams, you can likely solve it using the method of separation of variables.

Using the dot notation, these last two equations may be written as $a(t) = \dot{v}(t)$ and $a(t) = \ddot{x}(t)$, respectively.

The area bounded by a curve and the horizontal axis is the geometric definition of the **definite integral** of a function. Since the area bounded by a velocity-vs.-time graph gives the displacement, the definite integral of a velocity function gives the displacement. That is, given the equation $v = v(t)$ for the velocity v of an object as a function of time t, the integral of $v(t)$ from time $t = t_1$ to time $t = t_2$ equals the displacement during this time interval:

$$\text{displacement} = \int_{t_1}^{t_2} v(t)\,dt$$

For the velocity-vs.-time graph examined above, the slope of the line segment from $(t, v) = (0, 0)$ to $(t, v) = (2, 10)$ can be described by the equation $v(t) = 5t$. Therefore, the acceleration of the object from time $t = 0$ to $t = 2$ s is

$$a = \frac{dv}{dt} = \frac{d}{dt}(5t) = 5 \ \text{m/s}^2$$

as we found above.

The line segment from $(t, v) = (2, 10)$ to $(t, v) = (3, 0)$ can be described by the equation $v(t) = -10t + 30$. The displacement of the object from time $t = 0$ to $t = 3$ s is the value of the integral from $t = 0$ to $t = 3$ s, which we can get by adding the integral of $v(t) = 5t$ from $t = 0$ to $t = 2$ s to the integral of $v(t) = -10t + 30$ from $t = 2$ s to $t = 3$ s:

$$\begin{aligned}
\text{displacement} &= \int_0^2 5t\,dt + \int_2^3 (-10t + 30)\,dt \\
&= \left(\frac{5}{2}t^2\right)_0^2 + \left(-5t^2 + 30t\right)_2^3 \\
&= \frac{5}{2}\left(2^2 - 0^2\right) + \left[\left(-5 \cdot 3^2 + 30 \cdot 3\right) - \left(-5 \cdot 2^2 + 30 \cdot 2\right)\right] \\
&= 10 + [45 - 40] \\
&= 15 \text{ m}
\end{aligned}$$

which also agrees with the value computed above.

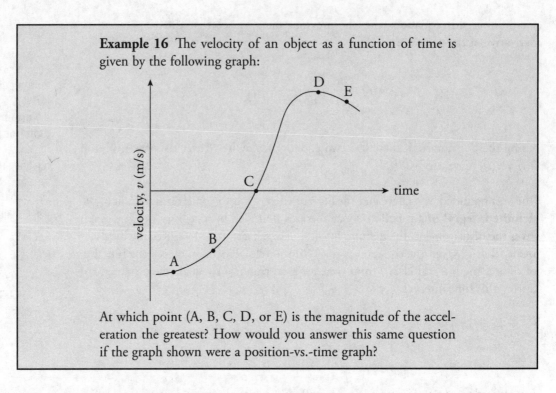

Example 16 The velocity of an object as a function of time is given by the following graph:

At which point (A, B, C, D, or E) is the magnitude of the acceleration the greatest? How would you answer this same question if the graph shown were a position-vs.-time graph?

Solution. The acceleration is the slope of the velocity-vs.-time graph, or *dv/dt*. Although this graph is not composed of straight lines, the concept of slope still applies; at each point, the slope of the curve is the slope of the tangent line to the curve. The slope is essentially zero at points A and D (where the curve is flat), small and positive at B, and small and negative at E. The slope at point C is large and positive, so this is where the object's acceleration is the greatest.

If the graph shown were a position-vs.-time graph, then the slope would be the velocity. The slope of the given graph starts at zero (around point A), slowly increases to a small positive value at B, continues to slowly increase to a large positive value at C, then, at around point D, this large positive slope decreases quickly to zero. Of the points designated on the graph, point D is the location of the greatest slope change, which means that this is the point of the greatest velocity change. Therefore, this is the point at which the magnitude of the acceleration is greatest.

Average vs. Instantaneous Quantities

We have a variety of methods to determine the velocity and the acceleration of an object. We also need to distinguish between average quantities and instantaneous quantities. Average velocity is the displacement over some time interval, while instantaneous velocity is how fast the object is traveling at a specific instant of time. The speedometer on a car gives us the instantaneous *speed* (it does not indicate the direction, so it does not indicate the instantaneous velocity). Typically, it is easiest to solve for average quantities using the definitions mentioned before. It is easiest to solve for instantaneous quantities by using one of the Big Five equations, or taking the slope of a given graph or derivative of a given function. Make sure you carefully read the question and understand whether it is asking for an average quantity or an instantaneous quantity.

FREE FALL

The simplest real-life example of motion under relatively constant acceleration is the motion of objects in Earth's gravitational field, ignoring any effects due to the air (mainly air resistance). Under these conditions, an object can fall *freely*, that is, it can fall experiencing only acceleration due to gravity. Near the surface of Earth, the gravitational acceleration has a constant magnitude of about 9.8 m/s²; this quantity is denoted g (for *gravitational* acceleration). On the AP Physics C Exams, you may use $g = 10$ m/s² as a simple approximation to $g = 9.8$ m/s². This is particularly helpful because, as we mentioned earlier, you will be permitted to use a calculator on the multiple-choice section. In this book, we will always use $g = 10$ m/s². And, of course, the gravitational acceleration vector, **g**, points *downward*.

> g is a positive constant because it is the magnitude of an acceleration. Anytime you see g in an equation, you will plug in 10 m/s² for it.

Since the acceleration is constant, we can use the Big Five with a replaced by $+g$ or $-g$. We will use y for displacement rather than s or x because the motion is vertical. To decide which of these two values to use for a, make a decision at the beginning of your calculations whether to call "down" the positive direction or the negative direction. If you call "down" the positive direction, then $a = +g$. If you call "down" the negative direction, then $a = -g$. We will always assume up is the positive direction unless all of the motion is downward.

In each of the following examples, we'll ignore effects due to the air.

Example 17 A rock is dropped from an 80-meter cliff. How long does it take to reach the ground?

Solution. Since all of the rock's motion is down, we call down the positive direction, so $a = +g$. We're given v_0, y, and a, and we are asked for t. So v is missing; it isn't given and it isn't asked for, and we use Big Five #3:

$$y = y_0 + v_0 t + \frac{1}{2}gt^2 \Rightarrow y = \frac{1}{2}gt^2 \text{ (since } y_0 = 0 \text{ and } v_0 = 0)$$

$$t = \sqrt{\frac{2y}{g}} = \sqrt{\frac{2(80)}{10}} = 4.0 \text{ s}$$

Example 18 A baseball is thrown straight upward with an initial speed of 20 m/s. How high will it go?

Solution. Since the ball travels upward, call up the positive direction. Therefore, down is the negative direction, so $a = -g$. The ball's velocity drops to zero at the instant the ball reaches its highest point, so we're given a, v_0, and v, and asked for y. Since t is missing, we use Big Five #5:

$$v^2 = v_0^2 + 2a\Delta y \Rightarrow 2a\Delta y = -v_0^2 \text{ (because } v = 0)$$

$$\Delta y = -\frac{v_0^2}{2a}$$

$$= -\frac{v_0^2}{2(-g)} = -\frac{(+20 \text{ m/s})^2}{2(-10 \text{ m/s}^2)} = 20 \text{ m}$$

Example 19 One second after being thrown straight down, an object is falling with a speed of 20 m/s. How fast will it be falling 2 seconds after it was traveling 20 m/s?

Solution. Because all the motion is downward, call down the positive direction, so $a = +g$ and $v_0 = +20$ m/s. We're given v_0, a, and t, and asked for v. Since y is missing, we use Big Five #2:

$$v = v_0 + at = (+20 \text{ m/s}) + (+10 \text{ m/s}^2)(2 \text{ s}) = 40 \text{ m/s}$$

Example 20 If an object is thrown straight upward with an initial speed of 8 m/s and takes 3 seconds to strike the ground, from what height was the object thrown?

Solution. The figure below shows the path of the ball. We will use up as positive because not all the motion is downward. Therefore, $a = -g$. We're given a, v_0, and t, and we need to find y, which is $y - y_0$. Since v is missing, we use Big Five #3:

Did you need to solve for the max height? Nope! It might be tempting to break up the motion into several parts, but it's not necessary.

In this example, the magnitude of the displacement equals the height from which the object was thrown.

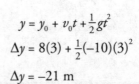

$$y = y_0 + v_0 t + \frac{1}{2}gt^2$$
$$\Delta y = 8(3) + \frac{1}{2}(-10)(3)^2$$
$$\Delta y = -21 \text{ m}$$

Notice that the displacement is −21 m. This means the object landed 21 m below where it started. Therefore, the height it started at was 21 m.

PROJECTILE MOTION

In general, an object that moves near the surface of Earth will not follow a straight-line path (for example, a baseball hit by a bat, a golf ball struck by a club, or a tennis ball hit from the baseline). If we launch an object at an angle other than straight upward and consider only the effect of acceleration due to gravity, then the object will travel along a parabolic trajectory.

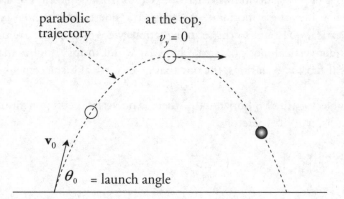

To simplify the analysis of parabolic motion, *we analyze the horizontal and vertical motions **separately***, using the Big Five. This is the key to doing projectile motion problems. Calling *down* the negative direction, we have the following:

Horizontal motion:

$$\Delta x = v_{0x}t$$

$$v_x = v_{0x} \text{ (constant!)}$$

$$a_x = 0$$

Vertical motion:

$$\Delta y = v_{0y}t + \frac{1}{2}(-g)t^2$$

$$v_y = v_{0y} + (-g)t$$

$$a_y = -g$$

$$v_y^2 = v_{0y}^2 + 2(-g)\Delta y$$

The quantity v_{0x}, which is the horizontal (or x) component of the initial velocity, is equal to $v_0 \cos \theta_0$, where θ_0 is the **launch angle**, the angle that the initial velocity vector, $\mathbf{v}_0$, makes with the horizontal. Similarly, the quantity v_{0y}, the vertical (or y) component of the initial velocity, is equal to $v_0 \sin \theta_0$.

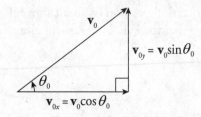

> **Example 21** An object is thrown horizontally off a cliff with an initial speed of 10 m/s. How far will it drop in 4 seconds assuming it does not hit the ground first?

Solution. The first step is to decide whether this is a horizontal question or a vertical question, since you must consider these motions separately. The question asks how far the object will drop. This is a vertical question, so the set of equations we will consider are those listed on the previous page under vertical motion. Next, "How far will it drop?" implies that we will use the first of the vertical-motion equations, the one that gives vertical displacement, y.

Now, since the object is thrown horizontally, there is no vertical component to its initial velocity vector $\mathbf{v}_0$; that is, $v_{0y} = 0$. Therefore,

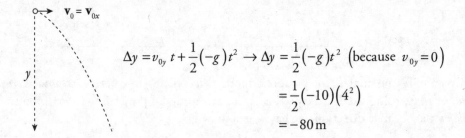

$$\Delta y = v_{0y}\, t + \frac{1}{2}\left(-g\right)t^2 \rightarrow \Delta y = \frac{1}{2}\left(-g\right)t^2 \ \left(\text{because } v_{0y}=0\right)$$
$$= \frac{1}{2}\left(-10\right)\left(4^2\right)$$
$$= -80\,\text{m}$$

The fact that Δy is negative means that the displacement is *down*. Also, notice that the information given about v_{0x} is irrelevant to the question.

> **Example 22** From a height of 100 m, a ball is thrown horizontally with an initial speed of 15 m/s. How far does it travel horizontally in the first 2 seconds?

Solution. The question, "How far does it travel horizontally?", immediately tells us that we should use the first of the horizontal-motion equations:

$$\Delta x = v_{0x}t = (15 \text{ m/s})(2 \text{ s}) = 30 \text{ m}$$

The information that the initial vertical position is 100 m above the ground is irrelevant (except for the fact that it's high enough that the ball doesn't strike the ground before the two seconds have elapsed).

Example 23 A projectile is traveling in a parabolic path for a total of 6 seconds. How does its horizontal velocity 1 s after launch compare to its horizontal velocity 4 s after launch?

Solution. The only acceleration experienced by the projectile is due to gravity, which is purely vertical, so there is no horizontal acceleration. If there's no horizontal acceleration, then the horizontal velocity cannot change during flight, and the projectile's horizontal velocity 1 s after it's launched is the same as its horizontal velocity 3 s later.

Example 24 An object is projected upward with a 30° launch angle and an initial speed of 40 m/s. How long will it take for the object to reach the top of its trajectory? How high is this?

Solution. When the projectile reaches the top of its trajectory, its velocity vector is momentarily horizontal; that is, $v_y = 0$. We are asked how long it will take the object to reach this point with $\theta = 30$ degrees. Using the vertical-motion equation for v_y, we can set it equal to 0 and solve for t:

$$v_{0y} + (-g)\, t = 0$$

$$t = \frac{v_{0y}}{g} = \frac{v_0 \sin \theta_0}{g} = \frac{(40 \text{ m/s}) \sin 30°}{10 \text{ m/s}^2} = 2 \text{ s}$$

At this time, the projectile's vertical displacement is

$$\Delta y = v_{0y} t + \frac{1}{2}(-g)\, t^2 = (v_0 \sin \theta_0)\, t + \frac{1}{2}(-g)\, t^2$$

$$= \left[(40 \text{ m/s}) \sin 30° \right](2 \text{ s}) + \frac{1}{2}\left(-10 \text{ m/s}^2\right)(2 \text{ s})^2$$

$$= 20 \text{ m}$$

Example 25 An object is projected upward with a 30° launch angle and an initial speed of 60 m/s. For how many seconds will it be in the air? How far will it travel horizontally before returning to its original height?

Solution. The total time the object spends in the air is equal to twice the time required to reach the top of the trajectory (because the parabola is symmetrical). So, as we did in the previous example, we find the time required to reach the top by setting v_y equal to 0, and now double that amount of time:

$$v_{0y} + (-g)t = 0$$

$$t = \frac{v_{0y}}{g} = \frac{v_0 \sin \theta_0}{g} = \frac{(60 \text{ m/s}) \sin 30^\circ}{10 \text{ m/s}^2} = 3 \text{ s}$$

Therefore, the *total* flight time (that is, up and down) is $T = 2t = 2 \times (3 \text{ s}) = 6 \text{ s}$.

Now, using the first horizontal-motion equation, we can calculate the horizontal displacement after 6 seconds:

$$\Delta x = v_{0x} T = (v_0 \cos \theta_0) T = \left[(60 \text{ m/s}) \cos 30^\circ \right] (6 \text{ s}) = 310 \text{ m}$$

By the way, the full horizontal displacement of a projectile to the same height is called the projectile's **range**.

A Note About Notation

We are trying to be as consistent as possible with the test by using equations identical to those on the AP Physics C Table of Information that you will be given to use during the exam. When the motion is vertical, we use y instead of x, and occasionally we use s to indicate when the motion is in two dimensions. It is important that you judge which variable best represents the values you are given and those you need to solve for.

Chapter 5 Drill

The answers and explanations can be found in Chapter 17.

Section I: Multiple Choice

1. An object that's moving with constant speed travels once around a circular path. Which of the following is/are true concerning this motion?

    I.   The displacement is zero.
    II. The average speed is zero.
    III. The acceleration is zero.

    (A) I only
    (B) I and II only
    (C) I and III only
    (D) III only
    (E) II and III only

2. At time $t = t_1$, an object's velocity is given by the vector $\mathbf{v}_1$ shown below:

    A short time later, at $t = t_2$, the object's velocity is the vector $\mathbf{v}_2$:

    If $v_2 = v_1$, which one of the following vectors best illustrates the object's average acceleration between $t = t_1$ and $t = t_2$?

    (A)

    (B)

    (C)

    (D)

    (E)

3. A rock is thrown off a 30-m cliff at a 45° angle above the horizontal. Which of the following is true regarding the acceleration of the rock?

    (A) The acceleration will be of magnitude g and have both horizontal and vertical components.
    (B) At the peak of the rock's path, the magnitude of acceleration will be half of what it was when the rock was initially thrown.
    (C) The acceleration will be of magnitude g and in a downward direction.
    (D) The acceleration will be downward during the rock's ascent and upward during the rock's descent.
    (E) The acceleration will increase throughout the rock's flight.

4. A baseball is thrown straight upward. What is the ball's acceleration at its highest point?

    (A) 0

    (B) $\frac{1}{2}g$, downward

    (C) g, downward

    (D) $\frac{1}{2}g$, upward

    (E) g, upward

5. An object's location, in meters, after t seconds have passed is given by the equation

    $$\mathbf{x}(t) = -3t^3 + t^2 + 6t$$

    What is the maximum velocity of the object?

    (A) 0.11 m/s
    (B) 2.73 m/s
    (C) 6.11 m/s
    (D) 7.39 m/s
    (E) 9.81 m/s

6. A rock is dropped off a cliff and strikes the ground with an impact speed of 30 m/s. How high was the cliff?

    (A) 15 m
    (B) 20 m
    (C) 30 m
    (D) 45 m
    (E) 60 m

7. A stone is thrown horizontally with an initial speed of 10 m/s from a bridge. If air resistance could be ignored, how long would it take the stone to strike the water 80 m below the bridge?

   (A) 1 s
   (B) 2 s
   (C) 4 s
   (D) 6 s
   (E) 8 s

8. A soccer ball, at rest on the ground, is kicked with an initial velocity of 10 m/s at a launch angle of 30°. Calculate its total flight time, assuming that air resistance is negligible.

   (A) 0.5 s
   (B) 1 s
   (C) 1.7 s
   (D) 2 s
   (E) 4 s

9. A stone is thrown horizontally with an initial speed of 30 m/s from a bridge. Find the stone's total speed when it enters the water 4 seconds later. (Ignore air resistance.)

   (A) 30 m/s
   (B) 40 m/s
   (C) 50 m/s
   (D) 60 m/s
   (E) 70 m/s

10. Which one of the following statements is true concerning the motion of an ideal projectile launched at an angle of 45° to the horizontal?

   (A) The acceleration vector points opposite to the velocity vector on the way up and in the same direction as the velocity vector on the way down.
   (B) The speed at the top of the trajectory is zero.
   (C) The object's total speed remains constant during the entire flight.
   (D) The horizontal speed decreases on the way up and increases on the way down.
   (E) The vertical speed decreases on the way up and increases on the way down.

11. An object is going to be launched from the ground with an initial velocity of **v**. It starts a distance d away from a wall of height h. Assume that the wall is close enough that any angle $\theta < 45°$ would not make it over the wall. Which of the following equations could be solved to find the angle θ that would maximize the horizontal distance the object travels while still ensuring it passes over the wall?

   (A) $h = d\tan\theta - \dfrac{1}{2}\left(\dfrac{gd^2}{v^2\cos^2\theta}\right)$

   (B) $h = d^2\cos\theta - \dfrac{1}{2}\left(\dfrac{gd^2}{v^2\cos^2\theta}\right)$

   (C) $h = d\tan\theta - \dfrac{1}{2}\left(\dfrac{gd^2}{v\cos\theta}\right)$

   (D) $h = d\sin\theta - \dfrac{1}{2}\left(\dfrac{gd^2}{v^2\cos^2\theta}\right)$

   (E) $h = d\sin\theta - \dfrac{1}{2}\left(\dfrac{gd^2}{v^2\cos\theta}\right)$

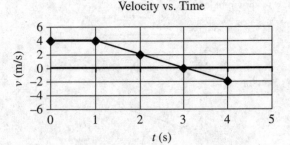

Velocity vs. Time

12. The velocity of an object moving in a straight line is graphed above. If $x = 3.0$ m at $t = 0$ s, what is the position of the particle at $t = 3.0$ s?

   (A) 6 m
   (B) 10 m
   (C) 8 m
   (D) 11 m
   (E) −5 m

Section II: Free Response

1. This question concerns the motion of a car on a straight track; the car's velocity as a function of time is plotted below.

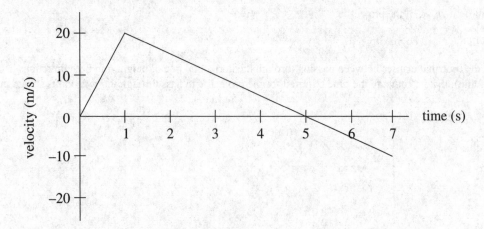

(a) Describe what happened to the car at time $t = 1$ s.

(b) How does the car's average velocity between time $t = 0$ and $t = 1$ s compare to its average velocity between times $t = 1$ s and $t = 5$ s?

(c) What is the displacement of the car from time $t = 0$ to time $t = 7$ s?

(d) Plot the car's acceleration during this interval as a function of time.

(e) Plot the object's position during this interval as a function of time. Assume that the car begins at $s = 0$.

2. Consider a projectile moving in a parabolic trajectory under constant gravitational acceleration. Its initial velocity has magnitude v_0, and its launch angle (with the horizontal) is θ_0. Solve the following in terms of given quantities and the acceleration of gravity, g.

(a) Calculate the maximum height, H, of the projectile.

(b) Calculate the (horizontal) range, R, of the projectile.

(c) For what value of θ_0 will the range be maximized?

(d) If $0 < h < H$, compute the time that elapses between passing through the horizontal line of height h in both directions (ascending and descending); that is, compute the time required for the projectile to pass through the two points shown in this figure:

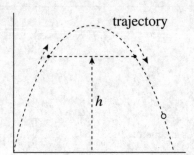

3. A cannonball is shot with an initial speed of 50 m/s at a launch angle of 40° toward a castle wall 220 m away. The height of the wall is 30 m. Assume that effects due to the air are negligible. (For this problem, use $g = 9.8$ m/s^2.)

    (a)   How long will it take the cannonball to reach the vertical plane of the wall?

    (b)   Will the cannonball strike the wall? If the cannonball strikes the wall, how far below the top of the wall does it strike? If the cannonball does not strike the wall, by how much does it clear the wall?

4. A particle moves along a straight axis in such a way that its acceleration at time t is given by the equation $a(t) = 6t$ (m/s^2). If the particle's initial velocity is 2 m/s and its initial position is $x = 4$ m, determine

    (a)   the time at which the particle's velocity is 14 m/s

    (b)   the particle's position at time $t = 3$ s

Summary

Definitions

○ Position is where an object is relative to a coordinate axis system.

○ Distance is the total measure of the ground traveled by an object.

○ Displacement is how far an object is from where it started. It can be represented as the final position minus the initial position: $\Delta s = s_f - s_i$.

○ average speed $= \dfrac{\text{distance}}{\text{time}}$

○ $\bar{v} = \dfrac{\text{displacement}}{\text{time}}$

○ $\bar{a} = \dfrac{\text{change in velocity}}{\text{time}}$

Constant Acceleration Equations

○ The Big 5 acceleration equations are:

$$\Delta x = x - x_0 = \bar{v}t$$

$$v = v_0 + at$$

$$x = x_0 + v_0 t + \frac{1}{2}at^2$$

$$x = x_0 + vt - \frac{1}{2}at^2$$

$$v^2 = v_0^2 + 2a(x - x_0)$$

Graphs

o   The slope of a position-vs.-time graph is the average velocity.

o   The area between the t-axis and the velocity function is the displacement.

o   The slope of a velocity-vs.-time graph is the average acceleration.

Motion Functions

o   The derivative of $x(t)$ is the velocity.

o   The integral of $v(t)$ is the displacement.

o   The derivative of $v(t)$ is the acceleration.

Free Fall and Projectiles

o   The acceleration due to gravity, g, is a constant 9.8 m/s^2 downward, for all objects close to the surface of Earth. Note that the test directions state that, to simplify calculations, you may use 10 m/s^2 in all problems.

o   Remember to analyze the vertical (constant acceleration) and horizontal (constant velocity) motions separately. You can use the constant acceleration equations with a replaced by g and x replaced by y to indicate the acceleration is in the vertical direction. The motion in the x direction has a constant velocity, so the only equation you need for that is $x = v_x t$.

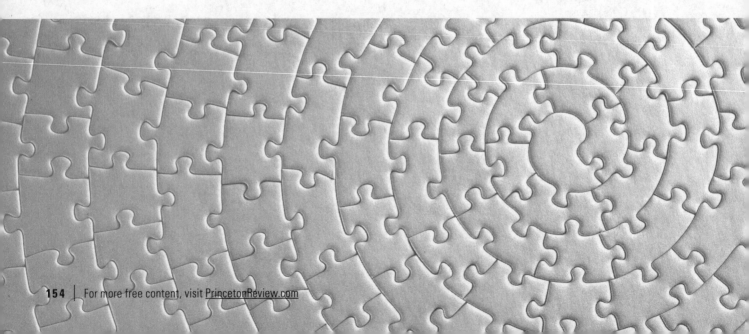

Chapter 6
Newton's Laws
of Motion

INTRODUCTION

In the previous chapter, we studied the vocabulary and equations that describe motion. Now we will learn why things move the way they do; this is the subject of **dynamics**.

An interaction between two bodies—a push or a pull—is called a **force**. If you lift a book, you exert an upward force (created by your muscles) on it. If you pull on a rope that's attached to a crate, you create a *tension* in the rope that pulls the crate. When a skydiver is falling through the air, Earth is exerting a downward pull called *gravitational force*, and the air exerts an upward force called *air resistance*. When you stand on the floor, the floor provides an upward, supporting force called the *normal force*. If you slide a book across a table, the table exerts a *frictional force* against the book, so the book slows down and then stops. Static cling provides a directly observable example of the *electrostatic force*. Protons and neutrons are held together in the nuclei of atoms by the *strong nuclear force* and radioactive nuclei decay through the action of the *weak nuclear force*.

The Englishman Sir Isaac Newton published a book in 1687 called *Philosophiae Naturalis Principia Mathematica (The Mathematical Principles of Natural Philosophy)*—referred to nowadays as simply *The Principia*—which began the modern study of physics as a scientific discipline. Three of the laws that Newton stated in *The Principia* form the basis for dynamics and are known simply as *Newton's Laws of Motion*.

THE FIRST LAW

Newton's First Law says that *an object will continue in its state of motion unless compelled to change by a net force impressed upon it*. That is, unless an unbalanced force acts on an object, the object's velocity will not change: If the object is at rest, then it will stay at rest; and if it is moving, then it will continue to move at a constant speed in a straight line.

Newton's First Law can be written as a set of relationships:
$\mathbf{F}_{net} = 0 \Rightarrow \mathbf{v}$ is constant
and
$\mathbf{v}$ is constant $\Rightarrow \mathbf{F}_{net} = 0$.

Basically, no net force means no change in velocity. This property of objects, their natural resistance to changes in their state of motion, is called **inertia**. In fact, the First Law is often referred to as the **Law of Inertia**.

THE SECOND LAW

Newton's Second Law predicts what will happen when a net force *does* act on an object: The object's velocity will change; the object will accelerate. More precisely, it says that its acceleration, **a**, will be directly proportional to the strength of the total—or *net*—force ($\mathbf{F}_{net}$) and inversely proportional to the object's mass, *m*:

Always keep this force equation in mind. This is the most important equation in mechanics!

$$\Sigma\mathbf{F} = \mathbf{F}_{net} = m\mathbf{a}$$

The **mass** of an object is the quantitative measure of its inertia; intuitively, it measures how much matter is contained in an object. Two boxes, one empty and one full, have different masses; the box that is full has the greater mass. Mass is measured in *kilograms*, abbreviated as kg.

(Note: An object whose mass is 1 kg weighs about 2.2 pounds.) It takes twice as much force to produce the same change in velocity of a 2 kg object than of a 1 kg object. Mass is a measure of an object's inertia, its resistance to acceleration.

Forces are represented by vectors; they have magnitude and direction. If several different forces act on an object simultaneously, then the net force, $\mathbf{F}_{net}$, is the vector sum of all these forces. (The phrase *resultant force* is also used to mean *net force*.)

Since $\mathbf{F}_{net} = m\mathbf{a}$, and m is a *positive* scalar, the direction of $\mathbf{a}$ always matches the direction of $\mathbf{F}_{net}$. Finally, since $F = ma$, the units for F equal the units of m times the units of a:

$$[F] = [m][a]$$
$$= \text{kg·m/s}^2$$

A force of 1 kg·m/s^2 is renamed 1 **newton** (abbreviated as N). A medium-size apple weighs about 1 N.

THE THIRD LAW

This is the law that's commonly remembered as *to every action, there is an equal, but opposite, reaction*. More precisely, if Object 1 exerts a force on Object 2, then Object 2 exerts a force back on Object 1, equal in strength but opposite in direction. These two forces, $\mathbf{F}_{1\text{-on-}2}$ and $\mathbf{F}_{2\text{-on-}1}$, are called an **action/reaction pair**.

While the forces in an action/reaction pair are equal in magnitude, the "reaction" (i.e., the resulting acceleration of the object) depends on the mass of each object. The greater the mass, the smaller the "reaction" (acceleration); the smaller the mass, the greater the "reaction" (acceleration).

> Newton's Third Law can be written as the following equation:
> $\mathbf{F}_{1\text{-on-}2} = -\mathbf{F}_{2\text{-on-}1}$

Example 1 What net force is required to maintain a 5,000 kg object moving at a constant velocity of magnitude 7,500 m/s?

Solution. The First Law says that any object will *continue* in its state of motion unless a force acts on it. Therefore, *no* net force is required to maintain a 5,000 kg object moving at a constant velocity of magnitude 7,500 m/s. Here's another way to look at it: Constant velocity means $\mathbf{a} = 0$, so the equation $\mathbf{F}_{net} = m\mathbf{a}$ immediately gives $\mathbf{F}_{net} = 0$.

Example 2 How much force is required to cause an object of mass 2 kg to have an acceleration of 4 m/s^2?

Solution. According to the Second Law, $\mathbf{F}_{net} = m\mathbf{a} = (2 \text{ kg})(4 \text{ m/s}^2) = 8 \text{ N}$.

Example 3 An object feels two forces: one of strength 8 N pulling to the left and one of strength 20 N pulling to the right. If the object's mass is 4 kg, what is its acceleration?

Solution. Forces are represented by vectors and can be added and subtracted. Therefore, an 8 N force to the left added to a 20 N force to the right yields a net force of 20 − 8 = 12 N to the right. Newton's Second Law gives $\mathbf{a} = \mathbf{F}_{net}/m$ = (12 N to the right)/(4 kg) = 3 m/s^2 to the right.

WEIGHT

Mass and weight are not the same thing—there is a clear distinction between them in physics—but they are often used interchangeably in everyday life. The **weight** of an object is the gravitational force exerted on it by a gravitational field. Mass, by contrast, is an intrinsic property of an object that measures its inertia. An object's mass does not change with location. An astronaut, for example, has the same mass on Earth as on the Moon. The astronaut's weight on the Moon, however, is 1/6 that on Earth because the gravitational force on the Moon is 1/6 that of Earth's.

Since weight is a force, we can use $\mathbf{F} = m\mathbf{a}$ to compute it, where the acceleration is the gravitational force imposed on an object. Therefore, setting $\mathbf{a} = \mathbf{g}$, the equation $\mathbf{F} = m\mathbf{a}$ becomes

$$\mathbf{F}_w = m\mathbf{g}$$

This is the equation for the weight of an object of mass m. ($\mathbf{F}_g$ and $\mathbf{F}_w$ are both commonly used to represent the force of gravity. Weight is often symbolized merely by $\mathbf{w}$, rather than $\mathbf{F}_w$.) Notice that mass and weight are proportional but not identical. Furthermore, mass is measured in kilograms, while weight is measured in newtons.

Example 4 What is the mass of an object that weighs 500 N?

Solution. We are given a weight, so we use the equation $F_w = mg$, where g is the acceleration due to gravity on Earth (10 m/s^2). Rearranging this equation gives

$$m = F_w/g = (500 \text{ N})/(10 \text{ m/s}^2) = 50 \text{ kg}$$

Example 5 A person weighs 150 pounds. Given that a pound is a unit of weight equal to 4.45 N, what is this person's mass?

Solution. This person's weight in newtons is (150 lb)(4.45 N/lb) = 667.5 N. Again, using the equation $F_w = mg$, where g = 10 m/s^2, gives

$$m = F_w/g = (667.5 \text{ N})/(10 \text{ m/s}^2) = 67 \text{ kg}$$

Example 6 A book whose mass is 2 kg rests on a table. Find the magnitude of the force exerted by the table on the book.

Solution. The book experiences two forces: The downward pull of Earth's gravity and the upward, supporting force exerted by the table. Since the book is at rest on the table, its acceleration is zero, so the net force on the book must be zero. Therefore, the magnitude of the upward force must equal the magnitude of the book's weight, which is $F_w = mg = (2 \text{ kg})(10 \text{ m/s}^2) = 20$ N.

Example 7 A can of paint with a mass of 6 kg hangs from a rope. If the can is to be pulled up to a rooftop with an acceleration of 1 m/s², what must the force of the tension be in the rope?

Solution. First, draw a picture. Represent the object of interest (the can of paint) as a heavy dot, and draw the forces that act on the object as arrows connected to the dot. This is called a **free-body** (or **force**) **diagram**.

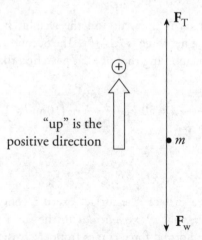

Tip
Always make the direction of acceleration the positive direction to ensure that the right-hand side of $\mathbf{F}_{net} = m\mathbf{a}$ is always positive.

We have the tension force in the rope, $\mathbf{F}_T$ (sometimes symbolized merely by $\mathbf{T}$), which is upward, and the downward force, $\mathbf{F}_w$. Calling *up* the positive direction, the net force is $F_T - F_w$. The Second Law, $F_{net} = ma$, becomes $F_T - F_w = ma$, so

$$F_T = F_w + ma = mg + ma = m(g + a) = 6(10 + 1) = 66 \text{ N}$$

Example 8 A can of paint with a mass of 6 kg hangs from a rope. If the can is to be pulled up to a rooftop with a constant velocity of 1 m/s, what must the tension in the rope be?

Solution. The phrase "constant velocity" automatically means $a = 0$ and, therefore, $F_{net} = 0$. In the diagram above, $\mathbf{F}_T$ would need to have the *same* magnitude as $\mathbf{F}_w$ in order to keep the can moving at a constant velocity. Thus, in this case, $F_T = F_w = mg = (6)(10) = 60$ N.

Example 9 How much tension must a rope have to lift a 50 N object with an acceleration of 10 m/s²?

Solution. First, draw a free-body diagram:

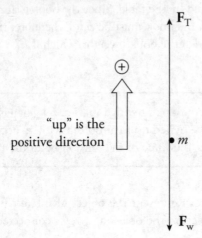

We have the tension force, **F**$_T$, which is upward, and the weight, **F**$_w$, which is downward. Calling *up* the positive direction, the net force is $F_T - F_w$. The Second Law, $F_{net} = ma$, becomes $F_T - F_w = ma$, so $F_T = F_w + ma$. Remembering that $m = F_w/g$, we find that

$$F_T = F_w + ma = F_w + \frac{F_w}{g}a = 50 \text{ N} + \frac{50 \text{ N}}{10 \text{ m/s}^2}\left(10 \text{ m/s}^2\right) = 100 \text{ N}$$

THE NORMAL FORCE

When an object is in contact with a surface, the surface exerts a contact force on the object. The component of the contact force that's *perpendicular* to the surface is called the **normal force** on the object. The normal force comes from electrostatic interactions among atoms. The normal force is what prevents objects from falling through tabletops or you from falling through the floor. The normal force is denoted by **F**$_N$, or simply by **N**. (If you use the latter notation, be careful not to confuse it with N, the abbreviation for the newton.)

> The word *normal* also means *perpendicular.*

Example 10 A book whose mass is 2 kg rests on a table. Find the magnitude of the normal force exerted by the table on the book.

Solution. The book experiences two forces: The downward gravitational pull and the upward, normal force exerted by the table. Since the book is at rest on the table, its acceleration is zero, so the net force on the book must be zero. Therefore, the magnitude of the normal force must equal the magnitude of the book's weight, which is $F_w = mg = (2)(10) = 20$ N. This means the normal force must be 20 N as well: $F_N = 20$ N. (Note that this is a repeat of Example 6, except now we have a name for the "upward, supporting force exerted by the table"; it's called the **normal force**.)

FRICTION

When an object is in contact with a surface, the surface exerts a contact force on the object. The component of the contact force that's *parallel* to the surface is called the **friction force** on the object. Friction, like the normal force, arises from electrostatic interactions between the object and the surface.

We'll look at two main categories of friction: (1) **static friction** and (2) **kinetic (sliding) friction**. If you attempt to push a heavy crate across a floor, at first you meet with resistance. Eventually, though, you can push hard enough to get the crate moving. The force that acted on the crate that canceled the force of your initial push was static friction, and the force that acts on the crate as it slides across the floor is kinetic friction. Static friction occurs when there is no relative motion between the object and the surface (no sliding); kinetic friction occurs when there *is* relative motion (when there's sliding).

The strength of the friction force depends, in general, on two things: The nature of the surfaces and the strength of the normal force. The nature of the surfaces is represented by the **coefficient of friction,** denoted by μ, which stands for the Greek letter *mu*. The greater the coefficient of friction, the stronger the friction force will be. For example, the coefficient of friction between rubber-soled shoes and a wooden floor is 0.7, but between rubber-soled shoes and ice is only 0.1. Also, since kinetic friction is generally weaker than static friction (it's easier to keep an object sliding once it's sliding than it is to start the object sliding in the first place), there are two coefficients of friction: one for static friction μ (μ_s) and one for kinetic friction (μ_k). For a given pair of surfaces, it's virtually always true that $\mu_k < \mu_s$. The strengths of these two types of friction forces are given by the following equations:

$$F_{\text{static friction, max}} = \mu_s F_N \quad \text{or} \quad F_{\text{static friction}} \leq \mu_s F_N$$

$$F_{\text{kinetic friction}} = \mu_k F_N$$

On the equation sheet for the free-response section, this information is represented as follows:

$$\boxed{F_{fric} \leq \mu N}$$

Note that the equation for the strength of the static friction force is for the maximum or lesser value. This is because static friction can vary, precisely counteracting weaker forces that attempt to move an object. For example, suppose an object feels a normal force of $F_N = 100$ N and the coefficient of static friction between it and the surface it's on is 0.5. Then, the *maximum* force that static friction can exert is (0.5)(100 N) = 50 N. However, if you push on the object with a force of, say, 20 N, then the static friction force will be 20 N (in the opposite direction), *not* 50 N; the object won't move. The net force on a stationary object must be zero. Static friction can take on all values, up to a certain maximum; you must overcome the maximum static friction force to get the object to slide. The direction of $\mathbf{F}_{\text{kinetic friction}} = \mathbf{F}_{f(\text{kinetic})}$ is opposite to that of motion (sliding), and the direction of $\mathbf{F}_{\text{static friction}} = \mathbf{F}_{f(\text{static})}$ is opposite to that of the intended motion.

Keep in mind that friction comes from electrostatic forces. This explains why both the normal force and the static friction force can change depending on how much force is applied.

Q: Can static friction act on an object that's moving?

Turn the page for the answer.

Example 11 A crate of mass 20 kg is sliding across a wooden floor. The coefficient of kinetic friction between the crate and the floor is 0.3.

   (a) Determine the strength of the friction force acting on the crate.

   (b) If the crate is being pulled by a force of 90 N (parallel to the floor), find the acceleration of the crate.

Solution. First, draw a free-body diagram.

In part (a), $F = 0$, and in part (b), $F = 90$ N for our free-body diagram. Reminder: Separate the horizontal and vertical forces and use $\Sigma F_x = ma_x$ and $\Sigma F_y = ma_y$.

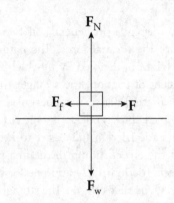

A: Yes, static friction can indeed act on an object that's moving! A few examples include a ball rolling down a hill, a person walking on a sidewalk, and a child sitting on top of and moving with a sled that is being pulled forward.

   (a) The normal force on the object balances the object's weight, so $F_N = mg = (20 \text{ kg})(10 \text{ m/s}^2) = 200$ N. Therefore, $F_{(\text{kinetic})} = \mu_k F_N = (0.3)(200 \text{ N}) = 60$ N.

   (b) The net horizontal force that acts on the crate is $F - F_f = 90 \text{ N} - 60 \text{ N} = 30$ N, so the acceleration of the crate is $a = F_{\text{net}}/m = (30 \text{ N})/(20 \text{ kg}) = 1.5$ m/s².

Example 12 A crate of mass 100 kg rests on the floor. The coefficient of static friction is 0.4. If a force of 250 N (parallel to the floor) is applied to the crate, what's the magnitude of the force of static friction on the crate?

Solution. The normal force on the object balances its weight, so $F_N = mg = (100 \text{ kg})(10 \text{ m/s}^2) = 1{,}000$ N. Therefore, $F_{\text{static friction, max}} = F_{f(\text{static}), \text{max}} = \mu_s F_N = (0.4)(1{,}000 \text{ N}) = 400$ N. This is the *maximum* force that static friction can exert, but in this case it's *not* the actual value of the static friction force. Since the applied force on the crate is only 250 N, which is less than the $F_{f(\text{static}), \text{max}}$, the force of static friction will be less also: $F_{f(\text{static})} = 250$ N, and the crate will not slide.

PULLEYS

Pulleys are devices that change the direction of the tension force in the cords that slide over them. Here, we'll consider each pulley to be frictionless and massless, which means that their masses are so much smaller than the objects of interest in the problem that they can be ignored.

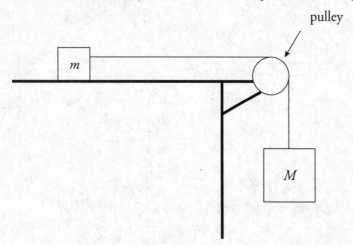

pulley

> **Example 13** In the diagram above, assume that the tabletop is frictionless. Determine the acceleration of the blocks once they're released from rest.

Solution. There are two blocks, so draw two free-body diagrams: The positive directions for each block must coincide. If the block on the table travels to the right then the hanging block travels down. This is why down is positive for the hanging block.

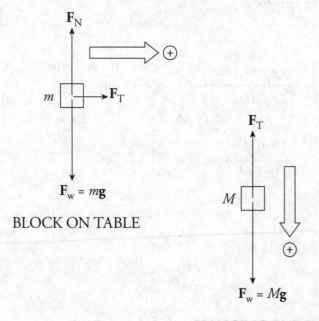

BLOCK ON TABLE

HANGING BLOCK

To get the acceleration of each one, we use Newton's Second Law, $\mathbf{F}_{net} = m\mathbf{a}$. Notice that while $\mathbf{F}_w = \mathbf{F}_N$ for the block on the table, $\mathbf{F}_w$ does not equal $\mathbf{F}_T$ for the hanging block, because that block is in motion in the downward direction.

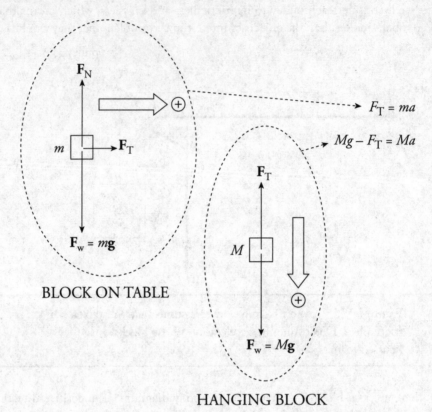

Note that there are two unknowns, F_T and a, but we can eliminate F_T by adding the two equations, and then we can solve for a.

$$F_T = ma \Bigg\}$$
$$\underline{Mg - F_T = Ma \Bigg\}}$$

Add the equations to eliminate F_T.

$$Mg = ma + Ma$$
$$= a(m + M)$$

$$\frac{Mg}{m + M} = a$$

Example 14 Using the same diagram as in the previous example, assume that m = 2 kg, M = 10 kg, and the coefficient of kinetic friction between the small block and the tabletop is 0.5. Compute the acceleration of the blocks.

Solution. Once again, draw a free-body diagram for each object. Note that the only difference between these diagrams and the ones in the previous example is the inclusion of the force of (kinetic) friction, $\mathbf{F}_f$, that acts on the block on the table.

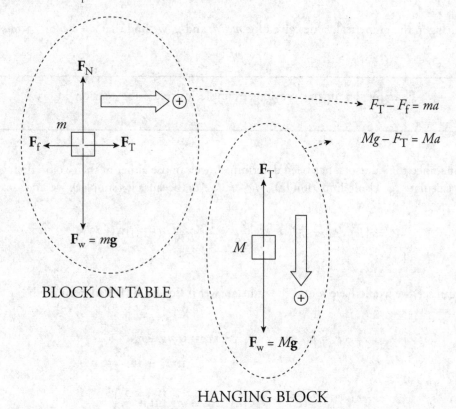

$$F_T - F_f = ma$$
$$Mg - F_T = Ma$$

BLOCK ON TABLE

HANGING BLOCK

As before, we have two equations that contain two unknowns (a and F_T):

$$F_T - F_f = ma \quad (1)$$
$$Mg - F_T = Ma \quad (2)$$

Add the equations (thereby eliminating F_T) and solve for a. Note that, by definition, $F_f = \mu F_N$, and from the free-body diagram for m, we see that $F_N = mg$, so $F_f = \mu mg$:

$$Mg - F_f = ma + Ma$$

$$Mg - \mu mg = a(m + M)$$

$$\frac{M - \mu m}{m + M} g = a$$

Substituting in the numerical values given for m, M, and μ, we find that $a = \frac{3}{4}g$ (or 7.5 m/s^2).

Example 15 In the previous example, calculate the tension in the cord.

Solution. Since the value of a has been determined, we can use either of the two original equations to calculate F_T. Using Equation (2), $Mg - F_T = Ma$ (because it's simpler), we find

$$F_T = Mg - Ma = Mg - M \cdot \frac{3}{4} g = \frac{1}{4} Mg = \frac{1}{4}(10)(10) = 25 \, \text{N}$$

As you can see, we would have found the same answer if Equation (1) had been used:

$$F_T - F_f = ma \Rightarrow F_T = F_f + ma = \mu mg + ma$$

$$= \mu mg + m \cdot \frac{3}{4} g$$

$$= mg \left(\mu + \frac{3}{4} \right)$$

$$= (2)(10)(0.5 + 0.75)$$

$$= 25 \, \text{N}$$

Bonus Tips and Tricks...

Check us out on YouTube for additional test-taking tips and must-know strategies at www.youtube.com/ThePrincetonReview

INCLINED PLANES

An **inclined plane** is basically a ramp. If an object of mass m is on the ramp, then the force of gravity on the object, $\mathbf{F}_w = m\mathbf{g}$, has two components: One that's parallel to the ramp ($mg \sin \theta$) and one that's normal to the ramp ($mg \cos \theta$), where θ is the incline angle. The force driving the block down the inclined plane is the component of the block's weight that's parallel to the ramp: $mg \sin \theta$.

When analyzing objects moving up or down inclined planes, it is almost always easiest to rotate the coordinate axes such that the x-axis is parallel to the incline and the y-axis is perpendicular to the incline, as shown in the diagram. The object would accelerate in both the x and y directions as it moved down along the incline if you did not rotate the axis. However, with the rotated axes, the acceleration in the y direction is zero. Now we only have to worry about the acceleration in the x direction.

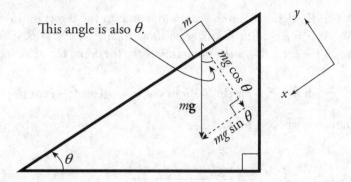

> **Example 16** A block slides down a frictionless inclined plane that makes a 30° angle with the horizontal. Find the acceleration of this block.

Solution. Let m denote the mass of the block, so the force that pulls the block down the incline is $mg \sin \theta$, and the block's acceleration down the plane is

$$a = \frac{F}{m} = \frac{mg \sin \theta}{m} = g \sin \theta = g \sin 30° = \frac{1}{2}g = 5 \text{ m/s}^2$$

> **Example 17** A block slides down an inclined plane that makes a 30° angle with the horizontal. If the coefficient of kinetic friction is 0.3, find the acceleration of the block.

Solution. First, draw a free-body diagram. Notice that, in the diagram shown below, the weight of the block, $\mathbf{F}_w = m\mathbf{g}$, has been written in terms of its scalar components: $F_w \sin\theta$ parallel to the ramp and $F_w \cos\theta$ normal to the ramp:

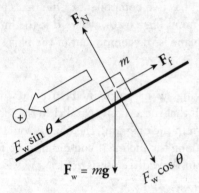

The force of friction, $\mathbf{F}_f$, that acts up the ramp (opposite to the direction in which the block slides) has magnitude $F_f = \mu F_N$. But the diagram shows that $F_N = F_w \cos\theta$, so $F_f = \mu(mg\cos\theta)$. Since the net force is the force down the ramp minus the frictional force, we have

$$F_w \sin\theta - F_f = mg\sin\theta - \mu mg\cos\theta = mg(\sin\theta - \mu\cos\theta)$$

Setting F_{net} equal to ma, we solve for a:

$$
\begin{aligned}
a = \frac{F_{net}}{m} &= \frac{mg(\sin\theta - \mu\cos\theta)}{m} \\
&= g(\sin\theta - \mu\cos\theta) \\
&= (10 \text{ m/s}^2)(\sin 30° - 0.3\cos 30°) \\
&= 2.4 \text{ m/s}^2
\end{aligned}
$$

UNIFORM CIRCULAR MOTION

In Chapter 5, we considered two types of motion: straight-line motion and parabolic motion. We will now look at motion that follows a circular path, such as a rock on the end of a string, a horse on a merry-go-round, and (to a good approximation) the Moon around Earth and Earth around the Sun.

If an object's speed around its path is a constant, then the object has **uniform circular motion**. You should remember that although the speed may be constant, the velocity is not, because the direction of the velocity is always changing. Since the velocity is changing, there must be acceleration. This acceleration does not change the speed of the object; it only changes the direction of the velocity to keep the object on its circular path. Also, in order to produce an acceleration, there must be a force; otherwise, the object would move off in a straight line (Newton's First Law).

The figure on the left shows an object moving counterclockwise along a circular trajectory, along with its velocity vectors at two nearby points. The vector $\mathbf{v}_1$ is the object's velocity at time $t = t_1$, and $\mathbf{v}_2$ is the object's velocity vector a short time later (at time $t = t_2$). The velocity vector is always tangential to the object's path (whatever the shape of the trajectory). Notice that since we are assuming constant speed, the lengths of $\mathbf{v}_1$ and $\mathbf{v}_2$ (their magnitudes) are the same.

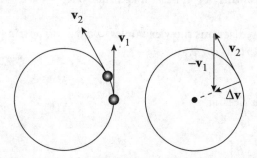

Notice that $\Delta\mathbf{v} = \mathbf{v}_2 - \mathbf{v}_1$ points toward the center of the circle (see the figure on the right). This means that the acceleration does also, since $\mathbf{a} = \Delta\mathbf{v}/\Delta t$. Because the acceleration vector points toward the center of the circle, it's called **centripetal acceleration**, or $\mathbf{a}_c$. The centripetal acceleration is what turns the velocity vector to keep the object traveling in a circle. The magnitude of the centripetal acceleration depends on the object's speed, v, and the radius, r, of the circular path according to the equation

$$a_c = \frac{v^2}{r}$$

When this is given on the free-response equation sheet, it is given in terms of angular velocity as well, which will be discussed later.

> **Example 18** An object of mass 5 kg moves at a constant speed of 6 m/s in a circular path of radius 2 m. Find the object's acceleration and the net force responsible for its motion.

Solution. By definition, an object moving at constant speed in a circular path is undergoing uniform circular motion. Therefore, it experiences a centripetal acceleration of magnitude v^2/r, always directed toward the center of the circle:

$$a_c = \frac{v^2}{r} = \frac{\left(6 \text{ m/s}\right)^2}{2 \text{ m}} = 18 \text{ m/s}^2$$

The force that produces the centripetal acceleration is given by Newton's Second Law, coupled with the equation for centripetal acceleration:

$$F_c = ma_c = m\frac{v^2}{r}$$

Centripetal force is an example of a non-real force. This means you don't draw it on your free-body diagrams. Instead, it will be the result of one or more forces that you do draw.

This equation gives the magnitude of the force. As for the direction, recall that because $\mathbf{F}_{net} = m\mathbf{a}$, the directions of $\mathbf{F}_{net}$ and $\mathbf{a}$ are always the same. Since centripetal acceleration points toward the center of the circular path, so does the force that produces it. Therefore, it's called **centripetal force**. Centripetal force is provided by everyday forces such as tension, friction, gravity, or normal forces. The centripetal force acting on this object has a magnitude of $F_c = ma_c = (5 \text{ kg})(18 \text{ m/s}^2) = 90$ N.

The following diagrams show examples of a ball on a string traveling in a *horizontal circle* and a *vertical circle*.

The figure on the right is an example of a "conical pendulum." Pendula will be explored in more detail in Chapter 10.

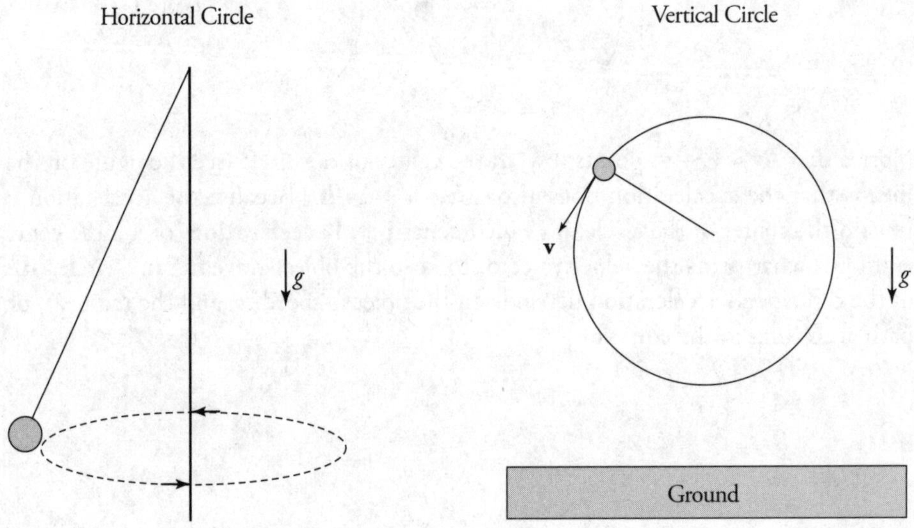

Horizontal Circle

Vertical Circle

Ground

Example 19 A 10 kg mass is attached to a string that has a breaking strength of 200 N. If the mass is whirled in a horizontal circle of radius 80 cm, what maximum speed can it have? Assume the string is horizontal.

Solution. The first thing to do in problems like this is to identify what force(s) provide the centripetal force. In this example, the tension in the string ($\mathbf{F}_T$) provides the centripetal force ($\mathbf{F}_c$):

$$\mathbf{F}_T \text{ provides } \mathbf{F}_c \Rightarrow F_T = m\frac{v^2}{r} \Rightarrow v = \sqrt{\frac{rF_T}{m}} \Rightarrow v_{max} = \sqrt{\frac{rF_{T, max}}{m}}$$

$$= \sqrt{\frac{(0.80 \text{ m})(200 \text{ N})}{10 \text{ kg}}}$$

$$= 4 \text{ m/s}$$

Example 20 An athlete who weighs 800 N is running around a curve at a speed of 5.0 m/s in an arc whose radius of curvature, r, is 5.0 m. Assuming his weight is equal to the force of static friction, find the centripetal force acting on him. What could happen to him if r were smaller?

Solution. Using the equation for the strength of the centripetal force, we find that

$$F_c = m\frac{v^2}{r} = \frac{F_w}{g} \cdot \frac{v^2}{r} = \frac{800 \text{ N}}{10 \text{ N/kg}} \cdot \frac{(5.0 \text{ m/s})^2}{5.0 \text{ m}} = 400 \text{ N}$$

In this case, static friction provides the centripetal force. Since his weight is equal to the force of static friction, this means the coefficient of static friction between his shoes and the ground is 1, so the maximum force that static friction can exert is $\mu_s F_N \approx F_N = F_w = 800$ N. Fortunately, 800 N is greater than 400 N. But notice that if the radius of curvature of the arc were much smaller, then F_c would become greater than F_w, and he would slip.

Example 21 A roller-coaster car enters the circular-loop portion of the ride. At the very top of the circle (where the people in the car are upside down), the speed of the car is 25 m/s, and the acceleration points straight down. If the diameter of the loop is 50 m and the total mass of the car (plus passengers) is 1,200 kg, find the magnitude of the normal force exerted by the track on the car at this point. Also find the normal force exerted by the track on the car when it is at the bottom of the loop. Assume it is still traveling 25 m/s at that location as well.

Solution. When analyzing circular motion, consider all forces pointing toward the center to be positive and all forces pointing away to be negative. There are two forces acting on the car at its topmost point: the normal force exerted by the track and the gravitational force, both of which point downward.

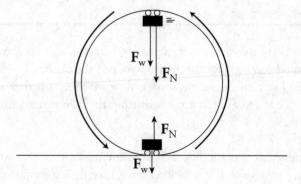

The combination of these two forces, $\mathbf{F}_N + \mathbf{F}_w$, provides the centripetal force:

$$F_N + F_w = m\frac{v^2}{r} \Rightarrow F_N = m\frac{v^2}{r} - F_w$$

$$= m\frac{v^2}{r} - mg$$

$$= m\left(\frac{v^2}{r} - g\right)$$

$$= (1,200\ \text{kg})\left(\frac{(25\ \text{m/s})^2}{(25\ \text{m})} - 10\ \text{m/s}^2\right)$$

$$= 1.8 \times 10^4\ \text{N}$$

Now we will determine the normal force exerted by the track on the car at the bottom of the loop.

$$F_c = m\frac{v^2}{r}$$

$$F_N - F_w = m\frac{v^2}{r} \Rightarrow F_N = m\frac{v^2}{r} + F_w$$

$$F_N = m\frac{v^2}{r} + mg$$

$$F_N = \frac{1,200(25)^2}{25} + 1,200(10)$$

$$F_N = 4.2 \times 10^4\ \text{N}$$

Notice that the normal force is much greater when the car is at the bottom of the track than when it is at the top. When it is at the top of the track, gravity is helping the car travel in a circle.

Example 22 In the previous example, if the net force on the car at its topmost point is straight down, why doesn't the car fall straight down?

Solution. Remember that force tells an object how to *accelerate*. If the car had zero velocity at this point, then it would certainly fall straight down, but the car has a nonzero velocity (to the left) at this point. The fact that the acceleration is downward means that, at the next moment, **v** will point down to the left at a slight angle, ensuring that the car remains on a circular path, in contact with the track.

The minimum centripetal acceleration of the car at the top of the track would be equal to the acceleration of gravity, $g = 9.8$ m/s^2. If a_c were less than g, then the car would fall off its circular path.

Chapter 6 Drill

The answers and explanations can be found in Chapter 17.

Section I: Multiple Choice

1. Consider a box being dragged across a table at a constant velocity by a string. Which of the following would be considered an action-reaction pair?

   I.   The force of gravity pulling the box down and the normal force pushing the box up
   II. The force of friction on the box and the tension force pulling the box
   III. The force of the box pushing down on the table and the normal force pushing the box up

   (A) I only
   (B) I and II only
   (C) II only
   (D) I and III only
   (E) III only

2. A person who weighs 800 N steps onto a scale that is on the floor of an elevator car. If the elevator accelerates upward at a rate of 5 m/s², what will the scale read?

   (A) 400 N
   (B) 800 N
   (C) 1,000 N
   (D) 1,200 N
   (E) 1,600 N

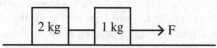

3. A force of 20 N is pulling on a 1.0 kg block, which is attached to a 2.0 kg block. Assuming the blocks are on a frictionless surface, find the tension in the rope between the blocks and the acceleration of the system of the blocks.

   (A) $F_T = 10$ N, $a = 10$ m/s²
   (B) $F_T = 13.3$ N, $a = 10$ m/s²
   (C) $F_T = 13.3$ N, $a = 20$ m/s²
   (D) $F_T = 13.3$ N, $a = 6.7$ m/s²
   (E) $F_T = 15$ N, $a = 6.7$ m/s²

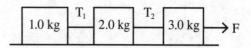

4. A force of 20 N is pulling on a 3.0 kg block, which is attached to a 2.0 kg block, which is attached to a 1.0 kg block. Assuming the blocks are on a frictionless surface, find the tension in the ropes between the blocks (F_{T1} between the 1 kg and 2 kg block and F_{T2} between the 2 kg and 3 kg block) and the acceleration of the system of the blocks.

   (A) $F_{T1} = 1$ N, $F_{T2} = 3.3$ N, $a = 1$ m/s²
   (B) $F_{T1} = 3.3$ N, $F_{T2} = 3.3$ N, $a = 3.3$ m/s²
   (C) $F_{T1} = 3.3$ N, $F_{T2} = 10$ N, $a = 3.3$ m/s²
   (D) $F_{T1} = 3.3$ N, $F_{T2} = 10$ N, $a = 6.7$ m/s²
   (E) $F_{T1} = 5$ N, $F_{T2} = 6.6$ N, $a = 6.7$ m/s²

5. The coefficient of static friction between a box and a ramp is 0.5. The ramp's incline angle is 30°. If the box is placed at rest on the ramp, the box will do which of the following?

   (A) Accelerate down the ramp
   (B) Accelerate briefly down the ramp but then slow down and stop
   (C) Move with constant velocity down the ramp
   (D) Not move
   (E) Cannot be determined from the information given

6.

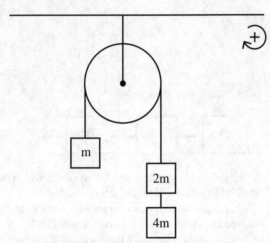

In the figure above, assume a frictionless, massless pulley system. Determine the acceleration of the masses once released from rest.

(A) $a = \frac{1}{3}(g)$ m/s²

(B) $a = \frac{1}{2}(g)$ m/s²

(C) $a = \frac{3}{5}(g)$ m/s²

(D) $a = \frac{5}{7}(g)$ m/s²

(E) $a = \frac{4}{3}(g)$ m/s²

7. Force of 20 N is pulling on a 2.0 kg block, which is attached to a 1.0 kg block. The coefficient of friction is equal to 0.2. Assume the surface is not frictionless, so you must account for friction when summing the net forces on each block. Find the tension in the rope between the blocks and the acceleration of the system of the blocks.

(A) $\mathbf{F}_T = 1$ N, $a = 4.7$ m/s²
(B) $\mathbf{F}_T = 3.3$ N, $a = 4.7$ m/s²
(C) $\mathbf{F}_T = 3.3$ N, $a = 4.7$ m/s²
(D) $\mathbf{F}_T = 6.7$ N, $a = 6.7$ m/s²
(E) $\mathbf{F}_T = 6.7$ N, $a = 4.7$ m/s²

8. An object of mass m is allowed to slide down a frictionless ramp of angle θ, and its speed at the bottom is recorded as v. If this same process was followed on a planet with twice the gravitational acceleration of Earth, what would be its final speed?

(A) $2v$

(B) $\sqrt{2}v$

(C) v

(D) $\dfrac{v}{\sqrt{2}}$

(E) $\dfrac{v}{2}$

9. An engineer is designing a loop for a roller coaster. If the loop has a radius of 25 m, how fast do the cars need to be moving at the top to ensure people would be safe even if the safety bars malfunctioned?

(A) 12.7 m/s
(B) 15.8 m/s
(C) 21.2 m/s
(D) 29.3 m/s
(E) 33.3 m/s

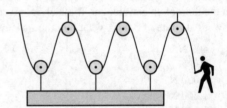

10. The pulley system above requires a person to pull on the rope with a minimum force F in order to lift the block of mass m. An inclined plane of angle θ requires that same force to push the block up the ramp. Assuming ideal conditions in both situations, what is the value of θ?

(A) 0°
(B) 9.6°
(C) 21°
(D) 47°
(E) 80°

Questions 11–12:

A 60 cm rope is tied to the handle of a bucket which is then whirled in a vertical circle. The mass of the bucket is 3 kg.

11. At the lowest point in its path, the tension in the rope is 50 N. What is the speed of the bucket?

(A) 1 m/s
(B) 2 m/s
(C) 3 m/s
(D) 4 m/s
(E) 5 m/s

12. What is the critical speed below which the rope would become slack when the bucket reaches the highest point in the circle?

(A) 0.6 m/s
(B) 1.8 m/s
(C) 2.4 m/s
(D) 3.2 m/s
(E) 4.8 m/s

13. An object moves at a constant speed in a circular path of radius r at a rate of 1 revolution per second. What is its acceleration?

(A) 0
(B) $2\pi^2 r$
(C) $2\pi^2 r^2$
(D) $4\pi^2 r$
(E) $4\pi^2 r^2$

Section II: Free Response

1. This question concerns the motion of a crate being pulled across a horizontal floor by a rope. In the diagram below, the mass of the crate is m, the coefficient of kinetic friction between the crate and the floor is μ, and the tension in the rope is $\mathbf{F}_T$.

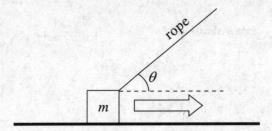

(a) Draw and label all of the forces acting on the crate.

(b) Compute the normal force acting on the crate in terms of m, F_T, θ, and g.

(c) Compute the acceleration of the crate in terms of m, F_T, θ, μ, and g.

(d) Assume that the magnitude of the tension in the rope is fixed but that the angle may be varied. For what value of θ would the resulting horizontal acceleration of the crate be maximized?

2. In the diagram below, a massless string connects two blocks—of masses m_1 and m_2, respectively—on a flat, frictionless tabletop. A force **F** pulls on Block #2, as shown:

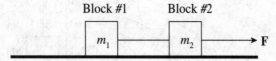

Solve for the following in terms of given quantities.

(a) Draw and label all of the forces acting on Block #1.

(b) Draw and label all of the forces acting on Block #2.

(c) What is the acceleration of Block #1?

(d) What is the tension in the string connecting the two blocks?

(e) If the string connecting the blocks were not massless, but instead had a mass of m, find

   (i)   the acceleration of Block #1, and

   (ii) the difference between the strength of the force that the connecting string exerts on Block #2 and the strength of the force that the connecting string exerts on Block #1.

3. In the figure shown, assume that the pulley is frictionless and massless.

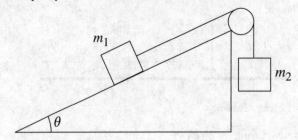

Solve for the following in terms of given quantities and the acceleration of gravity, g.

(a) If the surface of the inclined plane is frictionless, determine what value(s) of θ will cause the box of mass m_1 to

    (i) accelerate up the ramp

    (ii) slide up the ramp at constant speed

(b) If the coefficient of kinetic friction between the surface of the inclined plane and the box of mass m_1 is μ_k, derive (but do not solve) an equation satisfied by the value of θ which will cause the box of mass m_1 to slide up the ramp at constant speed.

4. A skydiver is falling with speed v_0 through the air. At that moment (time $t = 0$), she opens her parachute and experiences the force of air resistance whose strength is given by the equation $F = kv$, where k is a proportionality constant and v is her descent speed. The total mass of the skydiver and equipment is m. Assume that g is constant throughout her descent.

(a) Draw and label all the forces acting on the skydiver after her parachute opens.

(b) Determine the skydiver's acceleration in terms of m, v, k, and g.

(c) Determine the skydiver's terminal speed (that is, the eventual constant speed of descent).

(d) Sketch a graph of v as a function of time, starting at $t = 0$ and going until she lands, being sure to label important values on the vertical axis.

(e) Derive an expression for her descent speed, v, as a function of time t since opening her parachute in terms of m, k, and g.

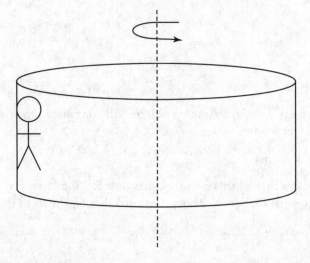

5. An amusement park ride consists of a large cylinder that rotates around its central axis as the passengers stand against the inner wall of the cylinder. Once the passengers are moving at a certain speed v, the floor on which they were standing is lowered. Each passenger feels pinned against the wall of the cylinder as it rotates. Let r be the inner radius of the cylinder.

Solve for the following in terms of given quantities and the acceleration of gravity, g.

(a) Draw and label all the forces acting on a passenger of mass m as the cylinder rotates with the floor lowered.

(b) Describe what conditions must hold to keep the passengers from sliding down the wall of the cylinder.

(c) Compare the conditions discussed in part (b) for an adult passenger of mass m and a child passenger of mass $m/2$.

6. A curved section of a highway has a radius of curvature of r. The coefficient of friction between standard automobile tires and the surface of the highway is μ_s.

(a) Draw and label all the forces acting on a car of mass m traveling along this curved part of the highway.

(b) Compute the maximum speed with which a car of mass m could make it around the turn without skidding in terms of μ_s, r, g, and m.

City engineers are planning on banking this curved section of highway at an angle of θ to the horizontal.

(c) Draw and label all of the forces acting on a car of mass m traveling along this banked turn. Do not include friction.

(d) The engineers want to be sure that a car of mass m traveling at a constant speed v (the posted speed limit) could make it safely around the banked turn even if the road were covered with ice (that is, essentially frictionless). Compute this banking angle θ in terms of r, v, g, and m.

Summary

Newton's Laws

o   Newton's First Law (Law of Inertia) states that objects will continue in their state of motion unless acted upon by an unbalanced force.

o   Newton's Second Law: $\mathbf{F}_{net} = m\mathbf{a}$

o   Newton's Third Law states that whenever two objects interact, the force that the first object exerts on the second object is equal to, but in the opposite direction of, the force the second object exerts on the first object.

Weight

o   The weight of an object is given by $\mathbf{F}_w = m\mathbf{g}$.

The Normal Force

o   The normal force is the component of the contact force exerted on an object in contact with a surface and is *perpendicular* to the surface.

o   The normal force can be represented by $\mathbf{F}_N$ or $\mathbf{N}$.

Friction

o   Friction is the component of the contact force exerted on an object in contact with a surface and is *parallel* to the surface.

o   Static friction occurs when there is no relative motion between the object and the surface. Its strength is given by the following equation: $F_{\text{static friction, max}} = \mu_s F_N$.

o   Kinetic friction occurs when there is relative motion between the two surfaces. Its strength is given by the following equation: $F_{\text{kinetic friction}} = \mu_k F_N$.

Inclined Planes

o   There are two components to the force of gravity on an object on an inclined plane: the force parallel to the ramp ($mg \sin \theta$) and the force normal to the ramp ($mg \cos \theta$).

o   To simplify analysis of an object moving up or down a ramp, rotate the coordinate axes so that the x-axis is parallel to the incline and the y-axis is perpendicular to the incline.

Uniform Circular Motion

o   The velocity vector is tangent to the circle.

o   The centripetal acceleration points toward the center of the circle, and therefore the centripetal force must also point to the center of the circle.

$$a_c = \frac{v^2}{r}$$

o   Any force, or component of a force, that points toward the center of the circle is positive and any force, or component of a force, that points away from the center is negative when using the following equation: $F_c = \frac{mv^2}{r}$.

Chapter 7
Work, Energy, and Power

INTRODUCTION

It wasn't until more than one hundred years after Newton that energy became incorporated into physics, but today its inclusion permeates every branch of the subject.

There are different forms of energy because there are different kinds of forces. There's gravitational energy (a meteor crashing into Earth), elastic energy (a stretched rubber band), thermal energy (an oven), radiant energy (sunlight), electrical energy (a lamp plugged into a wall socket), nuclear energy (nuclear power plants), and mass energy (the heart of Einstein's equation $E = mc^2$). Energy can come into a system or leave it through various interactions that produce changes. While **force** is the agent of change, **energy** is often defined as the ability to do work, and **work** is one way of transferring energy from one system to another. One of the most important laws in physics (the **Law of Conservation of Energy**, also known as the **First Law of Thermodynamics**) says that if you account for all its various forms, the total amount of energy in a given process will stay constant; that is, it will be *conserved*. For example, electrical energy can be converted into light and heat (this is how a light bulb works), but the amount of electrical energy *going into* the light bulb equals the total amount of light and heat *given off*.

> Energy cannot be created or destroyed; it can only be transferred (from one system to another) or transformed (from one form to another).

WORK

When you lift a book from the floor, you exert a force on it over a distance, and when you push a crate across a floor, you also exert a force on it over a distance. When you hold a book in your hand, you exert a force on the book (normal force), but since the book is at rest, its displacement = 0, so you do no work. If **F** is the force, and $d\mathbf{r}$ is an infinitesimal amount of displacement, then the work done is:

$$W = \int \mathbf{F} \cdot d\mathbf{r}$$

If the force is constant, it can be removed from the integral, and since $\mathbf{F} \cdot \mathbf{r} = (F \cos \theta)r$ for any angle θ, then $W = (F \cos \theta)r$. The unit of work is the newton-meter (N·m), which is also called a joule (J).

Notice that work depends on two vectors (**F** and **r**), but work itself is *not* a vector. Work is a *scalar* quantity. This is a result of the dot product, whereby two vectors are multiplied to produce a scalar. Another result of the dot product is that only the component of the force that is parallel (or antiparallel) to the displacement does any work. Any force or component of a force that is perpendicular to the direction that an object actually moves cannot do work because the displacement in that direction is zero.

Also, since an integral is the area under a curve, if a graph of force as a function of position or displacement is given, the work done by the force is the area under the curve whose boundaries correspond to the displacement, Δx.

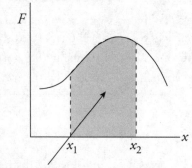

area = work done by **F**

> **Example 1** You slowly lift a book of mass 2 kg at constant velocity through a distance of 3 m. How much work did you do on the book?

Solution. In this case, the force you exert must balance the weight of the book (otherwise the velocity of the book wouldn't be constant), so $F = mg = (2\text{ kg})(10\text{ m/s}^2) = 20$ N. Since this force is straight upward and the displacement, **d**, of the book is also straight upward, **F** and **d** are parallel, so the work done by your lifting force is $W = Fd = (20\text{ N})(3\text{ m}) = 60\text{ N·m} = 60$ J.

> **Example 2** A 15 kg crate is moved along a horizontal floor by a warehouse worker who's pulling on it with a rope that makes a 30° angle with the horizontal. The tension in the rope is 200 N and the crate slides a distance of 10 m. How much work is done on the crate by the rope?

Solution. The figure below shows that $\mathbf{F}_\text{T}$ and **d** are not parallel. It's only the component of the force acting along the direction of motion, $\mathbf{F}_\text{T} \cos\theta$, that does work.

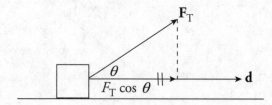

Therefore,

$$W = (F_\text{T} \cos\theta)d = (200\text{ N} \bullet \cos 30°)(10\text{ m}) = 1{,}730 \text{ J}$$

Example 3 In the previous example, assume that the coefficient of kinetic friction between the crate and the floor is 0.4.

(a) How much work is done by the normal force?

(b) How much work is done by the friction force?

Solution.

(a) Since the angle between $\mathbf{F}_N$ and $\mathbf{d}$ is 90° (by definition of *normal*) and cos 90° = 0, the normal force does zero work.

(b) The friction force, $\mathbf{F}_f$, is also not parallel to the motion; it's *antiparallel*. That is, the angle between $\mathbf{F}_f$ and $\mathbf{d}$ is 180°. Since cos 180° = –1, and since the strength of the normal force is $F_N = F_{T,y} - F_W = F_T \sin\theta - mg = 200 \text{ N} \cdot \sin 30° - (15 \text{ kg})(10 \text{ m/s}^2) = 100 \text{ N} - 150 \text{ N} = 50 \text{ N}$, the work done by the friction force is:

$$W = -F_f d = -\mu$$
$$\mu_k F_N d = -(0.4)(50 \text{ N})(10 \text{ m}) = -200 \text{ J}$$

The two previous examples show that work, which is a scalar quantity, may be positive, negative, or zero. If the angle between $\mathbf{F}$ and $\mathbf{d}$ (θ) is less than 90°, then the work is positive (because cos θ is positive in this case); if θ = 90°, the work is zero (because cos 90° = 0); and if $\theta > 90°$, then the work is negative (because cos θ is negative). Intuitively, if a force helps the motion, the work done by the force is positive, but if the force opposes the motion, then the work done by the force is negative.

Example 4 A box slides down an inclined plane (incline angle = 37°). The mass of the block, m, is 35 kg, the coefficient of kinetic friction between the box and the ramp, μ_k, is 0.3, and the length of the ramp, d, is 8 m.

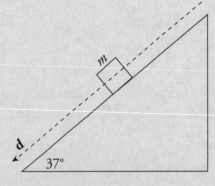

(a) How much work is done by gravity?

(b) How much work is done by the normal force?

(c) How much work is done by friction?

(d) What is the total work done?

Solution.

(a) Recall that the force that's directly responsible for pulling the box down the plane is the component of the gravitational force that's parallel to the ramp, so the angle between **F** and **d** is 0°:

$W = (F_{w\parallel} \cos(\theta)\, d = F_{w\parallel}\, mg \sin\theta)$ (where θ is the incline angle). This component is parallel to the motion, so the work done by gravity is

$$W_{\text{by gravity}} = (mg \sin\theta)d = (35 \text{ kg})(10 \text{ N/kg})(\sin 37°)(8 \text{ m}) = 1{,}680 \text{ J}$$

Note that the work done by gravity is positive, as we would expect it to be, since gravity is helping the motion. Also, be careful with the angle θ. The general definition of *work* reads $W = (F \cos\theta)d$, where θ is the angle between **F** and **d**. However, the angle between $\mathbf{F}_w$ and **d** is *not* 37° here, it is 0°, so the work done by gravity is not $(mg \cos 37°)d$. The angle θ used in the calculation above is the incline angle.

(b) Since the normal force is perpendicular to the motion, the work done by this force is zero.

(c) The strength of the normal force is $F_w \cos\theta$ (where θ is the incline angle), so the strength of the friction force is $F_f = \mu_k F_N = \mu_k F_w \cos\theta = \mu_k mg \cos\theta$. Since $\mathbf{F}_f$ is antiparallel to **d**, the cosine of the angle between these vectors (180°) is −1, so the work done by friction is

$$W_{\text{by friction}} = -F_f d = -(\mu_k mg \cos\theta)(d) = -(0.3)(35 \text{ kg})(10 \text{ N/kg})(\cos 37°)(8 \text{ m}) = -672 \text{ J}$$

Note that the work done by friction is negative, as we expect it to be, since friction is opposing the motion.

(d) The total work done is found simply by adding the values of the work done by each of the forces acting on the box:

$$W_{\text{total}} = \Sigma W = W_{\text{by gravity}} + W_{\text{by normal force}} + W_{\text{by friction}} = 1{,}680 + 0 + (-672) = 1{,}008 \text{ J}$$

Example 5 The force exerted by a spring when it's displaced by x from its natural length is given by the equation $F(x) = -kx$, where k is a positive constant. This equation is known as Hooke's law. What is the work done by a spring as it pushes out from $x = -x_2$ to $x = -x_1$ (where $x_2 > x_1$)?

Solution. Since the force is variable, we calculate the following definite integral:

$$W = \int_{-x_2}^{-x_1} F(x)\, dx = \int_{-x_2}^{-x_1} (-kx)\, dx = \left(-\tfrac{1}{2}kx^2\right)\Big|_{-x_2}^{-x_1}$$
$$= \left(\tfrac{1}{2}kx^2\right)\Big|_{x_1}^{x_2}$$
$$= \tfrac{1}{2}k\left(x_2^2 - x_1^2\right)$$

Another solution would involve sketching a graph of $F(x) = -kx$ and calculating the area under the graph from $x = -x_2$ to $x = -x_1$.

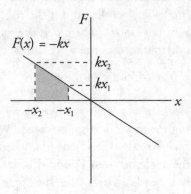

Here, the region is a trapezoid with area $A = \frac{1}{2}(\text{base}_1 + \text{base}_2) \times \text{height}$, so

$$
\begin{aligned}
W = A &= \tfrac{1}{2}(kx_2 + kx_1)(x_2 - x_1) \\
&= \tfrac{1}{2}k(x_2 + x_1)(x_2 - x_1) \\
&= \tfrac{1}{2}k\left(x_2^2 - x_1^2\right)
\end{aligned}
$$

KINETIC ENERGY

Consider an object at rest ($v_0 = 0$), and imagine that a steady force is exerted on it, causing it to accelerate. Let's be more specific; let the object's mass be m, and let $\mathbf{F}$ be the net force acting on the object, pushing it in a straight line. The object's acceleration is $a = F_{net}/m$, so after the object has traveled a distance Δx under the action of this force, its final speed, v, is given by Big Five #5:

$$
v^2 = v_0^2 + 2a(x - x_0) = 2a\Delta x = 2\frac{F_{net}}{m}\Delta x \quad \Rightarrow \quad F_{net}\Delta x = \tfrac{1}{2}mv^2
$$

But the quantity $F_{net}\Delta x$ is the total work done by the force, so $W_T = \frac{1}{2}mv^2$. The work done on the object has transferred energy to it, in the amount of $\frac{1}{2}mv^2$. The energy an object possesses by virtue of its motion is therefore defined as $\frac{1}{2}mv^2$ and is called **kinetic energy**:

$$
K = \tfrac{1}{2}mv^2
$$

THE WORK–ENERGY THEOREM

Kinetic energy is expressed in joules, just like work. The derivation on the previous page can be extended to an object with a nonzero initial speed, and the same analysis will show that the total work done on an object—or, equivalently, the work done by the net force—will equal its change in kinetic energy; this is known as the **work–energy theorem**:

$$W_{\text{total}} = \Delta K$$

Note that kinetic energy, like work, is a scalar quantity.

> A force that does positive work will increase an object's kinetic energy and speed. The opposite is true of a force that does negative work. A force that does no work will leave an object's kinetic energy and speed unchanged.

> **Example 6** What is the kinetic energy of a ball (mass = 0.10 kg) moving with a speed of 30 m/s?

Solution. From the definition,

$$K = \tfrac{1}{2}mv^2 = \tfrac{1}{2}(0.10 \text{ kg})(30 \text{ m/s})^2 = 45 \text{ J}$$

> **Example 7** A tennis ball (mass = 0.06 kg) is hit straight upward with an initial speed of 50 m/s. How high would it go if air resistance were negligible?

Solution. This could be done using the Big Five, but let's try to solve it using the concepts of work and energy. As the ball travels upward, gravity acts on it by doing negative work. [The work is negative because gravity is opposing the upward motion. $\mathbf{F}_w$ and $\mathbf{d}$ are in opposite directions, so $\theta = 180°$, which tells us that $W = (F_w \cos \theta)d = -F_w d$.] At the moment the ball reaches its highest point, its speed is 0, so its kinetic energy is also 0. The work–energy theorem says

$$W = \Delta K \quad \Rightarrow \quad -F_w d = 0 - \tfrac{1}{2}mv_0^2 \quad \Rightarrow \quad d = \frac{\tfrac{1}{2}mv_0^2}{F_w} = \frac{\tfrac{1}{2}mv_0^2}{mg} = \frac{\tfrac{1}{2}v_0^2}{g} = \frac{\tfrac{1}{2}(50 \text{ m/s})^2}{10 \text{ m/s}^2} = 125 \text{ m}$$

> **Example 8** Consider the box sliding down the inclined plane in Example 4. If it starts from rest at the top of the ramp, with what speed does it reach the bottom?

Solution. It was calculated in Example 4 that $W_{\text{total}} = 1,008$ J. According to the work–energy theorem,

$$W_{\text{total}} = \Delta K \quad \Rightarrow \quad W_{\text{total}} = K_{\text{fi}} - K = K_f = \tfrac{1}{2}mv^2 \quad \Rightarrow \quad v = \sqrt{\frac{2W_{\text{total}}}{m}} = \sqrt{\frac{2(1,008 \text{ J})}{35 \text{ kg}}} = 7.6 \text{ m/s}$$

> **Example 9** A pool cue striking a stationary billiard ball (mass = 0.25 kg) gives the ball a speed of 2 m/s. If the average force of the cue on the ball was 200 N, over what distance did this force act?

Solution. The kinetic energy of the ball as it leaves the cue is

$$K = \tfrac{1}{2}mv^2 = \tfrac{1}{2}(0.25 \text{ kg})(2 \text{ m/s})^2 = 0.50 \text{ J}$$

The work W done by the cue gave the ball this kinetic energy, so

$$W = \Delta K \quad \Rightarrow \quad W = K_f \quad \Rightarrow \quad Fd = K \quad \Rightarrow \quad d = \frac{K}{F} = \frac{0.50 \text{ J}}{200 \text{ N}} = 0.0025 \text{ m} = 0.25 \text{ cm}$$

POTENTIAL ENERGY

Kinetic energy is the energy an object has by virtue of its motion. Potential energy is independent of motion; it arises from the object's position (or the system's configuration). For example, a ball at the edge of a tabletop has energy that could be transformed into kinetic energy if it falls from the table. An arrow in an archer's pulled-back bow has energy that could be transformed into kinetic energy if the archer releases the arrow. Both of these are examples of *potential energy*, the energy an object or system has by virtue of its position or configuration. In each case, work was done on the object to put it in the given configuration (the ball was lifted to the tabletop, the bowstring was pulled back), and since work is the means of transferring energy, these things have *stored energy that can be converted into* kinetic energy. This is **potential energy**, denoted by U.

Because there are different types of forces, there are different types of potential energy. The ball at the edge of the tabletop provides an example of **gravitational potential energy**, U_{grav}, which is the energy stored by virtue of an object's position in a gravitational field. This energy would be converted to kinetic energy as gravity pulled the ball down to the floor. For now, let's concentrate on gravitational potential energy. Some textbooks label this $U_{grav} = PE_{grav}$.

Assume the ball has a mass m of 2 kg, and that the tabletop is $h = 1.5$ m above the floor. How much work did gravity do as the ball was lifted from the floor to the table? The strength of the gravitational force on the ball is $F_w = mg = (2 \text{ kg})(10 \text{ N/kg}) = 20$ N. The force $\mathbf{F}_w$ points downward, and the ball's motion was upward, so the work done by gravity during the ball's ascent was $W_{by \, gravity} = -F_w h = -mgh = -(20 \text{ N})(1.5 \text{ m}) = -30$ J.

So, someone performed +30 J of work to raise the ball from the floor to the tabletop. That energy is now stored and, if the ball was given a push to send it over the edge, by the time the ball reached the floor, it would release 30 J of kinetic energy. We therefore say that the change in the ball's gravitational potential energy in moving from the floor to the table was +30 J. That is,

Note that potential energy, like work (and kinetic energy), is expressed in joules.

$$\Delta U_g = -W_{\text{by gravity}}$$

In general, if an object of mass m is raised a height h (which is small enough that g stays essentially constant over this altitude change), then the increase in the object's gravitational potential energy is

$$\Delta U_g = mgh$$

In this equation, the work done by gravity as the object is raised does not depend on the path taken by the object. The ball could be lifted straight upward, or in some curvy path; it would make no difference. Gravity is said to be a **conservative** force because of this property.

If we decide on a reference level whereby $h = 0$, then we can say that the gravitational potential energy of an object of mass m at a height h is $U_{\text{grav}} = mgh$. Let's consider a passenger in an airplane reading a book. If the book is 1 m above the floor of the plane (our reference point), then, to the passenger, the gravitational potential energy of the book is mgh, where $h = 1$ m. However, to someone on the ground looking up, the floor of the plane may be 9,000 m above the ground. To this person, the gravitational potential energy of the book is mgH, where $H = 9,001$ m. Differences, or changes, in potential energy are unambiguous, but calculated values of potential energy are relative.

Example 10 A stuntwoman (mass = 60 kg) scales a 40-meter-tall rock face. What is her gravitational potential energy (relative to the ground)?

Solution. Calling the ground $h = 0$, we find

$$U_{\text{grav}} = mgh = (60 \text{ kg})(10 \text{ m/s}^2)(40 \text{ m}) = 24{,}000 \text{ J}$$

Example 11 If the stuntwoman in the previous example were to jump off the cliff, what would be her final speed as she landed on a large, air-filled cushion lying on the ground?

Solution. The gravitational potential energy would be transformed into kinetic energy. So

$$U \to K \quad \Rightarrow \quad K \to \tfrac{1}{2}mv^2 \quad \Rightarrow \quad v = \sqrt{\frac{2 \cdot U}{m}} = \sqrt{\frac{2(24{,}000 \text{ J})}{60 \text{ kg}}} = 28 \text{ m/s}$$

CONSERVATION OF MECHANICAL ENERGY

We have seen energy in its two basic forms: kinetic energy (K) and potential energy (U). The sum of an object's kinetic and potential energies is called its **mechanical energy**, E:

$$E = K + U$$

Assuming that no nonconservative forces (friction, for example) act on an object or system while it undergoes some change, then mechanical energy is conserved. That is, the initial mechanical energy, E_i, is equal to the final mechanical energy, E_f, or

$$K_i + U_i = K_f + U_f$$

> Note that because U is relative, so is E.

This is the simplest form of the Law of Conservation of Total Mechanical Energy, which we mentioned at the beginning of this section.

> **Example 12** A ball of mass 2 kg is dropped from a height of 5.0 m above the floor. Find the speed of the ball as it strikes the floor.

Solution. Ignoring the friction due to the air, we can apply Conservation of Mechanical Energy. Calling the floor our $h = 0$ reference level, we write

> Other ways of writing the Law of Conservation of Total Mechanical Energy include
> $\Delta K = -\Delta U$
> and
> $\Delta K + \Delta U = 0$.

$$K_i + U_i = K_f + U_f$$
$$0 + mgh = \tfrac{1}{2}mv^2 + 0$$
$$v = \sqrt{2gh}$$
$$= \sqrt{2(10 \text{ m/s}^2)(5.0 \text{ m})}$$
$$= 10 \text{ m/s}$$

Note that the ball's potential energy decreased, while its kinetic energy increased. This is the basic idea behind conservation of mechanical energy: One form of energy decreases while the other increases.

> **Example 13** A box is projected up a long ramp (incline angle with the horizontal = 37°) with an initial speed of 10 m/s. If the surface of the ramp is very smooth (essentially frictionless), how high up the ramp will the box go? What distance along the ramp will it slide?

Solution. Because friction is negligible, we can apply Conservation of Mechanical Energy. Calling the bottom of the ramp our $h = 0$ reference level, we write

$$K_i + U_i = K_f + U_f$$
$$\tfrac{1}{2}mv_0^2 + 0 = 0 + mgh$$
$$h = \frac{\tfrac{1}{2}v_0^2}{g}$$
$$= \frac{\tfrac{1}{2}(10 \text{ m/s})^2}{10 \text{ m/s}^2}$$
$$= 5 \text{ m}$$

Since the incline angle is $\theta = 37°$, the distance d it slides up the ramp is found in this way:

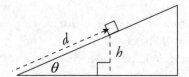

$$h = d \sin\theta$$
$$d = \frac{h}{\sin\theta} = \frac{5 \text{ m}}{\sin 37°} = 8.3 \text{ m}$$

Example 14 A skydiver jumps from a hovering helicopter that's 3,000 m above the ground. If air resistance can be ignored, how fast will he be falling when his altitude is 2,000 m?

Solution. Ignoring air resistance, we can apply Conservation of Mechanical Energy. Calling the ground our $h = 0$ reference level, we write

$$K_i + U_i = K_f + U_f$$
$$0 + mgH = \tfrac{1}{2}mv^2 + mgh$$
$$v = \sqrt{2g(H - h)}$$
$$= \sqrt{2(10 \text{ m/s}^2)(3,000 \text{ m} - 2,000 \text{ m})}$$
$$\approx 140 \text{ m/s}$$

That's over 300 mph! The terminal velocity of a human falling is about 100 mph, which shows that air resistance *does* play a role, even before the parachute is opened.

The equation $K_i + U_i = K_f + U_f$ holds if no nonconservative forces are doing work. However, if work is done by such forces during the process under investigation, then the equation needs to be modified to account for this work as follows:

$$K_i + U_i + W_{\text{other}} = K_f + U_f$$

Example 15 A crash test dummy (mass = 40 kg) falls off a 50-meter-high cliff. On the way down, the force of air resistance has an average strength of 100 N. Find the speed with which he crashes into the ground.

Solution. The force of air resistance opposes the downward motion, so it does negative work on the dummy as he falls: $W_r = -F_r h$. Calling the ground $h = 0$, we find that

$$K_i + U_i + W_r = K_f + U_f$$
$$0 + mgh + (-F_r h) = \tfrac{1}{2}mv^2 + 0$$
$$v = \sqrt{2h(g - F_r/m)} = \sqrt{2(50)(10 - 100/40)} = 27 \text{ m/s}$$

Example 16 A skier starts from rest at the top of a 20° incline and skis in a straight line to the bottom of the slope, a distance d (measured along the slope) of 400 m. If the coefficient of kinetic friction between the skis and the snow is 0.2, calculate the skier's speed at the bottom of the run.

Solution. The strength of the friction force on the skier is $F_f = \mu_k F_N = \mu_k (mg \cos \theta)$, so the work done by friction is $-F_f d = \mu_k (mg \cos \theta) d$. The vertical height of the slope above the bottom of the run (which we designate the $h = 0$ level) is $h = d \sin \theta$. Therefore, Conservation of Mechanical Energy (including the negative work done by friction) gives

$$K_i + U_i + W_{friction} = K_f + U$$
$$0 + mgh + (-\mu_k mg \cos \theta \cdot d) = \tfrac{1}{2}mv^2 + 0$$
$$mg(d \sin \theta) + (-\mu_k mg \cos \theta \cdot d) = \tfrac{1}{2}mv^2$$
$$gd(\sin \theta - \mu_k \cos \theta) = \tfrac{1}{2}v^2$$
$$v = \sqrt{2gd(\sin \theta - \mu_k \cos \theta)}$$
$$= \sqrt{2(10)(400)[\sin 20° - (0.2)\cos 20°]}$$
$$= 35 \text{ m/s}$$

POTENTIAL ENERGY CURVES

The behavior of a system can be analyzed if we are given a graph of its potential energy, $U(x)$, and its mechanical energy, E. Since $K + U = E$, we have $\frac{1}{2}mv^2 + U(x) = E$, which can be solved for v, the velocity at position x:

$$v = \pm\sqrt{\frac{2}{m}\left[E - U(x)\right]}$$

For example, consider the following potential energy curve:

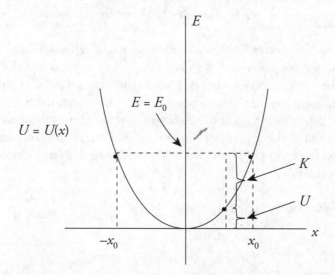

The graph shows how the potential energy, U, varies with position, x. A particular value of the total energy, $E = E_0$, is also shown. Motion of an object whose potential energy is given by $U(x)$ and which has a mechanical energy of E_0 is confined to the region $-x_0 \le x \le x_0$, because only in this range is $E_0 \ge U(x)$. At each position x in this range, the kinetic energy $K = E_0 - U(x)$ is positive. However, if $x > x_0$ (or if $x < -x_0$), then $U(x) > E_0$, which is physically impossible because the difference, $E_0 - U(x)$, which should give K, is negative.

This particular energy curve with $U(x) = \frac{1}{2}kx^2$, describes one of the most important physical systems: a simple harmonic oscillator. The force felt by the oscillator can be recovered from the potential energy curve. Recall that, in the case of gravitational potential energy, we defined $\Delta U_{grav} = -W_{by\ grav}$. In general, $\Delta U = -W$. If we account for a variable force of the form $F = F(x)$, which does the work W, then over a small displacement Δx, we have $\Delta U(x) = -W = -F(x)\ \Delta x$, so $F(x) = -\Delta U(x)/\Delta x$. In the limit as $\Delta x \to 0$, this last equation becomes

$$F(x) = -\frac{dU}{dx}$$

Therefore, in this case, we find $F(x) = -(d/dx)(\frac{1}{2}kx^2) = -kx$, which specifies a linear restoring force, a prerequisite for simple harmonic motion. This equation, $F(x) = -kx$, is called **Hooke's law** and is obeyed by ideal springs (see Example 5).

With this result, we can appreciate the oscillatory nature of the system whose energy curve is sketched above. If x is positive (and not greater than x_0), then $U(x)$ is increasing, so dU/dx is positive, which tells us that F is negative. So the oscillator feels a force—and an acceleration—in the negative direction, which pulls it back through the origin ($x = 0$). If x is negative (and not less than $-x_0$), then $U(x)$ is decreasing, so dU/dx is negative, which tells us that F is positive. So the oscillator feels a force—and an acceleration—in the positive direction, which pushes it back through the origin ($x = 0$).

Furthermore, the difference between E_0 and U, which is K, decreases as x approaches x_0 (or as x approaches $-x_0$), dropping to zero at these points. The fact that K decreases to zero at $\pm x_0$ tells us that the oscillator's speed decreases to zero as it approaches these endpoints. It then changes direction and heads back toward the origin—where its kinetic energy and speed are maximized—for another oscillation. By looking at the energy curve with these observations in mind, you can almost see the oscillator moving back and forth between the barriers at $x = \pm x_0$.

The origin is a point at which $U(x)$ has a minimum, so the tangent line to the curve at this point is horizontal; the slope is zero. Since $F = -dU/dx$, the force F is 0 at this point [which we also know from the equation $F(x) = -kx$]; this means that this is a point of **equilibrium**. If the oscillator is pushed from this equilibrium point in either direction, the force $F(x)$ will attempt to restore it to $x = 0$, so this is a point of **stable** equilibrium. However, a point where the $U(x)$ curve has a maximum is also a point of equilibrium, but it's an **unstable** one, because if the system were moved from this point in either direction, the force would accelerate it *away* from the equilibrium position.

Consider the following potential energy curve:

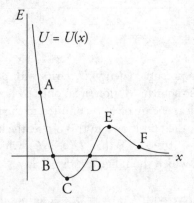

Point C is a position of stable equilibrium and E is a point of unstable equilibrium. Since $U(x)$ is decreasing at points A and F, $F(x)$ is positive, accelerating the system in the positive x direction. Points B and D mark the barriers of oscillation if the system has a mechanical energy E_0 of 0. To imagine an object moving in an area with this $U(x)$, imagine an object sliding on a frictionless hill of this shape. If released from rest at point B, it would accelerate toward point C, then slow down, stop at point D, and then return to point B. It would oscillate between these two points.

Example 17 An object of mass $m = 4$ kg has a potential energy function

$$U(x) = (x - 2) - (2x - 3)^3$$

where x is measured in meters and U in joules. The following graph is a sketch of the potential energy function.

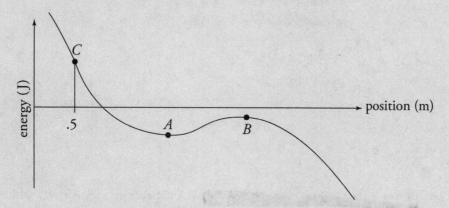

(a) Determine the positions of points A and B, the equilibrium points.

(b) If the object is released from rest at point B, can it reach point A or C? Explain.

(c) The particle is released from rest at point C. Determine its speed as it passes point A.

Solution.

(a) The points A and B are a local minimum and maximum, respectively, so the derivative of $U(x)$ will be zero at these locations.

$$\frac{dU}{dx} = 1 - 3(2x - 3)^2 (2) = 0$$

$$-24x^2 + 72x - 53 = 0$$

$$x = 1.30 \text{ m and } 1.70 \text{ m}$$

(b) The object has a negative total amount of mechanical energy at point B because all of its energy is potential energy. It will not be able to reach point C because that position has a potential energy well above zero. However, the object would be able to reach point A because the potential energy at A is less than the energy the object started with at point B. As the object moved from B to A, its potential energy would decrease (become more negative) and its kinetic energy would increase.

(c) First, we need to find how much potential energy the object has at point C, and this will define the total mechanical energy of the object. Then determine the potential energy at point A, $x = 1.3$ m, and use Conservation of Energy to determine the speed at point A.

$$U(0.5) = (0.5 - 2) - (2 \cdot (0.5) - 3)^3 = 6.5 \text{ J}$$
$$U(1.3) = -0.636 \text{ J}$$
$$E_C = E_A$$
$$6.5 = -0.636 + K_A$$
$$7.136 = \frac{1}{2}(4)v_A^2$$
$$v_A = 1.89 \text{ m/s}$$

POWER

Simply put, **power** is the rate at which work gets done (or energy gets transferred, which is the same thing). Suppose you and I each do 1,000 J of work, but I do the work in 2 minutes while you do it in 1 minute. We both did the same amount of work, but you did it more quickly; you were more powerful. Here's the definition of power:

$$\text{Power} = \frac{\text{Work}}{\text{time}} \qquad \text{in symbols} \rightarrow \qquad P = \frac{W}{t}$$

Since power is a rate, it can be represented as a derivative, too. The calculus (rather than algebra) equation for power is:

$$P = \frac{dW}{dt}$$

The unit of power is the joule per second (J/s), which is called the **watt**, and symbolized W (not to be confused with the symbol for work, W). One watt is 1 joule per second: 1 W = 1 J/s. Here in the United States, which still uses older English units like inches, feet, yards, miles, ounces, pounds, and so forth, you still hear of power ratings expressed in horsepower (particularly of engines). One horsepower is defined as the power output of a large horse. Horses can pull a 150-pound weight at a speed of $2\frac{1}{2}$ mph for quite a while. If the equation for work done by a constant force, namely $W = \mathbf{F} \cdot \mathbf{r}$, is inserted into the equation for power, where $\mathbf{r}$ is the displacement, the equation for power becomes $P = \frac{\mathbf{F} \cdot \mathbf{r}}{t}$. Since r/t (displacement over time) is velocity, another equation for power is:

$$P = \mathbf{F} \cdot \mathbf{v}$$

Since the theoretical horse mentioned above is pulling in the direction of its velocity,

$$P = Fv \quad \Rightarrow \quad 1 \text{ horsepower (hp)} = (150 \text{ lb})(2\tfrac{1}{2} \text{ mph})$$

Using unit conversions to convert horsepower to watts:

$$150 \text{ lb} = 150 \text{ lb} \times \frac{4.45 \text{ N}}{\text{lb}} = 667.5 \text{ N}$$

$$2\tfrac{1}{2} \text{ mph} = \frac{2\tfrac{1}{2} \text{ mi}}{\text{hr}} \times \frac{1,609 \text{ m}}{1 \text{ mi}} \times \frac{1 \text{ hr}}{3,600 \text{ s}} = 1.117 \text{ m/s}$$

Therefore,

$$1 \text{ hp} = (667.5 \text{ N})(1.117 \text{ m/s}) = 746 \text{ W}$$

By contrast, a human in good physical condition can do work at a steady rate of about 75 W (about 1/10 that of a horse!) but can attain power levels as much as twice this for short periods of time.

Example 18 A mover pushes a large crate (mass $m = 75$ kg) from the inside of the truck to the back end (a distance of 6 m), exerting a steady push of 300 N. If he moves the crate this distance in 20 s, what is his power output during this time?

Solution. The work done on the crate by the mover is $W = Fd = (300 \text{ N})(6 \text{ m}) = 1,800$ J. If this much work is done in 20 s, then the power delivered is $P = W/t = (1,800 \text{ J})/(20 \text{ s}) = 90$ W.

Example 19 What must be the power output of a rocket engine, which moves a 1,000 kg rocket at a constant speed of 8.0 m/s?

Solution. The equation $P = Fv$, with $F = mg$, yields

$$P = mgv = (1,000 \text{ kg})(10 \text{ N/kg})(8.0 \text{ m/s}) = 80,000 \text{ W} = 80 \text{ kW}$$

Chapter 7 Drill

The answers and explanations can be found in Chapter 17.

Section I: Multiple Choice

1. A force **F** of strength 20 N acts on an object of mass 3 kg as it moves a distance of 4 m. If **F** is perpendicular to the 4 m displacement, the work it does is equal to

    (A) 0 J
    (B) 60 J
    (C) 80 J
    (D) 600 J
    (E) 2,400 J

2. Under the influence of a force, an object of mass 4 kg accelerates from 3 m/s to 6 m/s in 8 s. How much work was done on the object during this time?

    (A) 27 J
    (B) 54 J
    (C) 72 J
    (D) 96 J
    (E) Cannot be determined from the information given

3. A student is asked to lift a box of mass *m* up to a height *h*. The student has the options of simply lifting the box, pushing it up an inclined plane of angle 30°, or using a pulley system in which the rope is looped around the pulley 4 times. If these are labeled as Option 1, Option 2, and Option 3, respectively, which of the following correctly shows the relationship of the work needed to lift the block in each situation? Assume ideal conditions.

    (A) Option 1 > Option 2 > Option 3
    (B) Option 1 > Option 2 = Option 3
    (C) Option 1 = Option 3 > Option 2
    (D) Option 1 = Option 2 = Option 3
    (E) Cannot be determined without additional information

4. A 4 kg box is pulled up a ramp of angle $\theta = 30$ degrees and height = 5 m at a constant velocity. How much work is done by the normal force?

    (A) 160 J
    (B) 80 J
    (C) 40 J
    (D) 20 J
    (E) 0 J

5. A man stands in an elevator as it begins to ascend. Does the normal force from the floor do work on the man?

    (A) Yes, and the work done will be positive.
    (B) Yes, and the work done will be negative.
    (C) Yes, but the sign can't be determined.
    (D) No, the normal force will do no work in any situation.
    (E) No, the normal force will do no work in this situation, but it can in others.

6. A block of mass 3.5 kg slides down a frictionless inclined plane of length 6.4 m that makes an angle of 30° with the horizontal. If the block is released from rest at the top of the incline, what is its speed at the bottom?

    (A) 5.0 m/s
    (B) 5.7 m/s
    (C) 6.4 m/s
    (D) 8.0 m/s
    (E) 10 m/s

7. A block of mass *m* slides from rest down an inclined plane of length *s* and height *h*. If *F* is the magnitude of the force of kinetic friction acting on the block as it slides, then the kinetic energy of the block when it reaches the bottom of the incline will be equal to

    (A) *mgh*
    (B) *mgh* – *Fh*
    (C) *mgs* – *Fh*
    (D) *mgh* – *Fs*
    (E) *mgs* – *Fs*

8. If a 1,000 kg car is accelerating at a rate of 4 m/s² and experiencing 300 N of drag force, how much force do the engines have to produce? Ignore any frictional effects with the road.

    (A) 16,300 N
    (B) 15,700 N
    (C) 4,300 N
    (D) 4,000 N
    (E) 3,700 N

9. An astronaut drops a rock from the top of a crater on the Moon. When the rock is halfway down to the bottom of the crater, its speed is what fraction of its final impact speed?

(A) $\dfrac{1}{4\sqrt{2}}$

(B) $\dfrac{1}{4}$

(C) $\dfrac{1}{2\sqrt{2}}$

(D) $\dfrac{1}{2}$

(E) $\dfrac{1}{\sqrt{2}}$

10. A block of mass m is at rest at the bottom of an inclined ramp of length d and angle θ. A string is attached to the block and connected to a motor that pulls on the string. The coefficient of friction between the block and the ramp as it slides up is μ. What is the minimum constant power P that the motor can expend to make the block reach the top of the ramp in t seconds?

(A) $m\left(\dfrac{d^2}{t^3} + \dfrac{gd\sin\theta}{t} + \dfrac{dg\mu\cos\theta}{t}\right)$

(B) $m\left(\dfrac{d^2}{t} + \dfrac{gd\sin\theta}{t} + \dfrac{dg\mu\cos\theta}{t}\right)$

(C) $m\left(\dfrac{d^2}{t} + \dfrac{gd\cos\theta}{t} \quad \dfrac{dg\mu\sin\theta}{t}\right)$

(D) $m\left(\dfrac{2d^2}{t^3} + \dfrac{gd\sin\theta}{t} + \dfrac{dg\mu\cos\theta}{t}\right)$

(E) $m\left(\dfrac{2d^2}{t^3} + \dfrac{gd\cos\theta}{t} + \dfrac{dg\mu\sin\theta}{t}\right)$

Section II: Free Response

1. A box of mass m is released from rest at point A, the top of a long, frictionless slide. Point A is at height H above the level of points B and C. Although the slide is frictionless, the horizontal surface from point B to C is not. The coefficient of kinetic friction between the box and this surface is μ_k, and the horizontal distance between point B and C is x.

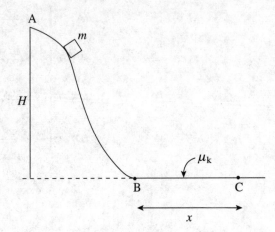

Solve for the following in terms of given quantities and the acceleration of gravity, g.

(a) Find the speed of the box when its height above point B is $\frac{1}{2}H$.

(b) Find the speed of the box when it reaches point B.

(c) Determine the value of μ_k so that the box comes to rest at point C.

(d) Now assume that points B and C were not on the same horizontal level. In particular, assume that the surface from B to C had a uniform upward slope so that point C was still at a horizontal distance of x from B but now at a vertical height of y above B. Answer the question posed in part (c).

2. The diagram below shows a roller-coaster ride which contains a circular loop of radius r. A car (mass m) begins at rest from point A and moves down the frictionless track from A to B, where it then enters the vertical loop (also frictionless), traveling once around the circle from B to C to D to E and back to B, after which it travels along the flat portion of the track from B to F (which is not frictionless).

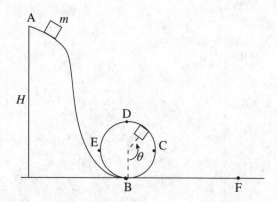

Solve for the following in terms of given quantities and the acceleration of gravity, g.

(a) Find the centripetal acceleration of the car when it is at point C.

(b) What is the minimum cut-off speed v_c that the car must have at point D to make it around the loop?

(c) What is the minimum height H necessary to ensure that the car makes it around the loop?

(d) If $H = 6r$ and the coefficient of friction between the car and the flat portion of the track from B to F is 0.5, how far along this flat portion of the track will the car travel before coming to rest at point F?

3. A ball $m = 3$ kg has the potential energy function

$$U(x) = 3(x - 1) - (x - 3)^3$$

where x is measured in meters and U in joules. The following graph is a sketch of this potential energy function.

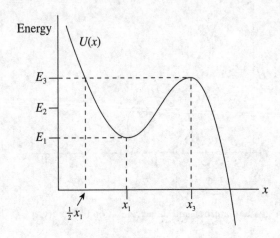

The energies indicated on the vertical axis are evenly spaced; that is, $E_3 - E_2 = E_2 - E_1$. The energy E_1 is equal to $U(x_1)$, and the energy E_3 is equal to $U(x_3)$.

(a) Determine the numerical values of x_1 and x_3.

(b) Describe the motion of the particle if its total energy is E_2.

(c) What is the particle's speed at $x = x_1$ if its total energy, E, equals 58 J?

(d) Sketch the graph of the particle's acceleration as a function of x. Be sure to indicate x_1 and x_3 on your graph.

(e) The particle is released from rest at $x = \frac{1}{2}x_1$. Find its speed as it passes through $x = x_1$.

4. The force on a 6 kg object is given by the equation: $F(x) = 3x + 5$, in newtons. The object is moving 2 m/s at the origin.

(a) Determine the work done on the object by the force when it is moved 4 m from the origin in the x direction.

(b) Determine the speed of the object when it has moved 4 m.

Summary

Work

o   Work is the dot product of force and displacement: $W = F \cdot x$

o   When force varies with x, the work is given by the following equation: $W = \int F(x)\,dx$

o   Work is positive when the force and displacement are parallel.

o   Work is negative when the force and displacement are antiparallel.

o   Work–Energy Theorem: $W = \Delta K = -\Delta U$

Energy

o   Kinetic energy is energy associated with motion and is given by the equation $K = \dfrac{1}{2}mv^2$.

o   Potential energy is stored energy. The potential energy due to gravity is given by the equation $U_g = mgh$. The potential energy due to springs is given by the equation $U_s = \dfrac{1}{2}kx^2$.

o   Work done by a conservative force only depends on the initial and final positions, and not on the path taken. Gravity and springs are examples of conservative forces.

o   Work done by a non-conservative force depends on the path taken, and mechanical energy is lost by heat, sound, and so on, when these forces act on a system. Friction and air resistance are examples of non-conservative forces.

o   Conservation of Mechanical Energy states that the total mechanical energy of a system is constant when there are no non-conservative forces acting on the system. It is usually written as $E_i = E_f$ or $K_i + U_i = K_f + U_f$.

Potential Energy Diagrams

o   The potential energy can be given as $U(x)$. Then $F = -\dfrac{dU}{dx}$.

o   If $\dfrac{dU}{dx} = 0$, then $F = 0$, and it is an equilibrium point.

o   Stable equilibrium occurs when the force restores the object back toward the equilibrium point after it is disturbed.

o   Unstable equilibrium occurs when the force moves the object further away from the equilibrium point after it is disturbed.

Power

o   Power is the rate at which work is done.

$$P = \frac{W}{t} = \frac{dW}{dt}$$

$$P = Fv$$

Chapter 8
Systems of Particles and Linear Momentum

INTRODUCTION

When Newton first expressed his Second Law, he didn't write $\mathbf{F}_{net} = m\mathbf{a}$. Instead, he expressed the law in the words, *The alteration of motion is...proportional to the... force impressed....* By "motion," he meant the product of mass and velocity, a vector quantity known as **linear momentum,** which is denoted by **p**:

$$\mathbf{p} = m\mathbf{v}$$

So Newton's original formulation of the Second Law read $\Delta\mathbf{p} \propto \mathbf{F}$, or, equivalently, $\mathbf{F} \propto \Delta\mathbf{p}$. But a large force that acts for a short period of time can produce the same change in linear momentum as a small force acting for a greater period of time. Knowing this, if we take the average force, $\overline{\mathbf{F}}$, that acts over the time interval Δt, we can turn the proportion above into an equation:

$$\overline{\mathbf{F}} = \frac{\Delta\mathbf{p}}{\Delta t}$$

This equation becomes $\mathbf{F} = m\mathbf{a}$, since $\Delta\mathbf{p}/\Delta t = \Delta(m\mathbf{v})/\Delta t = m(\Delta\mathbf{v}/\Delta t) = m\mathbf{a}$ (assuming that m remains constant). If we take the limit as $\Delta t \to 0$, then the equation above takes the form:

$$\mathbf{F} = \frac{d\mathbf{p}}{dt}$$

> **Example 1** A golfer strikes a golf ball of mass 0.05 kg, and the time of impact between the golf club and the ball is 1 ms. If the ball acquires a velocity of magnitude 70 m/s, calculate the average force exerted on the ball.

Solution. Using Newton's Second Law, we find

$$\overline{F} = \frac{\Delta p}{\Delta t} = \frac{\Delta(mv)}{\Delta t} = m\frac{v-0}{\Delta t} = (0.05 \text{ kg})\frac{70 \text{ m/s}}{10^{-3} \text{ s}} = 3{,}500 \text{ N} \quad [\approx 790 \text{ lb } (!)]$$

IMPULSE

The product of force and the time during which it acts is known as **impulse**; it's a vector quantity that's denoted by **J**:

$$\mathbf{J} = \overline{\mathbf{F}}\Delta t$$

In terms of impulse, Newton's Second Law can be written in yet another form:

$$\mathbf{J} = \Delta\mathbf{p}$$

Sometimes this is referred to as the **impulse–momentum theorem**, but it's just another way of writing Newton's Second Law. If **F** varies with time over the interval during which it acts, then the impulse delivered by the force **F** = **F**(*t*) from time *t* = t_1 to *t* = t_2 is given by the following definite integral:

$$\mathbf{J} = \int_{t_1}^{t_2} \mathbf{F}(t)\,dt$$

On the equation sheet for the free-response section, this information will be represented as follows:

$$\mathbf{J} = \int \mathbf{F}\,dt = \Delta \mathbf{p}$$

If a graph of force-versus-time is given, then the impulse of force **F** as it acts from t_1 to t_2 is equal to the area bounded by the graph of **F**, the *t*-axis, and the vertical lines associated with t_1 and t_2 as shown in the following graph.

Q: Can you think of how the impulse–momentum theorem can be used to make collisions safer?

Turn the page for the answer.

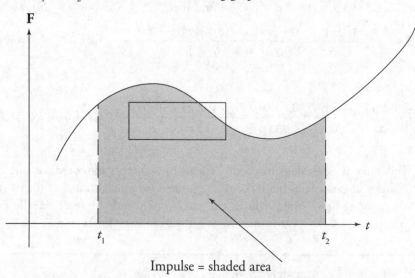

Impulse = shaded area

Example 2 A football team's kicker punts the ball (mass = 0.4 kg) and gives it a launch speed of 30 m/s. Find the impulse delivered to the football by the kicker's foot and the average force exerted by the kicker on the ball, given that the impact time is 8 ms.

Solution. Impulse is equal to change in linear momentum, so

$$J = \Delta p = p_\mathrm{f} - p_\mathrm{i} = p_\mathrm{f} = mv = (0.4 \text{ kg})(30 \text{ m/s}) = 12 \text{ kg·m/s}$$

Using the equation $\bar{F} = J/\Delta t$, we find that the average force exerted by the kicker is

$$\bar{F} = J/\Delta t = (12 \text{ kg·m/s})/(8 \times 10^{-3}\,\text{s}) = 1{,}500 \text{ N } [\approx 340 \text{ lb}]$$

Example 3 An 80 kg stuntman jumps out of a window that's 45 m above the ground.

(a) How fast is he falling when he reaches ground level?

(b) He lands on a large, air-filled target, coming to rest in 1.5 s. What average force does he feel while coming to rest?

(c) What if he had instead landed on the ground (impact time = 10 ms)?

Solution.

(a) His gravitational potential energy turns into kinetic energy: $mgh = \frac{1}{2}mv^2$, so

$$v = \sqrt{2gh} = \sqrt{2(10)(45)} = 30 \text{ m/s} \quad [\approx 70 \text{ mph}]$$

(You could also have answered this question using Big Five #5.)

(b) Using $\bar{\mathbf{F}} = \Delta\mathbf{p}/\Delta t$ and $\mathbf{p} = m\mathbf{v}$, we find that

$$\bar{\mathbf{F}} = \frac{\Delta\mathbf{p}}{\Delta t} = \frac{\mathbf{p}_f - \mathbf{p}_i}{\Delta t} = \frac{0 - m\mathbf{v}_i}{\Delta t} = \frac{-(80 \text{ kg})(30 \text{ m/s})}{1.5 \text{ s}} = 1{,}600 \text{ N} \quad \Rightarrow \quad \bar{F} = 1{,}600 \text{ N}$$

(c) In this case,

$$\bar{\mathbf{F}} = \frac{\Delta\mathbf{p}}{\Delta t} = \frac{\mathbf{p}_f - \mathbf{p}_i}{\Delta t} = \frac{0 - m\mathbf{v}_i}{\Delta t} = \frac{-(80 \text{ kg})(30 \text{ m/s})}{10 \times 10^{-3} \text{ s}} = 240{,}000 \text{ N} \Rightarrow \bar{F} = 240{,}000 \text{ N}$$

> A: For a constant change in momentum during a collision, a longer collision time will result in a smaller average force!

This force is equivalent to about 27 tons (!), which is more than enough to break bones and cause fatal brain damage. Notice how crucial impact time is: Increasing the slowing-down time reduces the acceleration and the force, ideally enough to prevent injury. This is the purpose of air bags in cars, for example.

Example 4 A small block of mass $m = 0.07$ kg, initially at rest, is struck by an impulsive force $\mathbf{F}$ of duration 10 ms whose strength varies with time according to the following graph:

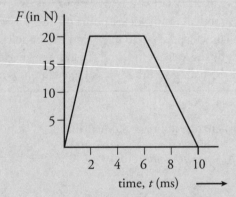

What is the resulting speed of the block?

Solution. The impulse delivered to the block is equal to the area under the F-vs.-t graph. The region is a trapezoid, so its area, $\frac{1}{2}(\text{base}_1 + \text{base}_2) \times \text{height}$, can be calculated as follows:

$$J = \frac{1}{2}(10 + 4) \times 20 = 0.14 \text{ N}$$

Now, by the impulse–momentum theorem,

$$J = \Delta p = p_f - p_i = p_f = m\,v_f \quad \Rightarrow \quad v_f = \frac{J}{m} = \frac{0.14 \text{ N} \cdot \text{s}}{0.07 \text{ kg}} = 2 \text{ m/s}$$

CONSERVATION OF LINEAR MOMENTUM

Newton's Third Law says that when one object exerts a force on a second object, the second object exerts an equal but opposite force on the first. Since Newton's Second Law says that the impulse delivered to an object is equal to the resulting change in its linear momentum, $\mathbf{J} = \Delta\mathbf{p}$, the two interacting objects experience equal but opposite momentum changes (assuming that there are no external forces), which implies that the total linear momentum of the system remains constant. In fact, given any number of interacting objects, each pair that comes in contact will undergo equal but opposite momentum changes, so the result described for two interacting objects will actually hold for any number of objects, given that the only forces they feel are from each other. This means that, in an isolated system, *the total linear momentum will remain constant*. This is the **Law of Conservation of Linear Momentum**. In equation form, for two objects colliding, we have $m_1\vec{v}_{1i} + m_2\vec{v}_{2i} = m_1\vec{v}_{1f} + m_2\vec{v}_{2f}$.

> **Example 5** An astronaut is floating in space near her shuttle when she realizes that the cord that's supposed to attach her to the ship has become disconnected. Her total mass (body + suit + equipment) is 91 kg. She reaches into her pocket, finds a 1 kg metal tool, and throws it out into space with a velocity of 9 m/s directly away from the ship. If the ship is 10 m away, how long will it take her to reach it?

Solution. Here, the astronaut + tool are the system. Because of Conservation of Linear Momentum,

$$m_{\text{astronaut}}\,\mathbf{v}_{\text{astronaut}} + m_{\text{tool}}\mathbf{v}_{\text{tool}} = 0$$

$$m_{\text{astronaut}}\,\mathbf{v}_{\text{astronaut}} = -m_{\text{tool}}\mathbf{v}_{\text{tool}}$$

$$\mathbf{v}_{\text{astronaut}} = -\frac{m_{\text{tool}}}{m_{\text{astronaut}}}\,\mathbf{v}_{\text{tool}}$$

$$= -\frac{1 \text{ kg}}{90 \text{ kg}}(-9 \text{ m/s}) = +0.1 \text{ m/s}$$

Using *distance = average speed × time*, we find

$$t = \frac{d}{v} = \frac{10 \text{ m}}{0.1 \text{ m/s}} = 100 \text{ s}$$

COLLISIONS

Conservation of Linear Momentum is routinely used to analyze **collisions**. The objects whose collision we will analyze form the *system*, and although the objects exert forces on each other during the impact, these forces are only *internal* (they occur within the system). The system's total linear momentum is conserved if there is no net external force on the system.

Collisions are classified into two major categories: (1) **elastic** and (2) **inelastic**. A collision is said to be *elastic* if kinetic energy is conserved. Ordinary macroscopic collisions are never truly elastic, because there is always a change in energy due to energy transferred as heat, deformation of the objects, and the sound of the impact. However, if the objects do not deform very much (for example, two billiard balls or a hard glass marble bouncing off a steel plate), then the loss of initial kinetic energy is small enough to be ignored, and the collision can be treated as virtually elastic. *Inelastic* collisions, then, are ones in which the total kinetic energy is different after the collision. An extreme example of inelasticism is **completely** (or **perfectly** or **totally**) **inelastic**. In this case, the objects stick together after the collision and move as one afterward. In all cases of isolated collisions (elastic or not), Conservation of Linear Momentum states that

$$\text{total } \mathbf{p}_{\text{before collision}} = \text{total } \mathbf{p}_{\text{after collision}}$$

> **Example 6** Two balls roll toward each other, one red and the other green. The red ball has a mass of 0.5 kg and a speed of 4 m/s just before impact. The green ball has a mass of 0.3 kg and a speed of 2 m/s. After the head-on collision, the red ball continues forward with a speed of 1.7 m/s. Find the speed of the green ball after the collision. Was the collision elastic?

Solution. First, remember that momentum is a vector quantity, so the direction of the velocity is crucial. Since the balls roll toward each other, one ball has a positive velocity while the other has a negative velocity. Let's call the red ball's velocity before the collision positive; then $\mathbf{v}_{\text{red}} = +4$ m/s, and $\mathbf{v}_{\text{green}} = -2$ m/s. Using a prime to denote *after the collision*, Conservation of Linear Momentum gives us the following:

$$\text{total } \mathbf{p}_{\text{before}} = \text{total } \mathbf{p}_{\text{after}}$$
$$m_{\text{red}}\mathbf{v}_{\text{red}} + m_{\text{green}}\mathbf{v}_{\text{green}} = m_{\text{red}}\mathbf{v}'_{\text{red}} + m_{\text{green}}\mathbf{v}'_{\text{green}}$$
$$(0.5)(+4) + (0.3)(-2) = (0.5)(+1.7) + (0.3)\mathbf{v}'_{\text{green}}$$
$$\mathbf{v}'_{\text{green}} = +1.83 \text{ m/s}$$

Notice that the green ball's velocity was reversed as a result of the collision; this typically happens when a lighter object collides with a heavier object. To see whether the collision was elastic, we need to compare the total kinetic energies before and after the collision. In this case, however, an explicit calculation is not needed since both objects experienced a decrease in speed as a result of the collision. Kinetic energy was lost, so the collision was inelastic; this is usually the case with macroscopic collisions. Most of the lost energy was transferred as heat; the two objects are both slightly warmer as a result of the collision.

Example 7 The balls in Example 6 roll toward each other. The red ball has a mass of 0.5 kg and a speed of 4 m/s just before impact. The green ball has a mass of 0.3 kg and a speed of 2 m/s. If the collision is completely inelastic, determine the velocity of the composite object after the collision.

Solution. If the collision is completely inelastic, then, by definition, the masses stick together after impact, moving with a velocity, $\mathbf{v}'$. Applying Conservation of Linear Momentum, we find

$$\text{total } \mathbf{p}_{\text{before}} = \text{total } \mathbf{p}_{\text{after}}$$
$$m_{\text{red}}\, \mathbf{v}_{\text{red}} + m_{\text{green}}\, \mathbf{v}_{\text{green}} = (m_{\text{red}} + m_{\text{green}})\mathbf{v}'$$
$$(0.5)(+4) + (0.3)(-2) = (0.5 + 0.3)\mathbf{v}'$$
$$\mathbf{v}' = +1.8 \text{ m/s}$$

Example 8 An object of mass m_1 is moving with velocity v_1 toward a target object of mass m_2, which is stationary ($v_2 = 0$). The objects collide head-on, and the collision is elastic. Show that the relative velocity before the collision, $v_2 - v_1$, has the same magnitude as $v_2' - v_1'$, the relative velocity after the collision.

Solution. Since the collision is elastic, both total linear momentum and kinetic energy are conserved. Therefore,

$$m_1 v_1 = m_1 v_1' + m_2 v_2' \qquad (1)$$
$$\tfrac{1}{2} m_1 v_1^2 = \tfrac{1}{2} m_1 v_1'^2 + \tfrac{1}{2} m_2 v_2'^2 \qquad (2)$$

Now for some algebra. Cancel the $\frac{1}{2}$'s in the second equation and factor to get the following pair of equations:

$$m_1(v_1 - v_1') = m_2 v_2' \qquad (1')$$
$$m_1(v_1 - v_1')(v_1 + v_1') = m_2 v_2'^2 \qquad (2')$$

Next, dividing the second equation by the first gives $v_1 + v_1' = v_2'$, so we can write

$$v_1 - v_1' = (m_2/m_1)v_2' \qquad (1'')$$
$$v_1 + v_1' = v_2' \qquad (2'')$$

Adding Equations $(1'')$ and $(2'')$ gives

$$2v_1 = \left(\frac{m_2}{m_1} + 1\right)v_2' \quad \Rightarrow \quad v_2' = \frac{2v_1}{\frac{m_2}{m_1} + 1} = \frac{2m_1}{m_1 + m_2} v_1$$

Substituting this result into Equation (2″) gives

$$v_1 + v_1' = \frac{2m_1}{m_1 + m_2} v_1 \quad \Rightarrow \quad v_1' = \left(\frac{2m_1}{m_1 + m_2} - 1 \right) v_1 = \frac{m_1 - m_2}{m_1 + m_2} v_1$$

Now we have calculated the final velocities, v_1 and v_2. To verify the claim made in the statement of the question, we notice that

$$v_2' - v_1' = \frac{2m_1}{m_1 + m_2} v_1 - \frac{m_1 - m_2}{m_1 + m_2} v_1 = \frac{m_1 + m_2}{m_1 + m_2} v_1 = v_1 - 0 = v_1 - v_2 = -(v_2 - v_1)$$

so the relative velocity after the collision, $v_2' - v_1'$, is equal (in magnitude) but opposite (in direction) to $v_2 - v_1$, the relative velocity before the collision. *This is a general property that characterizes elastic collisions.*

> If a problem involves an elastic collision between two moving masses, transform the problem into the reference frame of one of the moving masses. This will effectively turn the problem into the same one as Example 8. Just be sure to transform back into a stationary reference frame at the end!

Example 9 An object of mass m moves with velocity **v** toward a stationary object of mass $2m$. After impact, the objects move off in the directions shown in the following diagram:

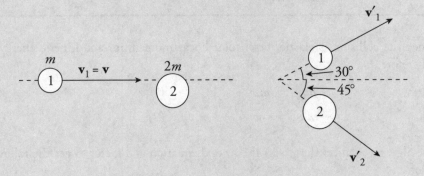

Before the collision   After the collision

(a) Determine the magnitudes of the velocities after the collision (in terms of v).

(b) Is the collision elastic? Explain your answer.

Solution.

(a) Conservation of Linear Momentum is a principle that establishes the equality of two vectors: $\mathbf{p}_{total}$ before the collision and $\mathbf{p}_{total}$ after the collision. Writing this single vector equation as two equations, one for the x-component and one for the y, we have

$$x\text{-component} : \quad mv = mv_1' \cos 30° + 2mv_2' \cos 45° \quad (1)$$

$$y\text{-component} : \quad 0 = mv_1' \sin 30° - 2mv_2' \sin 45° \quad (2)$$

Adding these equations eliminates v_2 because $\cos 45° = \sin 45°$.

$$mv = mv_1'(\cos 30° + \sin 30°)$$

and lets us determine v_1:

$$v_1' = \frac{v}{\cos 30° + \sin 30°} = \frac{2v}{1+\sqrt{3}}$$

Substituting this result into Equation (2) gives us

$$0 = m\frac{2v}{1+\sqrt{3}}\sin 30° - 2mv_2' \sin 45°$$

$$2mv_2' \sin 45° = m\frac{2v}{1+\sqrt{3}}\sin 30°$$

$$v_2' = \frac{\frac{2v}{1+\sqrt{3}}\sin 30°}{2\sin 45°} = \frac{v}{\sqrt{2}(1+\sqrt{3})}$$

(b) The collision is elastic only if kinetic energy is conserved. The total kinetic energy after the collision, K', is calculated as follows:

$$K' = \frac{1}{2} \cdot mv_1'^2 + \frac{1}{2} \cdot 2mv_2'^2$$

$$= \frac{1}{2}m\left(\frac{2v}{1+\sqrt{3}}\right)^2 + m\left(\frac{v}{\sqrt{2}(1+\sqrt{3})}\right)^2$$

$$= mv^2\left[\frac{4}{2(1+\sqrt{3})^2} + \frac{1}{2(1+\sqrt{3})^2}\right]$$

$$= \frac{5mv^2}{2(1+\sqrt{3})^2}$$

However, the kinetic energy before the collision is just $K = \frac{1}{2}mv^2$, so the fact that

$$\frac{5}{2(1+\sqrt{3})^2} < \frac{1}{2}$$

tells us that K' is less than K, so some kinetic energy is lost; the collision is inelastic.

CENTER OF MASS

The center of mass is the point where all of the mass of an object can be considered to be concentrated; it's the dot that represents the object of interest in a free-body diagram.

For a homogeneous body (that is, one for which the density is uniform throughout), the center of mass is where you intuitively expect it to be: at the geometric center. Thus, the center of mass of a uniform sphere or cube or box is at its geometric center.

If we have a collection of discrete particles, the center of mass of the system can be determined mathematically as follows. First consider the case where the particles all lie on a straight line. Call this the x-axis. Select some point to be the origin ($x = 0$) and determine the positions of each particle on the axis. Multiply each position value by the mass of the particle at that location, and get the sum for all the particles. Divide this sum by the total mass, and the resulting x-value is the center of mass:

$$x_{cm} = \frac{m_1 x_1 + m_2 x_2 + \cdots + m_n x_n}{m_1 + m_2 + \cdots + m_n}$$

On the AP Physics C Table of Information, this information will be represented as follows:

$$\mathbf{r}_{cm} = \sum m\mathbf{r} \,/\, \sum m$$

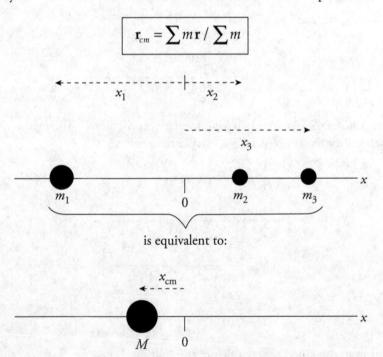

is equivalent to:

The system of particles behaves in many respects as if all its mass, $M = m_1 + m_2 + \cdots + m_n$, were concentrated at a single location, x_{cm}.

If the system consists of objects that are not confined to the same straight line, use the equation above to find the x-coordinate of their center of mass, and the corresponding equation,

$$y_{cm} = \frac{m_1 y_1 + m_2 y_2 + \cdots + m_n y_n}{m_1 + m_2 + \cdots + m_n}$$

to find the y-coordinate of their center of mass (and one more equation to calculate the z-coordinate, if they are not confined to a single plane).

From the equation

$$x_{cm} = \frac{m_1 x_1 + m_2 x_2 + \cdots + m_n x_n}{M}$$

we can derive

$$M v_{cm} = m_1 v_1 + m_2 v_2 + \cdots + m_n v_n$$

So, the total linear momentum of all the particles in the system $(m_1 v_1 + m_2 v_2 + \ldots + m_n v_n)$ is the same as $M v_{cm}$, the linear momentum of a single particle (whose mass is equal to the system's total mass) moving with the velocity of the center of mass.

We can also differentiate again and establish the following:

$$\mathbf{F}_{net} = M\mathbf{a}_{cm}$$

This says that the net (external) force acting on the system causes the center of mass to accelerate according to Newton's Second Law. In particular, *if the net external force on the system is zero, then the center of mass will not accelerate.*

Example 10 Two objects, one of mass m and one of mass $2m$, hang from light threads from the ends of a uniform bar of length $3L$ and mass $3m$. The masses m and $2m$ are at distances L and $2L$, respectively, below the bar. Find the center of mass of this system.

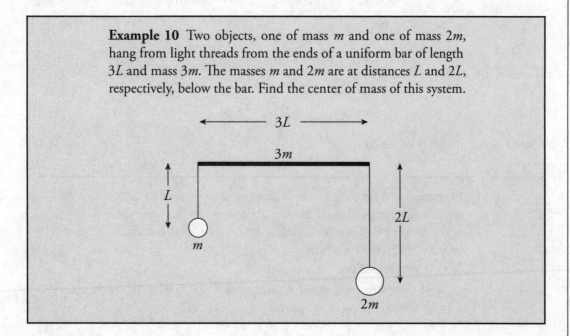

Solution. The center of mass of the bar alone is at its midpoint (because it is uniform), so we may treat the total mass of the bar as being concentrated at its midpoint. Constructing a coordinate system with this point as the origin, we now have three objects: one of mass m at $(-3L/2, -L)$, one of mass $2m$ at $(3L/2, -2L)$, and one of mass $3m$ at $(0, 0)$:

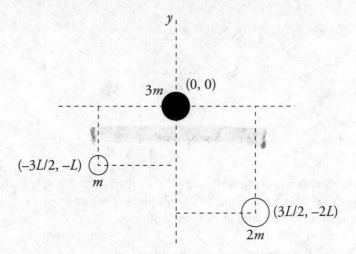

We figure out the x- and y-coordinates of the center of mass separately:

$$x_{cm} = \frac{m_1 x_1 + m_2 x_2 + m_3 x_3}{m_1 + m_2 + m_3} = \frac{(m)(-3L/2) + (2m)(3L/2) + (3m)(0)}{m + 2m + 3m} = \frac{3mL/2}{6m} = \frac{L}{4}$$

$$y_{cm} = \frac{m_1 y_1 + m_2 y_2 + m_3 y_3}{m_1 + m_2 + m_3} = \frac{(m)(-L) + (2m)(-2L) + (3m)(0)}{m + 2m + 3m} = \frac{-5mL}{6m} = -\frac{5L}{6}$$

Therefore, the center of mass is at

$$(x_{cm}, y_{cm}) = (L/4, -5L/6)$$

relative to the midpoint of the bar.

Example 11 A man of mass m is standing at one end of a stationary, floating barge of mass $3m$. He then walks to the other end of the barge, a distance of L meters. Ignore any frictional effects between the barge and the water.

(a) How far will the barge move?

(b) If the man walks at an average velocity of v, what is the average velocity of the barge?

Solution.

(a) Since there are no external forces acting on the man + barge system, the center of mass of the system cannot accelerate. In particular, since the system is originally at rest, the center of mass cannot move. Letting $x = 0$ denote the midpoint of the barge (which is its own center of mass, assuming it is uniform), we figure out the center of mass of the man + barge system:

$$x_{cm} = \frac{m_1 x_1 + m_2 x_2}{m_1 + m_2} = \frac{(m)(-L/2) + (3m)(0)}{m + 3m} = -\frac{L}{8}$$

So, the center of mass is a distance of $L/8$ from the midpoint of the barge, and since the mass is originally at the left end, the center of mass is a distance of $L/8$ to the left of the barge's midpoint.

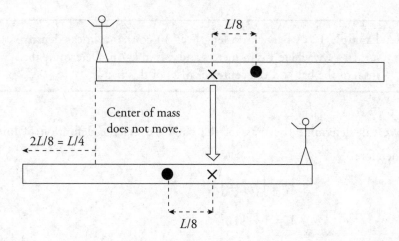

When the man reaches the other end of the barge, the center of mass will, by symmetry, be $L/8$ to the *right* of the midpoint of the barge. But, since the position of the center of mass cannot move, this means the barge itself must have moved a distance of

$$L/8 + L/8 = 2L/8 = L/4$$

to the left.

(b) Let the time it takes the man to walk across the barge be denoted by t; then $t = L/v$. In this amount of time, the barge moves a distance of $L/4$ in the *opposite* direction, so the velocity of the barge is

$$v_{barge} = \frac{-L/4}{t} = \frac{-L/4}{L/v} = -\frac{v}{4}$$

So far, we have dealt with objects that can be considered point masses, or masses with uniform density. Now we will learn how to find the center of mass of objects with non-uniform density.

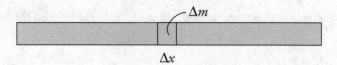

Take the example of a bar that becomes denser along its length. Here, we will deal with the linear density λ as a function of *x*, λ(*x*). Each small segment of the bar, Δ*x*, has a different mass, Δ*m*. We treat each Δ*m* as a point mass and then take the limit as Δ*x* approaches zero. Using the formula for calculating the center of mass of point masses, and replacing Δ*m* with *dm*, we get the integral shown below,

$$x_{cm} = \frac{1}{M} \int x \, dm$$

where *M* is the total mass, *x* is the distance to each *dm*, and you can substitute for *dm* in terms of *x*. Linear density is mass per length, so the equation is $\lambda = \frac{dm}{dx}$. Therefore, $M = \int dm = \int \lambda \, dx$.

> **Example 12** A bar with a length of 30 cm has a linear density λ = 10 + 6*x*, where *x* is in meters and λ is in kg/m. Determine the mass of the bar and the center of mass of this bar.

Solution. We can determine the mass of the bar by using the definition of linear density, $\lambda = \frac{dm}{dx}$. Therefore,

$$M = \int dm = \int \lambda \, dx$$

$$M = \int_0^{0.30} (10 + 6x) \, dx$$

$$M = (10x + 3x^2) \Big|_0^{0.30} = 3.27 \text{ kg}$$

To calculate the center of mass, we will use this equation:

$$x_{cm} = \frac{1}{M} \int x \, dm = \frac{1}{M} \int x \lambda \, dx$$

$$x_{cm} = \frac{1}{3.27} \int_0^{0.30} x(10 + 6x) dx$$

$$x_{cm} = \frac{1}{3.27} (5x^2 + 2x^3) \Big|_0^{0.30}$$

$$x_{cm} = \frac{1}{3.27} (0.504)$$

$$x_{cm} = 0.154 \text{ m}$$

This answer makes sense because the center of mass of the bar is beyond the midpoint.

Chapter 8 Drill

The answers and explanations can be found in Chapter 17.

Section I: Multiple Choice

1. An object of mass 2 kg has a linear momentum of magnitude 6 kg·m/s. What is this object's kinetic energy?

    (A) 3 J
    (B) 6 J
    (C) 9 J
    (D) 12 J
    (E) 18 J

2. The graph below shows the force on an object over time.

    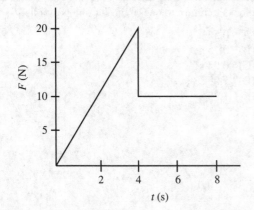

    If the object has a mass of 8 kg and is moving in a straight line, what is its change in speed?

    (A) 16 m/s
    (B) 14 m/s
    (C) 12 m/s
    (D) 10 m/s
    (E) 8 m/s

3. A box with a mass of 2 kg accelerates in a straight line from 4 m/s to 8 m/s due to the application of a force whose duration is 0.5 s. Find the average strength of this force.

    (A) 2 N
    (B) 4 N
    (C) 8 N
    (D) 12 N
    (E) 16 N

4. A ball of mass m traveling horizontally with velocity $\mathbf{v}$ strikes a massive vertical wall and rebounds back along its original direction with no change in speed. What is the magnitude of the impulse delivered by the wall to the ball?

    (A) 0
    (B) $\frac{1}{2}m\mathbf{v}$
    (C) $m\mathbf{v}$
    (D) $2m\mathbf{v}$
    (E) $4m\mathbf{v}$

5. Two objects, one of mass 3 kg moving with a speed of 2 m/s and the other of mass 5 kg and speed 2 m/s, move toward each other and collide head-on. If the collision is perfectly inelastic, find the speed of the objects after the collision.

    (A) 0.25 m/s
    (B) 0.5 m/s
    (C) 0.75 m/s
    (D) 1 m/s
    (E) 2 m/s

6. A student is trying to balance a meter stick on its midpoint. Given that $m_1 = 6$ kg and $m_2 = 2$ kg, how far from the left edge should the student hang a third mass, $m_3 = 10$ kg, to balance the meter stick?

    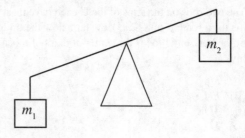

    (A) 40 cm
    (B) 50 cm
    (C) 60 cm
    (D) 70 cm
    (E) 80 cm

7. Two objects move toward each other, collide, and separate. If there was no net external force acting on the objects, but some kinetic energy was lost, then

(A) the collision was elastic and total linear momentum was conserved

(B) the collision was elastic and total linear momentum was not conserved

(C) the collision was not elastic and total linear momentum was conserved

(D) the collision was not elastic and total linear momentum was not conserved

(E) Cannot be determined

8. Three thin, uniform rods each of length L are arranged in the shape of an inverted U:

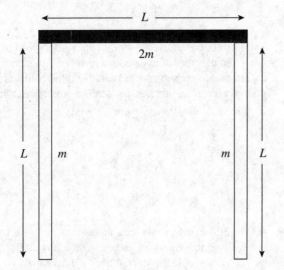

The two rods on the arms of the U each have mass m; the third rod has mass $2m$. How far below the midpoint of the horizontal rod is the center of mass of this assembly?

(A) $L/8$

(B) $L/4$

(C) $3L/8$

(D) $L/2$

(E) $3L/4$

9. A car of mass 1,000 kg collides head-on with a truck of mass 2,000 kg. Both vehicles are moving at a speed of 21 m/s, and the collision is perfectly inelastic. After the crash, the two vehicles skid to a halt. Assuming friction is the only force acting on the vehicles after the collision, how much work is done by friction after the crash?

(A) 73,500 J

(B) −73,500 J

(C) 147,000 J

(D) −147,000 J

(E) 220,500 J

10. A block of mass $m = 2$ kg has an initial speed of 3 m/s toward a second block. The second block, which starts 1 m away from the first block, has a mass of 5 kg and an initial speed of 2 m/s toward the first block. When the blocks collide, they experience a perfectly inelastic collision. How long after the collision does it take for the blocks to return to one of the starting positions?

(A) 0.55 s

(B) 0.75 s

(C) 0.95 s

(D) 1.05 s

(E) 1.25 s

Section II: Free Response

1. A steel ball of mass m is fastened to a light cord of length L and released when the cord is horizontal. At the bottom of its path, the ball strikes a hard plastic block of mass $M = 4m$, initially at rest on a frictionless surface. The collision is elastic.

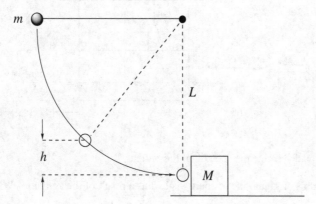

(a) Find the tension in the cord when the ball's height above its lowest position is $\frac{1}{2}L$. Write your answer in terms of m and g.

(b) Find the speed of the block immediately after the collision.

(c) To what height h will the ball rebound after the collision?

2. A *ballistic pendulum* is a device that may be used to measure the muzzle speed of a bullet. It is composed of a wooden block suspended from a horizontal support by cords attached at each end. A bullet is shot into the block, and as a result of the perfectly inelastic impact, the block swings upward. Consider a bullet (mass m) with velocity v as it enters the block (mass M). The length of the cords supporting the block each have length L. The maximum height to which the block swings upward after impact is denoted by y, and the maximum horizontal displacement is denoted by x.

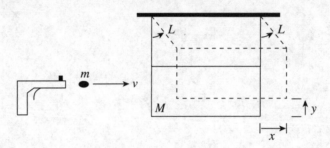

(a)   In terms of m, M, g, and y, determine the speed v of the bullet.

(b)   What fraction of the bullet's original kinetic energy is lost as a result of the collision? What happens to the lost kinetic energy?

(c)   If y is very small (so that y^2 can be neglected), determine the speed of the bullet in terms of m, M, g, x, and L.

(d)   Once the block begins to swing, does the momentum of the block remain constant? Why or why not?

3. An object of mass m moves with velocity $\mathbf{v}$ toward a stationary object of the same mass. After their impact, the objects move off in the directions shown in the following diagram:

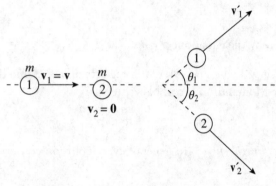

Before the collision   After the collision

Assume that the collision is elastic.

(a) If K_1 denotes the kinetic energy of object 1 before the collision, what is the kinetic energy of this object after the collision? Write your answer in terms of K_1 and θ_1.

(b) What is the kinetic energy of object 2 after the collision? Write your answer in terms of K_1 and θ_1.

(c) What is the relationship between θ_1 and θ_2?

Summary

Momentum

o   Linear momentum is given by the equation: $\mathbf{p} = m\mathbf{v}$.

o   Linear momentum is conserved when no external force acts on a system. This is known as the Law of Conservation of Linear Momentum. It can be written as:

$$\text{total } \mathbf{p}_{\text{before collision}} = \text{total } \mathbf{p}_{\text{after collision}}$$

o   Elastic collisions conserve *kinetic energy* (in general, every collision conserves energy but not necessarily kinetic energy).

o   Inelastic collisions do not conserve *kinetic energy*.

o   When the objects stick together, the collision is known as perfectly inelastic.

o   Anytime you are given a problem that involves a collision or separation, first consider whether you can use the Law of Conservation of Linear Momentum.

Impulse

o   Impulse is given by the equation: $J = \overline{F}\Delta t$

o   The Impulse–Momentum Theorem states that the impulse on an object is equal to the change in momentum of the object. The equation is: $J = \overline{F}\Delta t = \Delta p$.

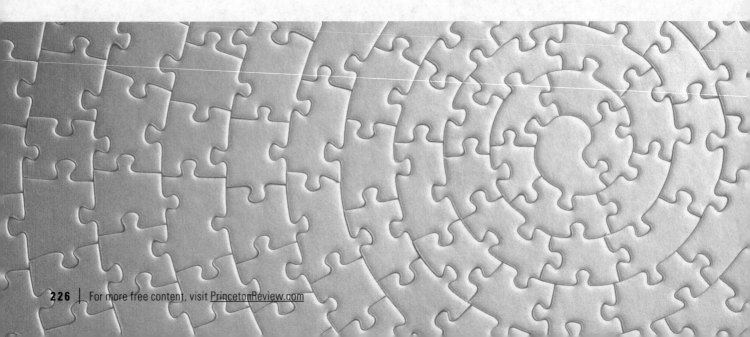

Center of Mass

o Usually, the motion of an object is describing the motion of the center of mass. When you use Newton's Second Law, $F = ma$, the acceleration you calculate is the acceleration of the center of mass.

o For point masses, $r_{cm} = \dfrac{\sum mr}{\sum m}$, where r is used for the position of each mass.

o For distributed mass (for example, a bar with non-uniform density), the equation is: $r_{cm} = \int r \, dm$.

o You usually use linear density, $\lambda \, dr = dm$, for dm and then integrate to solve for the center of mass.

Chapter 9
Rotation

INTRODUCTION

So far, we've studied only translational motion: objects sliding, falling, or rising, but in none of our examples have we considered spinning objects. We will now look at rotation, which will complete our study of motion. All motion is some combination of **translation** and **rotation**, which are illustrated in the figures below. Consider any two points in the object under study (on the left) and imagine connecting them by a straight line. If this line does not turn while the object moves, then the object is translating only. However, if this line does not always remain parallel to itself while the object moves, then the object is rotating.

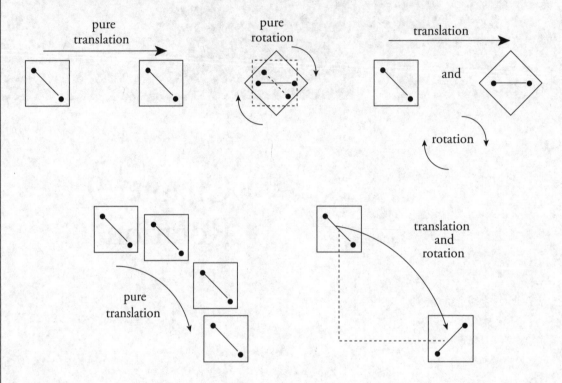

ROTATIONAL KINEMATICS

Mark several dots along a radius on a disk, and call this radius the *reference line*. If the disk rotates about its center, we can use the movement of these dots to talk about angular displacement, angular velocity, and angular acceleration.

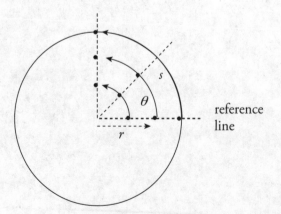

If the disk rotates as a rigid body, then all three dots shown have the same **angular displacement**, $\Delta\theta$. In fact, this is the definition of a **rigid body**: All points along a radial line always have the same angular displacement.

Just as the time rate-of-change of displacement gives velocity, the time rate-of-change of angular displacement gives angular velocity, denoted by ω (*omega*). The definition of the **average angular velocity** is:

$$\overline{\omega} = \frac{\Delta\theta}{\Delta t}$$

Note that if we let the time interval Δt approach 0, then the equation above leads to the definition of the instantaneous angular velocity:

$$\omega = \frac{d\theta}{dt}$$

And, finally, just as the time rate-of-change of velocity gives acceleration, the time rate-of-change of angular velocity gives angular acceleration, or α (*alpha*). The definition of the **average angular acceleration** is:

$$\overline{\alpha} = \frac{\Delta\omega}{\Delta t}$$

If we let the time interval Δt approach 0, then the equation above leads to the definition of the instantaneous angular acceleration:

$$\alpha = \frac{d\omega}{dt}$$

On the rotating disk illustrated on the previous page, we said that all points undergo the same angular displacement in any given time interval; this means that all points on the disk have the same angular velocity, ω, but not all points have the same linear velocity, v. This follows from the definition of **radian** measure. Expressed in radians, the angular displacement, $\Delta\theta$, is related to the arc length, Δs, by the equation

$$\Delta\theta = \frac{\Delta s}{r}$$

Rearranging this equation and dividing by Δt, we find that

$$\Delta s = r\,\Delta\theta \quad\Rightarrow\quad \frac{\Delta s}{\Delta t} = r\frac{\Delta\theta}{\Delta t} \quad\Rightarrow\quad \overline{v} = r\,\overline{\omega}$$

Or, using the equations $v = ds/dt$ and $\omega = d\theta/dt$,

$$\boxed{v = r\omega}$$

Therefore, the greater the value of r, the greater the value of v. Points on the rotating body farther from the rotation axis move more quickly than those closer to the rotation axis.

From the equation $v = r\omega$, we can derive the relationship that connects angular acceleration and linear acceleration. Differentiating both sides with respect to t (holding r constant), gives us

$$\frac{dv}{dt} = r\frac{d\omega}{dt} \quad \Rightarrow \quad a = r\alpha$$

(It's important to realize that the acceleration a in this equation is *not* centripetal acceleration; it's tangential acceleration, which arises from a change in speed caused by an angular acceleration. By contrast, centripetal acceleration does not produce a change in speed.) Often, tangential acceleration is written as a_t to distinguish it from centripetal acceleration (a_c).

Example 1 A rotating, rigid body makes one complete revolution in 2 s. What is its average angular velocity?

Solution. One complete revolution is equal to an angular displacement of 2π radians (that is, there are 2π radians in 360°), so the body's average angular velocity is

$$\bar{\omega} = \frac{\Delta\theta}{\Delta t} = \frac{2\pi \text{ rad}}{2 \text{ s}} = \pi \text{ rad/s}$$

Example 2 The angular velocity of a rotating disk increases from 2 rad/s to 5 rad/s in 0.5 s. What's the disk's average angular acceleration?

Solution. By definition,

$$\bar{\alpha} = \frac{\Delta\omega}{\Delta t} = \frac{(5-2) \text{ rad/s}}{0.5 \text{ s}} = 6 \text{ rad/s}^2$$

Example 3 A disk of radius 20 cm rotates at a constant angular velocity of 6 rad/s. How fast does a point on the rim of this disk travel (in m/s)?

Solution. The linear speed, v, is related to the angular speed, ω, by the equation $v = r\omega$. Therefore,

$$v = r\omega = (0.20 \text{ m})(6 \text{ rad/s}) = 1.2 \text{ m/s}$$

Note that, although we typically write the abbreviation *rad* when writing angular measurements, the radian is actually a dimensionless quantity, since, by definition, $\theta = s/r$. So, $\Delta\theta = 6$ means the same thing as $\Delta\theta = 6$ rad.

> **Example 4** The angular velocity of a rotating disk of radius 50 cm increases from 2 rad/s to 5 rad/s in 0.5 s. What is the linear tangential acceleration of a point on the rim of the disk during this time interval?

Solution. The linear acceleration a is related to the angular acceleration α by the equation $a = r\alpha$. Since $\alpha = 6$ rad/s^2 (as calculated in Example 2), we find that

$$a = r\alpha = (0.50 \text{ m})(6 \text{ rad/s}^2) = 3 \text{ m/s}^2$$

> **Example 5** Derive an expression for centripetal acceleration in terms of angular speed.

Solution. For an object revolving with linear speed v at a distance r from the center of rotation, the centripetal acceleration is given by the equation $a_c = v^2/r$. Using the fundamental equation $v = r\omega$, we find that

$$a_c = \frac{v^2}{r} = \frac{(r\omega)^2}{r} = \omega^2 r$$

On the equation sheet for the free-response section, this information will be represented as follows:

$$a_c = \frac{v^2}{r} = \omega^2 r$$

THE BIG FIVE FOR ROTATIONAL MOTION

The simplest type of rotational motion to analyze is motion in which the angular acceleration is *constant* (possibly equal to zero). Another restriction that will make our analysis easier (and which doesn't significantly diminish the power and applicability of our results) is to consider rotational motion around a *fixed* axis of rotation. In this case, there are only two possible directions for motion. One direction, counterclockwise, is called *positive* (+), and the opposite direction, clockwise, is called *negative* (−).

Let's review the quantities we've seen so far. The fundamental quantities for rotational motion are angular displacement ($\Delta\theta$), angular velocity (ω), and angular acceleration (α). Because we're dealing with angular acceleration, we know about changes in angular velocity, from initial velocity (ω_i or ω_0) to final velocity (ω_f or simply ω—with no subscript). And, finally, the motion takes place during some elapsed time interval, Δt. Therefore, we have five kinematics quantities: $\Delta\theta$, ω_0, ω, α, and Δt.

These five quantities are interrelated by a group of five equations which we call the *Big Five*. They work in cases in which the angular acceleration is uniform. These equations are identical to the Big Five we studied in Chapter 5 but, in these cases, the translational variables (*s*, *v*, or *a*) are replaced by the corresponding rotational variables (θ, ω, or α, respectively).

In Big Five #1, because angular acceleration is constant, the average angular velocity is simply the average of the initial angular velocity and the final angular velocity: $\bar{\omega} = \frac{1}{2}(\omega_0 + \omega)$. Also, if we set $t_i = 0$, then $\Delta t = t_f - t_i = t - 0 = t$, so we can just write "*t*" instead of "Δt" in the first four equations. This simplification in notation makes the equations a little easier to memorize.

Variable that's missing

| | Big Five #1: $\Delta \theta = \bar{\omega} t$ | α |

Notice that Big Five #1 and #2 are simply the definitions of $\bar{\omega}$ and $\bar{\alpha}$ written in forms that don't involve fractions.

Big Five #2: $\boxed{\omega = \omega_0 + \alpha t}$ — $\Delta \theta$

The two boxed equations are provided for you on the AP Physics C Table of Information, which you will be given on exam day—and in the Practice Tests in this book!

Big Five #3: $\boxed{\theta = \theta_0 + \omega_0 t + \frac{1}{2}\alpha t^2}$ — ω

Big Five #4: $\Delta \theta = \omega t - \frac{1}{2}\alpha(t)^2$ — ω_0

Big Five #5: $\omega^2 = \omega_0^2 + 2\alpha \Delta \theta$ — t

Each of the Big Five equations is missing exactly one of the five kinematics quantities and, as with the other Big Five you learned, the way you decide which equation to use is to determine which of the kinematics quantities is missing from the problem, and use the equation that's also missing that quantity. For example, if the problem never mentions the final angular velocity—ω is neither given nor asked for—then the equation that will work is the one that's missing ω; that's Big Five #3.

Example 6 An object with an initial angular velocity of 1 rad/s rotates with constant angular acceleration. Three seconds later, its angular velocity is 5 rad/s. Calculate its angular displacement during this time interval.

Solution. We're given ω_0, *t*, and ω, and asked for $\Delta\theta$. So α is missing, and we use Big Five #1:

$$\Delta\theta = \bar{\omega}\Delta t = \tfrac{1}{2}(\omega_0 + \omega)\Delta t = \tfrac{1}{2}(1 \text{ rad/s} + 5 \text{ rad/s})(3 \text{ s}) = 9 \text{ rad}$$

Example 7 Starting with zero initial angular velocity, a sphere begins to spin with constant angular acceleration about an axis through its center, achieving an angular velocity of 10 rad/s when its angular displacement is 20 rad. What is the value of the sphere's angular acceleration?

Solution. We're given ω_0, $\Delta\theta$, and ω, and asked for α. Since t is missing, we use Big Five #5:

$$\omega^2 = \omega_0^2 + 2\alpha\Delta\theta \quad \Rightarrow \quad \omega^2 = 2\alpha\Delta\theta \quad (\text{since} \quad \omega_0 = 0)$$

$$\alpha = \frac{\omega^2}{2\Delta\theta} = \frac{(10 \text{ rad/s})^2}{2(20 \text{ rad})} = 2.5 \text{ rad/s}^2$$

To summarize, here's a comparison of the fundamental quantities of translational and rotational motion and of the Big Five (assuming constant acceleration and a fixed axis of rotation):

	Translational	*Rotational*	*Connection*
displacement:	Δx	$\Delta\theta$	$\Delta x = r\Delta\theta$
velocity:	v	ω	$v = r\omega$
acceleration:	a	α	$a = r\alpha$
Big Five #1:	$\Delta x = x - x_0 = \bar{v}t$	$\Delta\theta = \bar{\omega}t$	
Big Five #2:	$v = v_0 + at$	$\omega = \omega_0 + \alpha t$	
Big Five #3:	$x = x_0 + v_0t + \frac{1}{2}at^2$	$\Delta\theta = \omega_0 t + \frac{1}{2}\alpha(t)^2$	
Big Five #4:	$x = x_0 + vt - \frac{1}{2}at^2$	$\Delta\theta = \omega t - \frac{1}{2}\alpha(t)^2$	
Big Five #5:	$v^2 = v_0^2 + 2a(x - x_0)$	$\omega^2 = \omega_0^2 + 2\alpha\Delta\theta$	

ROTATIONAL DYNAMICS

The dynamics of translational motion involve describing the acceleration of an object in terms of its mass (inertia) and the forces that act on it; $F_{net} = ma$. By analogy, the dynamics of rotational motion involve describing the angular (rotational) acceleration of an object in terms of its **rotational inertia** and the **torques** that act on it.

Torque

Intuitively, torque describes the effectiveness of a force in producing rotational acceleration. Consider a uniform rod that pivots around one of its ends, which is fixed. For simplicity, let's assume that the rod is at rest. What effect, if any, would each of the four forces in the figure below have on the potential rotation of the rod?

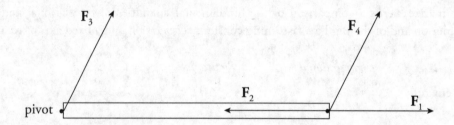

Our intuition tells us that $\mathbf{F}_1$, $\mathbf{F}_2$, and $\mathbf{F}_3$ would not cause the rod to rotate, but $\mathbf{F}_4$ would. What's different about $\mathbf{F}_4$? It has torque. Clearly, torque has something to do with rotation. Just like a force is a vector quantity that produces linear acceleration, a torque is a vector quantity that produces **angular acceleration**. Note that, just like for linear acceleration, an angular acceleration is something that either changes the direction of the angular velocity or changes the angular speed. A torque can be thought of as being positive if it produces counterclockwise rotation or negative if it produces clockwise rotation.

The torque of a force can be defined as follows. Let r be the distance from the pivot (axis of rotation) to the point of application of the force $\mathbf{F}$, and let θ be the angle between vectors $\mathbf{r}$ and $\mathbf{F}$.

> The pivot is often alternatively referred to as the hinge or the fulcrum.

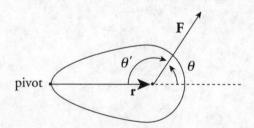

Then the torque of $\mathbf{F}$, denoted by τ (*tau*), is defined as:

$$\tau = rF \sin \theta$$

Imagine sliding **r** over so that its initial point is the same as that of **F**. *The angle between two vectors is the angle between them when they start at the same point.* However, the supplementary angle θ' can be used in place of θ in the definition of torque. This is because torque depends on $\sin\theta$, and the sine of an angle and the sine of its supplement are always equal. Therefore, when figuring out torque, use whichever of these angles is more convenient.

We will now see if this mathematical definition of torque supports our intuition about forces $\mathbf{F}_1$, $\mathbf{F}_2$, $\mathbf{F}_3$, and $\mathbf{F}_4$.

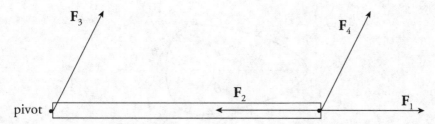

The angle between **r** and $\mathbf{F}_1$ is 0, and $\theta = 0$ implies $\sin\theta = 0$, so by the definition of torque, $\tau = 0$ as well. The angle between **r** and $\mathbf{F}_2$ is 180°, and $\theta = 180°$ gives us $\sin\theta = 0$, so $\tau = 0$. For $\mathbf{F}_3$, $r = 0$ (because $\mathbf{F}_3$ acts *at* the pivot, so the distance from the pivot to the point of application of $\mathbf{F}_3$ is zero); since $r = 0$, the torque is 0 as well. However, for $\mathbf{F}_4$, neither r nor $\sin\theta$ is zero, so $\mathbf{F}_4$ has a nonzero torque. Of the four forces shown in that figure, only $\mathbf{F}_4$ has torque and would produce rotational acceleration.

There's another way to determine the value of the torque. Of course, it gives the same result as the method given above, but this method is often easier to use. Look at the same object and force:

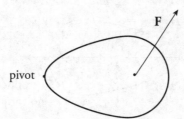

Instead of determining the distance from the pivot point to the point of application of the force, we will now determine the perpendicular distance from the pivot point to what's called the **line of action** of the force. This distance is the **lever arm** (or **moment arm**) of the force **F** relative to the pivot, and is denoted by l.

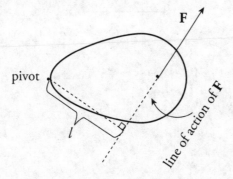

The torque of **F** is defined as the product

$$\tau = lF$$

Just as the lever arm is often called the moment arm, the torque is called the **moment** of the force. That these two definitions of torque, $\tau = rF \sin \theta$ and $\tau = lF$, are equivalent follows immediately from the fact that $l = r \sin \theta$:

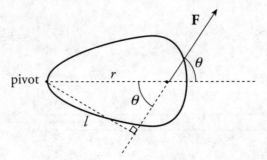

Since l is the component of **r** that's perpendicular to **F**, it is also denoted by $r_{\perp}$ ("r perp"). So the definition of torque can be written as $\tau = r_{\perp}F$.

These two equivalent definitions of torque make it clear that only the component of **F** that's perpendicular to **r** produces torque. The component of **F** that's parallel to **r** does not produce torque. Notice that $\tau = rF \sin \theta = rF_{\perp}$, where $F_{\perp}$ ("F perp") is the component of **F** that's perpendicular to **r**:

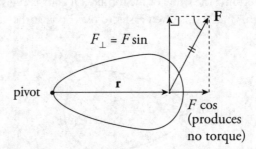

So the definition of torque can also be written as $\tau = rF_{\perp}$.

Also, since only the component of **F** perpendicular to **r** produces torque, the torque can be written as the cross product of **r** and **F**, and this is how it appears on the free-response equation sheet:

$$\boxed{\tau = \mathbf{r} \times \mathbf{F}}$$

Example 8 A student pulls down with a force of 40 N on a rope that winds around a pulley of radius 5 cm.

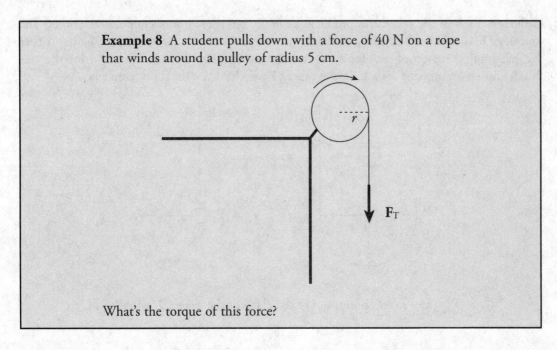

What's the torque of this force?

Solution. Since the tension force, $\mathbf{F}_T$, is tangent to the pulley, it is perpendicular to the radius vector **r** at the point of contact:

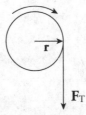

Therefore, the torque produced by this tension force is simply

$$\tau = rF_T = (0.05 \text{ m})(40 \text{ N}) = 2 \text{ N·m}$$

Example 9 What is the net torque on the cylinder shown below? The cylinder is pinned at its center.

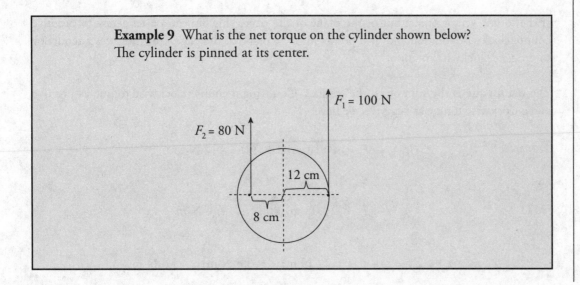

Solution. Each of the two forces produces a torque, but these torques oppose each other. The torque of $\mathbf{F}_1$ is counterclockwise, and the torque of $\mathbf{F}_2$ is clockwise. This can be visualized either by imagining the effect of each force, assuming that the other was absent, or by using the vector definition of torque, $\tau = \mathbf{r} \times \mathbf{F}$. In the case of $\mathbf{F}_1$, we have, by the right-hand rule,

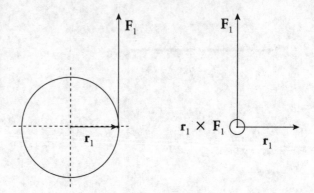

$\tau_1 = \mathbf{r}_1 \times \mathbf{F}_1$ points out of the plane of the page: $\odot$, while

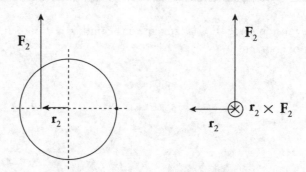

$\tau_2 = \mathbf{r}_2 \times \mathbf{F}_2$ points into the plane of the page: $\otimes$

These symbols are easy to remember if you think of a dart: $\odot$ is the point of the dart coming at you and $\otimes$ is the pattern of feathers you see at the back of the dart as it flies away from you toward the dartboard.

If the torque vector points out of the plane of the page, this indicates a tendency to produce counterclockwise rotation and, if it points into the plane of the page, this indicates a tendency to produce clockwise rotation.

The **net torque** is the sum of all the torques. Counting a counterclockwise torque as positive and a clockwise torque as negative, we have

$$\tau_1 = +r_1 F_1 = +(0.12 \text{ m})(100 \text{ N}) = +12 \text{ N·m}$$

and

$$\tau_2 = -r_2 F_2 = -(0.08 \text{ m})(80 \text{ N}) = -6.4 \text{ N·m}$$

so

$$\tau_{\text{net}} = \Sigma\tau = \tau_1 + \tau_2 = (+12 \text{ N·m}) + (-6.4 \text{ N·m}) = +5.6 \text{ N·m}$$

Rotational Inertia

Our goal is to develop a rotational analog of Newton's Second Law, $F_{net} = ma$. We're almost there; torque is the rotational analog of force and, therefore, τ_{net} is the rotational analog of F_{net}. The rotational analog of translational acceleration, a, is rotational (or angular) acceleration, α. We will now look at the rotational analog of inertial mass, m.

Consider a small point mass m at a distance r from the axis of rotation, being acted upon by a tangential force **F**.

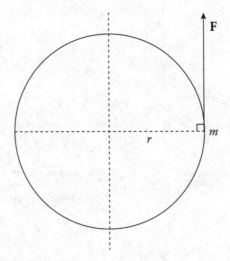

From Newton's Second Law, we have $F = ma$. Substituting $a = r\alpha$, this equation becomes $F = mr\alpha$. Now multiply both sides of this last equation by r to yield

$$rF = mr^2\alpha$$

or, since $rF = \tau$,

$$\tau = mr^2\alpha$$

In the equation $F = ma$, the quantity m is multiplied by the acceleration produced by the force F, while in the equation $\tau = mr^2\alpha$, the quantity mr^2 is multiplied by the rotational acceleration produced by the torque τ.

So for a point mass at a distance r from the axis of rotation, its **rotational inertia** (also called **moment of inertia**) is defined as mr^2. If we now take into account all the point masses that comprise the object under study, we can get the total rotational inertia, I, of the body by adding them:

$$I = \sum m_i r_i^2$$

For a continuous solid body, this sum becomes the integral

$$\boxed{I = \int r^2 \, dm}$$

This formula can be used to calculate expressions for the rotational inertia of cylinders (disks), spheres, slender rods, and hoops. Notice that the rotational inertia depends not only on m, but also on r; both the mass *and how it's distributed about the axis of rotation* determine I. By summing up all point masses and all external torques, the equation $\tau = mr^2\alpha$ becomes

$$\sum \tau_i = \left(\sum m_i r_i^2\right)\alpha \quad \text{or} \quad \tau_{net} = I\alpha$$

On the equation sheet for the free-response section, this information will be represented as follows:

$$\boxed{\sum \tau = \tau_{net} = I\alpha}$$

We've reached our goal:

Translational motion	*Rotational motion*
force, F	torque, τ
acceleration, a	rotational acceleration, α
mass, m	rotational inertia, I
$F_{net} = ma$	$\tau_{net} = I\alpha$

Example 10 Three beads, each of mass m, are arranged along a rod of negligible mass and length L. Figure out the rotational inertia of the assembly when the axis of rotation is through the center bead and when the axis of rotation is through one of the beads on the ends.

Solution.

(a) In the first case, both the left bead and the right bead are at a distance of $L/2$ from the axis of rotation, while the center bead is at distance zero from the axis of rotation. Therefore,

$$I = \sum m_i r_i^2 = m\left(\tfrac{L}{2}\right)^2 + m(0)^2 + m\left(\tfrac{L}{2}\right)^2 = \tfrac{1}{2}mL^2$$

(b) In the second case, the left bead is at distance zero from the rotation axis, the center bead is at distance $L/2$, and the right bead is at distance L. Therefore,

$$I' = \sum m_i r_i^2 = m(0)^2 + m\left(\tfrac{L}{2}\right)^2 + m(L)^2 = \tfrac{5}{4}mL^2$$

Note that, although both assemblies have the same mass (namely, $3m$), their rotational inertias are different, because of the different distribution of mass relative to the axis of rotation.

If the rotational inertia of a body is known relative to an axis that passes through the body's center of mass, then the rotational inertia, I', relative to any other rotation axis (parallel to the first one) can be calculated as follows. Let I_{cm} be the rotational inertia of a body relative to a rotation axis that passes through the body's center of mass, let the mass of the body be M, and let x be the distance from the axis through the center of mass to the rotation axis. Then

$$I = I_{cm} + mx^2$$

This is called the **parallel–axis theorem**. Let's use this result to calculate the rotational inertia of the three-bead assembly in part (b) from the value obtained in part (a), which is I_{cm}. Since $M = 3m$ and $x = L/2$, we have

$$I' = I_{cm} + Mx^2 = \tfrac{1}{2}mL^2 + (3m)\left(\tfrac{L}{2}\right)^2 = \tfrac{5}{4}mL^2$$

which agrees with the value calculated above.

Example 11 Show that the rotational inertia of a homogeneous cylinder of radius R and mass M, rotating about its central axis, is given by the equation $I = \frac{1}{2}MR^2$.

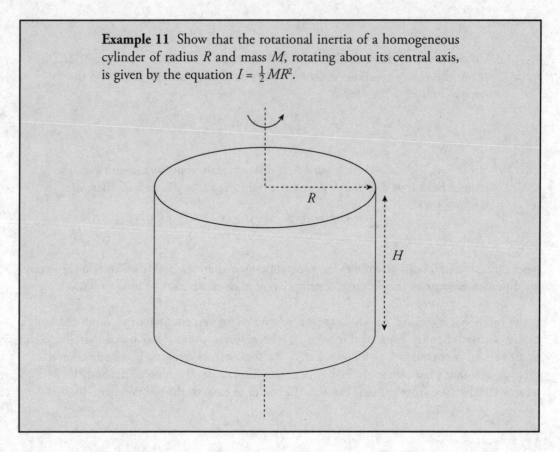

Solution. A solid is homogeneous if its density is constant throughout. Let ρ be the density of the cylinder. In order to compute I using the formula $I = \int r^2\,dm$, we choose our mass element to be an infinitesimally thin cylindrical ring of radius r:

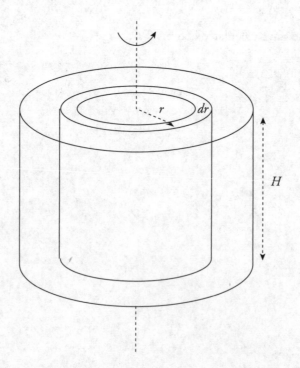

The mass of this ring is equal to its volume, $2\pi r \, dr \times H$, times the density; that is,

$$dm = \rho \times 2\pi r H \, dr$$

Therefore,

$$
\begin{aligned}
I = \int r^2 \, dm &= \int_{r=0}^{r=R} r^2 (\rho \cdot 2\pi r H \, dr) \\
&= 2\pi \rho H \int_0^R r^3 \, dr \\
&= 2\pi \rho H \left(\tfrac{1}{4} r^4 \right)_0^R \\
&= \tfrac{1}{2} \pi \rho H R^4
\end{aligned}
$$

To eliminate ρ, use the fact that the total mass of the cylinder is

$$M = \rho V = \rho \cdot \pi R^2 H$$

Putting this into the expression derived for I, we find that

$$I = \tfrac{1}{2} \pi \rho H R^4 = \tfrac{1}{2} (\rho \cdot \pi R^2 H) R^2 = \tfrac{1}{2} M R^2$$

The height of the cylinder (the dimension parallel to the axis of rotation) is irrelevant. Therefore, the formula $I = \tfrac{1}{2} M R^2$ gives the rotational inertia of any homogeneous solid cylinder revolving around its central axis. This includes a disk (which is just a really short cylinder).

Example 12 Again, consider the system of Example 9:

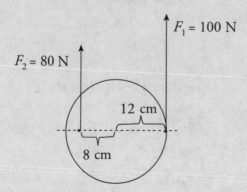

$F_2 = 80$ N

$F_1 = 100$ N

12 cm

8 cm

Assume that the mass of the cylinder is 50 kg. Given that the rotational inertia of a cylinder of radius R and mass M rotating about its central axis is given by the equation $I = \frac{1}{2}MR^2$, determine the rotational acceleration produced by the two forces shown.

Solution. In Example 9, we figured out that $\tau_{net} = +5.6$ N·m. The rotational inertia of the cylinder is

$$I = \tfrac{1}{2}MR^2 = \tfrac{1}{2}(50 \text{ kg})(0.12 \text{ m})^2 = 0.36 \text{ kg·m}^2$$

Therefore, from the equation $\tau_{net} = I\alpha$, we find that

$$\alpha = \frac{\tau_{net}}{I} = \frac{5.6 \text{ N·m}}{0.36 \text{ kg·m}^2} = 16 \text{ rad/s}^2$$

This angular acceleration will be counterclockwise because τ_{net} is counterclockwise.

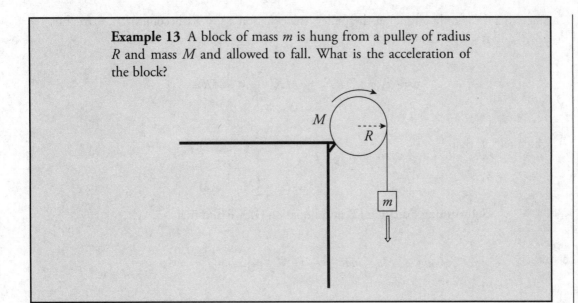

Example 13 A block of mass m is hung from a pulley of radius R and mass M and allowed to fall. What is the acceleration of the block?

Solution. Remember that, earlier, we treated pulleys as if they were massless, and no force was required to make them rotate. Now, however, we know how to take the mass of a pulley into account, by including its rotational inertia in our analysis. The pulley is a disk, so its rotational inertia is given by the formula $I = \frac{1}{2}MR^2$.

First, we draw a free-body diagram for the falling block:

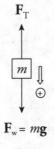

and apply Newton's Second Law:

$$mg - F_T = ma \quad (1)$$

Now the tension $\mathbf{F}_T$ in the cord produces a torque, $\tau = RF_T$, on the pulley:

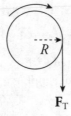

Since this is the only torque on the pulley, the equation $\tau_{\text{net}} = I\alpha$ becomes $RF_{\text{T}} = I\alpha$. But $I = \frac{1}{2}MR^2$ and $\alpha = a/R$. Therefore,

$$\tau = I\alpha \quad \Rightarrow \quad RF_{\text{T}} = \frac{1}{2}MR^2 \cdot \frac{a}{R} = \frac{1}{2}MRa \quad (1)$$

which tells us that

$$F_{\text{T}} = \frac{1}{2}Ma \quad (2)$$

> This last equation says that the cord doesn't slip as it slides over the pulley; the linear acceleration of a point on the rim, $a = R\alpha$, is equal to the acceleration of the connected block.

Substituting Equation (2) into Equation (1), we find that

$$mg - \frac{1}{2}Ma = ma \quad \Rightarrow \quad \left(\frac{1}{2}M + m\right)a = mg \quad \Rightarrow \quad a = \frac{m}{\frac{1}{2}M + m}g$$

KINETIC ENERGY OF ROTATION

A rotating object has rotational kinetic energy, just as a translating object has translational kinetic energy. The formula for kinetic energy is, of course, $K = \frac{1}{2}mv^2$, but this can't be directly used to calculate the kinetic energy of rotation because each point mass that makes up the body can have a different v. For this reason, we need a definition of $K_{\text{rotational}}$ that involves ω instead of v.

$$K_{\text{rotational}} = \sum K_i = \sum \frac{1}{2}(m_i r_i^2)\omega^2 = \frac{1}{2} \cdot \left(\sum m_i r_i^2\right) \cdot \omega^2 = \frac{1}{2}I\omega^2$$

In short, then:

$$K = \frac{1}{2}I\omega^2$$

Note that this expression for rotational kinetic energy follows the general pattern displayed by our previous results: I is the rotational analog of m and ω is the rotational analog of v. Therefore, the rotational analog of $\frac{1}{2}mv^2$ should be $\frac{1}{2}I\omega^2$.

Rolling Motion

One of the main types of motion associated with rotational motion is rolling motion. We will primarily deal with rolling motion without slipping.

Consider a disk rolling down an incline—without slipping:

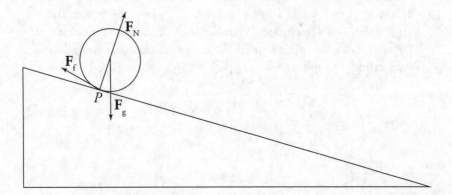

The point of contact of the object with the surface P is *instantaneously* at rest. If this were not the case, then the disk would be slipping down the incline, so the contact point must not be moving relative to the surface. In this case, the velocity of the center of mass of the disk is equal to the radius times the angular velocity of the disk.

You can take the torque around any point to determine the acceleration of the disk. It is often easiest to take it around the contact point P because then only gravity provides a torque about this point, and if you know the mass of the object, then you know the force of gravity. Make sure to use the parallel–axis theorem in this case since you are considering the rotational inertia about point P, not the center of mass. This will allow you to calculate the acceleration of the disk. Once you know the acceleration, you can calculate the necessary coefficient of friction to produce rolling without slipping using Newton's Second Law.

The total motion for an object that is rolling without slipping is the combined motion of the entire object translating with the velocity of the center of mass, and the object rotating about its center of mass, as shown below. This shows the object is instantaneously rotating about the contact point P.

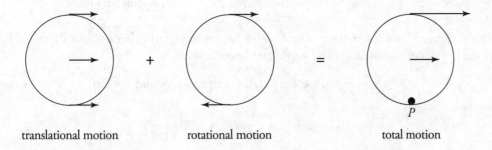

translational motion   rotational motion   total motion

For rolling motion, the total kinetic energy is the translational kinetic energy and the rotational kinetic energy.

$$K_{\text{rolling}} = K_{\text{translation}} + K_{\text{rotation}}$$
$$K_{\text{rolling}} = \frac{1}{2}mv_{\text{cm}}^2 + \frac{1}{2}I_{\text{cm}}\omega^2$$

Example 14 A cylinder of mass M and radius R rolls (without slipping) down an inclined plane whose incline angle with the horizontal is θ. Determine the acceleration of the cylinder's center of mass, and the minimum coefficient of friction that will allow the cylinder to roll without slipping on this incline.

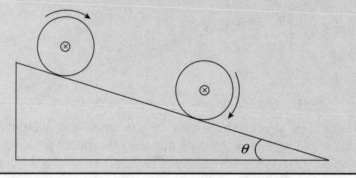

Solution. First, we draw a free-body diagram for the cylinder:

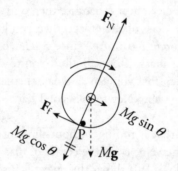

We know that the cylinder rolls without slipping, so the force of friction is not kinetic friction. Since the speed of the point on the cylinder in contact with the ramp is zero with respect to the ramp, *static* friction supplies the torque that allows the cylinder to roll smoothly.

Take the torque about the contact point to solve for the acceleration because the frictional force will not be part of the equation and we do not know it yet.

$$\sum \tau_P = I\alpha, \text{use the parallel–axis theorem and } a = R\alpha$$

$$(Mg \sin \theta)R = \left(\frac{1}{2}MR^2 + MR^2\right)\left(\frac{a}{R}\right)$$

$$Rg \sin \theta = \frac{3}{2}Ra$$

$$a = \frac{2}{3}g \sin \theta$$

Now we will use Newton's Second Law to solve for μ_s, since we know the acceleration.

$$\sum F = ma$$
$$Mg \sin \theta - F_f = Ma$$
$$Mg \sin \theta - \mu_s Mg \cos \theta = M \left(\frac{2}{3} g \sin \theta \right), \text{ cancel } M \text{ and } g.$$
$$\sin \theta - \mu_s \cos \theta = \frac{2}{3} \sin \theta$$
$$\frac{1}{3} \sin \theta = \mu_s \cos \theta$$
$$\mu_s = \frac{1}{3} \tan \theta$$

Example 15 A cylinder of mass M and radius R rolls (without slipping) down an inclined plane (of height h and length L) whose incline angle with the horizontal is θ. Determine the linear speed of the cylinder's center of mass when it reaches the bottom of the incline (assuming that it started from rest at the top).

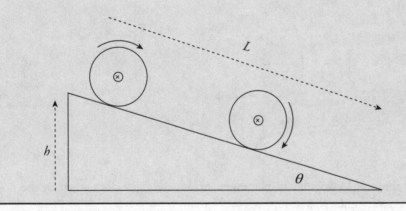

Solution. We will attack this problem using Conservation of Mechanical Energy. As the cylinder rolls down the ramp, its initial gravitational potential energy is converted into kinetic energy, which is a combination of translational kinetic energy (since the cylinder's center of mass is translating down the ramp) and rotational kinetic energy:

$$K_i + U_i = K_f + U_f$$
$$0 + Mgh = \left(\frac{1}{2} Mv_{cm}^2 + \frac{1}{2} I \omega^2 \right) + 0$$

Since $I = \frac{1}{2} MR^2$ and $\omega = v_{cm}/R$, this equation becomes

$$Mgh = \frac{1}{2} Mv_{cm}^2 + \frac{1}{2} \left(\frac{1}{2} MR^2 \right) \left(\frac{v_{cm}}{R} \right)^2 = \frac{1}{2} Mv_{cm}^2 + \frac{1}{4} Mv_{cm}^2 = \frac{3}{4} Mv_{cm}^2$$

Therefore,

$$v_{cm} = \sqrt{\frac{4}{3}gh}$$

We can verify this result using the result of the previous example. There, we found that the acceleration of the cylinder's center of mass as it rolled down the ramp was $a = \frac{2}{3}g\sin\theta$. Applying Big Five #5 gives us:

$$v^2 = v_0^2 + 2a\Delta s = 2aL = 2a\frac{h}{\sin\theta} = 2\cdot\frac{2}{3}g\sin\theta\cdot\frac{h}{\sin\theta} = \frac{4}{3}gh \quad \Rightarrow \quad v = \sqrt{\frac{4}{3}gh}$$

WORK AND POWER

Consider a small point mass m at distance r from the axis of rotation, acted upon by a tangential force **F**.

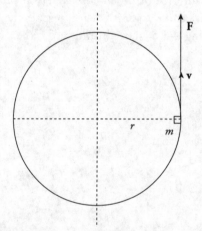

As it rotates through an angular displacement of $\Delta\theta$, the force does work on the point mass: $W = F\Delta s$, where $\Delta s = r\Delta\theta$. Therefore,

$$W = Fr\Delta\theta = \tau\Delta\theta$$

If the force is not purely tangential to the object's path, then only the tangential component of the force does work; the radial component does not (since it's perpendicular to the object's displacement). Therefore, for a general constant force F, the equation above would read $W = F_t r\Delta\theta = \tau\Delta\theta$, where F_t denotes the tangential component of F.

If we want to allow for a varying F—and a varying τ—then the work done is equal to the definite integral:

$$W = \int_{\theta_1}^{\theta_2} \tau \, d\theta$$

Again, notice the analogy between this equation and the one that defines work by F:

$$W = \int_{Fx_1}^{Fx_2} F\,dx$$

The work–energy theorem ($W = \Delta K$) also holds in the rotational case, where W is the work done by net torque and ΔK is the resulting change in the rotational kinetic energy.

The rate at which work is done, or the power (P), is defined by the equation

$$P = \frac{dW}{dt}$$

Over an infinitesimal angular displacement $d\theta$, the torque τ does an amount of work, dW, given by $dW = \tau\,d\theta$. This implies that

$$\frac{dW}{dt} = \tau\frac{d\theta}{dt} \quad \Rightarrow \quad P = \tau\omega$$

Once again, notice the parallel to $P = Fv$, the translational version of this equation.

Example 16 A block of mass $m = 5$ kg is hung from a pulley of radius $R = 15$ cm and mass $M = 8$ kg and then released from rest.

(a) What is the speed of the block as it strikes the floor, 2 m below its initial position?

(b) What is the rotational kinetic energy of the pulley just before the block strikes the floor?

(c) At what rate was work done on the pulley?

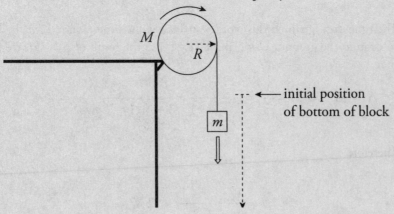

initial position
of bottom of block

Solution.

(a) Apply Conservation of Mechanical Energy. The initial gravitational potential energy of the block is transformed into the purely rotational kinetic energy of the pulley and translational kinetic energy of the falling block:

$$K_i + U_i = K_f + U_f$$

$$0 + mgh = (\tfrac{1}{2}mv^2 + \tfrac{1}{2}I\omega^2) + 0$$

$$= \tfrac{1}{2}mv^2 + \tfrac{1}{2} \cdot \tfrac{1}{2}MR^2 \cdot \left(\tfrac{v}{R}\right)^2$$

$$= \tfrac{1}{2}mv^2 + \tfrac{1}{4}Mv^2$$

$$= (\tfrac{1}{2}m + \tfrac{1}{4}M)v^2$$

$$v = \sqrt{\frac{mgh}{\tfrac{1}{2}m + \tfrac{1}{4}M}}$$

Substituting in the given numerical values, we get

$$v = \sqrt{\frac{(5)(10)(2)}{\tfrac{1}{2}(5) + \tfrac{1}{4}(8)}} = 4.7 \text{ m/s}$$

(b) The rotational kinetic energy of the pulley as the block strikes the floor is

$$K = \tfrac{1}{2}I\omega^2 = \tfrac{1}{2} \cdot \tfrac{1}{2}MR^2 \cdot \left(\tfrac{v}{R}\right)^2 = \tfrac{1}{4}Mv^2 = \tfrac{1}{4}(8)(4.7)^2 = 44 \text{ J}$$

(c) The rate at which work is done on the pulley is the power produced by the torque. One way to compute this is to first use the work–energy theorem to determine the work done by the torque and then divide this by the time during which the block fell. So,

$$W = \Delta K = K_f - K_i = K_f = 44 \text{ J}$$

The time during which this work was done is the time required for the block to drop to the ground. Using Big Five #1 and the result of part (a), we find that

$$\Delta s = \bar{v}t \quad \Rightarrow \quad t = \frac{\Delta s}{\bar{v}} = \frac{h}{\tfrac{1}{2}(v_0 + v)} = \frac{h}{\tfrac{1}{2}v} = \frac{2 \text{ m}}{\tfrac{1}{2}(4.7 \text{ m/s})} = 0.85 \text{ s}$$

Therefore,

$$P = \frac{W}{t} = \frac{44 \text{ J}}{0.85 \text{ s}} = 52 \text{ W}$$

ANGULAR MOMENTUM

So far, we've developed rotational analogs for displacement, velocity, acceleration, force, mass, and kinetic energy. We will finish by developing a rotational analog for linear momentum; it's called **angular momentum**.

Consider a small point mass m at distance r from the axis of rotation, moving with velocity $\mathbf{v}$ and acted upon by a tangential force $\mathbf{F}$.

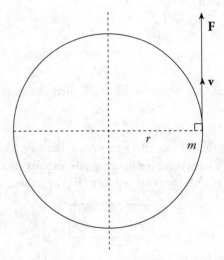

Then, by Newton's Second Law,

$$F = \frac{\Delta p}{\Delta t} = \frac{\Delta(mv)}{\Delta t}$$

If we multiply both sides of this equation by r and notice that $rF = \tau$, we get

$$\tau = \frac{\Delta(rm\,v)}{\Delta t}$$

Therefore, to form the analog of the law $F = \Delta p/\Delta t$ (force equals the rate of change of linear momentum), we say that torque equals the rate of change of angular momentum, and the angular momentum (denoted by L) of the point mass m is defined by the equation

$$L = rmv$$

If we now take into account all the point masses that comprise the object under study, we can get the angular momentum of the body by adding up all the individual contributions. This gives

$$L = I\omega$$

Note that this expression for angular momentum follows the general pattern we saw previously: I is the rotational analog of m, and ω is the rotational analog of v. Therefore, the rotational analog of mv should be $I\omega$.

If the point mass m does not move in a circular path, we can still define its angular momentum relative to any reference point.

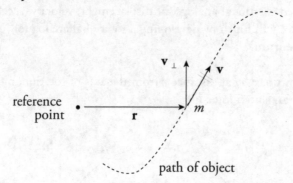

path of object

If $\mathbf{r}$ is the vector from the reference point to the mass, then the angular momentum is

$$L = rmv_\perp$$

where $v_\perp$ is the component of the velocity that's perpendicular to $\mathbf{r}$. This is the perpendicular component of $\mathbf{v}$ relative to $\mathbf{r}$ that's important for figuring out angular momentum, and this fact leads to the general vector definition with the cross product. The equation $L = (r)(mv_\perp) = (r)(p_\perp)$ becomes

$$\boxed{\mathbf{L} = \mathbf{r} \times \mathbf{p} = I\omega}$$

Example 17 A solid uniform sphere of mass $M = 8$ kg and radius $R = 50$ cm is revolving around an axis through its center at an angular speed of 10 rad/s. Given that the rotational inertia of the sphere is equal to $\frac{2}{5}MR^2$, what is the spinning sphere's angular momentum?

Solution. Apply the definition:

$$L = I\omega = \tfrac{2}{5}MR^2\omega = \tfrac{2}{5}(8 \text{ kg})(0.50 \text{ m})^2(10 \text{ rad/s}) = 8 \text{ kg} \cdot \text{m}^2/\text{s}$$

If you want to specify the direction of the angular momentum vector, $\mathbf{L}$, use the right-hand rule. Let the fingers of your right hand curl in the direction of rotation of the body. Your thumb gives the direction of $\mathbf{L}$, pointing along the rotation axis:

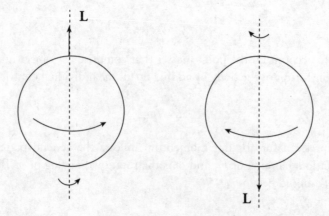

CONSERVATION OF ANGULAR MOMENTUM

Newton's Second Law says that

$$\mathbf{F}_{net} = \frac{d\mathbf{p}}{dt}$$

so if $F_{net} = 0$, then $\mathbf{p}$ is constant. This is Conservation of Linear Momentum.

The rotational analog of this is

$$\tau_{net} = \frac{d\mathbf{L}}{dt}$$

So if $\tau_{net} = 0$, then $\mathbf{L}$ is constant. This is **Conservation of Angular Momentum**. Basically, this says that if the torques on a body balance so that the net torque is zero, then the body's angular momentum can't change.

An often-cited example of this phenomenon is the spinning of a figure skater. As the skater pulls his or her arms inward, he or she moves more of his or her mass closer to the rotation axis and decreases his or her rotational inertia. Since the external torque on the skater is negligible, his or her angular momentum must be conserved. Since $L = I\omega$, a decrease in I causes an increase in ω, and the skater spins faster.

> **Example 18** A child of mass $m = 30$ kg stands at the edge of a small merry-go-round that's rotating at a rate of 1 rad/s. The merry-go-round is a disk of radius $R = 2.5$ m and mass $M = 100$ kg. If the child walks in toward the center of the disk and stops 0.5 m from the center, what will happen to the angular velocity of the merry-go-round (if friction can be ignored)?

Solution. The child walking toward the center of the merry-go-round does not provide an external torque to the child + disk system, so angular momentum is conserved. Let's denote the child as a point mass, and consider the following two views of the merry-go-round (looking down from above):

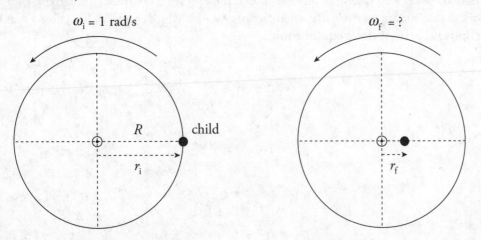

In the first picture, the total rotational inertia, I_i, is equal to the sum of the rotational inertia of the merry-go-round (MGR) and the child:

$$I_i = I_{MGR} + I_{child} = \tfrac{1}{2}MR^2 + mr_i^2 = \tfrac{1}{2}MR^2 + mR^2 = \left(\tfrac{1}{2}M + m\right)R^2$$

In the second picture, the total rotational inertia has decreased to

$$I_f = I_{MGR} + I'_{child} = \tfrac{1}{2}MR^2 + mr_f^2$$

So, by Conservation of Angular Momentum, we have

$$L_i = L_f$$
$$I_i\omega_i = I_f\omega_f$$
$$\left(\tfrac{1}{2}M + m\right)R^2\omega_i = \left(\tfrac{1}{2}MR^2 + mr_f^2\right)\omega_f$$
$$\omega_f = \frac{\left(\tfrac{1}{2}M + m\right)R^2}{\tfrac{1}{2}MR^2 + mr_f^2}\omega_i$$

and substituting the given numerical values gives us

$$\omega_f = \frac{\left(\tfrac{1}{2}M + m\right)R^2}{\tfrac{1}{2}MR^2 + mr_f^2}\omega_i$$
$$= \frac{\left(\tfrac{1}{2}\cdot 100 + 30\right)(2.5)^2}{\tfrac{1}{2}\cdot 100\cdot(2.5)^2 + 30\cdot(0.5)^2}(1\ \text{rad/s})$$
$$= 1.6\ \text{rad/s}$$

Notice that ω increased as I decreased, just as Conservation of Angular Momentum predicts.

EQUILIBRIUM

An object is said to be in **translational equilibrium** if the sum of the forces acting on it is zero; that is, if $F_{net} = 0$. Similarly, an object is said to be in **rotational equilibrium** if the sum of the torques acting on it is zero; that is, if $\tau_{net} = 0$. The term *equilibrium* by itself means both translational and rotational equilibrium. A body in equilibrium may be in motion; $F_{net} = 0$ does not mean that the velocity is zero; it only means that the velocity is constant. Similarly, $\tau_{net} = 0$ does not mean that the angular velocity is zero; it only means that it's constant. If an object is at rest, then it is said to be in **static equilibrium**.

Example 19 A uniform bar of mass m and length L extends horizontally from a wall. A supporting wire connects the wall to the bar's midpoint, making an angle of 55° with the bar. A sign of mass M hangs from the end of the bar.

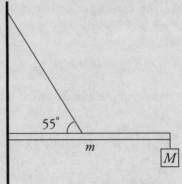

If the system is in static equilibrium, determine the tension in the wire and the strength of the force exerted on the bar by the wall if $m = 8$ kg and $M = 12$ kg.

Solution. Let $\mathbf{F}_C$ denote the (contact) force exerted by the wall on the bar. In order to simplify our work, we can write $\mathbf{F}_C$ in terms of its horizontal component, F_{Cx}, and its vertical component, F_{Cy}. Also, if $\mathbf{F}_T$ is the tension in the wire, then $F_{Tx} = F_T \cos 55°$ and $F_{Ty} = F_T \sin 55°$ are its components. This gives us the following force diagram:

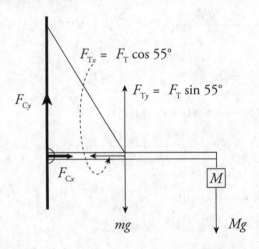

> The horizontal component of tension requires a horizontal force exerted by the wall on the bar to maintain equilibrium. Similarly, if the weights were too great for the vertical component of tension to support them, the wall would exert a vertical force upward on the bar to prevent the bar from falling.

The first condition for equilibrium requires that the sum of the horizontal forces is zero and the sum of the vertical forces is zero:

$$\Sigma F_x = 0: \qquad F_{Cx} - F_T \cos 55° = 0 \qquad (1)$$

$$\Sigma F_y = 0: \qquad F_{Cy} + F_T \sin 55° - mg - Mg = 0 \qquad (2)$$

We notice immediately that we have more unknowns (F_{Cx}, F_{Cy}, F_T) than equations, so this system cannot be solved as is. The second condition for equilibrium requires that the sum of the torques about any point is equal to zero. Choosing the contact point between the bar and the wall as our pivot, only three of the forces in the diagram on the previous page produce torque. $\mathbf{F}_{Ty}$ produces a counterclockwise torque, and both $m\mathbf{g}$ and $M\mathbf{g}$ produce clockwise torques, which must balance. From the definition $\tau = lF$, and taking counterclockwise torque as positive and clockwise torque as negative, we have

$$\Sigma\tau = 0: \qquad (L/2)F_{Ty} - (L/2)(mg) - LMg = 0 \quad (3)$$

This equation contains only one unknown and can be solved immediately:

$$\tfrac{L}{2}F_{Ty} = \tfrac{L}{2}mg + LMg$$
$$F_{Ty} = mg + 2Mg = (m + 2M)g$$

Since $F_{Ty} = F_T \sin 55°$, we can find that

$$F_T \sin 55° = (m + 2M)g \quad \Rightarrow \quad F_T = \frac{(m + 2M)g}{\sin 55°}$$
$$= \frac{(8 + 2 \cdot 12)(10)}{\sin 55°}$$
$$= 390 \text{ N}$$

Substituting this result into Equation (1) gives us F_{Cx}:

$$F_{Cx} = F_T \cos 55° = \frac{(m + 2M)g}{\sin 55°}\cos 55° = (8 + 2 \cdot 12)(10)\cot 55° = 220 \text{ N}$$

And finally, from Equation (2), we get

$$F_{Cy} = mg + Mg - F_T \sin 55°$$
$$= mg + Mg - \frac{(m + 2M)g}{\sin 55°}\sin 55°$$
$$= -Mg$$
$$= -(12)(10)$$
$$= -120 \text{ N}$$

Stay Up to Date!
For late-breaking information about test dates, exam formats, and any other changes pertaining to AP Physics C, make sure to check the College Board's website at apstudents. collegeboard.org/ courses

The fact that F_{Cy} turned out to be negative simply means that in our original force diagram, the vector $\mathbf{F}_{Cy}$ points in the direction opposite to how we drew it. That is, $\mathbf{F}_{Cy}$ points downward. Therefore, the magnitude of the total force exerted by the wall on the bar is

$$F_C = \sqrt{(F_{Cx})^2 + (F_{Cy})^2} = \sqrt{220^2 + 120^2} = 250 \text{ N}$$

Chapter 9 Drill

The answers and explanations can be found in Chapter 17.

Section I: Multiple Choice

1. A compact disc has a radius of 6 cm. If the disc rotates about its central axis at an angular speed of 5 rev/s, what is the linear speed of a point on the rim of the disc?

    (A) 0.3 m/s
    (B) 1.9 m/s
    (C) 7.4 m/s
    (D) 52 m/s
    (E) 83 m/s

2. A compact disc has a radius of 6 cm. If the disc rotates about its central axis at a constant angular speed of 5 rev/s, what is the total distance traveled by a point on the rim of the disc in 40 min?

    (A) 180 m
    (B) 360 m
    (C) 540 m
    (D) 720 m
    (E) 4.5 km

3. A disc starts at rest and experiences constant angular acceleration of 4 rad/s^2. When the disc has gone through a total angular displacement of 50 rad, what is its angular speed?

    (A) 12 rad/s
    (B) 15 rad/s
    (C) 16 rad/s
    (D) 20 rad/s
    (E) 23 rad/s

4. A rope of length 0.75 m has a small weight of mass $m =$ 0.1 kg attached to the end. If the rope is spun in a circle at a rate of 120 rpm, what acceleration is the weight experiencing?

    (A) 19,200 m/s^2
    (B) 10,800 m/s^2
    (C) 90 m/s^2
    (D) $12\pi^2$ m/s^2
    (E) 3π m/s^2

5. Joe and Alice are sitting on opposite ends of a seesaw. The fulcrum is beneath the center of the seesaw, which has a length of 4 m. Given that Joe's mass is 60 kg and Alice's mass is 30 kg, what will be the magnitude of the net torque on the seesaw when it is perfectly level?

    (A) 300 N·m
    (B) 600 N·m
    (C) 900 N·m
    (D) 1,200 N·m
    (E) 1,500 N·m

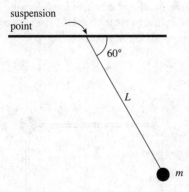

6. In the figure above, what is the torque about the pendulum's suspension point produced by the weight of the bob, given that the length of the pendulum, L, is 80 cm and $m = 0.50$ kg?

    (A) 0.5 N·m
    (B) 1.0 N·m
    (C) 1.7 N·m
    (D) 2.0 N·m
    (E) 3.4 N·m

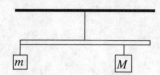

7. A uniform meter stick of mass 1 kg is hanging from a thread attached at the stick's midpoint. One block of mass $m = 3$ kg hangs from the left end of the stick, and another block, of unknown mass M, hangs below the 80 cm mark on the meter stick. If the stick remains at rest in the horizontal position shown above, what is the mass of M?

(A) 4 kg
(B) 5 kg
(C) 6 kg
(D) 8 kg
(E) 9 kg

8. What is the rotational inertia of the following body about the indicated rotation axis? (The masses of the connecting rods are negligible.)

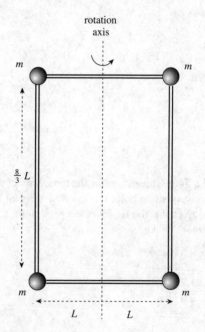

(A) $4mL^2$

(B) $\dfrac{32}{3}mL^2$

(C) $\dfrac{64}{9}mL^2$

(D) $\dfrac{128}{9}mL^2$

(E) $\dfrac{256}{9}mL^2$

9. A ramp of angle θ and height h has a block at the top. The block of mass m slides to the bottom without friction. A second ramp of the same angle and height H is constructed, and a solid cylinder also of mass m rolls down the ramp without slipping. If the cylinder and the block have equal amounts of kinetic energy upon reaching the bottoms of their respective ramps, what is the relationship between h and H?

(A) $h = \dfrac{1}{3}H$

(B) $h = \dfrac{1}{2}H$

(C) $h = \dfrac{2}{3}H$

(D) $h = \dfrac{3}{4}H$

(E) $h = H$

10. When a cylinder rolls down an inclined plane without slipping, which force is responsible for providing the torque that causes rotation?

(A) The force of gravity parallel to the plane
(B) The force of gravity perpendicular to the plane
(C) The normal force
(D) The force of static friction
(E) The force of kinetic friction

Section II: Free Response

1. In the figure below, the pulley is a solid disk of mass M and radius R, with rotational inertia $MR^2/2$. Two blocks, one of mass m_1 and one of mass m_2, hang from either side of the pulley by a light cord. Initially, the system is at rest, with Block 1 on the floor and Block 2 held at height h above the floor. Block 2 is then released and allowed to fall. Give your answers in terms of m_1, m_2, M, R, h, and g.

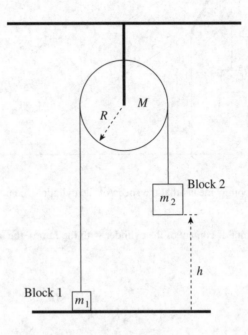

(a) What is the speed of Block 2 just before it strikes the ground?

(b) What is the angular speed of the pulley at this moment?

(c) What's the angular displacement of the pulley?

(d) How long does it take for Block 2 to fall to the floor?

2. The diagram below shows a solid uniform cylinder of radius R and mass M rolling (without slipping) down an inclined plane of incline angle θ. A thread wraps around the cylinder as it rolls down the plane and pulls upward on a block of mass m. Ignore the rotational inertia of the pulley.

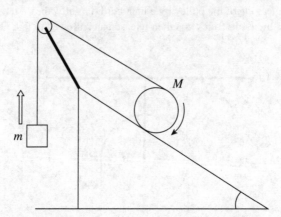

(a) Show that "rolling without slipping" means that the speed of the cylinder's center of mass, v_{cm}, is equal to $R\omega$, where ω is its angular speed.

(b) Show that, relative to P (the point of contact of the cylinder with the ramp), the speed of the top of the cylinder is $2v_{cm}$.

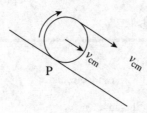

(c) What is the relationship between the magnitude of the acceleration of the block and the linear acceleration of the cylinder?

(d) What is the acceleration of the cylinder?

(e) What is the acceleration of the block?

3. Two slender uniform bars, each of mass M and length $2L$, meet at right angles at their midpoints to form a rigid assembly that's able to rotate freely about an axis through the intersection point, perpendicular to the page. Attached to each end of each rod is a solid ball of clay of mass m. A bullet of mass m_b is shot with velocity $\mathbf{v}$ as shown in the figure (which is a view from above of the assembly) and becomes embedded in the targeted clay ball.

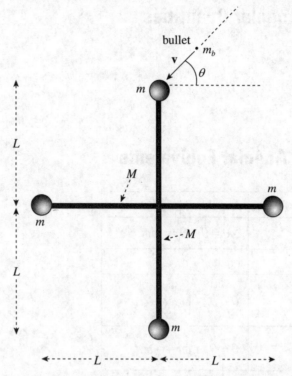

(a) Show that the moment of inertia of each slender rod about the given rotation axis, not including the clay balls, is $ML^2/3$.

(b) Determine the angular velocity of the assembly after the bullet has become lodged in the targeted clay ball.

(c) What is the resulting linear speed of each clay ball?

(d) Determine the ratio of the final kinetic energy of the assembly to the kinetic energy of the bullet before impact.

Summary

Relating Linear and Angular Quantities

$s = r\theta$

$v = r\omega$

$a_{\tan} = r\alpha$

Linear Equations and Angular Equivalents

Linear Equations (Big Five)	Angular Equivalent
$\Delta x = x - x_0 = \bar{v}\, t$	$\Delta\theta = \theta - \theta_0 = \bar{\omega} t$
$v = v_0 + at$	$\omega = \omega_0 + \alpha t$
$x = x_0 + v_0 t + \dfrac{1}{2}at^2$	$\theta = \theta_0 + \omega_0 t + \dfrac{1}{2}\alpha t^2$
$x = x_0 + vt - \dfrac{1}{2}at^2$	$\theta = \theta_0 + \omega t - \dfrac{1}{2}\alpha t^2$
$v^2 = v_0^2 + 2a(x - x_0)$	$\omega^2 = \omega_0^2 + 2\alpha(\theta - \theta_0)$
Basic Linear Equations	**Angular Equivalent**
$v = \dfrac{dx}{dt}$	$\omega = \dfrac{d\theta}{dt}$
$a = \dfrac{dv}{dt}$	$\alpha = \dfrac{d\omega}{dt}$

Basic Rotation Information

- **Rotational** inertia is the rotational analog of inertia, essentially a measure of how difficult it is to change an object's rotational motion.

  For point masses, $I = mr^2$.

  For distributed mass, $I = \int r^2 dm$.

- Parallel–Axis Theorem: $I = I_{cm} + mx^2$

- Torque is a force's ability to cause an object to rotate. The equation for torque is $\tau = \mathbf{r} \times \mathbf{F}$.

- Newton's Second Law for rotation: $\sum \tau = I\alpha$

- Rotating objects have rotational kinetic energy although the object does not necessarily translate. The equation for rotational kinetic energy is $K_{rotation} = \dfrac{1}{2}I\omega^2$.

  If the object is rolling, the rotational kinetic energy is

  $$K_{rolling} = K_{rotation} + K_{translation} =$$

  $$\frac{1}{2}I_{cm}\omega^2 + \frac{1}{2}mv_{cm}^2$$

Angular Momentum

- Angular momentum for a point particle is given by the equation $L = I\omega$.

- Angular momentum for a rigid object is given by the equation $\vec{L} = \vec{r} \times \vec{p}$.

- Angular momentum is conserved unless a net torque acts on the object. This is expressed by the equation $\sum \tau = dL/dt$.

- For an object to be in static equilibrium, the net force and the net torque must be zero: $\sum F = 0$, $\sum \tau = 0$.

Chapter 10
Oscillations

INTRODUCTION

In this chapter, we'll concentrate on a kind of periodic motion that's straightforward and that, fortunately, actually describes many real-life systems. This type of motion is called *simple harmonic motion*. The prototypical example of simple harmonic motion is a block that's oscillating on the end of a spring. What we learn about this simple system can be applied to many other oscillating systems.

SIMPLE HARMONIC MOTION (SHM): THE SPRING–BLOCK OSCILLATOR

When a spring is compressed or stretched from its natural length, a force is created. If the spring is displaced by x from its natural length, the force it exerts in response is given by the equation

$$\mathbf{F}_S = -k\mathbf{x}$$

This is known as **Hooke's law**. The proportionality constant, $k,$ is a positive number called the **spring** (or **force**) **constant** that indicates how stiff the spring is. The stiffer the spring, the greater the value of $k.$ The minus sign in Hooke's law tells us that $\mathbf{F}_S$ and $\mathbf{x}$ always point in opposite directions. For example, referring to the figure on the next page, when the spring is stretched (**x** is to the right), the spring pulls back (**F** is to the left); when the spring is compressed (**x** is to the left), the spring pushes outward (**F** is to the right). In all cases, the spring wants to return to its original length. As a result, the spring tries to restore the attached block to the **equilibrium position**, which is the position at which the net force on the block is zero. For this reason, we say that the spring provides a **restoring force**.

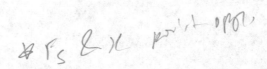

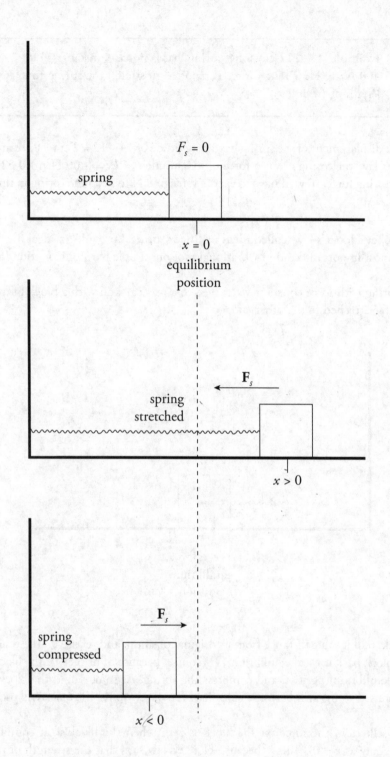

$F = -kx$ $.02$
$= 400 \cdot 0.\cancel{2}^{2}$

> **Example 1** A 12-cm-long spring has a force constant (k) of 400 N/m. How much force is required to stretch the spring to a length of 14 cm?

Solution. The displacement of the spring has a magnitude of $14 - 12 = 2$ cm = 0.02 m so, according to Hooke's law, the spring exerts a force of magnitude $F = kx = (400 \text{ N/m})(0.02 \text{ m}) = 8$ N. This is the restoring force, so we'd have to exert a force of 8 N to keep the spring in this stretched state.

Springs that obey Hooke's law (called **ideal** or **linear** springs) provide an ideal mechanism for defining the most important kind of vibrational motion: simple harmonic motion (SHM).

Consider a spring with force constant k attached to a vertical wall with a block of mass m on a frictionless table attached to the other end.

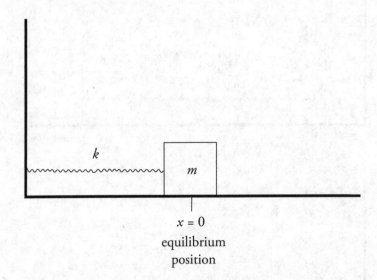

k

m

$x = 0$
equilibrium
position

Grab the block, pull it some distance from its original position, and release it. The restoring force will pull the block back toward equilibrium. Of course, because of its inertia, the block will pass through the equilibrium position and compress the spring. At some point, the block will stop, and the compressed spring will push the block back. In other words, the block will oscillate.

During the oscillation, the force on the block is zero when the block is at equilibrium (the point we designate as $x = 0$). This is because Hooke's law says that the strength of the spring's restoring force is given by the equation $F_S = kx$, so $F_S = 0$ at equilibrium. The acceleration of the block is also equal to zero at $x = 0$, since $F_S = 0$ at $x = 0$ and $a = F_S/m$. At the endpoints of the oscillation region, where the block's displacement, x, has the greatest magnitude, the restoring force and the magnitude of the acceleration are both at their maximum. Only the direction of the restoring force is different.

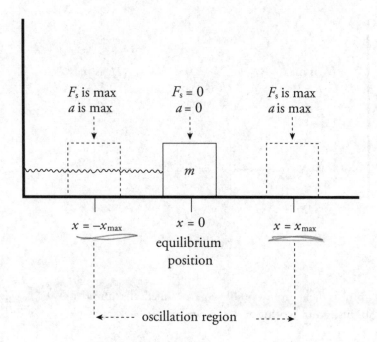

Simple Harmonic Motion in Terms of Energy

Another way to describe the block's motion is in terms of energy transfers. A stretched or compressed spring stores **elastic potential energy**, which is transformed into kinetic energy (and back again). This shuttling between potential and kinetic energy causes the oscillations. For a spring with spring constant k, the elastic potential energy it possesses—relative to its equilibrium position—is given by the equation

$$U_s = \tfrac{1}{2}kx^2$$

In terms of energy transfers, we can describe the block's oscillations as follows: When you initially pull the block out, you increase the elastic potential energy of the system. Upon releasing the block, this potential energy turns into kinetic energy, and the block moves. As it passes through its equilibrium position, $U_S = 0$. Then, as the block continues through that position, the spring is compressed, and the kinetic energy is transformed back into elastic potential energy.

By Conservation of Total Mechanical Energy, the sum $K + U_S$ is a constant. Therefore, when the block reaches the maximum displacement (that is, when $x = \pm x_{max}$), U_S is maximized, so K must be minimized; in fact, $K = 0$ at the endpoints of the oscillation region. As the block is passing through equilibrium, $x = 0$, so $U_S = 0$ and K is maximized.

> Notice that the farther you stretch or compress a spring, the more work you have to do, and, as a result, the more potential energy that's stored.

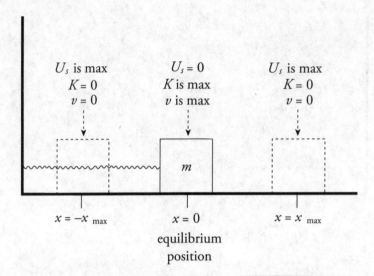

U_s is max
$K = 0$
$v = 0$

$U_s = 0$
K is max
v is max

U_s is max
$K = 0$
$v = 0$

m

$x = -x_{max}$ $x = 0$ $x = x_{max}$

equilibrium
position

The maximum displacement from equilibrium is called the **amplitude** of oscillation and is denoted by A. So, instead of writing $x = x_{max}$, we write $x = A$ (and $x = -x_{max}$ will be written as $x = -A$).

$$\tfrac{1}{2}kx^2 = E = \tfrac{1}{2}mv^2$$

Example 2 A block of mass $m = 0.05$ kg oscillates on a spring whose force constant k is 500 N/m. The amplitude of the oscillations is 4.0 cm. Calculate the maximum speed of the block.

Solution. First, let's get an expression for the maximum elastic potential energy of the system:

$$U_S = \tfrac{1}{2}kx^2 \quad \Rightarrow \quad U_{S,\,max} = \tfrac{1}{2}kx_{max}^2 = \tfrac{1}{2}kA^2$$

When all this energy has been transformed into kinetic energy—which occurs at the equilibrium position—the block will have maximum kinetic energy and maximum speed.

$$U_{S,\,max} \rightarrow K_{max} \quad \Rightarrow \quad \tfrac{1}{2}kA^2 = \tfrac{1}{2}mv_{max}^2$$

$$v_{max} = \sqrt{\frac{kA^2}{m}}$$

$$= \sqrt{\frac{(500 \text{ N/m})(0.04 \text{ m})^2}{0.05 \text{ kg}}}$$

$$= 4 \text{ m/s}$$

Example 3 Show why $U_S = \tfrac{1}{2}kx^2$ if $\mathbf{F}_S = -k\mathbf{x}$.

$$W_S = -\Delta U_S \qquad W_S = \int_{x_1}^{x_2} F(x)\, dx$$

Solution. By definition, $\Delta U_S = -W_S$, and, since $\mathbf{F}$ is not constant, the work done by $\mathbf{F}$ must be calculated using a definite integral. As the end of the spring moves from position $x = x_1$ to $x = x_2$, the work it does is equal to

$$\int_{x_1}^{x_2} -kx\ dx$$

$$
\begin{aligned}
W_S &= \int_{x_1}^{x_2} \mathbf{F}(x)\cdot d\mathbf{x} \\
&= \int_{x_1}^{x_2} -kx\ dx \qquad \left[-\tfrac{1}{2}kx^2\right]_{x_1}^{x_2} \\
&= \left(-\tfrac{1}{2}kx^2\right)_{x_1}^{x_2} \\
&= -\left(\tfrac{1}{2}kx_2^2 - \tfrac{1}{2}kx_1^2\right) \qquad = -\left[\tfrac{1}{2}kx_2^2 - \tfrac{1}{2}kx_1^2\right]
\end{aligned}
$$

Therefore,

$$\Delta U_S = U_{S2} - U_{S1} = -W_S = \tfrac{1}{2}kx_2^2 - \tfrac{1}{2}kx_1^2$$

So if we designate $U_{S1} = 0$ at $x_1 = 0$, the equation above yields $U_s = \tfrac{1}{2}kx^2$, as desired. (Even if a spring does not obey Hooke's law, this method can still be used to find the work done and the potential energy stored.)

$$E = K + U_S = 0 + \tfrac{1}{2}kx^2$$
$$= \tfrac{1}{2}\,500\cdot(0.08)^2$$

Example 4 A block of mass $m = 2.0$ kg is attached to an ideal spring of force constant $k = 500$ N/m. The amplitude of the resulting oscillations is 8.0 cm. Determine the total energy of the oscillator and the speed of the block when it's 4.0 cm from equilibrium.

Solution. The total energy of the oscillator is the sum of its kinetic and potential energies. By Conservation of Mechanical Energy, the sum $K + U_S$ is a constant. When the block is at its amplitude position, $x = 8$ cm, and its kinetic energy is zero:

$$E = K + U_S = 0 + \tfrac{1}{2}kA^2 = \tfrac{1}{2}(500\text{ N/m})(0.08\text{ m})^2 = 1.6\text{ J}$$

At any position x, we have

$$\tfrac{1}{2}mv^2 + \tfrac{1}{2}kx^2 = E$$

$$v = \sqrt{\frac{E - \tfrac{1}{2}kx^2}{\tfrac{1}{2}m}}$$

$$E = K + U_S$$
$$E = \tfrac{1}{2}mv^2 + \tfrac{1}{2}kx^2$$
$$1.6 = \tfrac{1}{2}(2)v^2 + \tfrac{1}{2}500\cdot(0.0\ )$$

Therefore, at $x = 4.0$ cm,

$$v = \sqrt{\frac{E - \tfrac{1}{2}kx^2}{\tfrac{1}{2}m}} = \sqrt{\frac{(1.6\text{ J}) - \tfrac{1}{2}(500\text{ N/m})(0.04\text{ m})^2}{\tfrac{1}{2}(2.0\text{ kg})}}$$

$$= 1.1\text{ m/s}$$

Example 5 A block of mass m = 3.0 kg is attached to an ideal spring of force constant k = 500 N/m. The block is at rest at its equilibrium position. An impulsive force acts on the block, giving it an initial speed of 2.0 m/s. Find the amplitude of the resulting oscillations.

Solution. When the impulsive force acts on the block at its equilibrium position, U_i = 0. Then, when all of the kinetic energy has been transformed into potential energy, the block is at its maximum displacement from equilibrium and it's at one of its amplitude positions. At this point, K_f = 0, and

$$K_i + U_i = K_f + U_f$$
$$\tfrac{1}{2}mv_i^2 + 0 = 0 + \tfrac{1}{2}kA^2$$
$$A = \sqrt{\frac{mv_i^2}{k}}$$
$$= \sqrt{\frac{(3.0\ \text{kg})(2.0\ \text{m/s})^2}{500\ \text{N/m}}}$$
$$= 0.15\ \text{m}$$

THE KINEMATICS OF SIMPLE HARMONIC MOTION

Now that we've explored the dynamics of the block's oscillations in terms of force and energy, let's talk about motion—or kinematics. As you watch the block oscillate, you should notice that it repeats each **cycle** of oscillation in the same amount of time. A cycle is a *round-trip*: for example, from position $x = A$ over to $x = -A$ and back again to $x = A$. The amount of time it takes to complete a cycle is called the **period** of the oscillations, or T. If T is short, the block is oscillating rapidly, and if T is long, the block is oscillating slowly.

Another way of indicating the rapidity of the oscillations is to count the number of cycles that can be completed in a given time interval: the greater the number of completed cycles, the more rapid the oscillations. The number of cycles that can be completed per unit time is called the **frequency** of the oscillations, or f, and is expressed in cycles per second or s^{-1}. One cycle per second is one **hertz** (abbreviated **Hz**).

One of the most basic equations of oscillatory motion expresses the inverse relationship between period and frequency:

$$\text{period} = \frac{\#\ \text{of seconds}}{\text{cycle}} \qquad \text{while} \qquad \text{frequency} = \frac{\#\ \text{of cycles}}{\text{second}}$$

Therefore:

$$T = \frac{1}{f} \qquad \text{and} \qquad f = \frac{1}{T}$$

> **Example 6** A block oscillating on the end of a spring moves from its position of maximum spring stretch to maximum spring compression in 0.25 s. Determine the period and frequency of this motion.

[handwritten: P = ½ 0.25s 0.5s]

[handwritten: f = 1/T]

[handwritten: 2 Hz]

Solution. The period is defined as the time required for one full cycle. Moving from one end of the oscillation region to the other is only half a cycle. Therefore, if the block moves from its position of maximum spring stretch to maximum spring compression in 0.25 s, the time required for a full cycle is twice as much; $T = 0.5$ s. Because frequency is the reciprocal of period, the frequency of the oscillations is $f = 1/T = 1/(0.5 \text{ s}) = 2$ Hz.

> **Example 7** A student observing an oscillating block counts 45.5 cycles of oscillation in one minute. Determine its frequency (in hertz) and period (in seconds).

[handwritten: 45.5/60 = 0.758 Hz]

Solution. The frequency of the oscillations in hertz (which is the number of cycles per second) is

$$f = \frac{45.5 \text{ cycles}}{\text{min}} \times \frac{1 \text{ min}}{60 \text{ s}} = \frac{0.758 \text{ cycles}}{\text{s}} = 0.758 \text{ Hz}$$

[handwritten: 1/0.758 = 1.385]

Therefore,

$$T = \frac{1}{f} = \frac{1}{0.758 \text{ Hz}} = 1.32 \text{ s}$$

One of the defining properties of the spring–block oscillator is that the frequency and period can be determined from the mass of the block and the force constant of the spring. The equations are as follows:

$$f = \frac{1}{2\pi}\sqrt{\frac{k}{m}} \quad \text{and} \quad T = \frac{1}{f} = 2\pi\sqrt{\frac{m}{k}}$$

[handwritten: $f = \frac{1}{2\pi}\sqrt{\frac{k}{m}}$]

[handwritten: $T = 2\pi\sqrt{\frac{m}{k}}$]

Let's analyze these equations. Suppose we had a small mass on a very stiff spring. Intuitively, we would expect that this strong spring would make the small mass oscillate rapidly, with a high frequency and a short period. Both of these predictions are substantiated by the equations above, because if m is small and k is large, then the ratio k/m is large (high frequency) and the ratio m/k is small (short period).

> **Example 8** A block of mass $m = 2.0$ kg is attached to a spring whose force constant, k, is 300 N/m. Calculate the frequency and period of the oscillations of this spring–block system.

[handwritten: $f = \frac{1}{2\pi}\sqrt{\frac{k}{m}} = \frac{1}{2\pi}\sqrt{\frac{300}{2}}$]

Solution. According to the equations,

$$f = \frac{1}{2\pi}\sqrt{\frac{k}{m}} = \frac{1}{2\pi}\sqrt{\frac{300 \text{ N/m}}{2.0 \text{ kg}}} = 1.9 \text{ Hz}$$

$$T = \frac{1}{f} = \frac{1}{1.9 \text{ Hz}} = 0.51 \text{ s}$$

Example 9 A block is attached to a spring and set into oscillatory motion, and its frequency is measured. If this block were removed and replaced by a second block with 1/4 the mass of the first block, how would the frequency of the oscillations compare to that of the first block?

$$f \propto \frac{1}{\sqrt{m}} \qquad \therefore \sqrt{\frac{m}{4}} = \sqrt{\frac{4}{m}} = 2$$

Solution. Since the same spring is used, k remains the same. According to the equation given above, f is inversely proportional to the square root of the mass of the block: $f \propto 1/\sqrt{m}$. Therefore, if m decreases by a factor of 4, then f increases by a factor of $\sqrt{4} = 2$.

The equations we saw above for the frequency and period of the spring–block oscillator do not contain A, the amplitude of the motion. In simple harmonic motion, *both the frequency and the period are independent of the amplitude*. The reason for this is that the strength of the restoring force is proportional to x, the displacement from equilibrium, as given by Hooke's law: $\mathbf{F}_S = -k\mathbf{x}$.

$$F_S = -kx$$

Example 10 A student performs an experiment with a spring–block simple harmonic oscillator. In the first trial, the amplitude of the oscillations is 3.0 cm, while in the second trial (using the same spring and block), the amplitude of the oscillations is 6.0 cm. Compare the values of the period, frequency, and maximum speed of the block between these two trials.

Solution. If the system exhibits simple harmonic motion, then the period and frequency for a trial using the same spring and block are independent of amplitude. However, the maximum speed of the block will be greater in the second trial than in the first. Since the amplitude is greater in the second trial, the system possesses more total energy ($E = \frac{1}{2}kA^2$). So when the block is passing through equilibrium (its position of greatest speed), the second system has more kinetic energy, meaning that the block will have a greater speed. In fact, from Example 2, we know that $v_{max} = A\sqrt{k/m}$, so A is twice as great in the second trial than in the first, and v_{max} will be twice as great in the second trial than in the first.

$$v_{max} = A\sqrt{\frac{k}{m}}$$

v_{max} is directly prop to A.

Example 11 For each of the following arrangements of two springs, determine the **effective spring constant**, k_{eff}. This is the force constant of a single spring that would produce the same force on the block as the pair of springs shown in each case.

(a)

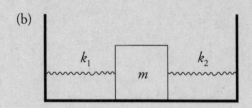

(b)

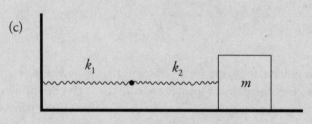

(c)

(d)   Determine k_{eff} in each of these cases if $k_1 = k_2 = k$.

Solution.

(a) Imagine that the block was displaced a distance x to the right of its equilibrium position. Then the force exerted by the first spring would be $F_1 = -k_1 x$ and the force exerted by the second spring would be $F_2 = -k_2 x$. The net force exerted by the springs would be

$$F_1 + F_2 = -k_1 x + -k_2 x = -(k_1 + k_2)x$$

Since $F_{eff} = -(k_1 + k_2)x$, we see that $k_{eff} = k_1 + k_2$.

(b) Imagine that the block was displaced a distance x to the right of its equilibrium position. Then the force exerted by the first spring would be $F_1 = -k_1 x$ and the force exerted by the second spring would be $F_2 = -k_2 x$. The net force exerted by the springs would be

$$F_1 + F_2 = -k_1 x + -k_2 x = -(k_1 + k_2)x$$

As in part (a), we see that since $F_{eff} = -(k_1 + k_2)x$, we get $k_{eff} = k_1 + k_2$.

(c) Imagine that the block was displaced a distance x to the right of its equilibrium position. Let x_1 be the distance that the first spring is stretched, and let x_2 be the distance that the second spring is stretched. Then $x = x_1 + x_2$. But $x_1 = -F/k_1$ and $x_2 = -F/k_2$, so

$$\frac{-F}{k_1} + \frac{-F}{k_2} = x$$

$$-F\left(\frac{1}{k_1} + \frac{1}{k_2}\right) = x$$

$$F = -\left(\frac{1}{\frac{1}{k_1} + \frac{1}{k_2}}\right)x$$

$$F = -\frac{k_1 k_2}{k_1 + k_2}x$$

Therefore,

$$k_{\text{eff}} = \frac{k_1 k_2}{k_1 + k_2}$$

(d) If the two springs have the same force constant, that is, if $k_1 = k_2 = k$, then in the first two cases the pairs of springs are equivalent to one spring that has twice their force constant: $k_{\text{eff}} = k_1 + k_2 = k + k = 2k$. In (c), the pair of springs is equivalent to a single spring with half their force constant:

$$k_{\text{eff}} = \frac{k_1 k_2}{k_1 + k_2} = \frac{kk}{k + k} = \frac{k^2}{2k} = \frac{k}{2}$$

THE SPRING–BLOCK OSCILLATOR: VERTICAL MOTION

So far, we've looked at a block sliding back and forth on a horizontal table, but the block could also oscillate vertically. In vertical oscillation, gravity causes the block to move downward to an equilibrium position where the spring is not at its natural length.

Consider a spring of negligible mass hanging from a stationary support. A block of mass m is attached to its end and allowed to come to rest, stretching the spring a distance d. At this point, the block is in equilibrium; the upward force of the spring is balanced by the downward force of gravity. Therefore,

$$kd = mg \quad \Rightarrow \quad d = \frac{mg}{k}$$

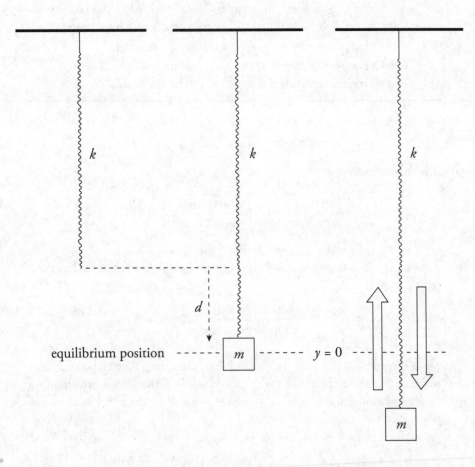

Next, imagine that the block is pulled down a distance A and released. The restoring force is greater than the block's weight, and, as a result, the block accelerates upward. As the block's momentum carries it up through the equilibrium position, F_S is less than the block's weight. As a result, the block decelerates, stops, and accelerates downward again, and the up-and-down motion repeats.

When the block is at a distance y below its equilibrium position, the spring is stretched a total distance of $d + y$, so the upward spring force is equal to $k(d + y)$. The downward force stays the same, mg. The net force on the block is

$$F = k(d + y) - mg$$

but this equation becomes $F = ky$, because $kd = mg$ (as we saw above). Since the resulting force on the block, $F = ky$, has the form of Hooke's law, we know that the vertical simple harmonic oscillations of the block have the same characteristics as the horizontal oscillations. The equilibrium position, $y = 0$, is not at the spring's natural length, but at the point where the hanging block is in equilibrium.

Example 12 A block of mass $m = 1.5$ kg is attached to the end of a vertical spring of force constant $k = 300$ N/m. After the block comes to rest, it is pulled down a distance of 2.0 cm and released.

(a) What is the frequency of the resulting oscillations?

(b) What are the minimum and maximum amounts of stretch of the spring during the oscillations of the block?

Solution.

(a) The frequency is given by

$$f = \frac{1}{2\pi}\sqrt{\frac{k}{m}} = \frac{1}{2\pi}\sqrt{\frac{300 \text{ N/m}}{1.5 \text{ kg}}} = 2.3 \text{ Hz}$$

(b) Before the block is pulled down, to begin the oscillations, it stretches the spring by a distance

$$d = \frac{mg}{k} = \frac{(1.5 \text{ kg})(10 \text{ N/kg})}{300 \text{ N/m}} = 0.05 \text{ m} = 5 \text{ cm}$$

Since the amplitude of the motion is 2 cm, the spring is stretched a maximum of 5 cm + 2 cm = 7 cm when the block is at the lowest position in its cycle, and a minimum of 5 cm − 2 cm = 3 cm when the block is at its highest position.

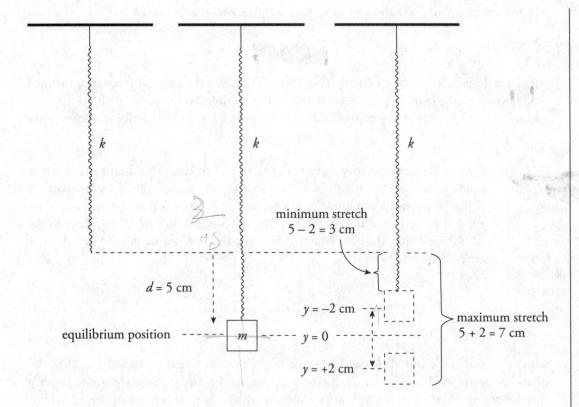

THE SINUSOIDAL DESCRIPTION OF SIMPLE HARMONIC MOTION

The position of the block during its oscillation can be written as a function of time. Take a look at the experimental setup below.

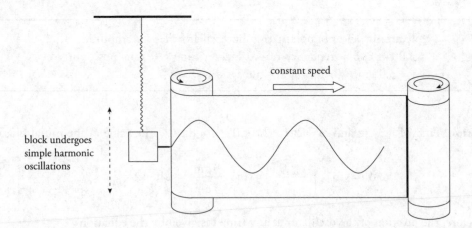

A small pen is attached to the oscillating block, and it makes a mark on the paper as the paper is pulled along by the roller on the right. Clearly, the simple harmonic motion of the block is sinusoidal.

The basic mathematical equation for describing simple harmonic motion is

$$\omega = 2\pi f$$

$$y = A \sin(\omega t)$$

where y is the position of the oscillator, A is the amplitude, ω is the **angular frequency** (defined as $2\pi f$, where f is the frequency of the oscillations), and t is time. Since $\sin(\omega t)$ oscillates between -1 and $+1$, the quantity $A \sin(\omega t)$ oscillates between $-A$ and $+A$; this describes the oscillation region.

> If $\phi_0 = \pi/2$, then $y = A\cos(\omega t)$. It's not uncommon to see the equation for the position of a simple harmonic oscillator given with a cosine instead of a sine function.

If $t = 0$, then the quantity $A \sin(\omega t)$ is also equal to zero. This means that $y = 0$ at time $t = 0$. However, what if $y \neq 0$ at time $t = 0$? For example, if the oscillator is pulled to one of its amplitude positions, say $y = A$, and released at time $t = 0$, then $y = A$ at $t = 0$. To account for the fact that the oscillator can begin anywhere in the oscillation region, the basic equation for the position of the oscillator given above is generalized as follows:

$$y = A \sin(\omega t + \phi_0)$$

where ϕ_0 is called the **initial phase**. The argument of the sine function, $\omega t + \phi_0$, is called the **phase** (or **phase angle**). By carefully choosing ϕ_0, we can be sure that the equation correctly specifies the oscillator's position no matter where it may have been at time $t = 0$. The value of ϕ_0 can be calculated from the equation

$$\phi_0 = \sin^{-1}\left(\frac{y_{\text{at } t=0}}{A}\right)$$

Example 13 A simple harmonic oscillator has an amplitude of 3.0 cm and a frequency of 4.0 Hz. At time $t = 0$, its position is $y = 3.0$ cm. Where is it at time $t = 0.3$ s?

Solution. First, $A = 3$ cm and $\omega = 2\pi f = 2\pi(4.0 \text{ s}^{-1}) = 8\pi \text{ s}^{-1}$. The value of the initial phase is

$$\phi_0 = \sin^{-1}\left(\frac{y_{\text{at } t=0}}{A}\right) = \sin^{-1}\frac{3\,\text{cm}}{3\,\text{cm}} = \sin^{-1} 1 = \tfrac{1}{2}\pi$$

Therefore, the position of the oscillator at any time t is given by the equation

$$y = (3\,\text{cm}) \cdot \sin\left[(8\pi\text{s}^{-1})\,t + \tfrac{1}{2}\pi\right]$$

So, at time $t = 0.3$ s, we find that $y = (3 \text{ cm}) \cdot \sin\left[(8\pi \text{ s}^{-1})(0.3 \text{ s}) + \tfrac{1}{2}\pi\right] = 0.93$. (Make sure your calculator is in *radian mode* when you evaluate this expression!)

Example 14 The position of a simple harmonic oscillator is given by the equation

$$y = (4 \text{ cm}) \cdot \sin[(6\pi \text{ s}^{-1})t - \tfrac{1}{2}\pi]$$

(a) Where is the oscillator at time $t = 0$?
(b) What is the amplitude of the motion?
(c) What is the frequency?
(d) What is the period?

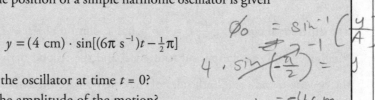

$\phi_0 = \sin^{-1}\left(\dfrac{y}{A}\right)$

$4 \cdot \sin\left(-\dfrac{\pi}{2}\right) = y$

$y = -4 \text{ cm}$

[handwritten: $3 \boxed{4\text{ cm}}{2}$]

[handwritten: $\dfrac{1}{3}$]

Solution.

(a) To determine y at time $t = 0$, we simply substitute $t = 0$ into the given equation and evaluate:

$$y_{\text{at } t=0} = (4 \text{ cm}) \cdot \sin(-\tfrac{1}{2}\pi) = -4 \text{ cm}$$

(b) The amplitude of the motion, A, is the coefficient in front of the $\sin(\omega t + \phi_0)$ expression. In this case, we read directly from the given equation that $A = 4$ cm.

(c) The coefficient of t is the angular frequency, ω. In this case, we see that $\omega = 6\pi \text{ s}^{-1}$. Since $\omega = 2\pi f$ by definition, we have $f = \omega/(2\pi) = 3$ Hz.

(d) Since $T = 1/f$, we calculate that $T = 1/f = 1/(3 \text{ Hz}) = 0.33$ s.

Instantaneous Velocity and Acceleration

If the position of a simple harmonic oscillator is given by the equation $y = A \sin(\omega t + \phi_0)$, its velocity and acceleration can be found by differentiation:

$$v(t) = \dot{y}(t) = \frac{d}{dt}[A\sin(\omega t + \phi_0)] = A\omega\cos(\omega t + \phi_0)$$

and

$$a(t) = \dot{v}(t) = \frac{d}{dt}[A\omega\cos(\omega t + \phi_0)] = -A\omega^2\sin(\omega t + \phi_0)$$

Note that both the velocity and acceleration vary with time.

Differential Equation for Simple Harmonic Motion

The differential equation, $\dfrac{d^2 y}{dt^2} = -\omega^2 y$ has a solution of $y(t) = A\sin(\omega t + \phi_0)$. To check that this function, $y(t)$, is a solution, substitute $y(t)$ and the second derivative of $y(t)$, which is $a(t)$, into the equation, and see if both sides of the equation are the same.

$$\frac{d^2 y}{dt^2} = -\omega^2 y$$

$$-A\omega^2 \sin(\omega t + \phi_0) = -\omega^2 (A\sin(\omega t + \phi_0))$$

check

$$-A\omega^2 \sin(\omega t + \phi_0) = -\omega^2 (A\sin(\omega t + \phi_0))$$

As you can see, the two sides of the equation are the same, so $y(t) = A\sin(\omega t + \phi_0)$ is a solution to the differential equation $\dfrac{d^2 y}{dt^2} = -\omega^2 y$. Therefore, if we can derive a differential equation that takes this form, we will know the solution to the equation and that the object is undergoing SHM.

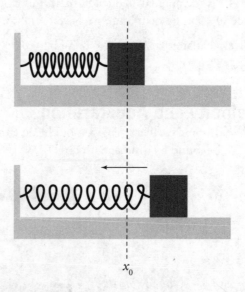

x_0

Let's do this for a spring-mass system sliding along a frictionless, horizontal floor, as shown above.

$$F = ma$$

$$-kx = m\frac{d^2 x}{dt^2}$$

$$-\frac{k}{m}x = \frac{d^2 x}{dt^2}$$

This is the same basic form of the differential equation from above, and for the spring-mass system:

$$\omega = \sqrt{\frac{k}{m}} \text{ and } x(t) = A \sin(\omega t)$$

Example 15 The position of a simple harmonic oscillator is given by the equation

$$y = 4\sin\left(6\pi t - \tfrac{1}{2}\pi\right)$$

where y is in cm and t is in seconds. What are the maximum speed and maximum acceleration of the oscillator?

Solution. Since the maximum value of $\cos(\omega t + \phi_0)$ is 1, the maximum value of v is $v_{max} = A\omega$; similarly, since the minimum value of $(\omega t + \phi_0)$ is -1, the maximum value of a is $a_{max} = A\omega^2$. From the equation for y, we see that $A = 4 \text{ cm} = 0.04 \text{ m}$ and $\omega = 6\pi \text{ s}^{-1}$. Therefore, $v_{max} = A\omega = (0.04 \text{ m})(6\pi \text{ s}^{-1}) = 0.75 \text{ m/s}$ and $a_{max} = A\omega^2 = (0.04 \text{ m})(6\pi \text{ s}^{-1})^2 = 14 \text{ m/s}^2$.

PENDULA

A **simple pendulum** consists of a weight of mass m attached to a massless rod that swings, without friction, about the vertical equilibrium position. The restoring force is provided by gravity and, as the figure below shows, the magnitude of the restoring force when the bob is θ to an angle to the vertical is given by the equation:

$$F_{restoring} = mg \sin \theta$$

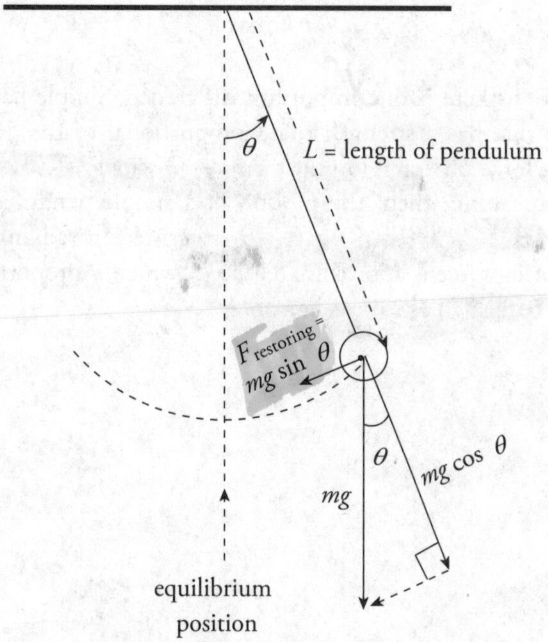

Although the displacement of the pendulum is measured by the angle that it makes with the vertical, rather than by its linear distance from the equilibrium position (as was the case for the spring–block oscillator), the simple pendulum shares many of the important features of the spring–block oscillator. For example,

- Displacement is zero at the equilibrium position.
- At the endpoints of the oscillation region (where $\theta = \pm\theta_{max}$), the restoring force and the tangential acceleration (a_t) have their greatest magnitudes, the speed of the pendulum is zero, and the potential energy is maximized.
- As the pendulum passes through the equilibrium position, its kinetic energy and speed are maximized.

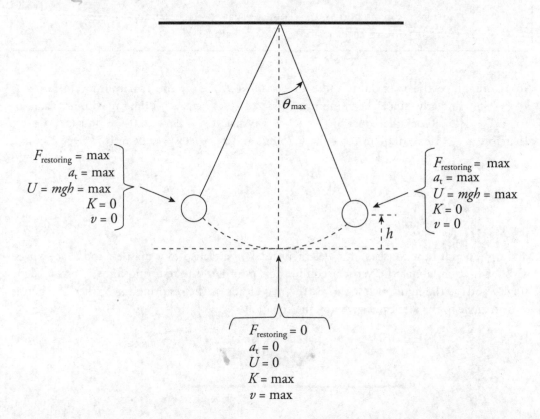

Despite these similarities, there is one important difference. Simple harmonic motion results from a restoring force that has a strength that's proportional to the displacement. The magnitude of the restoring force on a pendulum is $mg \sin \theta$, which is *not* proportional to the displacement θ. Strictly speaking, then, the motion of a simple pendulum is not really simple harmonic. However, if θ is small, then $\sin \theta \approx \theta$ (measured in radians). So, in this case, the magnitude of the restoring force is approximately $mg\theta$, which *is* proportional to θ. So if θ_{max} is small, the motion can be treated as simple harmonic.

If the restoring force is given by $mg\theta$, rather than $mg \sin \theta$, then the frequency and period of the oscillations depend only on the length of the pendulum and the value of the gravitational acceleration, according to the following equations:

$$f = \frac{1}{2\pi}\sqrt{\frac{g}{L}} \qquad \text{and} \qquad T = 2\pi\sqrt{\frac{L}{g}}$$

Differential Equation for a Pendulum

We have derived a differential equation for a spring-mass system that has a solution of $y(t) = A \sin(\omega t + \phi_o)$. Now we will do the same for a pendulum with small amplitude oscillations, which allows us to approximate $\sin \theta = \theta$. We will use s for the arc length displacement.

Consider a pendulum of length L and mass m undergoing small amplitude oscillations.

$$F = ma$$

$$-mg \sin\theta = m\frac{d^2 s}{dt^2}$$

$$-g\theta = \frac{d^2 s}{dt^2}$$

$$-g\theta = L\frac{d^2\theta}{dt^2} \quad \text{because } s = L\theta$$

$$-\frac{g}{L}\theta = \frac{d^2\theta}{dt^2}$$

This is the same basic form of the differential equation for the spring-mass system. Therefore,

for a pendulum: $\omega = \sqrt{\frac{g}{L}}$ and $\theta(t) = \theta_{max} \sin(\omega t + \phi_0)$.

Note that neither frequency nor period depends on the amplitude (the maximum angular displacement, θ_{max}); this is a characteristic feature of simple harmonic motion. Also notice that neither depends on the mass of the weight.

$T = 1s$

Example 16 A simple pendulum has a period of 1 s on Earth. What would its period be on the Moon (where g is one-sixth of its value here)?

Solution. The equation $T = 2\pi\sqrt{L/g}$ shows that T is inversely proportional to $\sqrt{g}$, so if g decreases by a factor of 6, then T increases by a factor of $\sqrt{6}$. That is,

$$T_{\text{on Moon}} = \sqrt{6} \times T_{\text{on Earth}} = \sqrt{6}\,(1\,\text{s}) = 2.4\ \text{s}$$

$$T = 2\pi\sqrt{\frac{L}{g}}$$

$$T \propto \frac{1}{\sqrt{g}} = \frac{1}{\sqrt{\frac{g}{6}}}$$

$$= \sqrt{6} \cdot T_E = \sqrt{6} \cdot 1s$$

$$= 2.4s$$

Chapter 10 Drill

The answers and explanations can be found in Chapter 17.

Section I: Multiple Choice

1. Which of the following is/are characteristics of simple harmonic motion?

    I.   The acceleration is constant.
    II. The restoring force is proportional to the displacement.
    III. The frequency is independent of the amplitude.

    (A) II only
    (B) I and II only
    (C) I and III only
    (D) II and III only
    (E) I, II, and III

2. What combination of values will result in a spring–block system with the greatest frequency?

    (A) $A = 50$ cm; $k = 80$ N/m; $m = 2$ kg
    (B) $A = 50$ cm; $k = 100$ N/m; $m = 2$ kg
    (C) $A = 50$ cm; $k = 100$ N/m; $m = 4$ kg
    (D) $A = 75$ cm; $k = 80$ N/m; $m = 2$ kg
    (E) $A = 75$ cm; $k = 100$ N/m; $m = 4$ kg

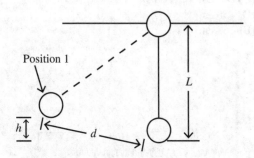

3. A pendulum of length L is pulled to Position 1, as shown in the figure, and released from rest. The straight line distance from Position 1 to the equilibrium position of the pendulum is d. The vertical height change of the pendulum from Position 1 to the equilibrium position (the lowest point of the swing) is h. What is the maximum speed of the pendulum in terms of h, d, L, and gravity (g)?

    (A) $\sqrt{\dfrac{2gh}{L}}$

    (B) $\sqrt{2gh}$

    (C) $\sqrt{2gd}$

    (D) $\sqrt{2gL}$

    (E) $\sqrt{\dfrac{2gd}{L}}$

4. A student measures the maximum speed of a block undergoing simple harmonic oscillations of amplitude A on the end of an ideal spring. If the block is replaced by one with twice the mass but the amplitude of its oscillations remains the same, then the maximum speed of the block will

    (A) decrease by a factor of 4
    (B) decrease by a factor of 2
    (C) decrease by a factor of $\sqrt{2}$
    (D) remain the same
    (E) increase by a factor of 2

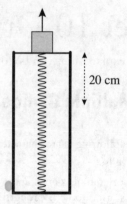

5. A spring with a natural length of 40 cm and a spring constant of 400 N/m is hung vertically with a 10 kg mass attached to the end. Assuming the spring's mass is negligible, what will be the final length of the spring when it reaches equilibrium?

(A) 25 cm
(B) 35 cm
(C) 40 cm
(D) 50 cm
(E) 65 cm

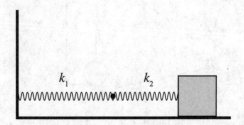

6. The spring-block system above features two connected springs of different, nonzero spring constants, k_1 and k_2. For what value of k_1 would the system above have the same effective spring constant as if the spring with constant k_2 were the only spring in the system?

(A) $k_1 = \dfrac{1}{k_2}$

(B) $k_1 = k_2$

(C) $k_1 = 2k_2$

(D) $k_1 = \dfrac{1}{2k_2}$

(E) It is not possible.

7. A block of mass $m = 4$ kg on a frictionless, horizontal table is attached to one end of a spring of force constant $k = 400$ N/m and undergoes simple harmonic oscillations about its equilibrium position ($x = 0$) with amplitude $A = 6$ cm. If the block is at $x = 6$ cm at time $t = 0$, then which of the following equations (with x in centimeters and t in seconds) gives the block's position as a function of time?

(A) $x = 6\sin(10t + \frac{1}{2}\pi)$

(B) $x = 6\sin(10\pi t + \frac{1}{2}\pi)$

(C) $x = 6\sin(10\pi t - \frac{1}{2}\pi)$

(D) $x = 6\sin(10t)$

(E) $x = 6\sin(10t - \frac{1}{2}\pi)$

8. A vertical spring with spring constant $k = \pi^2$ N/m has a platform of mass $m = 1$ kg attached to the top. The spring is placed inside a tube so that it rests 10 cm below the top of the tube. A small block of negligible mass is placed on the platform, and the spring is compressed 10 cm. When released, it launches the block into the air such that it has an initial speed of 10 m/s when it exits the tube. If the resting position of the spring is considered $x = 0$ cm, at what position will the spring platform and block make contact when the block falls back down?

(A) $x = -10$ cm
(B) $x = 0$ cm
(C) $x = 10$ cm
(D) -10 cm $< x < 0$ cm
(E) 0 cm $< x < 10$ cm

9. A simple pendulum swings about the vertical equilibrium position with a maximum angular displacement of 5° and period T. If the same pendulum is given a maximum angular displacement of 10°, then which of the following best gives the period of the oscillations?

(A) $T/2$
(B) $T/\sqrt{2}$
(C) T
(D) $T\sqrt{2}$
(E) $2T$

10. Which of the following best describes the relationship between the tension force, F_T, in the string of a pendulum and the component of gravity that pulls antiparallel to the tension, F_G? Assume that the pendulum is only displaced by a small amount.

(A) $F_T > F_G$
(B) $F_T \geq F_G$
(C) $F_T = F_G$
(D) $F_T \leq F_G$
(E) $F_T < F_G$

Section II: Free Response

1. The figure below shows a block of mass m (Block 1) that's attached to one end of an ideal spring of force constant k and natural length L. The block is pushed so that it compresses the spring to 3/4 of its natural length and then released from rest. Just as the spring has extended to its natural length L, the attached block collides with another block (also of mass m) at rest on the edge of the frictionless table. When Block 1 collides with Block 2, half of its kinetic energy is lost to heat; the other half of Block 1's kinetic energy at impact is divided between Block 1 and Block 2. The collision sends Block 2 over the edge of the table, where it falls a vertical distance H, landing at a horizontal distance R from the edge.

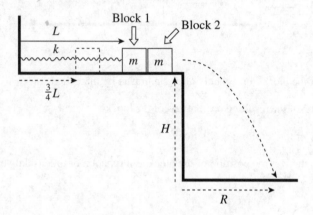

(a) What is the acceleration of Block 1 at the moment it's released from rest from its initial position? Write your answer in terms of k, L, and m.

(b) If v_1 is the velocity of Block 1 just before impact, show that the velocity of Block 1 just after impact is $\frac{1}{2}v_1$.

(c) Determine the amplitude of the oscillations of Block 1 after Block 2 has left the table. Write your answer in terms of L only.

(d) Determine the period of the oscillations of Block 1 after the collision, writing your answer in terms of T_0, the period of the oscillations that Block 1 would have had if it did not collide with Block 2.

(e) Find an expression for R in terms of H, k, L, m, and g.

2. A bullet of mass m is fired horizontally with speed v into a block of mass M initially at rest, at the end of an ideal spring on a frictionless table. At the moment the bullet hits, the spring is at its natural length, L. The bullet becomes embedded in the block, and simple harmonic oscillations result.

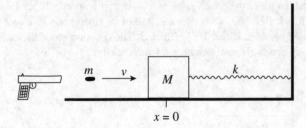

(a) Determine the speed of the block immediately after the impact by the bullet.

(b) Determine the amplitude of the resulting oscillations of the block.

(c) Compute the frequency of the resulting oscillations.

(d) Derive an equation which gives the position of the block as a function of time (relative to $x = 0$ at time $t = 0$).

3. A block of mass M oscillates with amplitude A on a frictionless horizontal table, connected to an ideal spring of force constant k. The period of its oscillations is T. At the moment when the block is at position $x = \frac{1}{2}A$ and moving to the right, a ball of clay of mass m dropped from above lands on the block.

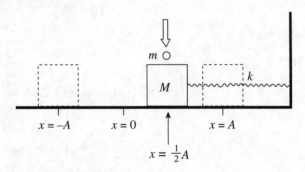

(a)   What is the velocity of the block just before the clay hits?

(b)   What is the velocity of the block just after the clay hits?

(c)   What is the new period of the oscillations of the block?

(d)   What is the new amplitude of the oscillations? Write your answer in terms of A, k, M, and m.

(e)   Would the answer to part (c) be different if the clay had landed on the block when it was at a different position? Support your answer briefly.

(f)   Would the answer to part (d) be different if the clay had landed on the block when it was at a different position? Support your answer briefly.

4. An object of total mass M is allowed to swing around a fixed suspension point P. The object's moment of inertia with respect to the rotation axis perpendicular to the page through P is denoted by I. The distance between P and the object's center of mass C, is d.

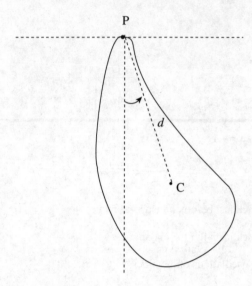

(a) Compute the torque τ produced by the weight of the object when the line PC makes an angle θ with the vertical. (Take the counterclockwise direction as positive for both θ and τ.)

(b) If θ is small, so that $\sin \theta$ may be replaced by θ, write the restoring torque τ computed in part (a) in the form $\tau = -\kappa\theta$.

A simple harmonic oscillator whose displacement from equilibrium, z, satisfies an equation of the form $\dfrac{d^2z}{dt^2} = -bz$ has a period of oscillation given by the formula $T = \dfrac{2\pi}{\sqrt{b}}$.

(c) Setting z equal to θ in the equation above, use the result of part (b) to derive an expression for the period of small oscillations of the object shown above.

(d) Answer the question posed in part (c) if the object were a uniform bar of mass M and length L (whose moment of inertia about one of its ends is given by the equation $I = \dfrac{1}{3}ML^2$.)

Summary

Simple Harmonic Motion

o Simple harmonic motion will occur when there is a restoring force on an object that is proportional to the displacement from equilibrium.

o The mathematical equation of an object undergoing SHM is $y(t) = A \sin(\omega t + \phi)$.

o This equation is a solution to the differential equation for SHM: $\dfrac{d^2 y}{dt^2} = -\omega^2 y$.

o The mathematical equation for angular frequency is $\omega = 2\pi f$.

o The period is the length of time it takes the object to complete one cycle:

$$T = \frac{\text{time}}{\text{cycles}}$$

o The period can be expressed as an equation proportional to the angular frequency: $T = \dfrac{2\pi}{\omega}$.

o The frequency is the number of cycles the object completes in one unit of time:

$$f = \frac{1}{T} = \frac{\text{cycles}}{\text{time}}$$

o An object undergoing SHM typically transforms potential energy to kinetic energy and back to potential energy in a repeating cycle.

o The two examples of SHM that we have studied thus far are a mass oscillating on a spring and a simple pendulum.

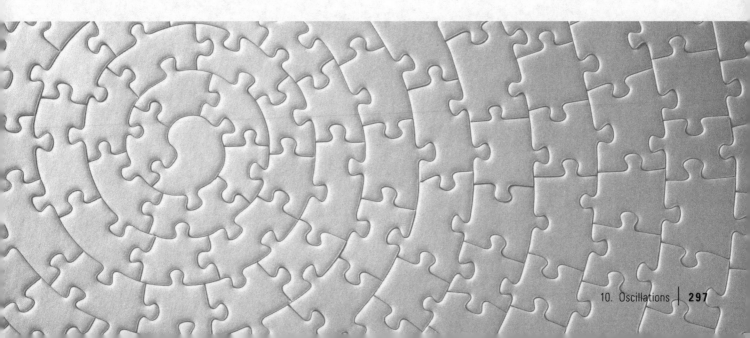

Chapter 11
Gravitation

KEPLER'S LAWS

Johannes Kepler spent years of exhaustive study distilling volumes of data collected by his mentor, Tycho Brahe, into three simple laws that describe the motion of planets.

Kepler's First Law

Every planet moves in an elliptical orbit, with the Sun at one focus.

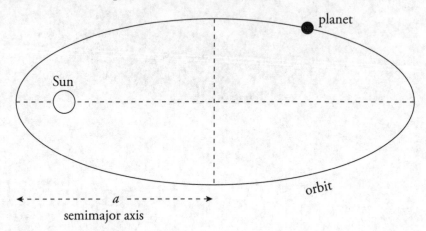

Kepler's Second Law

As a planet moves in its orbit, a line drawn from the Sun to the planet sweeps out equal areas in equal time intervals. Physically, this is a consequence of conservation of angular momentum.

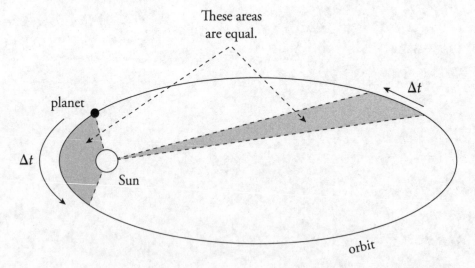

Kepler's Third Law

If T is the period, the time required to make one revolution, and a is the length of the semimajor axis of a planet's orbit, then the ratio T^2/a^3 is the same for all planets orbiting the same star.

NEWTON'S LAW OF GRAVITATION

Newton eventually proved that Kepler's first two laws imply a law of gravitation: Any two objects in the universe exert an attractive force on each other—called the **gravitational force**—whose strength is proportional to the product of the objects' masses and inversely proportional to the square of the center-to-center distance between them. If we let G be the **universal gravitational constant**, then the strength of the gravitational force is given by the equation:

$$\mathbf{F}_G = -\frac{Gm_1m_2}{r^2}\hat{\mathbf{r}}$$

Recall that $\hat{\mathbf{r}}$ is the unit vector from m_1 to m_2. Consider a mass, m_1, close to the surface of Earth. We will use Newton's Law of Gravitation and Newton's Second Law to show that the gravitational acceleration of m_1 is independent of the mass of the object, as shown below.

$$F_g = G\frac{m_1 M_E}{R_E^{\,2}} = m_1 a_g$$

$$\frac{GM_E}{R_E^{\,2}} = a_g = g$$

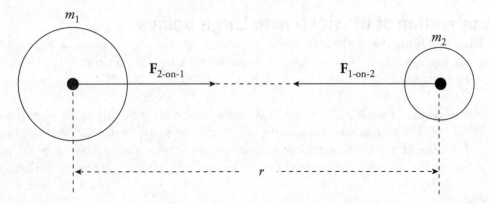

The forces $\mathbf{F}_{1\text{-on-2}}$ and $\mathbf{F}_{2\text{-on-1}}$ act along the line that joins the bodies and form an action/reaction pair.

The first reasonably accurate numerical value for G was determined by Cavendish more than one hundred years after Newton's Law was published. To three decimal places, the currently accepted value of G is

$$G = 6.67 \times 10^{-11} \text{ N·m}^2/\text{kg}^2$$

Kepler's Third Law then follows from Newton's Law of Gravitation. We'll show how this works for the case of a circular orbit of radius R (which can be considered an elliptical orbit with eccentricity zero). If the orbit is circular, then Kepler's Second Law says that the planet's orbit speed, v, must be constant. Therefore, the planet executes uniform circular motion, and centripetal force is provided by the gravitational attraction of the Sun. If we let M be the mass of the Sun and m be the mass of the planet, then this last statement can be expressed mathematically as:

> Note that this formula makes sense when we consider that each object is contributing an m/r term to the total force of gravity.

$$\frac{mv^2}{R} = G\frac{Mm}{R^2} \quad (1)$$

The period of a planet's orbit is the time it requires to make one revolution around the Sun, so dividing the distance covered, $2\pi R$, by the planet's orbit speed, v, we have

$$T = \frac{2\pi R}{v} \quad (2)$$

Equation (1) implies that $v^2 = GM/R$. Squaring both sides of Equation (2) and then substituting $v^2 = GM/R$, we find that

$$T^2 = \frac{4\pi^2 R^2}{v^2} = \frac{4\pi^2 R^2}{GM/R} = \frac{4\pi^2}{GM} R^3$$

Therefore,

$$\frac{T^2}{R^3} = \frac{4\pi^2}{GM}, \text{ a constant}$$

which is Kepler's Third Law for a circular orbit of radius R.

Acceleration of Gravity Due to Large Bodies

We have been using the acceleration due to gravity $g = 10 \text{ m/s}^2$ for all objects falling near the surface of Earth. We have assumed that the mass does not affect the acceleration of gravity and now we will show why.

Q: Are astronauts in the International Space Station actually "weightless"?

Turn the page for the answer.

Assume a small mass m is located near a large body (i.e., a planet or star) of mass M. The gravitational force on the object near the surface will equal the mass of the object times the acceleration of gravity a_g. The equation below shows how the mass of the object cancels out, and the acceleration of gravity is independent of that mass of the object.

$$F_g = ma_g$$
$$\frac{GmM}{r^2} = ma_g, \text{ cancel out the } m\text{'s}$$
$$\frac{GM}{r^2} = a_g$$

From this expression, we can see that the acceleration due to gravity for any object on a planet would be related to the mass and radius of the planet, but not the mass of the object.

THE GRAVITATIONAL ATTRACTION DUE TO AN EXTENDED BODY

Newton's Law of Gravitation is really a statement about the force between two point particles: objects that are very small in comparison to the distance between them. Newton also proved that a uniform sphere attracts another body as if all of the sphere's mass were concentrated at its center.

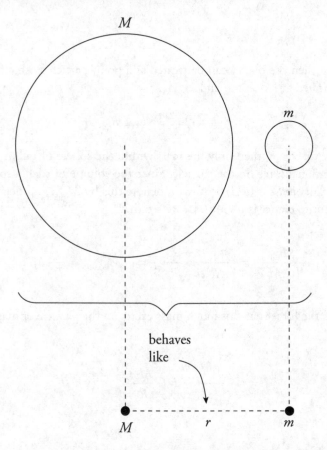

For this reason, we can apply Newton's Law of Gravitation to extended bodies, that is, to objects that are not small relative to the distance between them.

Additionally, a uniform shell of mass does not exert a gravitational force on a particle inside it. This means that if a spherical planet is uniform, then as we descend into it, only the mass of the sphere *underneath* us exerts a gravitational force; the shell *above* exerts no force because we're inside it.

Example 1 What is the gravitational force on a particle of mass m at a distance x from the center of a spherically symmetric planet of uniform density ρ, total mass M, and radius R for

(a) $x \geq R$

(b) $x < R$

Solution.

(a) If $x \geq R$, then the planet can be treated as a point particle with all its mass concentrated at its center, and

$$F = G\frac{Mm}{x^2} \quad (x \geq R)$$

(b) However, if $x < R$, then only the mass within the sphere of radius x exerts a gravitational force on the particle. Since the volume of such a sphere is $(4/3)\pi x^3$, its mass is $(4/3)\pi x^3 \rho$; we'll denote this by $M_{\text{within } x}$. Since the mass of the entire planet is $(4/3)\pi R^3 \rho$, we see that

$$\frac{M_{\text{within } x}}{M} = \frac{\frac{4}{3}\pi x^3 \rho}{\frac{4}{3}\pi R^3 \rho} = \frac{x^3}{R^3} \quad \Rightarrow \quad M_{\text{within } x} = \frac{x^3}{R^3}M$$

Therefore, the force that this much mass exerts on the particle of mass m is

$$F = G\frac{\frac{x^3}{R^3}M \cdot m}{x^2} = G\frac{Mm}{R^3}x \quad (x < R)$$

In summary then,

$$F_{\text{grav}} = \begin{cases} G\dfrac{Mm}{R^3}x & \text{for } x < R \\[2mm] G\dfrac{Mm}{x^2} & \text{for } x \geq R \end{cases}$$

A: No, the astronauts still have weight; they simply weigh less because they are farther from the Earth.

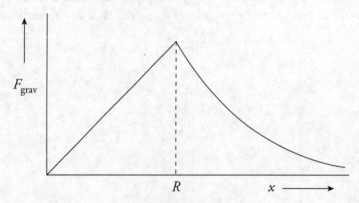

Example 2 Given that the radius of Earth is 6.37×10^6 m, determine the mass of Earth.

Solution. Consider a small object of mass m near the surface of Earth (mass M). Its weight is mg, but its weight is just the gravitational force it feels due to Earth, which is GMm/R^2. Therefore,

$$mg = G\frac{Mm}{R^2} \implies M = \frac{gR^2}{G}$$

Since we know that $g = 10$ m/s^2 and $G = 6.67 \times 10^{-11}$ N·m^2/kg^2, we can substitute to find

$$M = \frac{gR^2}{G} = \frac{(10 \text{ m/s}^2)(6.37 \times 10^6 \text{ m})^2}{6.67 \times 10^{-11} \text{ N} \cdot \text{m}^2/\text{kg}^2} = 6.0 \times 10^{24} \text{ kg}$$

Example 3 We can derive the expression GM/R^2 by equating mg and GMm/R^2 (as we did in the previous example), and this gives the magnitude of the *absolute gravitational acceleration*, a quantity that's sometimes denoted g_0. The notation g is acceleration, but with the spinning of Earth taken into account. Show that if an object is at the equator, its *measured weight* (the weight that a scale would measure), mg, is less than its *true weight*, mg_0, and compute the weight difference for a person of mass $m = 60$ kg.

Solution. Imagine looking down at Earth from above the North Pole.

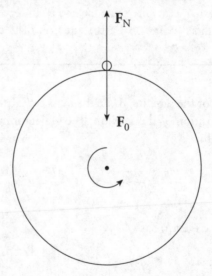

The net force toward the center of Earth is $\mathbf{F}_0 - \mathbf{F}_N$, which provides the centripetal force on the object. Therefore,

$$F_0 - F_N = \frac{mv^2}{R}$$

Since $v = 2\pi R/T$, where T is Earth's rotation period, we have

$$F_0 - F_N = \frac{m}{R}\left(\frac{2\pi R}{T}\right)^2 = \frac{4\pi^2 mR}{T^2}$$

or, since $F_0 = mg_0$ and $F_N = mg$,

$$mg_0 - mg = \frac{4\pi^2 mR}{T^2}$$

Since the quantity $4\pi^2 mR/T^2$ is positive, mg must be less than mg_0. The difference between mg_0 and mg, for a person of mass $m = 60$ kg, is only:

$$\frac{4\pi^2 mR}{T^2} = \frac{4\pi^2 (60 \text{ kg})(6.37\times10^6 \text{ m})}{\left(24 \text{ hr}\times\frac{60 \text{ min}}{\text{hr}}\times\frac{60 \text{ s}}{\text{min}}\right)^2} = 2.0 \text{ N}$$

and the difference between g_0 and g is

$$g_0 - g = \frac{mg_0 - mg}{m} = \frac{4\pi^2 R}{T^2} = \frac{4\pi^2 (6.37\times10^6 \text{ m})}{\left(24 \text{ hr}\times\frac{60 \text{ min}}{\text{hr}}\times\frac{60 \text{ s}}{\text{min}}\right)^2} = 0.034 \text{ m/s}^2$$

> **Example 4** Communications satellites are often parked in geosynchronous orbits above Earth's surface. Such satellites have orbit periods that are equal to Earth's rotation period, so they remain above the same position on Earth's surface. Determine the altitude and the speed that a satellite must have to be in a geosynchronous orbit above a fixed point on Earth's equator. (The mass of Earth is 5.98×10^{24} kg.)

Solution. Let m be the mass of the satellite, M be the mass of Earth, and R be the distance from the center of Earth to the position of the satellite. The gravitational pull of Earth provides the centripetal force on the satellite, so

$$G\frac{Mm}{R^2} = \frac{mv^2}{R} \quad \Rightarrow \quad G\frac{M}{R} = v^2$$

The orbit speed of the satellite is $2\pi R/T$, so

$$G\frac{M}{R} = \left(\frac{2\pi R}{T}\right)^2$$

which implies that

$$G\frac{M}{R} = \frac{4\pi^2 R^2}{T^2} \quad \Rightarrow \quad 4\pi^2 R^3 = GMT^2 \quad \Rightarrow \quad R = \sqrt[3]{\frac{GMT^2}{4\pi^2}}$$

Now, the key feature of a geosynchronous orbit is that its period matches Earth's rotation period, $T = 24$ hr. Substituting the numerical values of G, M, and T into this expression, we find that

$$R = \sqrt[3]{\frac{GMT^2}{4\pi^2}} = \sqrt[3]{\frac{(6.67\times10^{-11})(5.98\times10^{24})(24\cdot60\cdot60)^2}{4\pi^2}}$$
$$= 4.23\times10^7 \text{ m}$$

Therefore, if r_E is the radius of Earth, then the satellite's altitude above Earth's surface must be

$$h = R - r_E = (4.23\times10^7 \text{ m}) - (6.37\times10^6 \text{ m}) = 3.59\times10^7 \text{ m}$$

which is equal to $5.6r_E$. The speed of the satellite in this orbit is, from the first equation in our calculations,

$$v = \sqrt{G\frac{M}{R}} = \sqrt{(6.67\times10^{-11} \text{ N}\cdot\text{m}^2/\text{kg}^2)\frac{5.98\times10^{24} \text{ kg}}{4.23\times10^7 \text{ m}}} = 3{,}070 \text{ m/s}$$

regardless of the mass of the satellite (as long as $m \ll M$).

Example 5 A uniform, slender bar of mass M has length L. Determine the gravitational force it exerts on the point particle of mass m shown below:

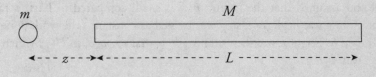

Solution. Since the bar is an extended body (and not spherically symmetric), we must calculate F using an integral. Select an arbitrary segment of length dx and mass dM in the bar, at a distance x from its left-hand end.

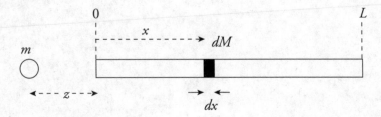

Then, since the bar is uniform, $dM = (M/L)dx$, so the gravitational force between m and dM is

$$dF = G\frac{m \cdot dM}{(z+x)^2} = G\frac{m \cdot \dfrac{M}{L}dx}{(z+x)^2} = G\frac{mM}{L(z+x)^2}dx$$

Now, by adding (that is, by integrating) all of the contributions dF, we get the total gravitational force, F:

$$\begin{aligned}
F &= \int dF \\
&= \int_{x=0}^{x=L} G\frac{mM}{L(z+x)^2}dx \\
&= G\frac{mM}{L}\int_{x=0}^{x=L}\frac{dx}{(z+x)^2} \\
&= G\frac{mM}{L}\left(\frac{-1}{z+x}\right)_{x=0}^{x=L} \\
&= G\frac{mM}{L}\left(\frac{-1}{z+L}+\frac{1}{z}\right) \\
&= G\frac{mM}{L}\frac{L}{z(z+L)} \\
&= G\frac{mM}{z(z+L)}
\end{aligned}$$

GRAVITATIONAL POTENTIAL ENERGY

When we developed the equation $U = mgh$ for the gravitational potential energy of an object of mass m at height h above the surface of Earth, we took the surface of Earth to be our $U = 0$ reference level and assumed that the height, h, was small compared to Earth's radius. In that case, the variation in g was negligible, so g was treated as constant. The work done by gravity as an object was raised to height h was then simply $-F_{grav} \times \Delta s = -mgh$, so U_{grav}, which by definition equals $-W_{by\ grav}$, was mgh.

But now we'll take variations in g into account and develop a general equation for gravitational potential energy, one that isn't restricted to small altitude changes.

Consider an object of mass m at a distance r_1 from the center of Earth (or any spherical body) moving by some means to a position r_2:

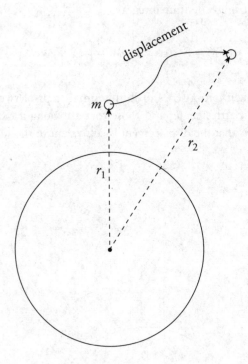

How much work did the gravitational force perform during this displacement? The answer is given by the equation:

$$W_{\text{by grav}} = GMm\left(\frac{1}{r_2} - \frac{1}{r_1}\right)$$

> If you're unfamiliar with this equation, don't worry! We'll be proving this one on the next page!

Therefore, since $\Delta U_{\text{grav}} = -W_{\text{by grav}}$, we get

$$U_2 - U_1 = -GMm\left(\frac{1}{r_2} - \frac{1}{r_1}\right)$$

Let's choose our $U = 0$ reference at infinity. That is, we decide to allow $U_2 \to 0$ as $r_2 \to \infty$. Then this equation becomes

$$U = -\frac{GMm}{r}$$

Notice that, according to this equation (and our choice of $U = 0$ when $r = \infty$), the gravitational potential energy is always negative. This just means that energy has to be added to bring an object (mass m) bound to the gravitational field of M to a point very far from M, at which $U = 0$.

PROVING THE EQUATION

Because the gravitational force is not constant over the displacement, the work done by this force must be calculated using a definite integral:

$$dW = \mathbf{F} \cdot d\mathbf{r} \quad \Rightarrow \quad W = \int_{r=r_1}^{r=r_2} \mathbf{F} \cdot d\mathbf{r}$$

Displacement can be broken into a series of infinitesimal steps of two types: Those that are at a constant distance from Earth's center and those that are along a radial line going away from Earth's center. The result is that the displacement is equivalent to the sum of two curves $C_1 + C_2$:

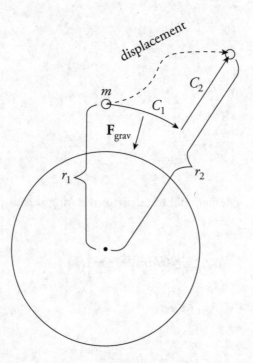

The gravitational force along C_1 does no work, because it's always perpendicular to the displacement. Along C_2, the inward gravitational force is in the opposite direction from the outward displacement, so along this vector,

$$\begin{aligned}
\mathbf{F} \cdot d\mathbf{r} &= F \cdot dr \cdot \cos 180° \\
&= -F \, dr \\
&= -\frac{GMm}{r^2} dr
\end{aligned}$$

Therefore,

$$W_{\text{by grav}} = \int_{r_1}^{r_2} \mathbf{F} \cdot d\mathbf{r}$$

$$= \int_{r_1}^{r_2} -\frac{GMm}{r^2} dr$$

$$= GMm \int_{r_1}^{r_2} -\frac{1}{r^2} dr$$

$$= GMm \left(\frac{1}{r}\right)_{r_1}^{r_2}$$

$$= GMm \left(\frac{1}{r_2} - \frac{1}{r_1}\right)$$

Example 6 With what minimum speed must an object of mass m be launched in order to escape Earth's gravitational field? (This is called **escape speed**, v_{esc}.)

Solution. When launched, the object is at the surface of Earth ($r_i = r_E$) and has an upward, initial velocity of magnitude v_i. To get it far away from Earth, we want to bring its gravitational potential energy to zero, but to find the *minimum* launch speed, we want the object's final speed to be zero by the time it gets to this distant location. So, by Conservation of Energy,

$$K_i + U_i = K_f + U_f$$

$$\frac{1}{2}mv_i^2 + \frac{-GM_E m}{r_E} = 0 + 0$$

> Note: This is the minimum speed needed to escape *Earth's* gravitational field but, in this situation, a projectile would also have to contend with the *Sun's* gravitational field.

which gives

$$\frac{1}{2}mv_i^2 = \frac{GM_E m}{r_E} \quad \Rightarrow \quad v_i = \sqrt{\frac{2GM_E}{r_E}}$$

Substituting the known numerical values for G, M_E, and r_E gives us:

$$v_i = v_{\text{esc}} = \sqrt{\frac{2GM_E}{r_E}} = \sqrt{\frac{2(6.67 \times 10^{-11})(5.98 \times 10^{24})}{6.37 \times 10^6}}$$

$$= 1.12 \times 10^4 \text{ m/s}$$

Example 7 A satellite of mass m is in a circular orbit of radius R around Earth (radius r_E, mass M).

(a) What is its total mechanical energy (where U_{grav} is considered zero as R approaches infinity)?

(b) How much work would be required to move the satellite into a new orbit, with radius $2R$?

Solution.

(a) The mechanical energy, E, is the sum of the kinetic energy, K, and potential energy, U. You can calculate the kinetic energy since you know that the centripetal force on the satellite is provided by the gravitational attraction of Earth:

$$\frac{mv^2}{R} = \frac{GMm}{R^2} \quad \Rightarrow \quad mv^2 = \frac{GMm}{R} \quad \Rightarrow \quad K = \tfrac{1}{2}mv^2 = \frac{GMm}{2R}$$

Therefore,

$$E = K + U = \frac{GMm}{2R} + \frac{-GMm}{R} = -\frac{GMm}{2R}$$

(b) From the equation $K_i + U_i + W = K_f + U_f$, we see that

$$W = (K_f + U_f) - (K_i + U_i)$$
$$= E_f - E_i$$

Therefore, the amount of work necessary to effect the change in the satellite's orbit radius from R to $2R$ is

$$W = E_{at\,2R} - E_{at\,R}$$
$$= -\frac{GMm}{2(2R)} - \left(\frac{-GMm}{2R}\right)$$
$$= \frac{GMm}{4R}$$

A Note on Elliptical Orbits

The expression for the total energy of a satellite in a circular orbit of radius R [derived in Example 7(a)] is:

$$E = -\frac{GMm}{2R} \quad \text{(circular orbit)}$$

And this also holds for a satellite traveling in an elliptical orbit if the radius R is replaced by a, (the length of the semimajor axis):

$$E = -\frac{GMm}{2a} \quad \text{(elliptical orbit)}$$

Example 8 An asteroid of mass m is in an elliptical orbit around the Sun (mass M). Assume that $m \ll M$.

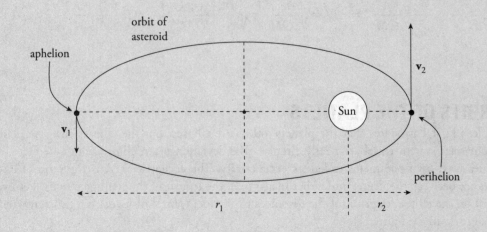

(a) What is the total energy of the asteroid?

(b) What is the ratio of v_1 (the asteroid's speed at aphelion) to v_2 (the asteroid's speed at perihelion)? Write your answer in terms of r_1 and r_2.

(c) What is the time necessary for the asteroid to make a complete orbit around the Sun?

Solution.

(a) The total energy of the asteroid is equal to $-GMm/2a$, where a is the semimajor axis. However, notice in the figure above that $2a = r_1 + r_2$. Therefore, the total energy of the asteroid is

$$E = -\frac{GMm}{r_1 + r_2}$$

(b) One way to answer this question is to invoke Conservation of Angular Momentum. When the asteroid is at aphelion, its angular momentum (with respect to the center of the Sun) is $L_1 = r_1 m v_1$. When the asteroid is at perihelion, its angular momentum is $L_2 = r_2 m v_2$. Therefore,

$$L_1 = L_2 \quad \Rightarrow \quad r_1 m v_1 = r_2 m v_2 \quad \Rightarrow \quad \frac{v_1}{v_2} = \frac{r_2}{r_1}$$

This tells us that the asteroid's speed at aphelion is less than its speed at perihelion (because $r_2/r_1 < 1$), as implied by Kepler's Second Law (a line drawn from the Sun to the asteroid must sweep out equal areas in equal time intervals). The closer the asteroid is to the Sun, the faster it has to travel to make this true.

(c) As you know, the time necessary for the asteroid to make a complete orbit around the Sun is the orbit period, T. Using Kepler's Third Law, with $a = \frac{1}{2}(r_1 + r_2)$, we find

$$\frac{T^2}{a^3} = \frac{4\pi^2}{GM} \quad \Rightarrow \quad T = \sqrt{\frac{4\pi^2 a^3}{GM}} = \sqrt{\frac{4\pi^2 \left[\frac{1}{2}(r_1 + r_2)\right]^3}{GM}} = \sqrt{\frac{\pi^2 (r_1 + r_2)^3}{2GM}}$$

ORBITS OF THE PLANETS

Kepler's First Law states that the planets' orbits are ellipses, but the ellipses that the planets in our solar system travel are nearly circular. The deviation of an ellipse from a perfect circle is measured by a parameter called its **eccentricity**. The eccentricity, e, is the ratio of c (the distance between the center and either focus) to a, the length of the semimajor axis. For every point on the ellipse, the sum of the distances to the foci (plural of focus) is a constant (and is equal to $2a$ in the figure below).

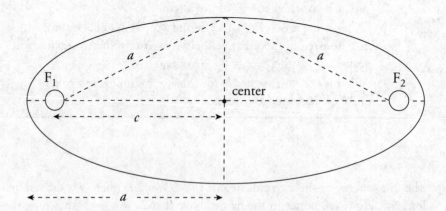

Kepler's First Law also states that one of the foci of a planet's elliptical orbit is located at the position of the Sun. Actually, the focus is at the center of mass of the Sun-planet system, because when one body orbits another, both bodies orbit around their center of mass, a point called the **barycenter**.

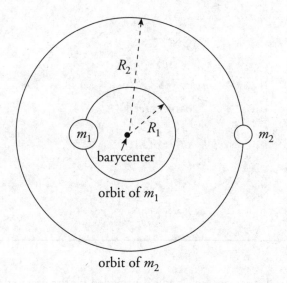

For most of the planets, which are much less massive than the Sun, this correction to Kepler's First Law has little significance, because the center of mass of the Sun and the planet system is close enough to the Sun's center. For example, let's figure out the center of mass of the Sun-Earth system. The mass of Earth is $m = 5.98 \times 10^{24}$ kg, the mass of the Sun is $M = 1.99 \times 10^{30}$ kg, and the Sun-Earth distance averages $R = 1.496 \times 10^{11}$ m. Therefore, letting $x = 0$ be at the Sun's center, we have

$$x_{cm} = \frac{Mx_{Sun} + mx_{Earth}}{M+m} = \frac{M(0) + (5.98 \times 10^{24}\text{ kg})(1.496 \times 10^{11}\text{ m})}{(1.99 \times 10^{30}\text{ kg}) + (5.98 \times 10^{24}\text{ kg})} = 450\text{ km}$$

So the center of mass of the Sun-Earth system is only 450 km from the center of the Sun, a distance of less than 0.1% of the Sun's radius.

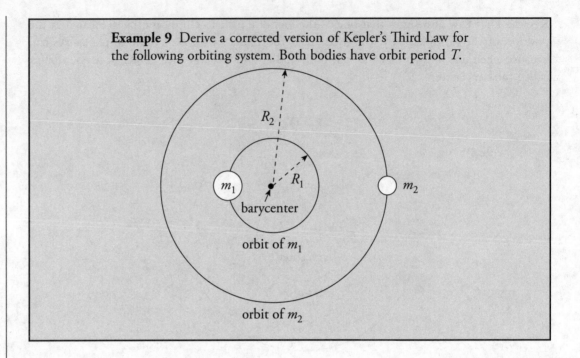

Example 9 Derive a corrected version of Kepler's Third Law for the following orbiting system. Both bodies have orbit period T.

Solution. The centripetal force on each body is provided by the gravitational pull of the other body, so

$$\frac{m_1 v_1^2}{R_1} = G\frac{m_1 m_2}{(R_1 + R_2)^2} \qquad \text{and} \qquad \frac{m_2 v_2^2}{R_2} = G\frac{m_1 m_2}{(R_1 + R_2)^2}$$

which imply

$$\frac{v_1^2}{R_1} = G\frac{m_2}{(R_1 + R_2)^2} \qquad \text{and} \qquad \frac{v_2^2}{R_2} = G\frac{m_1}{(R_1 + R_2)^2}$$

But, since both bodies have the same orbit period, T, we have

$$v_1 = \frac{2\pi R_1}{T} \qquad \text{and} \qquad v_2 = \frac{2\pi R_2}{T}$$

Substituting these results into the preceding pair of equations gives us

$$\frac{(2\pi R_1/T)^2}{R_1} = G\frac{m_2}{(R_1 + R_2)^2} \qquad \text{and} \qquad \frac{(2\pi R_2/T)^2}{R_2} = G\frac{m_1}{(R_1 + R_2)^2}$$

which simplify to

$$\frac{4\pi^2 R_1}{T^2} = G\frac{m_2}{(R_1 + R_2)^2} \qquad \text{and} \qquad \frac{4\pi^2 R_2}{T^2} = G\frac{m_1}{(R_1 + R_2)^2}$$

Adding this last pair of equations gives us the desired result:

$$\frac{4\pi^2(R_1+R_2)}{T^2} = G\frac{m_1+m_2}{(R_1+R_2)^2}$$

$$\frac{4\pi^2}{T^2} = \frac{G(m_1+m_2)}{(R_1+R_2)^3}$$

$$\frac{T^2}{(R_1+R_2)^3} = \frac{4\pi^2}{G(m_1+m_2)}$$

Note that this final equation is a general version of Kepler's Third Law for a circular orbit derived earlier, $T^2/R^3 = 4\pi^2/GM$, where it was assumed that the planet orbited at a distance R from the center of the Sun.

Example 10 An artificial satellite of mass m travels at a constant speed in a circular orbit of radius R around Earth (mass M). What is the speed of the satellite?

Solution. The centripetal force on the satellite is provided by Earth's gravitational pull. Therefore,

$$\frac{mv^2}{R} = G\frac{Mm}{R^2}$$

Solving this equation for v yields

$$v = \sqrt{G\frac{M}{R}}$$

Notice that the satellite's speed doesn't depend on its mass; even if it were a baseball, if its orbit radius were R, then its orbit speed would still be $\sqrt{GM/R}$.

Chapter 11 Drill

The answers and explanations can be found in Chapter 17.

Section I: Multiple Choice

1. An object has an altitude of 2 times the Earth's radius, r_E, and experiences some force of gravity, $F_{g,0}$. If the object's altitude is doubled, then the new force of gravity will be

   (A) $4F_{g,0}$

   (B) $\left(\dfrac{25}{9}\right)F_{g,0}$

   (C) $2F_{g,0}$

   (D) $\left(\dfrac{1}{2}\right)F_{g,0}$

   (E) $\left(\dfrac{9}{25}\right)F_{g,0}$

2. At the surface of Earth, an object of mass m has weight w. If this object is transported to a height above the surface that's twice the radius of Earth, then, at the new location,

   (A) its mass is $m/2$ and its weight is $w/2$
   (B) its mass is m and its weight is $w/2$
   (C) its mass is $m/2$ and its weight is $w/4$
   (D) its mass is m and its weight is $w/4$
   (E) its mass is m and its weight is $w/9$

3. A moon of mass m orbits a planet of mass $100m$. Let the strength of the gravitational force exerted by the planet on the moon be denoted by F_1, and let the strength of the gravitational force exerted by the moon on the planet be F_2. Which of the following is true?

   (A) $F_1 = 100F_2$
   (B) $F_1 = 10F_2$
   (C) $F_1 = F_2$
   (D) $F_2 = 10F_1$
   (E) $F_2 = 100F_1$

4. Humans cannot survive for long periods under gravity more than 4 times what we experience on Earth. If a planet were discovered with the same mass as Earth, what is the smallest radius it could have (in terms of Earth's radius, r_E) without being dangerous to humans?

   (A) $\dfrac{1}{8}r_E$

   (B) $\dfrac{1}{4}r_E$

   (C) $\dfrac{1}{2}r_E$

   (D) $2r_E$

   (E) $4r_E$

5. In terms of the Earth's radius R_E, how far above the surface of the Earth would an object have to be to experience a gravitational force equivalent to what it would feel if it were buried halfway to the center of the Earth?

   (A) $(\sqrt{2}+1)R_E$

   (B) $\sqrt{2}\,R_E$

   (C) R_E

   (D) $R_E/2$

   (E) $(\sqrt{2}-1)R_E$

6. A moon of Jupiter has a nearly circular orbit of radius R and an orbit period of T. Which of the following expressions gives the mass of Jupiter?

   (A) $2\pi R/T$
   (B) $4\pi^2 R/T^2$
   (C) $2\pi R^3/(GT^2)$
   (D) $4\pi R^2/(GT^2)$
   (E) $4\pi^2 R^3/(GT^2)$

7. Two large bodies, Body A of mass m and Body B of mass $4m$, are separated by a distance R. At what distance from Body A, along the line joining the bodies, would the gravitational force on an object be equal to zero? (Ignore the presence of any other bodies.)

   (A) $R/16$
   (B) $R/8$
   (C) $R/5$
   (D) $R/4$
   (E) $R/3$

8. The mean distance from Saturn to the Sun is 9 times greater than the mean distance from Earth to the Sun. Approximately how long is a Saturn year?

   (A) 18 Earth years
   (B) 27 Earth years
   (C) 81 Earth years
   (D) 243 Earth years
   (E) 729 Earth years

9. A planet of mass M and radius r requires a minimum speed of v to escape its gravitational field. If the planet's radius were doubled (but the density remained constant), the new escape speed would be

   (A) $\dfrac{1}{2}v$

   (B) v

   (C) $\sqrt{2}\,v$

   (D) $2v$

   (E) $2\sqrt{2}\,v$

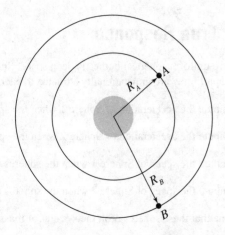

10. Two satellites, A and B, orbit a planet in circular orbits having radii R_A and R_B, respectively, as shown above. If $R_B = 3R_A$, the velocities v_A and v_B of the two satellites are related by which of the following?

    (A) $v_B = v_A$

    (B) $v_B = 3v_A$

    (C) $v_B = 9v_A$

    (D) $v_B = v_A\sqrt{3}$

    (E) $v_B = \dfrac{v_A}{\sqrt{3}}$

Section II: Free Response

1. Consider two uniform spherical bodies in deep space. Sphere 1 has mass m_1 and Sphere 2 has mass m_2. Starting from rest from a distance R apart, they are gravitationally attracted to each other.

   (a) Compute the acceleration of Sphere 1 when the spheres are a distance $R/2$ apart.

   (b) Compute the acceleration of Sphere 2 when the spheres are a distance $R/2$ apart.

   (c) Compute the speed of Sphere 1 when the spheres are a distance $R/2$ apart.

   (d) Compute the speed of Sphere 2 when the spheres are a distance $R/2$ apart.

   Now assume that these spheres orbit their center of mass with the same orbit period, T.

   (e) Determine the radii of their orbits. Write your answer in terms of m_1, m_2, T, and fundamental constants.

2. A satellite of mass m is in the elliptical orbit shown below around Earth (radius r_E, mass M). Assume that $m \ll M$.

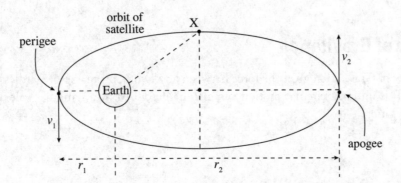

(a) Determine v_1, the speed of the satellite at perigee (the point of the orbit closest to Earth). Write your answer in terms of r_1, r_2, M, and G.

(b) Determine v_2, the speed of the satellite at apogee (the point of the orbit farthest from Earth). Write your answer in terms of r_1, r_2, M, and G.

(c) Express the ratio v_1/v_2 in simplest terms.

(d) What is the satellite's angular momentum (with respect to Earth's center) when it's at apogee?

(e) Determine the speed of the satellite when it's at the point marked X in the figure.

(f) Determine the period of the satellite's orbit. Write your answer in terms of r_1, r_2, M, and fundamental constants.

Summary

Newton's Law of Gravitation

o Newton's Law of Gravitation gives the force between any two point masses, regardless of their mass or location. This uniform circular motion can also be described by an angular velocity and a centripetal acceleration.

$$F_g = \frac{GM_1 M_2}{r^2}$$

$$a_g = \frac{GM_1}{r^2}$$

Circular Orbits

$$F_g = F_c$$

$$\frac{GM_1 M_2}{r^2} = \frac{M_2 v^2}{r}$$

$$\frac{GM_1}{r} = v^2$$

$$v = \frac{2\pi r}{T}$$

General Orbits

o The gravitational potential energy is represented by the following formula:

$$U_g = -\frac{GM_1 M_2}{r}$$

o   Both *mechanical energy* and *angular momentum* are conserved for orbits.

$$E_{total} = 0$$

o   Systems are bound when the total mechanical energy is less than zero. This means that the gravitational potential energy is greater than the kinetic energy of the system.

$$\frac{1}{2} M_1 v_{esc}^{\ 2} - \frac{GM_1 M_2}{d} = 0$$

o   For an object to escape, its kinetic energy must be greater than or equal to its gravitational potential energy. So

$$v_{esc} = \sqrt{\frac{2GM_2}{d}} \text{ for } M_1 \text{ to escape the pull of } M_2.$$

Gravity of Spheres and Shells

o   The gravity due to a spherical shell of mass M and radius R where the mass is at a distance r away from the center of the shell is represented by the equations below.

   Outside the shell is $F_g = \dfrac{GMm}{r^2}$ and the gravity inside the shell is $F_g = 0$.

o   The gravity due to a uniform, solid sphere where a small mass m is at a distance r away from the center of the sphere is represented by the equations below.

   Outside the sphere is $F_g = \dfrac{GMm}{r^2}$ and the gravity inside the shell is $F_g = \dfrac{GMm}{R^3} r$.

o   The gravity of a spherically symmetric sphere of mass M and radius R (that is, $\rho(r)$) is represented by the equation

$$\vec{F}_g = -Gm \int \frac{dM}{r^2} \hat{r} \text{ where } dM = \rho(r) dV$$

Chapter 12
Electrostatics

INTRODUCTION

The basic components of atoms are protons, neutrons, and electrons. Protons and neutrons form the nucleus (and are referred to collectively as *nucleons*), while the electrons occupy regions of space outside of the nucleus. Most of an atom consists of empty space. In fact, if a nucleus were the size of the period at the end of this sentence, then the nearest electrons would be 5 meters away. So what holds such an apparently tenuous structure together? One of the most powerful forces in nature: the *electromagnetic force*. Protons and electrons are characterized by their **electric charge** that gives them an attractive force. Electric charge can be either positive, negative, or neutral. A positive particle always attracts a negative particle, and particles of the same charge always repel each other. Protons are positively charged, and electrons are negatively charged.

> An even stronger force than the electromagnetic force is the strong nuclear force. This force keeps the protons and neutrons bound together in the nucleus.

Protons and electrons are intrinsically charged, but bulk matter is not. Since neutral atoms contain an equal number of protons and electrons, their overall electric charge is 0, because the negative charges cancel out the positive charges. Therefore, in order for matter to be **charged**, an imbalance between the numbers of protons and electrons must exist. This can be accomplished by either the removal or addition of electrons (that is, by the **ionization** of some of the object's atoms). If you remove electrons, then the object becomes positively charged, while if you add electrons, then it becomes negatively charged. Furthermore, charge is **conserved**. For example, if you rub a glass rod with a piece of silk, then the silk will acquire a negative charge and the glass will be left with an *equal* positive charge. *Net charge cannot be created or destroyed.* (*Charge* can be created or destroyed—it happens all the time—but *net* charge cannot.)

The magnitude of charge on an electron (and therefore on a proton) is denoted by e. This stands for **elementary charge** because it's the basic unit of electric charge. The charge of an ionized atom must be a whole number times e because charge can be added or subtracted only in units of e. For this reason, we say that charge is **quantized**. To remind us of the quantized nature of electric charge, the charge of a particle (or object) is denoted by the letter q. In the SI system of units, charge is expressed in **coulombs** (abbreviated **C**). One coulomb is a tremendous amount of charge; the value of e is about 1.6×10^{-19} C.

COULOMB'S LAW

The electric force between two charged particles obeys a law that is very similar to that describing the gravitational force between two masses: they are both inverse-square laws. The **electric force** between two particles with charges of q_1 and q_2, separated by a distance r, is given by the equation

$$F_E = k \frac{q_1 q_2}{r^2}$$

> Coulomb's Law can instead be used to find the magnitude of $\mathbf{F}_E$ by ignoring the signs of q_1 and q_2. The correct sign for $\mathbf{F}_E$, and hence the direction of $\mathbf{F}_E$, can be added on by assessing if the charges get forced in a positive or negative direction via attraction (opposite) or repulsion (similar).

This is **Coulomb's law**. We interpret a negative F_E as an attraction between the charges (q_1 and q_2 have opposite signs) and a positive F_E as a repulsion. The value of the proportionality constant, k, called **Coulomb's constant**, is approximately 9×10^9 N·m²/C². For reasons that will become clear later in this chapter, k is usually written in terms of a fundamental constant known as the **permittivity of free space**, denoted ε_0, whose numerical value is approximately 8.85×10^{-12} $C^2/N \cdot m^2$. The equation that gives k in terms of ε_0 is:

$$k = \frac{1}{4\pi\varepsilon_0}$$

Coulomb's law for the force between two point charges is then written as

$$F_E = \frac{1}{4\pi\varepsilon_0} \frac{q_1 q_2}{r^2}$$

Let's compare the electric force to the gravitational force between two protons that are a distance r apart.

$$F_E = k\frac{ee}{r^2} \ \& \ F_g = G\frac{m_p m_p}{r^2}$$

$$\frac{F_E}{F_g} = \frac{ke^2}{Gm_p{}^2} = 1.29 \times 10^{36}$$

Example 1 Consider two small spheres, one carrying a charge of +1.5 nC and the other a charge of –2.0 nC, separated by a distance of 1.5 cm. Find the electric force between them. ("n" is the abbreviation for "nano," which means 10^{-9}.)

Solution. The electric force between the spheres is given by Coulomb's law:

$$F_E = \frac{1}{4\pi\varepsilon_0}\frac{q_1 q_2}{r^2} = (9\times10^9 \ \text{N}\cdot\text{m}^2/\text{C}^2)\frac{(1.5\times10^{-9} \ \text{C})(-2.0\times10^{-9} \ \text{C})}{(1.5\times10^{-2} \ \text{m})^2} = -1.8 \times 10^{-6} \ \text{N}$$

The fact that F_E is negative means that the force is one of *attraction*, which we naturally expect, since one charge is positive and the other is negative. The force between the spheres is along the line that joins the charges, as we've illustrated below. The two forces shown form an action/reaction pair.

Superposition

Consider three point charges: q_1, q_2, and q_3. The total electric force acting on, say, q_2 is simply the sum of $\mathbf{F}_{\text{1-on-2}}$ (the electric force on q_2 due to q_1) and $\mathbf{F}_{\text{3-on-2}}$ (the electric force on q_2 due to q_3):

$$\mathbf{F}_{\text{on 2}} = \mathbf{F}_{\text{1-on-2}} + \mathbf{F}_{\text{3-on-2}}$$

The fact that electric forces can be added in this way is known as **superposition**.

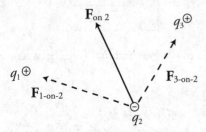

Example 2 Consider four equal, positive point charges that are situated at the vertices of a square. Find the net electric force on a negative point charge placed at the square's center.

Solution. Refer to the following diagram. The attractive forces due to the two charges on each diagonal cancel out: $\mathbf{F}_1 + \mathbf{F}_3 = \mathbf{0}$, and $\mathbf{F}_2 + \mathbf{F}_4 = \mathbf{0}$, because the distances between the negative charge and the positive charges are all the same and the positive charges are all equivalent. Therefore, by symmetry, the net force on the center charge is zero.

Taking a step back from your problem to assess symmetry can save you time and effort on test day.

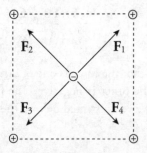

Example 3 If the two positive charges on the bottom side of the square in the previous example were removed, what would be the net electric force on the negative charge? Assume that each side of the square is 4.0 cm, each positive charge is 1.5 μC, and the negative charge is –6.2 nC. (μ is the symbol for "micro-," which equals 10^{-6}.)

Solution. If we break down $\mathbf{F}_1$ and $\mathbf{F}_2$ into horizontal and vertical components, then by symmetry, the two horizontal components will cancel each other out, and the two vertical components will add:

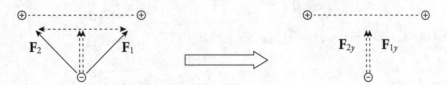

Since the diagram on the left shows the components of $\mathbf{F}_1$ and $\mathbf{F}_2$ making a right triangle with legs each of length 2 cm, it must be that $F_{1y} = F_1 \sin 45°$ and $F_{2y} = F_2 \sin 45°$. Also, the magnitude of $\mathbf{F}_1$ equals that of $\mathbf{F}_2$. So the net electric force on the negative charge is $F_{1y} + F_{2y} = 2F \sin 45°$, where F is the strength of the force between the negative charge and each of the positive charges. If s is the length of each side of the square, then the distance r between each positive charge and the negative charge is $r = \frac{1}{2}s\sqrt{2}$ and

$$F_E = 2F \sin 45° = 2 \frac{1}{4\pi\varepsilon_0} \frac{qQ}{r^2} \sin 45°$$

$$= 2(9\times10^9 \text{ N}\cdot\text{m}^2/\text{C}^2) \frac{(1.5\times10^{-6}\text{ C})(6.2\times10^{-9}\text{ C})}{(\frac{1}{2}\cdot 4.0\times10^{-2}\cdot\sqrt{2}\text{ m})^2} \sin 45°$$

$$= 0.15 \text{ N}$$

The direction of the net force is straight upward, toward the center of the line that joins the two positive charges.

> **Example 4** Two pith balls of mass m are each given a charge of $+q$. They are hung side-by-side from two threads each of length L and move apart as a result of their electrical repulsion. Find the equilibrium separation distance x in terms of m, q, and L. (Use the fact that if θ is small, then $\tan\theta \approx \sin\theta$.)

Solution. Three forces act on each ball: weight, tension, and electrical repulsion:

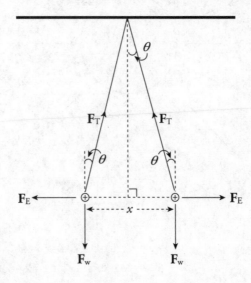

When the balls are in equilibrium, the net force each feels is zero. Therefore, the vertical component of $\mathbf{F}_T$ must cancel out $\mathbf{F}_w$ and the horizontal component of $\mathbf{F}_T$ must cancel out $\mathbf{F}_E$:

$$F_T \cos \theta = F_w \qquad \text{and} \qquad F_T \sin \theta = F_E$$

Dividing the second equation by the first, we get $\tan \theta = F_E/F_w$. Therefore,

$$\tan \theta = \frac{k\dfrac{q^2}{x^2}}{mg} = \frac{kq^2}{mgx^2}$$

Now, to approximate: If θ is small, the $\tan \theta \approx \sin \theta$ and, from the diagram, $\sin \theta = \frac{1}{2}x/L$. Therefore, the equation above becomes

$$\frac{\frac{1}{2}x}{L} = \frac{kq^2}{mgx^2} \quad \Rightarrow \quad \frac{1}{2}mgx^3 = kq^2L \quad \Rightarrow \quad x = \sqrt[3]{\frac{2kq^2L}{mg}}$$

THE ELECTRIC FIELD

The presence of a massive body such as Earth causes objects to experience a gravitational force directed toward Earth's center. For objects located outside Earth, this force varies inversely with the square of the distance and directly with the mass of the gravitational source. A vector diagram of the gravitational field surrounding Earth looks like this:

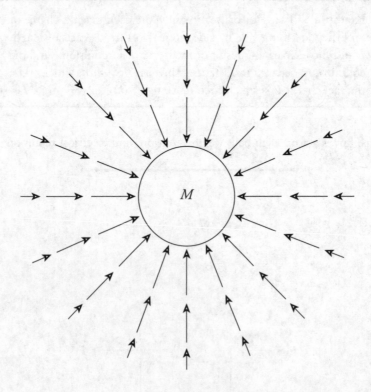

We can think of the space surrounding Earth as permeated by a **gravitational field** that's created by Earth. Any mass that's placed in this field then experiences a gravitational force due to this field.

The same process is used to describe the electric force. Rather than having two charges reach out across empty space to each other to produce a force, we can instead interpret the interaction in the following way: The presence of a charge creates an **electric field** in the space that surrounds it. Another charge placed in the field created by the first will experience a force due to the field.

Consider a point charge Q in a fixed position and assume that it's positive. Now imagine moving a tiny positive test charge q around to various locations near Q. At each location, measure the force that the test charge experiences, and call it **F**. Divide this force by the test charge q; the resulting vector is the **electric field vector**, **E**, at that location:

> By virtue of existing, a mass alters the space around itself to cause other nearby masses to "fall" in toward the original mass. Similarly, a charge alters the space around itself to cause other nearby charges to "fall" in toward or away from the original charge. These alterations of space are represented by fields.

$$\mathbf{E} = \frac{\mathbf{F}_e}{q}$$

The reason for dividing by the test charge is simple. If we were to use a different test charge with, say, twice the charge of the first one, then each of the forces **F** we'd measure would be twice as much as before. But when we divided this new, stronger force by the new, greater test charge, the factors of 2 would cancel, leaving the same ratio as before. So this ratio tells us the intrinsic strength of the field due to the source charge, independent of whatever test charge we may use to measure it.

What would the electric field of a positive charge Q look like? Since the test charge used to measure the field is positive, every electric field vector would point radially away from the source charge. *If the source charge is positive, the electric field vectors point away from it; if the source charge is negative, then the field vectors point toward it.* And, since the force decreases as we get farther away from the charge (as $1/r^2$), so does the electric field. This is why the electric field vectors farther from the source charge are shorter than those that are closer.

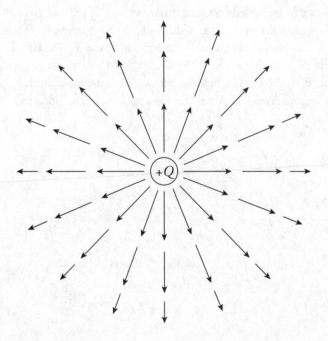

Since the force on the test charge q has a strength of $qQ/4\pi\varepsilon_0 r^2$, when we divide this by q, we get the expression for the strength of the electric field created by a point-charge source of magnitude Q:

$$E = \frac{1}{4\pi\varepsilon_0}\frac{Q}{r^2}$$

To make it easier to sketch an electric field, lines are drawn through the vectors such that the electric field vector is tangent to the line everywhere it's drawn.

The strength of the field can be figured out by looking at the density of the field lines. Where the field lines are denser, the field is stronger.

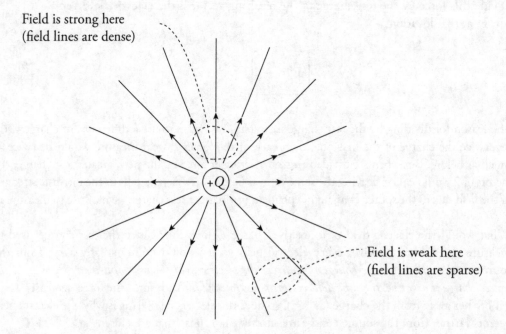

Field is strong here
(field lines are dense)

Field is weak here
(field lines are sparse)

Electric field vectors can be added like any other vectors. If we had two source charges, their fields would overlap and effectively add; a third charge wandering by would feel the effect of the combined field. This means that at each position in space, $\mathbf{E}_{total} = \mathbf{E}_1 + \mathbf{E}_2$. (This is superposition again.) In the diagram below, $\mathbf{E}_1$ is the electric field vector at a particular location due to the charge $+Q$, and $\mathbf{E}_2$ is the electric field vector at that same location due to the other charge, $-Q$. Adding these vectors gives the overall field vector $\mathbf{E}_{total}$ at that location.

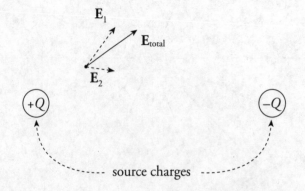

If this is done at enough locations, the electric field lines can be sketched.

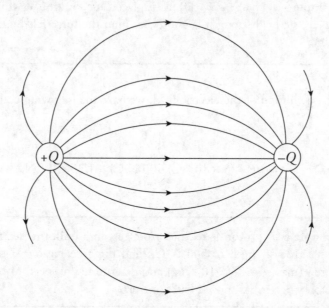

Note that, like electric field vectors, electric field lines always point away from positive source charges and toward negative ones. Two equal but opposite charges, like the ones shown in the diagram above, form a pair called an **electric dipole**.

If a positive charge $+q$ were placed in the electric field above, it would experience a force that is tangent to, and in the same direction as, the field line passing through $+q$'s location. After all, electric fields are sketched from the point of view of what a positive test charge would do. On the other hand, if a negative charge $-q$ were placed in the electric field, it would experience a force that is tangent to, but in the direction opposite from, the field line passing through $-q$'s location.

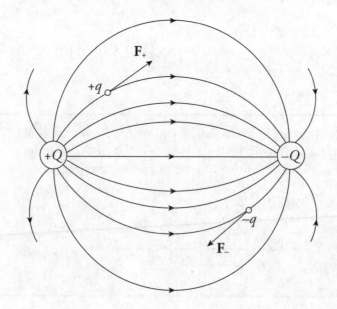

Notice that electric field lines never cross.

Example 5 A charge q = +3.0 nC is placed at a location at which the electric field strength is 400 N/C. Find the force felt by the charge q.

Solution. From the definition of the electric field, we have the following equation:

$$\mathbf{F}_{on\ q} = q\mathbf{E}$$

Therefore, in this case, $F_{on\ q} = qE = (3 \times 10^{-9}\ \text{C})(400\ \text{N/C}) = 1.2 \times 10^{-6}\ \text{N}$.

Example 6 A proton is released from rest in a uniform electric field with a strength of 500 N/C. Find the acceleration of the proton (mass = 1.67×10^{-27} kg) and determine the speed of the proton when it has moved half a meter.

Solution. From the definition of the electric field, we can calculate the force and then use Newton's Second Law to determine the acceleration.

$$F_{on\ q} = qE = ma$$
$$(1.6 \times 10^{-19}\ \text{C})(500\ \text{N/C}) = (1.67 \times 10^{-27}\ \text{kg})a$$
$$a = 4.79 \times 10^{10}\ \text{m/s}^2$$

We can use the constant acceleration equations to solve for the speed of the proton after traveling 1 m.

$$v_2^2 = v_1^2 + 2a\Delta x \quad \Rightarrow \quad v_2^2 = 2(4.79 \times 10^{10})(0.5)$$
$$v_2 = 2.19 \times 10^5\ \text{m/s}$$

Example 7 A dipole is formed by two point charges, each of magnitude 4.0 nC, separated by a distance of 6.0 cm. What is the strength of the electric field at the point midway between them?

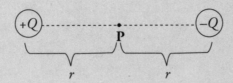

Solution. Let the two source charges be denoted $+Q$ and $-Q$. At point P, the electric field vector due to $+Q$ would point directly away from $+Q$, and the electric field vector due to $-Q$ would point directly toward $-Q$. Therefore, these two vectors point in the same direction (from $+Q$ to $-Q$), so their magnitudes would add.

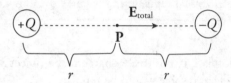

Using the equation for the electric field strength due to a single point charge, we find that

$$E_{total} = \frac{1}{4\pi\varepsilon_0}\frac{Q}{r^2} + \frac{1}{4\pi\varepsilon_0}\frac{Q}{r^2} = 2\frac{1}{4\pi\varepsilon_0}\frac{Q}{r^2}$$

$$= 2(9\times10^9 \ \text{N}\cdot\text{m}^2/\text{C}^2)\frac{4.0\times10^{-9} \ \text{C}}{[\frac{1}{2}(6.0\times10^{-2} \ \text{m})]^2}$$

$$= 8.0\times10^4 \ \text{N/C}$$

Example 8 If a charge $q = -5.0$ pC were placed at the midway point described in the previous example, describe the force it would feel. ("p" is the abbreviation for "pico-," which means 10^{-12}.)

Solution. Since the field **E** at this location is known, the force felt by q is easy to calculate:

$$\mathbf{F}_{on \ q} = q\mathbf{E} = (-5.0 \times 10^{-12} \ \text{C})(8.0 \times 10^4 \ \text{N/C to the right}) = 4.0 \times 10^{-7} \ \text{N to the } \textit{left}$$

Example 9 What can you say about the electric force that a charge would feel if it were placed at a location at which the electric field was zero?

Solution. Remember that $\mathbf{F}_{on \ q} = q\mathbf{E}$. So if $\mathbf{E} = 0$, then $\mathbf{F}_{on \ q} = 0$. Zero field means zero force.

Example 10 Positive charge is distributed uniformly over a large, horizontal plate, which then acts as the source of a vertical electric field. An object of mass 5 g is placed at a distance of 2 cm above the plate. If the strength of the electric field at this location is 10^6 N/C, how much charge would the object need to have in order for the electrical repulsion to balance the gravitational pull?

Solution. Clearly, since the plate is positively charged, the object would also have to carry a positive charge so that the electric force would be repulsive.

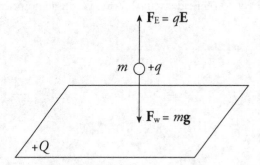

Let q be the charge on the object. Then, in order for F_E to balance mg, we must have

$$qE = mg \quad \Rightarrow \quad q = \frac{mg}{E} = \frac{(5 \times 10^{-3} \text{ kg})(10 \text{ N/kg})}{10^6 \text{ N/C}} = 5 \times 10^{-8} \text{ C} = 50 \text{ nC}$$

Example 11 A thin, nonconducting rod that carries a uniform linear charge density λ is bent into a semicircle of radius R. Find the electric field at the center of curvature of the semicircle.

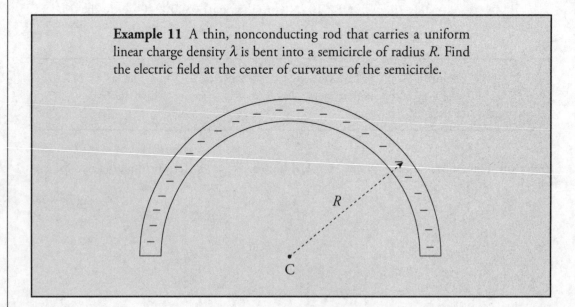

Solution. Refer to the figure below. Consider a small section of the rod subtended by an angle $d\theta$. The length of this section is (by the definition of radian measure) $R\,d\theta$, so the charge it carries is equal to $\lambda R\,d\theta$. The field it creates at C is $(1/4\pi\varepsilon_0)(\lambda R\,d\theta/R^2)$, pointing directly toward this section.

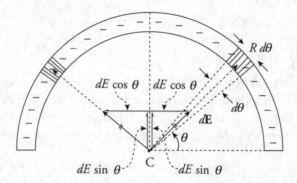

However, another section of the rod creates a field of the same strength and, as the figure shows, cancels out the horizontal component of the field created by the first section. Only the two vertical components, $dE\sin\theta$, remain. The total electric field at C, due to all sections of the rod, is found by adding up the individual contributions:

$$E = \int 2\,dE\sin\theta$$
$$= \int_{\theta=0}^{\theta=\pi/2} 2\left(\frac{1}{4\pi\varepsilon_0}\cdot\frac{\lambda R\,d\theta}{R^2}\right)\sin\theta$$
$$= \frac{\lambda}{2\pi\varepsilon_0 R}\int_0^{\pi/2}\sin\theta\,d\theta$$
$$= \frac{\lambda}{2\pi\varepsilon_0 R}\left[-\cos\theta\right]_0^{\pi/2}$$
$$= \frac{\lambda}{2\pi\varepsilon_0 R}$$

Since the rod is negatively charged, the field **E,** at C, points upward, toward the rod.

Electric Field Drill

Answers and explanations can be found in Chapter 17.

The electric field defined in the figure above has a magnitude of 5.0 N/C. For each particle described below, when the particle is placed in the electric field, find the force and direction exerted on the particle by the electric field, along with the acceleration of the particle.

1. A proton at rest
2. A positron at rest
3. An electron at rest
4. A proton traveling into the page ($\otimes$) at 3 m/s
5. An electron moving right ($\rightarrow$) at 3 m/s

6. An electron moving with the electric field (↑) at 3 m/s
7. A positron moving against the electric field (↓) at 3 m/s
8. An electron moving down and to the right, at an angle of 30 degrees to the horizontal, at 3 m/s

GAUSS'S LAW

For individual point charges, we can compute the electric field they produce by using the equation $E = Q/4\pi\varepsilon_0 r^2$ for each source charge, Q, then add up the resulting electric field vectors to obtain the net field. However, if the source charge is spread over a wire or a plate or throughout a cylinder or a sphere, then we need another method for determining the resulting electric field.

First, let's discuss the concept of **flux**. Consider the flow of a fluid through a pipe and imagine a small rectangular surface perpendicular to the flow. Multiplying the area of this rectangle (A) by the speed of the flow (v) gives the volume of fluid flowing through the surface per unit time. This is called the **fluid-volume flux**, symbolized Φ (the uppercase Greek letter *phi*):

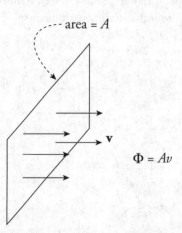

area = A

$\mathbf{v}$

$\Phi = Av$

In general, the fluid-volume flux through the surface is given by the equation $\Phi = Av \cos\theta$, where θ is the angle between the velocity $\mathbf{v}$ of the flow and the normal to the surface. If this rectangular surface were tilted at an angle θ to the flow, then less fluid would flow through it per unit time. In fact, if θ were 90°, then no fluid would flow through the surface. (You might also recognize this as the dot product, $\Phi = \mathbf{A} \cdot \mathbf{v}$, where $\mathbf{A}$ is the vector whose magnitude is A and whose direction is normal to the surface.)

All the fluid that used to cross the surface down here (below the dotted line) now no longer passes through the tilted surface. So, the flux through the surface has been decreased. It's now the same as the flux through the rectangular portion perpendicular to $\mathbf{v}$ that lies above the dotted line. This area is $A \cos\theta$.

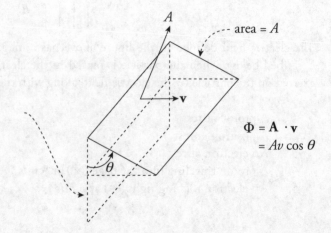

A

area = A

$\mathbf{v}$

θ

$\Phi = \mathbf{A} \cdot \mathbf{v}$
$\quad = Av \cos\theta$

Let's now apply this same idea to an electric field. Consider a source charge $+Q$ and imagine a tiny patch (area = A) of a spherical surface of radius r. The following diagram shows a cross-section of the situation.

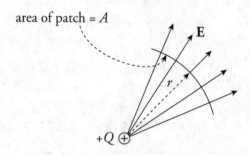

area of patch = A

Then we can think of the electric field vectors flowing through the area A and, since $\mathbf{E}$ is everywhere perpendicular to this patch area A, the **electric flux** (Φ_E) would simply be the product: $\Phi_E = EA$. Although the electric field isn't really flowing like the fluid in the previous discussion, we can still use the same terminology. It might be more appropriate to think of the electric flux as a measure of the amount of electric field passing through the surface.

In general, the electric flux through a patch of area dA is given by the equation:

$$d\Phi_E = \mathbf{E} \cdot d\mathbf{A}$$

The total electric flux is then found by adding (that is, integrating) all of the contributions:

$$\Phi_E = \int d\Phi_E = \int \mathbf{E} \cdot d\mathbf{A}$$

There's More Where That Came From
For more review with Gauss's Law, flip over to page 480.

Let's now imagine a *closed* surface—a sphere, for simplicity—surrounding the charge $+Q$. Place this closed spherical surface at a distance r from $+Q$.

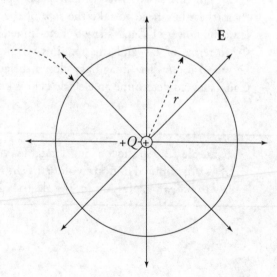

imaginary spherical surface surrounding the charge $+Q$

The total flux outward through this closed surface is

$$\Phi_E = EA = E(4\pi r^2)$$

where $4\pi r^2$ is the surface area of a sphere.

We know what the electric field is at the position of this imaginary sphere; it's $E = Q/4\pi\varepsilon_0 r^2$. Therefore, the previous equation becomes

$$\Phi_E = \frac{1}{4\pi\varepsilon_0}\frac{Q}{r^2} \cdot 4\pi r^2 = \frac{Q}{\varepsilon_0}$$

It can be shown that the previous equation holds true for *any closed surface*; the total electric flux through the surface is equal to $1/\varepsilon_0$ times the charge enclosed:

$$\Phi_E = \frac{Q_{enclosed}}{\varepsilon_0}$$

On the equation sheet for the free-response section, this information will be represented as follows:

$$\oint \mathbf{E} \cdot d\mathbf{A} = \frac{Q}{\varepsilon_0}$$

where the integral is for a closed surface, $\int$, which is the same as $\oint$.

Now you can finally see why we write Coulomb's constant, k_0, as $1/(4\pi\varepsilon_0)$. It makes the 4π's cancel out in this equation.

This is **Gauss's law**. The quantity $Q_{enclosed}$ is the algebraic sum of the charges inside. For example, if the imaginary closed surface we construct encloses, say, charges of $+Q$ and $-3Q$, then the total charge enclosed is $-2Q$, the algebraic sum, so this is $Q_{enclosed}$ in this case. In fact, many people place a Σ in front of $Q_{enclosed}$ to remind them that they need to take the *sum* of the charges enclosed when using Gauss's law. The power of Gauss's law to determine electric fields is evident if we can write Φ_E in terms of E—and this requires symmetry—because then the equation can be solved for E. The imaginary closed surface that we will construct when using Gauss's law to determine an electric field is known as a **Gaussian surface**.

Example 12 Consider a long, straight wire carrying a total charge of $+Q$, distributed uniformly along its entire length, L. Use Gauss's law to find an expression for the electric field created by the wire.

Solution. By symmetry, the electric field must point radially away from the wire. To take advantage of this symmetry, we will construct a cylindrical Gaussian surface—of length x and radius r—around the wire:

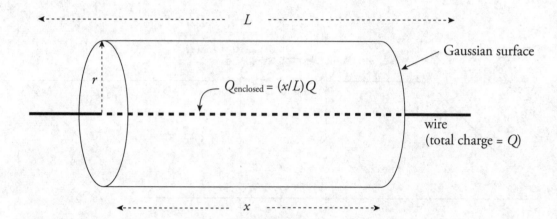

The electric field is always perpendicular to the lateral surface area, A, of the cylinder. Therefore, the electric flux through the cylinder is equal to $EA = E(2\pi rx)$. The flux through each of the two lids of the cylinder is zero, because $\mathbf{E}$ is parallel to the end surfaces (so $\mathbf{E} \cdot d\mathbf{A} = E \cdot dA \cos 90° = 0$). So the total electric flux through the Gaussian surface is $\Phi_E = 0 + E(2\pi rx) + 0 = E(2\pi rx)$. Next we need to know the charge enclosed by the Gaussian surface. Since a total charge of Q is distributed uniformly along the entire length L of the wire, the amount of charge on a section of length x is $(x/L)Q$. Therefore, Gauss's law becomes

$$\Phi_E = \frac{Q_{\text{enclosed}}}{\varepsilon_0}$$

$$E \cdot 2\pi rx = \frac{(x/L)Q}{\varepsilon_0}$$

$$E = \frac{Q/L}{2\pi\varepsilon_0 r}$$

The ratio Q/L (charge per unit length) is called the *linear charge density*, and is denoted by λ; using this notation, the electric field at a distance r from the wire is given by the equation $E = \lambda/(2\pi\varepsilon_0 r)$. (If the charge on the wire was $-Q$, then the field would point radially inward.)

> **Example 13** A very large rectangular plate has a surface charge density of $+\sigma$ (this is the charge per unit area). Use Gauss's law to determine an expression for the electric field it creates.

Solution. By symmetry, the electric field lines must be perpendicular to the plate. We will construct a cylindrical Gaussian surface, with cross-sectional area A and total length $2r$, perpendicular to the plate:

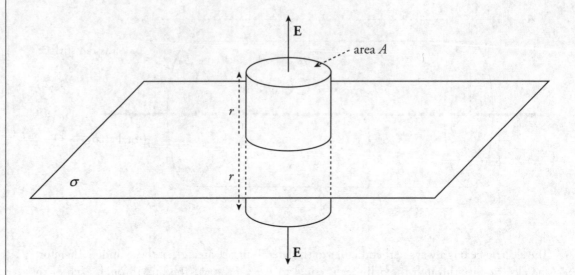

The electric flux through the lateral surface area is now zero, since **E** is parallel to the cylindrical surface here. However, there is flux through the end caps: $EA + EA = 2EA$. (Flux outward from the closed surface is counted as positive; flux inward is counted as negative.) Therefore,

$$\Phi_E = \frac{Q_{enclosed}}{\varepsilon_0}$$

$$2EA = \frac{\sigma A}{\varepsilon_0}$$

$$E = \frac{\sigma}{2\varepsilon_0}$$

The electric field does not depend on the distance r from the plate. This would only be true if the plate had infinite area. In practical terms, this result means that if we are close to a very large plate (and away from its edges), then the plate looks infinite, and moving a small amount toward or away from the plate won't make much difference; the electric field will remain essentially constant.

Example 14 A nonconducting sphere of radius a has excess charge distributed throughout its volume so that the volume charge density ρ as a function of r (the distance from the sphere's center) is given by the equation $\rho(r) = \rho_0(r/a)^2$, where ρ_0 is a constant. Determine the electric field at points inside and outside the sphere.

Solution. In order to apply Gauss's law, we need to find the total charge enclosed by a Gaussian surface of radius r. Since the charge density depends on r only, consider a thin spherical shell of radius R and thickness dR.

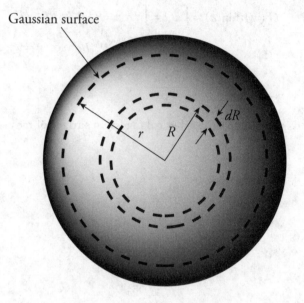

Gaussian surface

The shell's volume is $dV = 4\pi R^2 \, dR$, so the charge contained in the shell is

$$dQ = \rho \, dV = \rho_0 (R/a)^2 \cdot 4\pi R^2 \, dR$$

The total charge enclosed within a sphere of radius r is found by adding up all the charges on these shells, from $R = 0$ to $R = r$. Call this Equation 1.

$$Q\,(\text{within } r) = \int dQ = \int_{R=0}^{R=r} \rho_0 (R/a)^2 \cdot 4\pi R^2 \, dR$$
$$= (4\pi\rho_0 / a^2) \int_{R=0}^{R=r} R^4 \, dR$$
$$= \frac{4\pi\rho_0}{5a^2} r^5$$

The electric flux through the spherical Gaussian surface of radius r is

$$\Phi_E = EA = E(4\pi r^2)$$

so, by Gauss's law,

$$\Phi_E = \frac{Q_{\text{enclosed}}}{\varepsilon_0}$$
$$E(4\pi r^2) = \frac{1}{\varepsilon_0} \frac{4\pi\rho_0}{5a^2} r^5$$
$$E = \frac{\rho_0}{5\varepsilon_0 a^2} r^3$$

This gives the electric field at all points within the given sphere. For points outside, we first figure out the charge on the entire sphere. To do so, we can refer back to the integral we used in Equation 1 to find the total charge.

$$Q \, (\text{within } a) = \int dQ = \int_{R=0}^{R=a} \rho_0 (R/a)^2 \cdot 4\pi R^2 \, dR$$
$$= (4\pi\rho_0/a^2)\int_{R=0}^{R=a} R^4 \, dR$$
$$= \frac{4\pi\rho_0 a^3}{5}$$

Then, applying Gauss's law again on a spherical Gaussian surface of radius $r > a$, we get

$$\Phi_E = \frac{Q_{\text{enclosed}}}{\varepsilon_0}$$
$$E(4\pi r^2) = \frac{1}{\varepsilon_0}\frac{4\pi\rho_0 a^3}{5}$$
$$E = \frac{\rho_0 a^3}{5\varepsilon_0 r^2}$$

To summarize, we have found

$$E(r) = \begin{cases} \dfrac{\rho_0}{5\varepsilon_0 a^2}\, r^3 & \text{for } r \le a \\[2ex] \dfrac{\rho_0 a^3}{5\varepsilon_0 r^2} & \text{for } r \ge a \end{cases}$$

Example 15 Use Gauss's law to show that the electric field inside a charged hollow metal sphere is zero.

Solution. Recall that for a conductor, any excess charge resides on its surface. Therefore, if a Gaussian surface is constructed anywhere *within* the metal, it contains no net charge, so **E** = 0 everywhere inside the sphere.

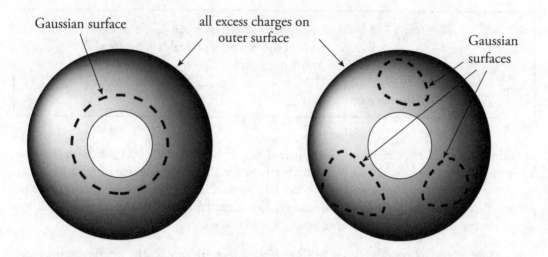

Gaussian surface

all excess charges on
outer surface

Gaussian
surfaces

Example 16 Consider a neutral, hollow metal sphere. Imagine that a charge of $+q$ could be introduced into the cavity, insulated from the inner wall. What will happen?

Solution. *A static electric field cannot be sustained within the body of a conductor.* Therefore, enough electrons will migrate to the inner wall of the cavity to form a *shield* of charge $-q$ around the inside charge $+q$.

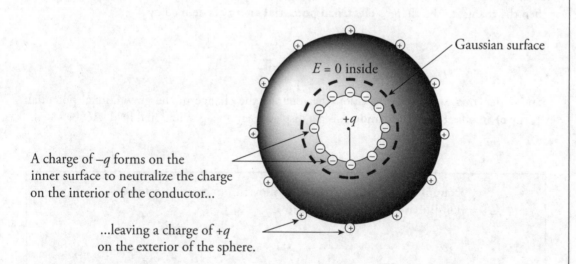

$E = 0$ inside

Gaussian surface

$+q$

A charge of $-q$ forms on the inner surface to neutralize the charge on the interior of the conductor...

...leaving a charge of $+q$ on the exterior of the sphere.

To see why this must be so, if any Gaussian surface surrounding the cavity is drawn within the conductor, then, because E *must* be 0 inside, the electric flux must be 0, so $Q_{enclosed}$ must be 0. The only way to ensure that $Q_{enclosed} = 0$ is to have a charge of $-q$ on the inner wall of the cavity. Since the sphere was originally neutral, an excess charge of $+q$ will then appear on the outer surface.

Example 17 A solid metal sphere of radius *a*, carrying a charge of +*q*, is placed inside and concentric with a neutral hollow metal sphere of inner radius *b* and outer radius *c*. Determine the electric field for *r* < *a*, *a* < *r* < *b*, *b* < *r* < *c*, and *r* > *c*. Also, describe the charge distributions.

Solution. For *r* < *a*, the electric field is 0, since there can be no electric field within the body of a conductor. For *r* between *a* and *b*, the inner sphere acts like a point charge of +*q* at its center, so $E = q/(4\pi\varepsilon_0 r^2)$. For *r* between *b* and *c*, the electric field is again zero, since there can be no electric field within the body of a conductor. For *r* > *c*, the ensemble behaves as if all its charge were concentrated at its center, so once again, $E = q/(4\pi\varepsilon_0 r^2)$.

The charge on the sphere of radius *a* will be at its surface. To protect the interior of the outer spherical shell from an electric field, a charge of –*q* will reside on the surface of radius *b* (shielding the charge of +*q* that's introduced into the cavity), leaving +*q* to appear on the outer surface (radius *c*).

ELECTRICAL POTENTIAL ENERGY

When a charge moves in an electric field, unless its displacement is always perpendicular to the field, the electric force does work on the charge. If W_E is the work done by the electric force, then the change in the charge's **electrical potential energy** is defined by

$$\Delta U_E = -W_E$$

Notice that this is the same equation that defined the change in the gravitational potential energy of an object of mass *m* undergoing a displacement in a gravitational field ($\Delta U_G = -W_G$).

Example 18 A positive charge +*q* moves from position A to position B in a uniform electric field **E**:

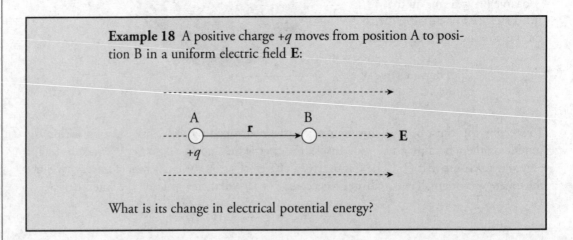

What is its change in electrical potential energy?

Solution. Since the field is uniform, the electric force that the charge feels, $\mathbf{F}_E = q\mathbf{E}$, is constant. Since q is positive, $\mathbf{F}_E$ points in the same direction as $\mathbf{E}$, and, as the figure shows, they point in the same direction as the displacement, $\mathbf{r}$. This makes the work done by the electric field equal to $W_E = F_E r = qEr$, so the change in the electrical potential energy is

$$\Delta U_E = -qEr$$

Note that the change in potential energy is negative, which means that potential energy has decreased; this always happens when the field does positive work. It's just like when you drop a rock to the ground: Gravity does positive work, and the rock loses gravitational potential energy.

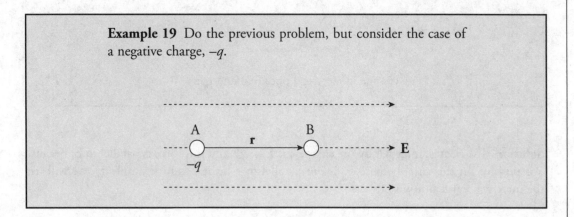

Example 19 Do the previous problem, but consider the case of a negative charge, $-q$.

Solution. In this case, an outside agent must be pushing the charge to make it move, because the electric force *naturally* pushes negative charges against field lines. Therefore, we expect that the work done by the electric field is negative. The electric force, $\mathbf{F}_E = (-q)\mathbf{E}$, points in the direction opposite to the displacement, so the work it does is $W_E = -F_E r = -qEr$, and the change in electrical potential energy is positive: $\Delta U_E = -W_E = -(-qEr) = qEr$. Since the change in potential energy is positive, the potential energy increased; this always happens when the field does negative work. It's like when you lift a rock off the ground: Gravity does negative work, and the rock gains gravitational potential energy.

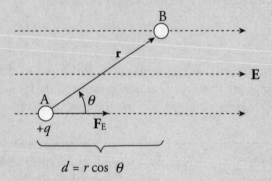

Example 20 A positive charge $+q$ moves from position A to position B in a uniform electric field **E**:

What is its change in electrical potential energy?

Solution. The electric force felt by the charge q is $\mathbf{F}_E = q\mathbf{E}$ and this force is parallel to **E,** because q is positive. In this case, because $\mathbf{F}_E$ is not parallel to **r** (as it was in Example 1), we will use the more general definition of work:

$$W_E = \mathbf{F}_E \bullet \mathbf{r} = F_E\, r \cos \theta = qEr \cos \theta$$

But $r \cos \theta = d$, so

$$W_E = qEd$$

and

$$\Delta U_E = -W_E = -qEd$$

Because the electric force is a conservative force, which means that the work done does not depend on the path that connects the positions A and B, the work calculated above could have been figured out by considering the path from A to B composed of the segments $\mathbf{r}_1$ and $\mathbf{r}_2$:

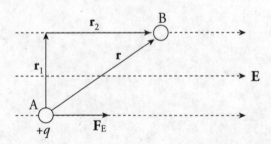

Along $\mathbf{r}_1$, the electric force does no work since this displacement is perpendicular to the force. Thus, the work done by the electric field as q moves from A to B is simply equal to the work it does along $\mathbf{r}_2$. And since the length of $\mathbf{r}_2$ is $d = r \cos \theta$, we have $W_E = F_E d = qEd$, just as before.

Example 21 A positive charge $+q$ moves from position A to B in the electric field, **E**, created by the source charge $+Q$ (only a portion of the electric field is drawn):

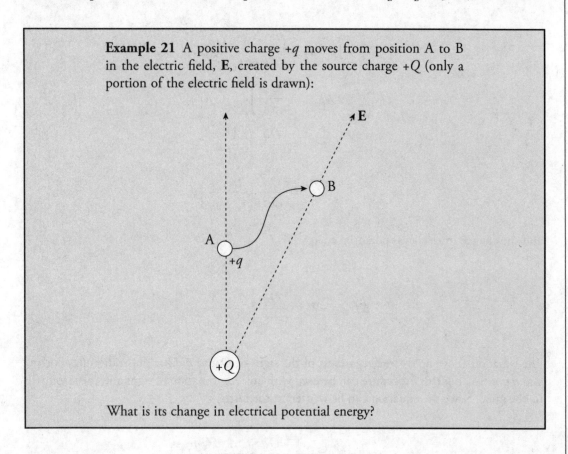

What is its change in electrical potential energy?

Solution. As you know, movement that's perpendicular to the field lines causes the electric field to do no work. So replace the path pictured by another path, composed of $\mathbf{r}_1$ and $\mathbf{r}_2$:

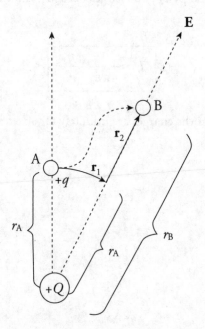

At every point along the arc r_1, the electric force is radial and thus perpendicular to the displacement, so $\mathbf{F}_E$ does no work on the charge as it moves along $\mathbf{r}_1$. But along radial line $\mathbf{r}_2$, the electric force is parallel to the displacement. The work done by the electric force is

$$W_E = \int \mathbf{F}_E \cdot d\mathbf{r} = \int_{r=r_A}^{r=r_B} \frac{1}{4\pi\varepsilon_0}\frac{Qq}{r^2}\,dr$$

$$= \frac{Qq}{4\pi\varepsilon_0}\int_{r=r_A}^{r=r_B}\frac{1}{r^2}\,dr$$

$$= \frac{Qq}{4\pi\varepsilon_0}\left(-\frac{1}{r}\right)_{r_A}^{r_B}$$

$$= \frac{Qq}{4\pi\varepsilon_0}\left(\frac{1}{r_A}-\frac{1}{r_B}\right)$$

and the change in electrical potential energy is

$$\Delta U_E = -W_E = \frac{Qq}{4\pi\varepsilon_0}\left(\frac{1}{r_B}-\frac{1}{r_A}\right)$$

The equation above holds true regardless of the signs of Q and q. Our illustration has both Q and q positive, but this procedure can be used with any combination of signs and the result will be the same. Since the equation can be written in the form

$$U_B - U_A = \frac{1}{4\pi\varepsilon_0}\frac{Qq}{r_B} - \frac{1}{4\pi\varepsilon_0}\frac{Qq}{r_A}$$

we can define the potential energy of two charges (let's just call them q_1 and q_2), separated by a distance r, to be

$$U = \frac{1}{4\pi\varepsilon_0}\frac{q_1 q_2}{r}$$

This definition says that when the charges are infinitely far apart, their potential energy is zero.

Example 22 A positive charge $q_1 = +2 \times 10^{-6}$ C is held stationary, while a negative charge, $q_2 = -1 \times 10^{-8}$ C, is released from rest at a distance of 10 cm from q_1. Find the kinetic energy of charge q_2 when it's 1 cm from q_1.

Solution. The gain in kinetic energy is equal to the loss in potential energy; you know this from Conservation of Energy. The change in electrical potential energy is

$$\Delta U = \frac{q_1 q_2}{4\pi\varepsilon_0}\left(\frac{1}{r_B} - \frac{1}{r_A}\right)$$

$$= (9\times10^9 \ \text{N}\cdot\text{m}^2/\text{C}^2)(+2\times10^{-6} \ \text{C})(-1\times10^{-8} \ \text{C})\left(\frac{1}{0.01 \ \text{m}} - \frac{1}{0.10 \ \text{m}}\right)$$

$$= -0.016 \ \text{J}$$

So the gain in kinetic energy is +0.016 J. Since q_2 started from rest (with no kinetic energy), this is the kinetic energy of q_2 when it's 1 cm from q_1.

Example 23 Two positive charges, q_1 and q_2, are held in the positions shown below. How much work would be required to bring (from infinity) a third positive charge, q_3, and place it so that the three charges form the corners of an equilateral triangle of side length s?

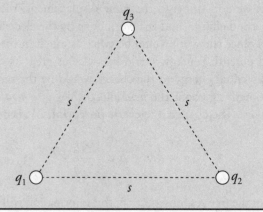

Solution. An external agent would need to do positive work, equal in magnitude to the negative work done by the electric force on q_3 as it is brought into place, so let's first compute this quantity. Let's first compute the work done *by the electric force* as q_3 is brought in. Since q_3 is fighting against both q_1's and q_2's electric fields, the total work done on q_3 by the electric force, W_E, is equal to the work done on q_3 by q_1 ($W_{1\text{-}3}$) plus the work done on q_3 by q_2 ($W_{2\text{-}3}$).

Using the equation $W_E = -\Delta U_E$ and the one we derived above for ΔU_E, we have

$$W_{1\text{-}3} + W_{2\text{-}3} = -\Delta U_{1\text{-}3} + -\Delta U_{2\text{-}3}$$

$$= -\frac{q_1 q_3}{4\pi\varepsilon_0}\left(\frac{1}{s} - 0\right) + -\frac{q_2 q_3}{4\pi\varepsilon_0}\left(\frac{1}{s} - 0\right)$$

$$= -\left(\frac{1}{4\pi\varepsilon_0}\frac{q_1 q_3}{s} + \frac{1}{4\pi\varepsilon_0}\frac{q_2 q_3}{s}\right)$$

Therefore, the work that an external agent must do to bring q_3 into position is

$$-W_E = \frac{1}{4\pi\varepsilon_0}\frac{q_1 q_3}{s} + \frac{1}{4\pi\varepsilon_0}\frac{q_2 q_3}{s}$$

In general, the work required by an external agent to assemble a collection of point charges q_1, $q_2, \dots, q_n$ (bringing each one from infinity), such that the final fixed distance between q_i and q_j is r_{ij}, is equal to the total electrical potential energy of the arrangement:

$$W_{\text{by external agent}} = U_{\text{total}} = \frac{1}{4\pi\varepsilon_0}\sum_{i<j}\frac{q_i q_j}{r_{ij}}$$

Electric Potential

Let W_E be the work done by the electric field on a charge q as it undergoes a displacement. If another charge, say $2q$, were to undergo the same displacement, the electric force would be twice as great on this second charge, and the work done by the electric field would be twice as much, $2W_E$. Since the work would be twice as much in the second case, the change in electrical potential energy would be twice as great as well, but the ratio of the change in potential energy to the charge would be the same: $U_E/q = 2U_E/2q$. This ratio says something about the *field* and the *displacement*, but not the charge that made the move. The change in **electric potential**, ΔV, is defined as this ratio:

> Sometimes, electric potential is represented by φ, and change in electric potential, Δφ, is represented by *V* or voltage.

$$\Delta V = \frac{\Delta U_E}{q}$$

On the equation sheet for the free-response section, this information will be represented as follows:

$$\boxed{U_E = qV}$$

Electric potential is electrical potential energy *per unit charge*; the units of electric potential are joules per coulomb. One joule per coulomb is called one **volt** (abbreviated V); so 1 J/C = 1 V.

Consider the electric field that's created by a point source charge Q. If a charge q moves from a distance r_A to a distance r_B from Q, then the change in the potential energy is

$$U_B - U_A = \frac{Qq}{4\pi\varepsilon_0}\left(\frac{1}{r_B} - \frac{1}{r_A}\right)$$

The difference in electric potential between positions A and B in the field created by Q is

$$V_B - V_A = \frac{U_B - U_A}{q} = \frac{Q}{4\pi\varepsilon_0}\left(\frac{1}{r_B} - \frac{1}{r_A}\right)$$

If we designate $V_A \to 0$ as $r_A \to \infty$ (an assumption that's stated on the AP Physics Exam), then the electric potential at a distance r from Q is

$$V = \frac{1}{4\pi\varepsilon_0}\frac{Q}{r}$$

Note that the potential depends on the source charge making the field and the distance from it.

Since electric potential is a scalar, if there is more than one charge, the electric potential at a given location is the sum of the potentials due to each of the charges:

$$V = \frac{1}{4\pi\varepsilon_0}\sum_i \frac{q_i}{r_i}$$

Q: Can you derive an expression for the electric potential at point C in Example 11 of Chapter 12?

Turn the page for the solution.

Example 24 Let $Q = 2 \times 10^{-9}$ C. What is the potential at a point P that is 2 cm from Q?

Solution. Relative to $V = 0$ at infinity, we have

$$V = \frac{1}{4\pi\varepsilon_0}\frac{Q}{r} = \left(9\times10^9 \text{ N}\cdot\text{m}^2/\text{C}^2\right)\frac{2\times10^{-9} \text{ C}}{0.02 \text{ m}} = 900 \text{ V}$$

This means that the work done by the electric field on a charge of q coulombs brought from infinity to a point 2 cm from Q would be $-900q$ joules.

Note that, like potential energy, potential is a *scalar*. In the preceding example, we didn't have to specify the direction of the vector from the position of Q to the point P because it didn't matter. At *any* point on a sphere that's 2 cm from Q, the potential will be 900 V. These spheres around Q are called **equipotential surfaces**, and they're surfaces of constant potential. Their cross sections in any plane are circles and are (therefore) perpendicular to the electric field lines. The equipotentials are always perpendicular to the electric field lines.

Example 25 How much work is done by the electric field as a charge moves along an equipotential surface?

Solution. If the charge always remains on a single equipotential, then, by definition, the potential, V, never changes. Therefore, $\Delta V = 0$, so $\Delta U_E = 0$. Since $W_E = -\Delta U_E$, the work done by the electric field is zero.

Example 26 If charges $q_1 = 4 \times 10^{-9}$ C and $q_2 = -6 \times 10^{-9}$ C are stationary, calculate the potential at point A in the figure below:

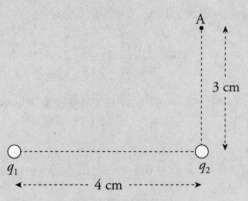

Solution. Potentials add like ordinary numbers. Therefore, the potential at A is just the sum of the potentials at A due to q_1 and q_2. Note that the distance from q_1 to A is 5 cm.

A: The electric potential at point C in Example 11 of Chapter 12 is $V = \dfrac{\lambda}{4\varepsilon_0}$.

$$
\begin{aligned}
V &= \frac{1}{4\pi\varepsilon_0}\frac{q_1}{r_{1A}} + \frac{1}{4\pi\varepsilon_0}\frac{q_2}{r_{2A}} \\
&= \frac{1}{4\pi\varepsilon_0}\left(\frac{q_1}{r_{1A}} + \frac{q_2}{r_{2A}}\right) \\
&= (9\times10^9 \ \text{N}\cdot\text{m}^2/\text{C}^2)\left(\frac{4\times10^{-9}\ \text{C}}{0.05\ \text{m}} + \frac{-6\times10^{-9}\ \text{C}}{0.03\ \text{m}}\right) \\
&= -1{,}080\ \text{V}
\end{aligned}
$$

Example 27 Using the charges from Example 26, how much work would it take to move a charge $q = +1 \times 10^{-2}$ C from point A to point B (the point midway between q_1 and q_2)?

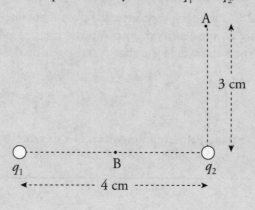

Solution. $\Delta U_E = q\Delta V$, so if we calculate the potential difference between points A and B and multiply by q, we will have found the change in the electrical potential energy: $\Delta U_{A \to B} = q\Delta V_{A \to B}$. Then, since the work by the electric field is $-\Delta U$, the work required by an external agent is ΔU. In this case, the potential at point B is

$$
\begin{aligned}
V_B &= \frac{1}{4\pi\varepsilon_0}\left(\frac{q_1}{r_{1B}} + \frac{q_2}{r_{2B}}\right) \\
&= (9\times10^9 \ \text{N}\cdot\text{m}^2/\text{C}^2)\left(\frac{4\times10^{-9} \ \text{C}}{0.02 \ \text{m}} + \frac{-6\times10^{-9} \ \text{C}}{0.02 \ \text{m}}\right) \\
&= -900 \ \text{V}
\end{aligned}
$$

In the preceding example, we calculated the potential at point A: $V_A = -1{,}080$ V, so $\Delta V_{A \to B} = V_B - V_A = (-900 \ \text{V}) - (-1{,}080 \ \text{V}) = +180$ V. This means that the change in electrical potential energy as q moves from A to B is

$$
\Delta U_{A \to B} = qV_{A \to B} = (+1 \times 10^{-2} \ \text{C})(+180 \ \text{V}) = 1.8 \ \text{J}
$$

This is the work required by an external agent to move q from A to B.

The Potential in a Uniform Field

Example 28 Consider a very large, flat plate that contains a uniform surface charge density σ. At points that are not too far from the plate, the electric field is uniform and given by the equation

$$E = \frac{\sigma}{2\varepsilon_0}$$

What is the potential at a point which is a distance d from the sheet, relative to the potential of the sheet itself?

Solution. Let A be a point on the plate and let B be a point a distance d from the sheet. Then

$$V_B - V_A = \frac{-W_{E,\,A \to B} \text{ on } q}{q}$$

Since the field is constant, the force that a charge q would feel is also constant, and is equal to

$$F_E = qE = q\frac{\sigma}{2\varepsilon_0}$$

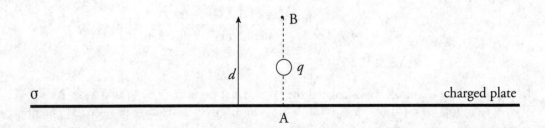

Therefore,

$$W_{E,\,A \to B} = F_E d$$
$$= \frac{q\sigma}{2\varepsilon_0}d$$

so applying the definition gives us

$$V_B - V_A = \frac{-W_{E,\,A \to B}}{q} = -\frac{\sigma}{2\varepsilon_0}d$$

This says that for a positive σ, the potential decreases linearly as we move away from the plate.

Example 29 Two large flat plates—one carrying a charge of $+Q$, the other $-Q$—are separated by a distance d. The electric field between the plates, **E**, is uniform. Determine the potential difference between the plates.

Solution. Imagine a positive charge q moving from the positive plate to the negative plate:

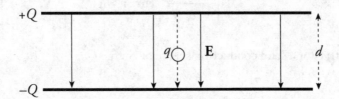

Since the work done by the electric field is

$$W_{E,+\to-} = F_E d = qEd$$

the potential difference between the plates is

$$V_- - V_+ = \frac{-W_{E,+\to-}}{q} = \frac{-qEd}{q} = -Ed$$

This tells us that the potential of the positive plate is greater than the potential of the negative plate by the amount Ed. This equation can also be written as

$$E = -\frac{V_- - V_+}{d}$$

Therefore, if the potential difference and the distance between the plates are known, then the magnitude of the electric field can be determined quickly.

THE POTENTIAL OF A SPHERE

Think about a conducting spherical shell of radius R. If the shell carries an excess charge Q, we know that this excess charge will be found on the outer surface and that the electric field inside the sphere will be zero. What is the potential due to the shell? At points outside the shell, the field exists as if all of the charge of the sphere were concentrated at the center. That is, for points outside the shell, the potential is the same as it would be due to a single point charge Q:

$$V = \frac{1}{4\pi\varepsilon_0}\frac{Q}{r} \quad \text{for } r > R$$

What about at points inside the shell? Since **E** = 0 everywhere inside, there would be no electric force and no work done on a charge moving inside the shell. *The potential is constant within the sphere.* So if we moved a charge q from the surface of the spherical shell to its inside, the potential wouldn't change. The potential everywhere inside the shell is equal to the potential on the surface, which is $(1/4\pi\varepsilon_0)(Q/R)$. Therefore,

$$V = \begin{cases} \dfrac{1}{4\pi\varepsilon_0}\dfrac{Q}{R} & \text{for } r \leq R \\[2mm] \dfrac{1}{4\pi\varepsilon_0}\dfrac{Q}{r} & \text{for } r > R \end{cases}$$

The same is also true for a solid conducting sphere of charge Q.

Example 30 The figure below shows two concentric, conducting, thin spherical shells. The inner shell has a radius of a and carries a charge of q. The outer shell has a radius of b and carries a charge of Q. The inner shell is supported on an insulating stand.

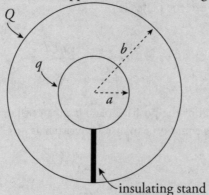

insulating stand

(a) What would the potential of the inner shell be if the outer shell were absent?

(b) What is the potential of the inner shell with the outer shell present?

(c) Show that the potential difference between the inner shell and the outer shell, $V_a - V_b$, does not depend on the charge on the outer shell.

Solution.

(a) The potential on (and within) a sphere of radius a containing a charge q is given by the equation $V_a = (1/4\pi\varepsilon_0)(q/a)$.

(b) The potential inside the outer sphere (if the inner sphere were absent) is equal to $(1/4\pi\varepsilon_0)(Q/b)$. The potential of the inner sphere if the outer sphere were absent is $(1/4\pi\varepsilon_0)(q/a)$. So the potential of the inner sphere with the outer sphere present is the sum: $(1/4\pi\varepsilon_0)(Q/b) + (1/4\pi\varepsilon_0)(q/a)$.

Here's another way to arrive at this result: Outside (and on) the outer sphere, the potential is the same as if the total charge were concentrated at the center, $(1/4\pi\varepsilon_0)(q + Q)/r$. So at a point *on* the outer sphere, the potential is found by substituting b for r in this expression; this gives $V_b = (1/4\pi\varepsilon_0)(q + Q)/b$. Now, the electric field between the spheres is, by Gauss's law, simply equal to $(1/4\pi\varepsilon_0)(q/r^2)$. Therefore, the work required *by an external agent* to move a charge (let's call it q_0) from infinity to the inner sphere is equal to:

$$W_{\text{from } \infty \text{ to } a} = W_{\text{from } \infty \text{ to } b} + W_{\text{from } b \text{ to } a}$$

$$= q_0 V_b + -\int_{r=b}^{r=a} q_0 E \, dr$$

$$= q_0 V_b - q_0 \int_{r=b}^{r=a} \frac{1}{4\pi\varepsilon_0} \frac{q}{r^2} \, dr$$

$$= q_0 V_b - q_0 \frac{q}{4\pi\varepsilon_0} \left[-\frac{1}{r} \right]_b^a$$

$$= q_0 V_b - q_0 \frac{q}{4\pi\varepsilon_0} \left(\frac{1}{b} - \frac{1}{a} \right)$$

Therefore,

$$V_a = \frac{W_{\text{from } \infty \text{ to } a}}{q_0}$$

$$= V_b - \frac{q}{4\pi\varepsilon_0} \left(\frac{1}{b} - \frac{1}{a} \right)$$

$$= \left[\frac{1}{4\pi\varepsilon_0} \frac{q+Q}{b} \right] - \frac{q}{4\pi\varepsilon_0} \left(\frac{1}{b} - \frac{1}{a} \right)$$

$$= \frac{1}{4\pi\varepsilon_0} \left(\frac{Q}{b} + \frac{q}{a} \right)$$

(c) At points outside the outer sphere, the two spheres behave as if all their charge was concentrated at the center. So at points outside the sphere ($r > b$), the potential is equal to $(1/4\pi\varepsilon_0)(q + Q)/r$. This must also give the potential at the surface of the outer sphere when $r = b$. Therefore, the potential difference between the inner sphere and the outer sphere is

$$V_a - V_b = \left[\frac{1}{4\pi\varepsilon_0} \frac{q}{a} + \frac{1}{4\pi\varepsilon_0} \frac{Q}{b} \right] - \frac{1}{4\pi\varepsilon_0} \frac{q+Q}{b}$$

$$= \frac{1}{4\pi\varepsilon_0} \frac{q}{a} - \frac{1}{4\pi\varepsilon_0} \frac{q}{b}$$

$$= \frac{q}{4\pi\varepsilon_0} \left(\frac{1}{a} - \frac{1}{b} \right)$$

This expression does not depend on Q, which is what we were asked to show.

Example 31 A *nonconducting* sphere of radius R has an excess charge of Q distributed uniformly throughout its volume (that is, the volume charge density, ρ, is a constant). The electric field at a distance $r < R$ from the center is given by the equation

$$E = \frac{1}{4\pi\varepsilon_0} \frac{Q}{R^3} r$$

What is the potential inside the sphere? That is, what is the potential at a distance r from the sphere's center?

Solution. The potential at some distance, let's call it r_0, from the center is equal to the negative of the work done by the electric field as a charge q is brought to r_0 from infinity, divided by q. (By definition, $V = U_E/q = -W_E/q$, where we take $V = 0$ at infinity.) So our first step (the big one) is to figure out the work done by the electric field in bringing a charge q in from infinity. At points outside the sphere, the electric field is simply $(1/4\pi\varepsilon_0)(Q/r^2)$, since the sphere behaves as if all of its charge were concentrated at its center. Once we get inside, however, the electric field is given by the formula in the question. Therefore,

$$W_E = W_{E, \infty \text{ to } R} + W_{E, R \text{ to } r_0}$$

$$= \int_{\infty}^{R} q \cdot \frac{1}{4\pi\varepsilon_0} \frac{Q}{r^2} dr + \int_{R}^{r_0} q \cdot \frac{1}{4\pi\varepsilon_0} \frac{Q}{R^3} r \, dr$$

$$= \frac{qQ}{4\pi\varepsilon_0} \int_{\infty}^{R} \frac{1}{r^2} dr + \frac{qQ}{4\pi\varepsilon_0 R^3} \int_{R}^{r_0} r \, dr$$

$$= \frac{qQ}{4\pi\varepsilon_0} \left[-\frac{1}{r} \right]_{\infty}^{R} + \frac{qQ}{4\pi\varepsilon_0 R^3} \left[\frac{r^2}{2} \right]_{R}^{r_0}$$

$$= -\frac{qQ}{4\pi\varepsilon_0} \frac{1}{R} + \frac{qQ}{8\pi\varepsilon_0 R^3} (r_0^2 - R^2)$$

$$= \frac{qQ}{4\pi\varepsilon_0} \left(\frac{r_0^2 - 3R^2}{2R^3} \right)$$

Taking the negative of this result, dividing by q, and replacing r_0 with r gives us our answer:

$$V = \frac{-W_E}{q} = \frac{Q}{4\pi\varepsilon_0}\left(\frac{3R^2 - r^2}{2R^3}\right)$$

Note how this result differs from the potential of a *conducting* sphere. In that case, the potential was constant throughout the interior of the sphere (and was equal to the value of the potential on the surface). In this case, the potential is not constant; it depends on r.

THE POTENTIAL OF A CYLINDER

Example 32 Consider a very long conducting cylinder of radius R, which carries a uniform linear charge density λ. The electric field at a distance r, where $r > R$, from the center of the cylinder is given by the equation

$$E = \frac{\lambda}{2\pi\varepsilon_0 r}$$

Determine a formula for the potential at a point outside the cylinder, relative to the potential on the cylinder.

Solution. Let A be a point on the cylinder (at $r = R$) and let B be a point outside the cylinder (at $r = r_0 > R$). Then, by definition,

$$V_B - V_A = \frac{-W_{E, A\to B} \text{ on } q}{q}$$

Next, we figure out the work done by the electric field as a charge q moves from A to B:

$$
\begin{aligned}
W_{E, A\to B} \text{ on } q &= \int_A^B \mathbf{F}_E \cdot d\mathbf{r} \\
&= \int_{r=R}^{r=r_0} q \cdot \frac{\lambda}{2\pi\varepsilon_0 r} dr \\
&= \frac{q\lambda}{2\pi\varepsilon_0} \int_{r=R}^{r=r_0} \frac{dr}{r} \\
&= \frac{q\lambda}{2\pi\varepsilon_0} \left[\ln r\right]_R^{r_0} \\
&= \frac{q\lambda}{2\pi\varepsilon_0} (\ln r_0 - \ln R) \\
&= \frac{q\lambda}{2\pi\varepsilon_0} \ln\frac{r_0}{R}
\end{aligned}
$$

Replacing r_0 with r, we get

$$V_B - V_A = \frac{-W_{E, A \to B} \text{ on } q}{q}$$

$$= -\frac{\lambda}{2\pi\varepsilon_0} \ln \frac{r}{R}$$

$$= \frac{\lambda}{2\pi\varepsilon_0} \ln \frac{R}{r}$$

Deriving the Field from the Potential

From the definition of potential,

$$V_B - V_A = \frac{-W_{E, A \to B} \text{ on } q}{q}$$

$$= -\frac{1}{q} \int_A^B F_E \, dr$$

$$= -\frac{1}{q} \int_A^B qE \, dr$$

$$= \int_A^B (-E) \, dr$$

If A and B are separated by an infinitesimal distance dr, then

$$V_B - V_A = \int_A^B dV$$

and this gives us

$$\int_A^B (-E) \, dr = \int_A^B dV$$

from which we can conclude that

$$(-E) \, dr = dV$$

$$\boxed{E = -\frac{dV}{dr}}$$

So, if we know how the potential varies as a function of r, we can determine the electric field variation with r.

Example 33 If the potential at a distance r from a source point charge Q is given by the equation $V(r) = (1/4\pi\varepsilon_0)(Q/r)$, determine a formula for the electric field.

Solution. Using the relationship derived above,

$$E = -\frac{dV}{dr} = -\frac{d}{dr}\left(\frac{1}{4\pi\varepsilon_0}\frac{Q}{r}\right) = -\frac{Q}{4\pi\varepsilon_0}\frac{d}{dr}\left(\frac{1}{r}\right) = -\frac{Q}{4\pi\varepsilon_0}\left(-\frac{1}{r^2}\right) = \frac{1}{4\pi\varepsilon_0}\frac{Q}{r^2}$$

which is a result we know well.

Chapter 12 Drill

The answers and explanations can be found in Chapter 17.

Section I: Multiple Choice

1. What will happen to the magnitude of electric force between two particles if the distance between them is doubled and each charge is tripled?

    (A) It will be multiplied by a factor of $\dfrac{9}{4}$.

    (B) It will be multiplied by a factor of $\dfrac{3}{2}$.

    (C) It will be multiplied by a factor of $\dfrac{3}{4}$.

    (D) It will be multiplied by a factor of $\dfrac{2}{3}$.

    (E) It will be multiplied by a factor of $\dfrac{4}{9}$.

70 cm

2. The particle shown above is initially at rest inside a uniform electric field $E = 50$ N/C toward the left. The field area is a square with 1-meter-long sides. If the particle has a mass of 10^{-10} kg and a charge of $q = -5 \times 10^{-13}$ C, how long will it take the particle to escape the electric field?

    (A) 1.10 s
    (B) 1.23 s
    (C) 1.55 s
    (D) 2.31 s
    (E) 2.59 s

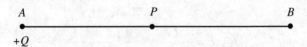

3. Points A and B are equidistant from point P. At point A, there is a particle of charge $+Q$. This results in an electric field, E, at point P. If we want to triple the electric field at P, what charge should be placed at point B?

    (A) $+3Q$
    (B) $+2Q$
    (C) $-Q$
    (D) $-2Q$
    (E) $-3Q$

4. A sphere of charge $+Q$ is fixed in position. A smaller sphere of charge $+q$ is placed near the larger sphere and released from rest. The small sphere will move away from the large sphere with

    (A) decreasing velocity and decreasing acceleration
    (B) decreasing velocity and increasing acceleration
    (C) decreasing velocity and constant acceleration
    (D) increasing velocity and decreasing acceleration
    (E) increasing velocity and increasing acceleration

5. A particle of negligible mass and charge $q = 1$ μC is fixed in place. A small object of mass $m = 10^{-3}$ kg and charge $q = 1$ μC is released from rest from a position 1 m directly above the particle. How far does the object fall before the electric force manages to push it away?

    (A) 0.1 m
    (B) 0.3 m
    (C) 0.5 m
    (D) 0.7 m
    (E) 0.9 m

6. The figure below shows four point charges and the cross section of a Gaussian surface:

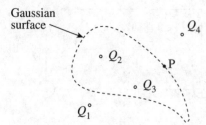

Which of the following statements is true concerning the situation depicted?

(A) The net electric flux through the Gaussian surface depends on all four charges shown, but the electric field at point P depends only on charges Q_2 and Q_3.

(B) The net electric flux through the Gaussian surface depends only on charges Q_2 and Q_3, but the electric field at point P depends on all four charges.

(C) The net electric flux through the Gaussian surface depends only on charges Q_2 and Q_3, and the electric field at point P depends only on charges Q_2, Q_3, and Q_4.

(D) The net electric flux through the Gaussian surface depends only on charges Q_1 and Q_4, and the electric field at point P depends only on charges Q_2 and Q_3.

(E) Both the net electric flux through the Gaussian surface and the electric field at point P depend on all four charges.

7. A nonconducting sphere of radius R contains a total charge of $-Q$ distributed uniformly throughout its volume (that is, the volume charge density, ρ, is constant).

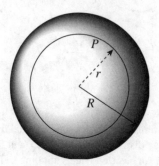

The magnitude of the electric field at point P, at a distance $r < R$ from the sphere's center, is equal to

(A) $\dfrac{1}{4\pi\varepsilon_0}\dfrac{Q}{R^3}r$

(B) $\dfrac{1}{4\pi\varepsilon_0}\dfrac{Q}{R^2}r^2$

(C) $\dfrac{1}{4\pi\varepsilon_0}\dfrac{Q}{R^3}r^3$

(D) $\dfrac{1}{4\pi\varepsilon_0}\dfrac{Q}{R^3r^2}$

(E) $\dfrac{1}{4\pi\varepsilon_0}\dfrac{Q}{r^2}$

8. Calculate the electric flux through a Gaussian surface of area A enclosing an electric dipole where each charge has magnitude q.

(A) 0
(B) $Aq/(4\pi\varepsilon_0)$
(C) $Aq^2/4\pi\varepsilon_0$
(D) $Aq/(4\pi\varepsilon_0 r)$
(E) $Aq/(4\pi\varepsilon_0 r^2)$

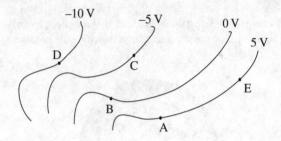

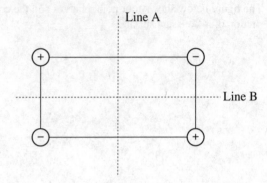

9. The diagram above shows equipotential lines produced by a charge distribution. A, B, C, D, and E are points in the plane. An electron begins at point A. The electron is then moved to point E and then from point E to point C. Which of the following correctly describes the work done *by the field* for each part of the movement?

	Movement from A to E	Movement from E to C
(A)	Negative	Positive
(B)	Zero	Positive
(C)	Zero	Negative
(D)	Negative	Zero
(E)	Positive	Positive

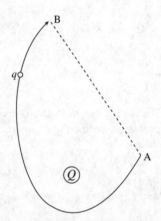

10. How much work would the electric field (created by the stationary charge Q) perform as a charge q is moved from point A to B along the curved path shown? $V_A = 200$ V, $V_B = 100$ V, $q = -0.05$ C, the length of line segment AB = 10 cm, and the length of the curved path = 20 cm.

(A) −10 J
(B) −5 J
(C) +5 J
(D) +10 J
(E) +2 J

11. The picture above shows 4 charges fixed in position at the corners of a rectangle measuring 2 cm by 4 cm. Assuming the charges are all of equal magnitude, how many locations on either Line A or Line B would be places with 0 net electric potential?

(A) 1
(B) 5
(C) All of Line A
(D) All of Line B
(E) All of both Line A and Line B

12. A particle with a charge of +q and mass m starts at rest and moves linearly from a position of high potential, A, to a position of low potential, B. Which of the following expressions will give the particle's speed at position B?

(A) $\sqrt{\dfrac{2q(V_A - V_B)}{m}}$

(B) $\sqrt{\dfrac{2q(V_B - V_A)}{m}}$

(C) $\sqrt{\dfrac{q(V_A - V_B)}{m}}$

(D) $\sqrt{\dfrac{q(V_A - V_B)}{2m}}$

(E) $\sqrt{\dfrac{2q(V_B - V_A)}{2m}}$

Section II: Free Response

1. In the figure shown, all four charges ($+Q$, $+Q$, $-q$, and $-q$) are situated at the corners of a square. The net electric force on each charge $+Q$ is zero.

    (a)   Express the magnitude of q in terms of Q.

    (b)   Is the net electric force on each charge $-q$ also equal to zero? Justify your answer.

    (c)   Determine the electric field at the center of the square.

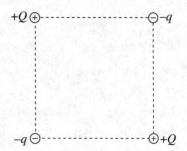

2. Two charges, $+Q$ and $+2Q$, are fixed in place along the y-axis of an x-y coordinate system as shown in the figure below. Charge 1 is at the point $(0, a)$, and Charge 2 is at the point $(0, -2a)$.

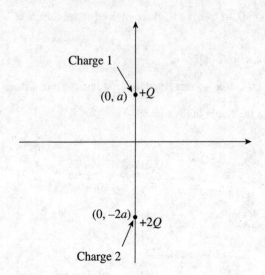

(a)   Find the electric force (magnitude and direction) felt by Charge 1 due to Charge 2.

(b)   Find the electric field (magnitude and direction) at the origin created by both Charges 1 and 2.

(c)   Is there a point on the x-axis where the total electric field is zero? If so, where? If not, explain briefly.

(d)   Is there a point on the y-axis where the total electric field is zero? If so, where? If not, explain briefly.

(e)   If a small negative charge, $-q$, of mass m were placed at the origin, determine its initial acceleration (magnitude and direction).

3. A positively charged, thin, nonconducting rod of length ℓ lies along the y-axis with its midpoint at the origin. The linear charge density within the rod is uniform and denoted by λ. Points P_1 and P_2 lie on the positive x-axis, at distances x_1 and x_2, respectively, from the rod.

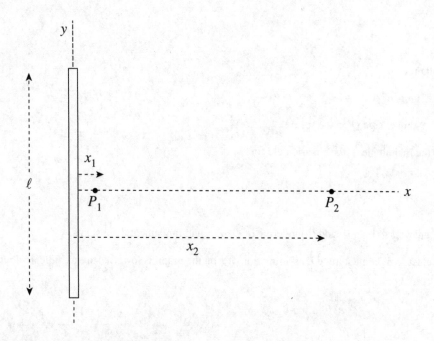

(a) Use Gauss's law to approximate the electric field at point P_1, given that x_1 is very small compared to ℓ. Write your answer in terms of λ, x_1, and fundamental constants.

(b) What is the total charge Q on the rod?

(c) Compute the electric field at point P_2, given that x_2 is not small compared to ℓ. For $x_2 = \ell$, write your answer in terms of Q, ℓ, and fundamental constants. You may use the fact that

$$\int (x^2 + y^2)^{-3/2}\, dy = \frac{y}{x^2\sqrt{x^2 + y^2}} + c$$

4. A solid glass sphere of radius a contains excess charge distributed throughout its volume such that the volume charge density depends on the distance r from the sphere's center according to the equation

$$\rho(r) = \rho_s(r/a)$$

where ρ_s is a constant.

(a) What are the units of ρ_s?

(b) Compute the total charge Q on the sphere.

(c) Determine the magnitude of the electric field for

(i) $r < a$

(ii) $r \geq a$

Write your answers to both (i) and (ii) in terms of Q, a, r, and fundamental constants.

(d) Sketch the electric field magnitude E as a function of r on the graph below. Be sure to indicate on the vertical axis the value of E at $r = a$.

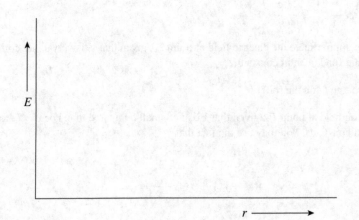

Summary

General Information

- There are positive and negative electric charges. Opposite charges attract and like charges repel.

- Conductors allow charge to flow and insulators do not allow charge to flow.

Electric Forces

- The force between two charges is given by Coulomb's law: $F = k\dfrac{q_1 q_2}{r^2}$.

Electric Fields

- The electric field indicates the direction of the force on a positive test charge, q.

- The strength of the electric field is given by $E = \dfrac{F_{on\,q}}{q}$.

- For a point charge, Q, the electric field is

$$E = k\frac{Q}{r^2}$$

Electric Flux and Gauss's Law

- Electric flux is given by the number of electric field lines that pass through a surface.

- It is given by the dot product of the electric field and area vector: $\Phi_E = \mathbf{E} \cdot \mathbf{A}$.

- For a variable electric field, the flux is given by $\Phi_E = \displaystyle\int \mathbf{E} \cdot d\mathbf{A}$.

o Gauss's law is given by the equation

$$\Phi_E = \oint E \cdot dA = \frac{Q_{enc}}{\varepsilon_0}$$

It can also be shown as this:

$$\int_{\substack{closed \\ surface}} E \cdot dA = \frac{Q_{enclosed}}{\varepsilon_0}$$

o Gauss's law is generally used to solve for the electric field created by symmetric charge distributions (for example, spheres, spherical shells, lines of charge, and a sheet of charge).

Electric Potential

o When a positive test charge q_0 moves through an electric field, its electric potential energy changes. The change in electric potential energy is: $\Delta U = -q_0 \int E \cdot ds$. Note that this is a dot product between the electric field and the path, ds.

o When the electric field is uniform, then $\Delta U = -q_0 E d$, where d is the distance along the electric field that the charge moved.

o The electric potential is the electric potential energy divided by the test charge. Therefore, $\Delta V = \frac{\Delta U}{q_0} = -\int \mathbf{E} \cdot ds$. This can also be rearranged as $-\frac{dV}{dr} = E$. Either one of these equations can be used to relate the electric potential to the electric field strength. The units of electric potential are volts (V), and it is often called *voltage*.

o When the electric field is uniform, the electric potential is given by the strength of the electric field and the distance the object moves along the electric field:

$$\Delta V = -E d$$

o The electric potential due to a point charge, Q, is $V = k\frac{Q}{r}$.

o The electric potential energy between a pair of charges is $U = k\frac{q_1 q_2}{r}$.

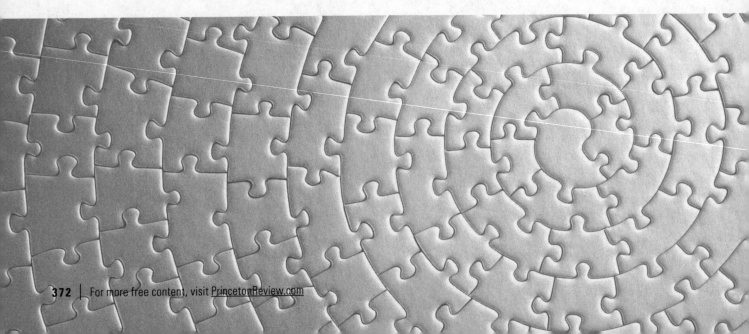

Chapter 13
Conductors, Capacitors, and Dielectrics

INTRODUCTION

When an object moves in a gravitational field, it usually experiences a change in kinetic energy and in gravitational potential energy due to the work done on the object by gravity. Similarly, when a charge moves in an electric field, it generally experiences a change in kinetic energy and in electrical potential energy due to the work done on it by the electric field. By exploring the idea of electric potential, we can simplify our calculations of work and energy changes within electric fields.

CONDUCTORS AND INSULATORS

Materials can be classified into broad categories based on their ability to permit the flow of charge. Materials that permit the flow of charge are called **conductors**; they conduct electricity. Metals are the best examples of conductors, but other conductors are aqueous solutions that contain dissolved electrolytes (such as salt water). Metals conduct electricity because they consist of positively charged nuclei with outermost electrons that are not tightly bound to any one atom. Instead, the electrons are distributed uniformly throughout the structure and are not bound to a single nucleus. In the presence of an electric field, the electrons are free to move about the lattice, creating a sort of sea of mobile (or conduction) electrons. This freedom allows excess charge to flow freely.

The outermost electrons in an **insulator**, on the other hand, have no such mobility. There is no pathway by which electrons can move through the solid because the energy required to mobilize the electrons is prohibitively large. Examples of insulators are glass, wood, rubber, and plastic.

Midway between conductors and insulators is a class of materials known as **semiconductors**. That is, they are less conducting than most metals, but more conducting than most insulators. Examples of semiconducting materials are silicon and germanium.

An extreme example of a conductor is the **superconductor**. This is a material that offers absolutely no resistance to the flow of charge; it is a *perfect* conductor of electric charge because an electrical current can flow forever. Many metals and ceramics become superconducting when they are brought to extremely low temperatures.

Example 1 A solid sphere of copper is given a negative charge. Discuss the electric field inside and outside the sphere.

Solution. The excess electrons that are deposited on the sphere move quickly to the outer surface (copper is a great conductor). *Any excess charge on a conductor resides entirely on the outer surface.*

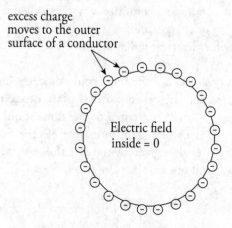

excess charge
moves to the outer
surface of a conductor

Electric field
inside = 0

Once these excess electrons establish a uniform distribution on the outer surface of the sphere, there will be no net electric field within the sphere. Why not? Since there is no additional excess charge inside the conductor, there are no excess charges to serve as a source or sink of an electric field line cutting down into the sphere, because field lines begin or end on excess charges.

> *There can be no electrostatic field within the body of a conductor.*

In fact, you can shield yourself from electric fields simply by surrounding yourself with metal. Charges may move around on the outer surface of your cage, but within the cage, the electric field will be zero. For points outside the sphere, it can be shown that the sphere behaves as if all its excess charge were concentrated at its center. (Remember that this is just like the gravitational field due to a uniform spherical mass.) Also, *the electric field is always perpendicular to the surface, no matter what shape the surface may be.* See the diagram below.

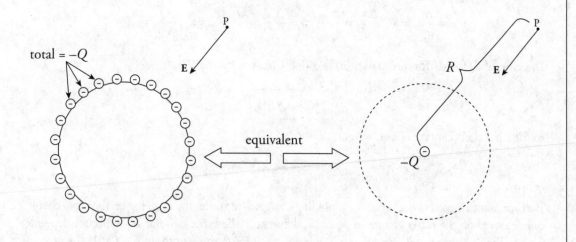

total = –Q

P

E

equivalent

R

P

E

–Q

CAPACITORS

Consider two conductors, separated by some distance, that carry equal but opposite charges, $+Q$ and $-Q$. Such a pair of conductors comprise a system called a **capacitor**. Work must be done to create this separation of charge, and, as a result, potential energy is stored. Capacitors are basically storage devices for electrical potential energy.

The conductors may have any shape, but the most common capacitors are parallel metal plates or sheets. These types of capacitors are called **parallel-plate capacitors**. We'll assume that the distance d between the plates is small compared to the dimensions of the plates since, in this case, the electric field between the plates is uniform. The electric field due to *one* such plate, if its surface charge density is $\sigma = Q/A$, is given by the equation $E = \sigma/(2\varepsilon_0)$, with **E** pointing away from the sheet if σ is positive and toward the plate if σ is negative.

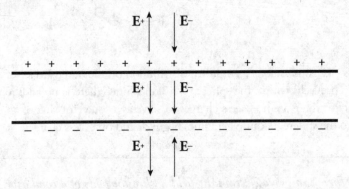

Therefore, with two plates, one with surface charge density $+\sigma$ and the other, $-\sigma$, the electric fields combine to give a field that's zero outside the plates and that has the magnitude

$$E_{\text{total}} = \frac{\sigma}{2\varepsilon_0} + \frac{\sigma}{2\varepsilon_0} = \frac{\sigma}{\varepsilon_0}$$

in between. In Example 29 in Chapter 12, we learned that the magnitude of the potential difference, ΔV, between the plates satisfies the relationship $\Delta V = Ed$, so combining this with the previous equation, we get

$$E = \frac{\sigma}{\varepsilon_0} \implies \frac{\Delta V}{d} = \frac{\sigma}{\varepsilon_0} \implies \frac{\Delta V}{d} = \frac{Q/A}{\varepsilon_0} \implies \frac{Q}{\Delta V} = \frac{\varepsilon_0 A}{d}$$

The ratio of Q to ΔV, for *any* capacitor, is called its **capacitance** (C),

$$\boxed{C = Q/\Delta V}$$

so for a parallel-plate capacitor, we get

$$C = \frac{\varepsilon_0 A}{d}$$

The capacitance measures the capacity for holding charge. The greater the capacitance, the more charge can be stored on the plates at a given potential difference. The capacitance of any capacitor depends only on the size, shape, and separation of the conductors. From the definition, $C = Q/\Delta V$, the units of C are coulombs per volt. One coulomb per volt is renamed one **farad** (abbreviated F): 1 C/V = 1 F.

Example 2 A 10-nanofarad parallel-plate capacitor holds a charge of magnitude 50 μC on each plate.

(a) What is the potential difference between the plates?

(b) If the plates are separated by a distance of 0.2 mm, what is the area of each plate?

Solution.

(a) From the definition, $C = Q/\Delta V$, we find that

$$\Delta V = \frac{Q}{C} = \frac{50 \times 10^{-6} \text{ C}}{10 \times 10^{-9} \text{ F}} = 5{,}000 \text{ V}$$

(b) From the equation $C = \varepsilon_0 A/d$, we can calculate the area, A, of each plate:

$$A = \frac{Cd}{\varepsilon_0} = \frac{(10 \times 10^{-9} \text{ F})(0.2 \times 10^{-3} \text{ m})}{8.85 \times 10^{-12} \text{ C}^2/\text{N} \cdot \text{m}^2} = 0.23 \text{ m}^2$$

Capacitors of Other Geometries

Example 3 A long cable consists of a solid conducting cylinder of radius a, which carries a linear charge density of $+\lambda$, concentric with an outer cylindrical shell of radius b, which carries a linear charge density of $-\lambda$. This is a **coaxial** cable. Determine the capacitance of the cable.

Solution. We first apply the definition, $C = Q/\Delta V$. In Chapter 12, Example 32, we saw that the electric field outside of a conducting cylinder is given by the equation

$$E = \frac{\lambda}{2\pi\varepsilon_0 r}$$

Therefore, we can get the potential difference between the inner cylinder and the outer cylindrical shell by integrating the electric field. Since

$$V_B - V_A = \Delta V_{A \to B} = \frac{-W_{E,\,A \to B} \text{ on } q}{q} = -\frac{1}{q}\int_A^B q\mathbf{E} \cdot d\mathbf{r} = \int_B^A \mathbf{E} \cdot d\mathbf{r}$$

we have

$$V_{\text{inner}} - V_{\text{outer}} = \int_{r=a}^{r=b} \frac{\lambda}{2\pi\varepsilon_0 r}\,dr = \frac{\lambda}{2\pi\varepsilon_0}[\ln r]_a^b = \frac{\lambda}{2\pi\varepsilon_0}\ln\frac{b}{a}$$

Therefore, over a length ℓ of cable, the magnitude of the charge Q on each cylinder is $\lambda \ell$, so by the definition of capacitance, we get

$$C = \frac{Q}{\Delta V} = \frac{\lambda \ell}{\frac{\lambda}{2\pi\varepsilon_0} \ln \frac{b}{a}} = \frac{2\pi\varepsilon_0 \ell}{\ln \frac{b}{a}}$$

Example 4 A spherical conducting shell of radius a, which carries a charge of $+Q$, is concentric with an outer spherical shell of radius b, which carries a charge of $-Q$. What is the capacitance of this spherical capacitor?

Solution. We first apply the definition, $C = Q/\Delta V$. We know that, in the region between the shells, the field is due to the inner shell alone:

$$E = \frac{1}{4\pi\varepsilon_0} \frac{Q}{r^2}$$

Therefore, the potential difference between the inner spherical shell and the outer spherical shell is obtained by integrating the electric field as follows. Since

$$V_B - V_A = \Delta V_{A \to B} = \frac{-W_{E, A \to B} \text{ on } q}{q} = -\frac{1}{q}\int_A^B q\mathbf{E} \cdot d\mathbf{r} = \int_B^A \mathbf{E} \cdot d\mathbf{r}$$

we have

$$V_{\text{inner}} - V_{\text{outer}} = \int_{r=a}^{r=b} \frac{1}{4\pi\varepsilon_0} \frac{Q}{r^2} dr = \frac{Q}{4\pi\varepsilon_0} \int_{r=a}^{r=b} \frac{1}{r^2} dr = \frac{Q}{4\pi\varepsilon_0}\left[-\frac{1}{r}\right]_a^b = \frac{Q}{4\pi\varepsilon_0}\left(\frac{1}{a} - \frac{1}{b}\right)$$

So, by definition of capacitance, we get

$$C = \frac{Q}{\Delta V} = \frac{Q}{\frac{Q}{4\pi\varepsilon_0}\left(\frac{1}{a} - \frac{1}{b}\right)} = \frac{4\pi\varepsilon_0}{\frac{1}{a} - \frac{1}{b}} = 4\pi\varepsilon_0 \frac{ab}{b-a}$$

The capacitance of a single, isolated conductor can also be defined; we can just think of the other conductor as being infinitely far away. In this case, if a conductor holds a charge Q at a potential of ΔV, then its capacitance is defined as $C = Q/\Delta V$. For a single sphere of radius a, we determined earlier in this chapter that the potential at its surface is $V = (1/4\pi\varepsilon_0)Q/a$, so its capacitance is

$$C = \frac{Q}{\Delta V} = \frac{Q}{\frac{1}{4\pi\varepsilon_0}\frac{Q}{a}} = 4\pi\varepsilon_0 a$$

which says that C is proportional to the radius of the sphere. This makes sense; a larger sphere would have the capacity to hold more charge and should therefore have a greater capacitance. This same result for the capacitance of a single sphere can be obtained from the equation we derived on the previous page for the spherical shell capacitor simply by letting the radius b of the outer sphere go to infinity:

$$\lim_{b \to \infty} \frac{4\pi\varepsilon_0}{\frac{1}{a} - \frac{1}{b}} = \frac{4\pi\varepsilon_0}{\frac{1}{a} - 0} = 4\pi\varepsilon_0 a$$

Combinations of Capacitors

Capacitors are often arranged in combination in electric circuits. Here, we'll look at two types of arrangements, the parallel combination and the series combination.

A collection of capacitors are said to be in **parallel** if they all share the same potential difference. The following diagram shows two capacitors wired in parallel:

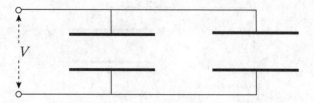

The top plates are connected by a wire and form a single equipotential; the same is true for the bottom plates. Therefore, the potential difference across one capacitor is the same as the potential difference across the other capacitor.

We want to find the capacitance of a *single* capacitor that would perform the same function as this combination. If the capacitances are C_1 and C_2, then the charge on the first capacitor is $Q_1 = C_1 \Delta V$ and the charge on the second capacitor is $Q_2 = C_2 \Delta V$. The total charge on the combination is $Q_1 + Q_2$, so the equivalent capacitance, C_P, must be

$$C_P = \frac{Q}{\Delta V} = \frac{Q_1 + Q_2}{\Delta V} = \frac{Q_1}{\Delta V} + \frac{Q_2}{\Delta V} = C_1 + C_2$$

On the equation sheet for the free-response section, this information will be represented as follows:

$$C_p = \sum_i C_i$$

So, the **equivalent** capacitance of a collection of capacitors in parallel is found by adding the individual capacitances.

A collection of capacitors are said to be in **series** if they all share the same charge magnitude. The following diagram shows two capacitors wired in series:

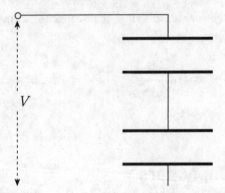

When a potential difference is applied, as shown, negative charge will be deposited on the bottom plate of the bottom capacitor; this will push an equal amount of negative charge away from the top plate of the bottom capacitor toward the bottom plate of the top capacitor. When the system has reached equilibrium, the charges on all the plates will have the same magnitude:

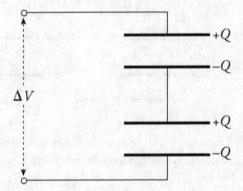

If the top and bottom capacitors have capacitances of C_1 and C_2, respectively, then the potential difference across the top capacitor is $\Delta V_1 = Q/C_1$, and the potential difference across the bottom capacitor is $\Delta V_2 = Q/C_2$. The total potential difference across the combination is $\Delta V_1 + \Delta V_2$, which must equal ΔV. Therefore, the equivalent capacitance, C_S, must be

$$C_S = \frac{Q}{\Delta V} = \frac{Q}{\Delta V_1 + \Delta V_2} = \frac{Q}{\frac{Q}{C_1} + \frac{Q}{C_2}} = \frac{1}{\frac{1}{C_1} + \frac{1}{C_2}}$$

We can write this in another form:

$$\frac{1}{C_S} = \frac{1}{C_1} + \frac{1}{C_2}$$

On the equation sheet for the free-response section, this information will be represented as follows:

$$\frac{1}{C_S} = \sum_i \frac{1}{C_i}$$

In other words, the *reciprocal* of the capacitance of a collection of capacitors in series is found by adding the reciprocals of the individual capacitances.

Example 5 Given that $C_1 = 2\ \mu F$, $C_2 = 4\ \mu F$, and $C_3 = 6\ \mu F$, calculate the equivalent capacitance for the following combination:

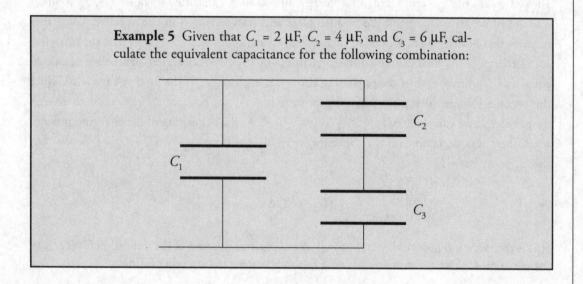

Solution. Notice that C_2 and C_3 are in series, and they are in parallel with C_1. That is, the capacitor equivalent to the series combination of C_2 and C_3 (which we'll call $C_{2\text{-}3}$) is in parallel with C_1. We can represent this as follows:

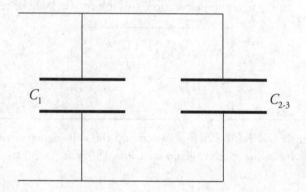

So, the first step is to find $C_{2\text{-}3}$:

$$\frac{1}{C_{2\text{-}3}} = \frac{1}{C_2} + \frac{1}{C_3} \quad \Rightarrow \quad C_{2\text{-}3} = \frac{C_2 C_3}{C_2 + C_3}$$

Now this is in parallel with C_1, so the overall equivalent capacitance ($C_{1\text{-}2\text{-}3}$) is

$$C_{1\text{-}2\text{-}3} = C_1 + C_{2\text{-}3} = C_1 + \frac{C_2 C_3}{C_2 + C_3}$$

Substituting in the given numerical values, we get

$$C_{1\text{-}2\text{-}3} = (2\ \mu F) + \frac{(4\ \mu F)(6\ \mu F)}{(4\ \mu F) + (6\ \mu F)} = 4.4\ \mu F$$

The Energy Stored in a Capacitor

To figure out the electrical potential energy stored in a capacitor, imagine taking a small amount of negative charge off the positive plate and transferring it to the negative plate. This requires that positive work is done by an external agent, and this is the reason that the capacitor stores energy. If the final charge on the capacitor is Q, then we transferred an amount of charge equal to Q, fighting against the prevailing voltage at each stage. If the final voltage is ΔV, then the average voltage during the charging process is $\frac{1}{2}\Delta V$; so, because ΔU_E is equal to charge times voltage, we can write $\Delta U_E = Q \cdot \frac{1}{2}\Delta V = \frac{1}{2}Q\Delta V$. At the beginning of the charging process, when there was no charge on the capacitor, we had $U_i = 0$, so $\Delta U_E = U_f - U_i = U_f - 0 = U_f$; therefore, we have

$$U_E = \tfrac{1}{2}Q\Delta V$$

This is the electrical potential energy stored in a capacitor. Because of the definition $C = Q/\Delta V$, the equation for the stored potential energy can be written in two other forms:

$$U_E = \tfrac{1}{2}(C\Delta V)\cdot \Delta V = \tfrac{1}{2}C(\Delta V)^2 \quad \text{or} \quad U_E = \tfrac{1}{2}Q\cdot\frac{Q}{C} = \frac{Q^2}{2C}$$

On the equation sheet for the free-response section, this information will be represented as follows:

$$U_C = \frac{1}{2}Q\Delta V = \frac{1}{2}C\Delta V^2$$

Example 6 Find the charge stored and the voltage across each capacitor in the following circuit, given that $\mathcal{E} = 180$ V, $C_1 = 30\ \mu\text{F}$, $C_2 = 60\ \mu\text{F}$, and $C_3 = 90\ \mu\text{F}$.

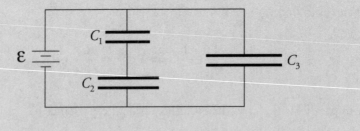

Solution. Once the charging currents stop, the voltage across C_3 is equal to the voltage across the battery, so $V_3 = 180$ V. This gives us $Q_3 = C_3 V_3 = (90 \ \mu\text{F})(180 \ \text{V}) = 16.2$ mC. Since C_1 and C_2 are in series, they must store identical amounts of charge and, from the diagram, the sum of their voltages must equal the voltage of the battery. So, if we let Q be the charge on each of these two capacitors, then $Q = C_1 V_1 = C_2 V_2$, and $V_1 + V_2 = 180$ V. The equation $C_1 V_1 = C_2 V_2$ becomes $(30 \ \mu\text{F}) V_1 = (60 \ \mu\text{F}) V_2$, so $V_1 = 2V_2$. Substituting this into $V_1 + V_2 = 180$ V gives us $V_1 = 120$ V and $V_2 = 60$ V. The charge stored on each of these capacitors is

$$(30 \ \mu\text{F})(120 \ \text{V}) = C_1 V_1 = C_2 V_2 = (60 \ \mu\text{F})(60 \ \text{V}) = 3.6 \ \text{mC}$$

Example 7 In the diagram below, $C_1 = 2$ mF and $C_2 = 4$ mF. When Switch S is open, a battery (which is not shown) is connected between points a and b and charges capacitor C_1 so that $V_{ab} = 12$ V. The battery is then disconnected.

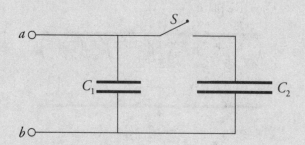

After the switch is closed, what will be the common voltage across each of the parallel capacitors (once electrostatic conditions are reestablished)?

Solution. When C_1 is fully charged, the charge on (each of the plates of) C_1 has the magnitude $Q = C_1 V = (2 \ \text{mF})(12 \ \text{V}) = 24$ mC. After the switch is closed, this charge will be redistributed in such a way that the resulting voltages across the two capacitors, V', are equal. This happens because the capacitors are in parallel. So if Q_1' is the new charge magnitude on C_1 and Q_2' is the new charge magnitude on C_2, we have $Q_1' + Q_2' = Q$, so $C_1 V' + C_2 V' = Q$, which gives us:

$$V' = \frac{Q}{C_1 + C_2} = \frac{24 \ \text{mC}}{2 \ \text{mF} + 4 \ \text{mF}} = 4 \ \text{V}$$

DIELECTRICS

One method of keeping the plates of a capacitor apart, which is necessary to maintain charge separation and store potential energy, is to insert an insulator (called a **dielectric**) between the plates.

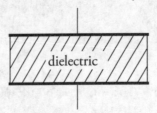

A dielectric always increases the capacitance of a capacitor. Let's see why this happens. Imagine charging a capacitor to a potential difference of V with charge $+Q$ on one plate and $-Q$ on the other. Now disconnect the capacitor from the charging source and insert a dielectric. What happens? Although the dielectric is not a conductor, the electric field that existed between the plates causes the molecules within the dielectric material to polarize; there is more electron density on the side of the molecule near the positive plate.

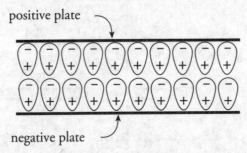

The effect of this is to form a layer of negative charge along the top surface of the dielectric and a layer of positive charge along the bottom surface; this separation of charge induces its own electric field ($\mathbf{E}_i$), within the dielectric, that opposes the original electric field, $\mathbf{E}$, within the capacitor:

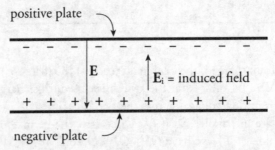

So the overall electric field has been reduced from its previous value: $\mathbf{E}_{total} = \mathbf{E} + \mathbf{E}_i$, and $E_{total} = E - E_i$. Let's say that the electric field has been reduced by a factor of κ (the Greek letter *kappa*) from its original value

$$E_{\text{with dielectric}} = E_{\text{without dielectric}} - E_i = \frac{E}{\kappa}$$

Since $\Delta V = Ed$ for a parallel-plate capacitor, we see that ΔV must have decreased by a factor of κ. But $C = Q/\Delta V$, so if ΔV decreases by a factor of κ, then C increases by a factor of κ:

$$C_{\text{with dielectric}} = \kappa C_{\text{without dielectric}}$$

The value of κ, called the **dielectric constant**, varies from material to material, but it's always greater than 1.

This implies that the full equation for the capacitance of a parallel-plate capacitor is

$$C = \frac{\kappa \varepsilon_0 A}{d}$$

Although the description for why the capacitance increases with the insertion of a dielectric assumed that the source that charged the capacitor was disconnected (so that Q remains constant), the same result holds if the source of potential remains connected to the capacitor. The capacitance still increases, but because ΔV must now remain constant, the equation $Q = C\Delta V$ tells us that more charge will appear on the plates (because C increases).

The presence of a dielectric also limits the potential difference that can be applied across the plates. If ΔV gets too high, then $E = \Delta V/d$ gets so strong that electrons in the dielectric material can be ejected from their atoms and propelled toward the positive plate. This discharges the capacitor (and typically burns a hole through the dielectric). This event is called **dielectric breakdown**.

The capacitance formulas derived in this chapter have assumed that no dielectric was present; the permittivity constant that appears in the formulas is ε_0, the permittivity of *free space* (vacuum). If a dielectric is present, then the permittivity increases to $\varepsilon = \kappa \varepsilon_0$, so the occurrence of ε_0 in each formula is simply replaced by $\varepsilon = \kappa \varepsilon_0$.

Example 8 The plates of a parallel-plate capacitor are separated by a distance of 2.0 mm and each has an area of 10 cm^2. If a layer of polystyrene (whose dielectric constant is 2.6) is sandwiched between the plates, calculate

(a) the capacitance
(b) the maximum amount of charge that can be placed on the plates, given that polystyrene suffers dielectric breakdown if the electric field exceeds 20 million volts per meter.

Solution.

(a) The capacitance of the parallel-plate capacitor, with a dielectric, is

$$C = \frac{\kappa \varepsilon_0 A}{d} = \frac{(2.6)(8.85 \times 10^{-12} \text{ F/m}) \left[10 \text{ cm}^2 \cdot \left(\frac{1 \text{ m}}{100 \text{ cm}} \right)^2 \right]}{2.0 \times 10^{-3} \text{ m}} = 1.2 \times 10^{-11} \text{ F}$$

Notice the units for ε_0; in calculations that involve capacitance, it is usually easier to write F/m rather than $C^2/N \cdot m^2$.

(b) First, note the units for the electric field: V/m. These units follow from the equation $\Delta V = Ed$. In the past, we've written the units of E as N/C. These two units are actually equivalent: 1 V/m = 1 N/C. We were asked to determine Q_{max} if E_{max} is 20×10^6 V/m and, from the equations $\Delta V = Ed$ and $Q = C \Delta V$, we get

$$\begin{aligned} Q_{max} &= C \Delta V_{max} = C E_{max} d \\ &= (1.2 \times 10^{-11} \text{ F})(20 \times 10^6 \text{ V/m})(2 \times 10^{-3} \text{ m}) \\ &= 4.8 \times 10^{-7} \text{ C} \end{aligned}$$

Chapter 13 Drill

The answers and explanations can be found in Chapter 17.

Section I: Multiple Choice

1. A capacitor is fully charged by a battery. The battery is disconnected, and a dielectric is inserted into the capacitor. Which of the following statements is/are true?

    I. The voltage will remain the same.
    II. The potential energy of the capacitor will increase.
    III. The capacitance of the capacitor will increase.

    (A) I only
    (B) I and II only
    (C) I and III only
    (D) II and III only
    (E) III only

2. If the electric field does negative work on a negative charge as the charge undergoes a displacement from Position A to Position B within an electric field, then the electrical potential energy

    (A) is negative
    (B) is positive
    (C) increases
    (D) decreases
    (E) cannot be determined from the information given

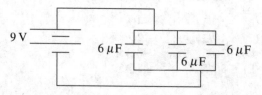

3. Three 6 μF capacitors are connected in parallel to a 9 V battery as shown above. Determine the energy stored in each capacitor.

    (A) 243 J
    (B) 7.29×10^{-4} J
    (C) 8.10×10^{-5} J
    (D) 2.43×10^{-4} J
    (E) 27 J

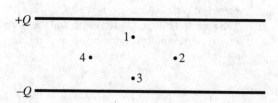

4. Which points in this uniform electric field (between the plates of the capacitor) shown above lie on the same equipotential?

    (A) 1 and 2 only
    (B) 1 and 3 only
    (C) 2 and 4 only
    (D) 3 and 4 only
    (E) 1, 2, 3, and 4 all lie on the same equipotential since the electric field is uniform.

5. Two isolated and widely separated conducting spheres each carry a charge of $-Q$. Sphere 1 has a radius of a and Sphere 2 has a radius of $4a$. If the spheres are now connected by a conducting wire, what will be the final charge on each sphere?

	Sphere 1	Sphere 2
(A)	$-Q$	$-Q$
(B)	$-2Q/3$	$-4Q/3$
(C)	$-4Q/3$	$-2Q/3$
(D)	$-2Q/5$	$-8Q/5$
(E)	$-8Q/5$	$-2Q/5$

6. A parallel-plate capacitor is charged to a potential difference of ΔV; this results in a charge of $+Q$ on one plate and a charge of $-Q$ on the other. The capacitor is disconnected from the charging source, and a dielectric is then inserted. What happens to the potential difference and the stored electrical potential energy?

(A) The potential difference decreases, and the stored electrical potential energy decreases.
(B) The potential difference decreases, and the stored electrical potential energy increases.
(C) The potential difference increases, and the stored electrical potential energy decreases.
(D) The potential difference increases, and the stored electrical potential energy increases.
(E) The potential difference decreases, and the stored electrical potential energy remains unchanged.

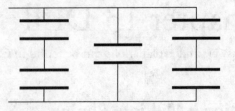

7. If each of the capacitors in the array shown above is C, what is the capacitance of the entire combination?

(A) $C/2$
(B) $2C/3$
(C) $5C/6$
(D) $2C$
(E) $5C/3$

Section II: Free Response

1. The figure below shows a parallel-plate capacitor. Each rectangular plate has length L and width w, and the plates are separated by a distance d.

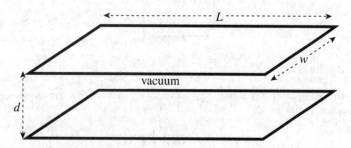

(a) Determine the capacitance.

An electron (mass m, charge $-e$) is shot horizontally into the empty space between the plates, midway between them, with an initial velocity of magnitude v_0. The electron just barely misses hitting the end of the top plate as it exits. (Ignore gravity.)

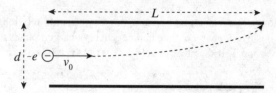

(b) In the diagram, sketch the electric field vector at the position of the electron when it has traveled a horizontal distance of $L/2$.

(c) In the diagram, sketch the electric force vector on the electron at the same position as in part (b).

(d) Determine the strength of the electric field between the plates. Write your answer in terms of L, d, m, e, and v_0.

(e) Determine the charge on the top plate.

(f) How much potential energy is stored in the capacitor?

2. In the figure shown, you have two conducting plates of area A with a distance D between them.

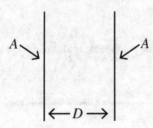

If the conductance of this capacitor is C_0:

(a) what happens to the conductance of this capacitor if a dielectric, with a dielectric constant of 2, is inserted halfway between the plates, as shown below?

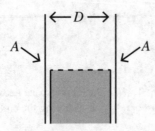

(b) Now, imagine the plates no longer have an area of A each, but they are infinitely long, as shown below. For points R, S, and T in the figure, with the positively charged plate and negatively charged plates as shown, draw the vector representing the electric field at points R, S, and T. If the electric field is 0, write "E = 0" next to the vector.

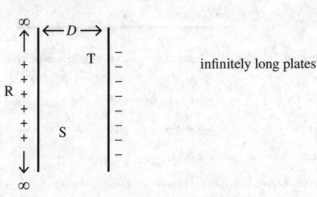

infinitely long plates

(i)   R

(ii) S

(iii) T

3. In the figure shown, you have a 12V battery and 6 different capacitors in the circuit, some in series and some in parallel.

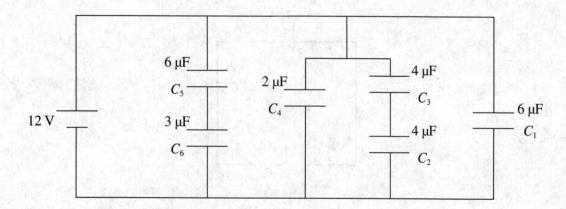

(a) What is the charge on C_1 as noted on the capacitor on the far right?

(b) What is the equivalent capacitance for the entire circuit?

(c) What are the voltages across each of the capacitors, C_1 through C_6?

4. In the figure shown, you have a simple circuit with a battery of 12 V and a capacitance of 1 F.

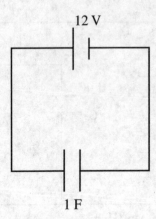

In each of the following four scenarios, describe what happens to C, the capacitance, Q, the charge, V, the voltage, E, the Electric Field, and U, the energy.

(a) The battery remains attached, and a material is inserted in the capacitor that has a dielectric constant of 2.

(b) The battery is detached (so the capacitor is isolated), and a material with a dielectric constant of 2 is inserted.

(c) The battery remains attached, and the plates of the capacitor are pulled apart so that the distance between the plates is doubled from d to $2d$.

(d) The battery is detached (so the capacitor is isolated), and the plates of the capacitor are pulled apart so that the distance between the plates is doubled from d to $2d$.

Summary

Capacitance

o Capacitors store electric potential energy by separating opposite charges. Capacitance is measured by the ratio of the charge stored to the potential difference across the capacitor: $C = \dfrac{Q}{\Delta V}$. The units are farads (F).

o The capacitance of a parallel-plate capacitor is $C = \dfrac{\varepsilon_0 A}{d}$.

o When capacitors are in parallel, the equivalent capacitance is the sum of the capacitances: $C_{eq} = C_1 + C_2 + \ldots = \sum_i C_i$.

o When capacitors are in series, the reciprocal of the equivalent capacitance is the sum of the reciprocals of each capacitance: $\dfrac{1}{C_{eq}} = \dfrac{1}{C_1} + \dfrac{1}{C_2} + \ldots = \sum_i \dfrac{1}{C_i}$.

o The energy stored in a capacitor is

$$U = \frac{1}{2} C \Delta V^2$$

o Dielectrics are used in capacitors to induce an opposing electric field that decreases the total electric field between the plates. This decreases the electric potential and therefore increases the capacitance.

Chapter 14
Electric Circuits

INTRODUCTION

In the previous chapter, when we studied electrostatic fields, we learned that within a conductor, an electrostatic field cannot be sustained; the source charges move to the surface and the conductor forms a single equipotential. We will now look at conductors within which an electric field can be sustained because a power source maintains a potential difference across the conductor, allowing charges to continually move through it. This ordered motion of charge through a conductor is called **electric current**.

ELECTRIC CURRENT

Picture a piece of metal wire. Within the metal, electrons are zooming around at speeds of about a million m/s in random directions, colliding with other electrons and positive ions in the lattice. This constitutes charge in motion, but it doesn't constitute *net* movement of charge because the electrons move randomly. If there's no net motion of charge, there's no current. However, if we were to create a potential difference between the ends of the wire (if we set up an electric field), the electrons would experience an electric force, and they would start to drift through the wire.

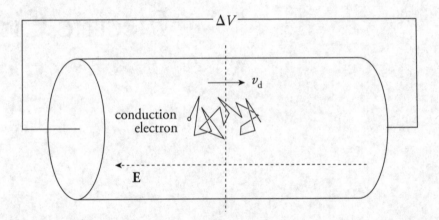

This is current. Although the electric field would travel through the wire at nearly the speed of light, the electrons themselves would still have to make their way through a crowd of atoms and other free electrons, so their **drift speed**, v_d, would be relatively slow: on the order of a millimeter per second.

To measure the current, we have to measure how much charge crosses a plane per unit of time. If an amount of charge of magnitude ΔQ crosses an imaginary plane in a time interval Δt, then the **current** is

$$I = \frac{\Delta Q}{\Delta t}$$

If the amount of current is changing during the time interval Δt, then the equation from the previous page gives the average current. We can define the instantaneous current as

$$I = \frac{dQ}{dt}$$

Because current is charge per unit time, it's expressed in coulombs per second. One coulomb per second is an **ampere** (abbreviated as **A**), or amp. So, 1 C/s = 1 A.

Although the charge carriers that constitute the current within a metal are electrons, the direction of the current is taken to be the direction in which *positive* charge carriers would move. So, if the conduction electrons drift to the right, we'd say the current points toward the left.

> This convention dates back to Benjamin Franklin's experiments before the electron had been discovered. It is the accepted convention and is used on the AP Physics C Exam.

Resistance

Let's say we had a copper wire and a glass fiber that had the same length and cross-sectional area, and we hooked up the ends of the metal wire to a source of potential difference and measured the resulting current. If we were to do the same thing with the glass fiber, the current would probably be too small to measure, but why? Well, the glass provided more resistance to the flow of charge. If the potential difference is ΔV and the current is I, then the **resistance** is

$$R = \frac{\Delta V}{I}$$

On the equation sheet for the free-response section, this information will be represented as follows:

$$\Delta V = IR$$

This is known as **Ohm's law**. Ohm's law is not a "law" in the sense that all circuit elements follow it. Only certain conductors do, and these devices are called "ohmic." Devices for which the ratio $\Delta V/I$ is not constant are called "non-ohmic." A light bulb is an example of a non-ohmic device. Notice that if the current is large, the resistance is low, and if the current is small, then resistance is high. The Δ in the equation above is often omitted, but you should always assume that, in this context, $V = \Delta V$ = potential difference, also called voltage.

Because resistance is voltage divided by current, it is expressed in volts per amp. One volt per amp is one **ohm** (Ω, *omega*). So, 1 V/A = 1 Ω.

Resistivity

The resistance of an object depends on two things: what the material it is made of and its shape. For example, again think of the copper wire and glass fiber of the same length and area. They have the same shape, but their resistances are different because they're made of different materials. Glass has a much greater intrinsic resistance than copper does; it has a greater **resistivity**. Each material has its own characteristic resistivity. Moreover, resistance depends on how the material is shaped. For a wire of length L and cross-sectional area A made of a material with resistivity ρ, resistance is given by

> The resistivity of most materials has a dependence on temperature: the higher the temperature, the higher the resistivity. However, for typical ambient temperatures, resistivity is essentially constant for a given material.

$$R = \frac{\rho L}{A}$$

The resistivity of copper is around 10^{-8} $\Omega \cdot$m, while the resistivity of glass is *much* greater, around 10^{12} $\Omega \cdot$m.

> **Example 1** A cylindrical wire of radius 1 mm and length 2 m is made of platinum (resistivity = 1×10^{-7} $\Omega \cdot$m). If a voltage of 9 V is applied between the ends of the wire, what will be the resulting current?

Solution. First, the resistance of the wire is given by the equation

$$R = \frac{\rho L}{A} = \frac{\rho L}{\pi r^2} = \frac{(1 \times 10^{-7} \ \Omega \cdot \text{m})(2 \ \text{m})}{\pi (0.001 \ \text{m})^2} = 0.064 \ \Omega$$

Then, from $I = V/R$, we get

$$I = \frac{V}{R} = \frac{9 \ \text{V}}{0.064 \ \Omega} = 140 \ \text{A}$$

ELECTRIC CIRCUITS

An electric current is maintained when the terminals of a voltage source (a battery, for example) are connected by a conducting pathway in what's called a **circuit**. If the current always travels in the same direction through the pathway, it's called a **direct current**.

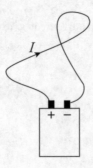

The job of the voltage source is to provide a potential difference called an **electromotive force,** or **emf,** which drives the flow of charge. The emf isn't really a force; it's the work done per unit charge and it's measured in volts.

To try to imagine what's happening in a circuit in which a steady-state current is maintained, let's follow one of the charge carriers that's drifting through the pathway. (Remember, we're pretending that the charge carriers are positive.) The charge is introduced by the positive terminal of the battery and enters the wire, where it's pushed by the electric field. It encounters resistance, bumping into the relatively stationary atoms that make up the metal's lattice and setting them into greater motion. So, the electrical potential energy that the charge had when it left the battery is turning into heat. By the time the charge reaches the negative terminal, all of its original electrical potential energy is lost. In order to keep the current going, the voltage source must do positive work on the charge, forcing it to move from the negative terminal toward the positive terminal. The charge is now ready to make another journey around the circuit.

Energy and Power

When a carrier of positive charge q drops by an amount V in potential, it loses potential energy in the amount qV. If this happens in time t, then the rate at which this energy is transformed is equal to $(qV)/t = (q/t)V$. But q/t is equal to the current, I, so the rate at which electrical energy is transferred is given by the equation

$$P = IV$$

This equation works for the power delivered by a battery to the circuit as well as for resistors. The power dissipated in a resistor, as electrical potential energy is turned into heat, is given by $P = IV$, but because of the relationship $V = IR$, we can express this in two other ways:

$$P = IV = I(IR) = I^2 R$$

or

$$P = IV = \frac{V}{R} \cdot V = \frac{V^2}{R}$$

Resistors become hot when current passes through them; the thermal energy generated is called **joule heat**.

Circuit Analysis

We will now develop a way of specifying the current, voltage, and power associated with each element in a circuit. Our circuits will contain three basic elements: batteries, resistors, and connecting wires. As we've seen, the resistance of an ordinary metal wire is negligible; resistance is provided by devices that control the current: **resistors**. All the resistance of the system is concentrated in the resistors, which are symbolized in a circuit diagram by this symbol:

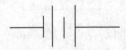

Batteries are denoted by the symbol:

where the longer vertical line represents the **positive** (higher potential) terminal, and the shorter line is the **negative** (lower potential) terminal. Sometimes, a battery is denoted by more than one pair of such lines:

Here's a simple circuit diagram:

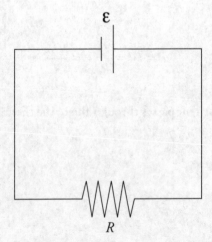

The emf ($\mathcal{E}$) of the battery is indicated, as is the resistance (R) of the resistor. Determining the current in this case is straightforward, because there's only one resistor. The equation $V = IR$, with voltage given by $\mathcal{E}$, gives us

$$I = \frac{\mathcal{E}}{R}$$

Combinations of Resistors

Two common ways of combining resistors within a circuit are to place them either in **series** (one after the other):

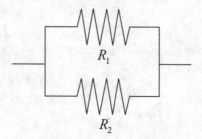

or in **parallel**:

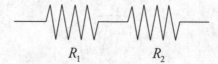

In order to simplify the circuit, our goal is to find the equivalent resistance of combinations. Resistors are said to be in series if they all share the same current and if the total voltage drop across them is equal to the sum of the individual voltage drops.

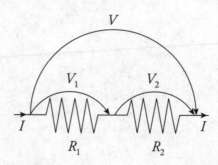

In this case, then, if V denotes the voltage drop across the combination, we have

$$R_{\text{equiv}} = \frac{V}{I} = \frac{V_1 + V_2}{I} = \frac{V_1}{I} + \frac{V_2}{I} = R_1 + R_2$$

This idea can be applied to any number of resistors in series (not just two):

$$R_S = \sum_i R_i$$

Resistors are said to be in parallel if they all share the same voltage drop, and the total current entering the combination is split among the resistors. Imagine that a current I enters the combination. It splits; some of the current, I_1, would go through R_1, and the remainder, I_2, would go through R_2.

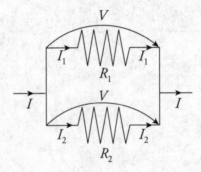

So if V is the voltage drop across the combination, we have

$$I = I_1 + I_2 \quad \Rightarrow \quad \frac{V}{R_{\text{equiv}}} = \frac{V}{R_1} + \frac{V}{R_2} \quad \Rightarrow \quad \frac{1}{R_{\text{equiv}}} = \frac{1}{R_1} + \frac{1}{R_2}$$

Resistors and capacitors follow opposite rules. Remember that capacitors in parallel simply add, and capacitors in series add as reciprocals. We see here that the rules have been switched for resistors. Remember RAS CAP: Resistors Add directly in Series (RAS) and Capacitors Add directly in Parallel (CAP).

This idea can be applied to any number of resistors in parallel (not just two): The reciprocal of the equivalent resistance for resistors in parallel is equal to the sum of the reciprocals of the individual resistances:

$$\frac{1}{R_P} = \sum_i \frac{1}{R_i}$$

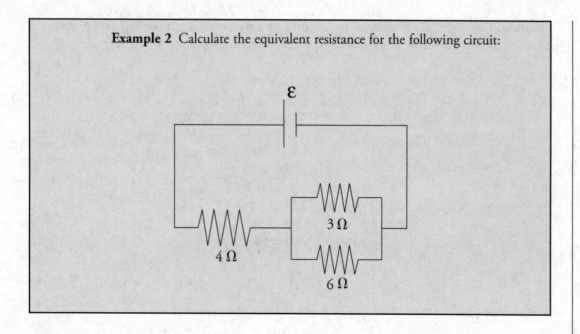

Example 2 Calculate the equivalent resistance for the following circuit:

Solution. First, find the equivalent resistance of the two parallel resistors:

$$\frac{1}{R_P} = \frac{1}{3 \ \Omega} + \frac{1}{6 \ \Omega} \quad \Rightarrow \quad \frac{1}{R_P} = \frac{1}{2 \ \Omega} \quad \Rightarrow \quad R_P = 2 \ \Omega$$

This resistance is in series with the 4 Ω resistor, so the overall equivalent resistance in the circuit is $R = 4 \ \Omega + 2 \ \Omega = 6 \ \Omega$.

Example 3 Determine the current through each resistor, the voltage drop across each resistor, and the power given off (dissipated) as heat in each resistor of this circuit:

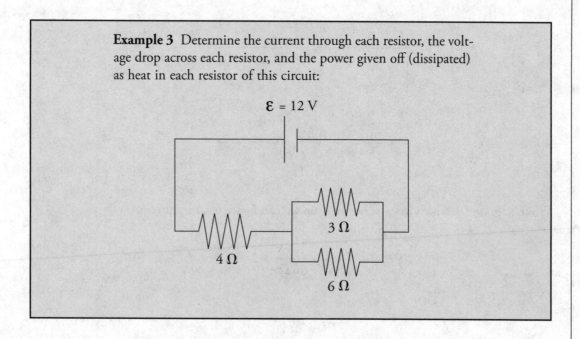

Solution. You might want to redraw the circuit each time we replace a combination of resistors by its equivalent resistance. From our work in the preceding example, we have

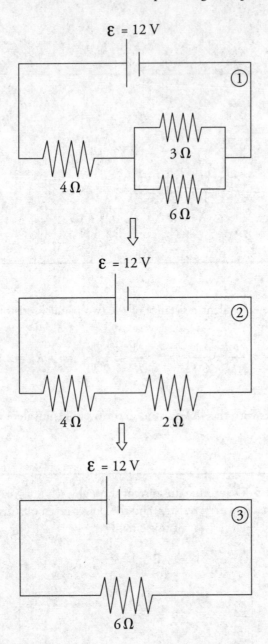

From diagram ③, which has just one resistor, we can figure out the current:

$$I = \frac{\varepsilon}{R} = \frac{12 \text{ V}}{6 \, \Omega} = 2 \text{ A}$$

Now we can work our way back to the original circuit (diagram ①). In going from ③ to ②, we are going back to a series combination, and what do resistors in series share? That's right, the same current. So, we take the current, $I = 2$ A, back to diagram ②. The current through each resistor in diagram ② is 2 A.

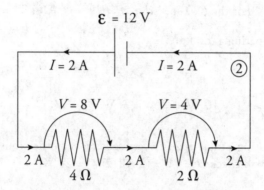

Since we know the current through each resistor, we can figure out the voltage drop across each resistor using the equation $V = IR$. The voltage drop across the 4 Ω resistor is (2 A)(4 Ω) = 8 V, and the voltage drop across the 2 Ω resistor is (2 A)(2 Ω) = 4 V. Notice that the total voltage drop across the two resistors is 8 V + 4 V = 12 V, which matches the emf of the battery—this is to be expected.

Now for the last step: going from diagram ② back to diagram ①. Nothing needs to be done with the 4 Ω resistor; nothing about it changes in going from diagram ② to ①, but the 2 Ω resistor in diagram ② goes back to the parallel combination. And what do resistors in parallel share? The same voltage drop. So we take the voltage drop, $V = 4$ V, back to diagram ①. The voltage drop across each of the two parallel resistors in diagram ① is 4 V.

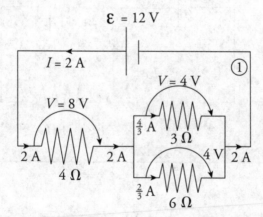

Since we know the voltage drop across each resistor, we can figure out the current through each resistor by using the equation $I = V/R$. The current through the 3 Ω resistor is (4 V)/(3 Ω) = $\frac{4}{3}$ A, and the current through the 6 Ω resistor is (4 V)/(6 Ω) = $\frac{2}{3}$ A. Note that the current entering the parallel combination (2 A) equals the total current passing through the individual resistors $\left(\frac{4}{3} \text{ A} + \frac{2}{3} \text{ A}\right)$. Again, this is to be expected.

Finally, we will calculate the power dissipated as heat by each resistor. We can use any of the equivalent formulas: $P = IV$, $P = I^2R$, or $P = V^2/R$.

$$\text{For the 4 } \Omega \text{ resistor: } P = IV = (2 \text{ A})(8 \text{ V}) = 16 \text{ W}$$

$$\text{For the 3 } \Omega \text{ resistor: } P = IV = (\tfrac{4}{3} \text{ A})(4 \text{ V}) = \tfrac{16}{3} \text{ W}$$

$$\text{For the 6 } \Omega \text{ resistor: } P = IV = (\tfrac{2}{3} \text{ A})(4 \text{ V}) = \tfrac{8}{3} \text{ W}$$

So, the resistors are dissipating a total of

$$16 \text{ W} + \tfrac{16}{3} \text{ W} + \tfrac{8}{3} \text{ W} = 24 \text{ W}$$

If the resistors are dissipating a total of 24 J every second, then they must be provided with that much power. This is easy to check: $P = IV = (2 \text{ A})(12 \text{ V}) = 24 \text{ W}$.

Example 4 Answer the following questions for the circuit below.

(a) In which direction will current flow and why?

(b) What's the overall emf?

(c) What's the current in the circuit?

(d) At what rate is energy consumed by, and provided to, this circuit?

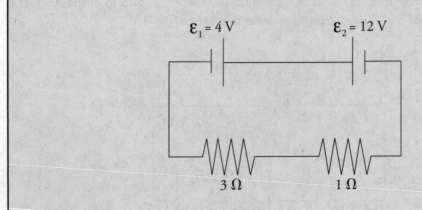

Solution.

(a) The battery whose emf is ε_1 wants to send current clockwise, while the battery whose emf is ε_2 wants to send current counterclockwise. Since $\varepsilon_2 > \varepsilon_1$, the battery whose emf is ε_2 is the more powerful battery, and the current will flow counterclockwise.

(b) Charges forced through ε_1 will lose, rather than gain, 4 V of potential, so the overall emf of this circuit is $\varepsilon_2 - \varepsilon_1 = 8$ V.

(c) Since the total resistance is $3 \ \Omega + 1 \ \Omega = 4 \ \Omega$, the current will be
$I = (8 \ \text{V})/(4 \ \Omega) = 2 \ \text{A}$.

(d) ε_2 will provide energy at a rate of $P_2 = IV_2 = (2 \ \text{A})(12 \ \text{V}) = 24 \ \text{W}$, while ε_1 will absorb at a rate of $P_1 = IV_1 = (2 \ \text{A})(4 \ \text{V}) = 8 \ \text{W}$.
Finally, energy will be dissipated in these resistors at a rate of $I^2 R_1 + I^2 R_2 = (2 \ \text{A})^2(3 \ \Omega) + (2 \ \text{A})^2(1 \ \Omega) = 16 \ \text{W}$. Once again, energy is conserved; the power delivered (24 W) equals the power taken (8 W + 16 W = 24 W).

Example 5 All real batteries contain **internal resistance**, r. Determine the current in the following circuit when the switch S is closed.

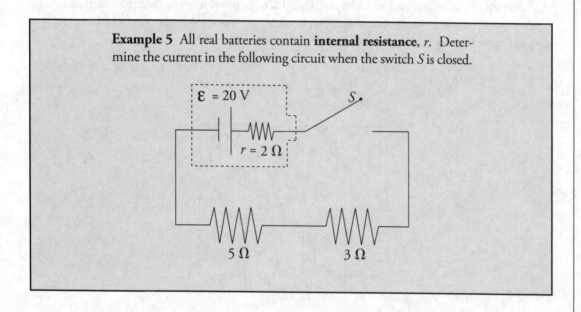

Solution. Before the switch is closed, there is no complete conducting pathway from the positive terminal of the battery to the negative terminal, so no current flows through the resistors. However, once the switch is closed, the resistance of the circuit is $2 \ \Omega + 3 \ \Omega \ + 5 \ \Omega = 10 \ \Omega$, so the current in the circuit is $I = (20 \ \text{V})/(10 \ \Omega) = 2 \ \text{A}$. Often, the battery and its internal resistance are enclosed in a dashed box in the shape of a battery.

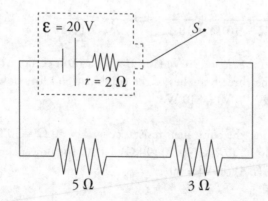

In this case, a distinction can be made between the emf of the battery and the actual voltage it provides once the current has begun. Since $I = 2 \ \text{A}$, the voltage drop across the internal resistance is $Ir = (2 \ \text{A})(2 \ \Omega) = 4 \ \text{V}$, so the effective voltage provided by the battery to the rest of the circuit—called the **terminal voltage**—is lower than the ideal emf. It is $V = \varepsilon - Ir = 20n - 4 \ \text{V} = 16 \ \text{V}$.

Example 6 A student has three 30 Ω resistors and an ideal 90 V battery. (A battery is *ideal* if it has a negligible internal resistance.) Compare the current drawn from—and the power supplied by—the battery when the resistors are arranged in parallel versus in series.

Solution. Resistors in series always provide an equivalent resistance that's greater than any of the individual resistances, and resistors in parallel always provide an equivalent resistance that's smaller than their individual resistances. So, hooking up the resistors in parallel will create the smallest resistance and draw the greatest total current:

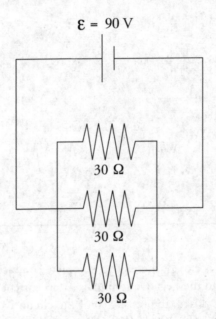

In this case, the equivalent resistance is

$$\frac{1}{R_P} = \frac{1}{30\ \Omega} + \frac{1}{30\ \Omega} + \frac{1}{30\ \Omega} \quad \Rightarrow \quad \frac{1}{R_P} = \frac{1}{10\ \Omega} \quad \Rightarrow \quad R_P = 10\ \Omega$$

and the total current is $I = \mathcal{E}/R_P = (90\ \text{V})/(10\ \Omega) = 9$ A. (You could verify that 3 A of current would flow in each of the three branches of the combination.) The power supplied by the battery will be $P = IV = (9\ \text{A})(90\ \text{V}) = 810$ W.

If the resistors are in series, the equivalent resistance is $R_S = 30\ \Omega + 30\ \Omega + 30\ \Omega = 90\ \Omega$, and the current drawn is only $I = \mathcal{E}/R_S = (90\ \text{V})/(90\ \Omega) = 1$ A. The power supplied by the battery in this case is just $P = IV = (1\ \text{A})(90\ \text{V}) = 90$ W.

Example 7 A **voltmeter** is a device that's used to measure the voltage between two points in a circuit. An **ammeter** is used to measure current. Determine the readings on the voltmeter (denoted —Ⓥ—) and the ammeter (denoted —Ⓐ—) in the circuit below.

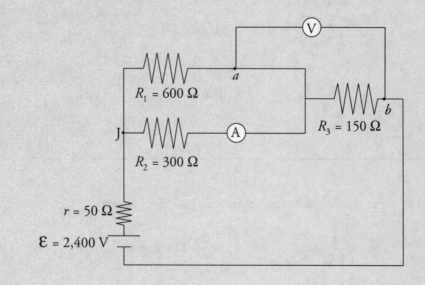

An ammeter is added in series with the resistor whose current is to be measured. If an ammeter has non-negligible resistance, it will alter the equivalent resistance of the circuit and thus the current that is meant to be measured!

A voltmeter is added in parallel with a resistor whose voltage drop is to be measured. If a voltmeter has a low resistance, some current will go through the voltmeter instead of the resistor and alter the voltage drop that is meant to be measured!

Solution. We consider the ammeter to be ideal; this means it has negligible resistance, so it doesn't alter the current that it's trying to measure. Similarly, we consider the voltmeter to have an extremely high resistance, so it draws negligible current away from the circuit.

Our first goal is to find the equivalent resistance in the circuit. The $600\,\Omega$ and $300\,\Omega$ resistors are in parallel; they're equivalent to a single $200\,\Omega$ resistor. This is in series with the battery's internal resistance, r, and R_3. The overall equivalent resistance is therefore $R = 50\,\Omega + 200\,\Omega + 150\,\Omega = 400\,\Omega$, so the current supplied by the battery is $I = \mathcal{E}/R$ (2,400 V)/(400 Ω) = 6 A. At the junction marked J, this current splits. Since R_1 is twice R_2, half as much current will flow through R_1 as through R_2; the current through R_1 is $I_1 = 2$ A, and the current through R_2 is $I_2 = 4$ A. The voltage drop across each of these resistors is $I_1 R_1 = I_2 R_2 = 1{,}200$ V (matching voltages verify the values of currents I_1 and I_2). Since the ammeter is in the branch that contains R_2, it will read $I_2 = 4$ A.

The voltmeter will read the voltage drop across R_3, which is $V_3 = IR_3 = (6\text{ A})(150\,\Omega) = 900$ V. The potential at point b is 900 V lower than at point a. Ammeters measure current, and since current is the same in series, ammeters should be connected in series in a circuit. By the same logic, voltmeters measure potential, which is the same in parallel. Thus, we connect them in parallel.

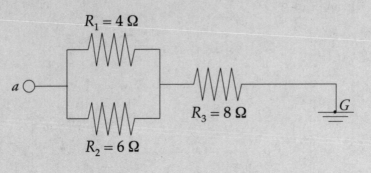

Example 8 The diagram below shows a point *a* at potential $V = 20$ V connected by a combination of resistors to a point (denoted *G*) that is **grounded**. *The ground is considered to be at potential zero.* If the potential at point *a* is maintained at 20 V, what is the current through R_3?

$R_1 = 4\ \Omega$

a

$R_3 = 8\ \Omega$

G

$R_2 = 6\ \Omega$

Solution. R_1 and R_2 are in parallel; their equivalent resistance is R_p, where

$$\frac{1}{R_P} = \frac{1}{4\ \Omega} + \frac{1}{6\ \Omega} \quad \Rightarrow \quad R_P = \frac{24}{10}\ \Omega = 2.4\ \Omega$$

R_p is in series with R_3, so the equivalent resistance is

$$R = R_p + R_3 = (2.4\ \Omega) + (8\ \Omega) = 10.4\ \Omega$$

and the current that flows through R_3 is

$$I_3 = \frac{V}{R} = \frac{20\text{ V}}{10.4\ \Omega} = 1.9\text{ A}$$

Kirchhoff's Rules

When the resistors in a circuit cannot be classified as either in series or in parallel, we need another method for analyzing the circuit. The rules of Gustav Kirchhoff (pronounced "Keer-koff") can be applied to any circuit:

The Loop Rule. The sum of the potential differences (positive and negative) that traverse any closed loop in a circuit must be zero.

The Junction Rule. The total current that enters a junction must equal the total current that leaves the junction. (This is also known as the **Node Rule**.)

The Loop Rule just says that, starting at any point, by the time we get back to that same point by following any closed loop, we have to be back to the same potential. Therefore, the total drop in potential must equal the total rise in potential. Put another way, the Loop Rule says that all the decreases in electrical potential energy (for example, caused by resistors in the direction of the current) must be balanced by all the increases in electrical potential energy (for example, caused by a source of emf from the negative to positive terminal). So the Loop Rule is basically a statement of the Law of Conservation of Energy.

Similarly, the Junction Rule simply says that the charge (per unit time) that goes into a junction must equal the charge (per unit time) that comes out. This is basically a re-statement of the Law of Conservation of Charge.

In practice, the Junction Rule is straightforward to apply. The most important things to remember about the Loop Rule can be summarized as follows:

- When going across a resistor in the *same* direction as the current, the potential *drops* by *IR*.
- When going across a resistor in the *opposite* direction from the current, the potential *increases* by *IR*.
- When going from the negative to the positive terminal of a source of emf, the potential *increases* by ε.
- When going from the positive to the negative terminal of a source of emf, the potential *decreases* by ε.

Circuit Drill

Answers and explanations can be found in Chapter 17.

For each of the following circuits, identify the current across each of the resistors.

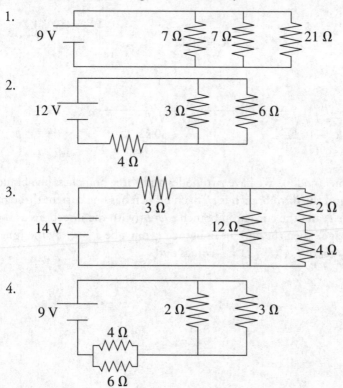

1. 9 V   7 Ω 7 Ω   21 Ω

2. 12 V   3 Ω   6 Ω   4 Ω

3. 14 V   3 Ω   12 Ω   2 Ω   4 Ω

4. 9 V   4 Ω   6 Ω   2 Ω   3 Ω

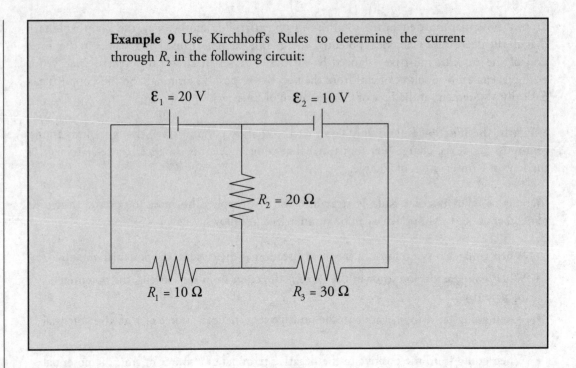

Solution. First, let's label some points in the circuit.

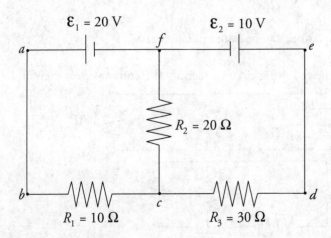

The points c and f are junctions (nodes). We have two nodes and three branches: one branch is *fabc*, another branch is *cdef*, and the third branch is *cf*. Each branch has one current throughout. If we label the current in *fabc* I_1 and the current in branch *cdef* I_2 (with the directions as shown in the diagram on the next page), then the current in branch *cf* must be $I_1 - I_2$, by the Junction Rule: I_1 comes into c, and a total of $I_2 + (I_1 - I) = I_1$ comes out.

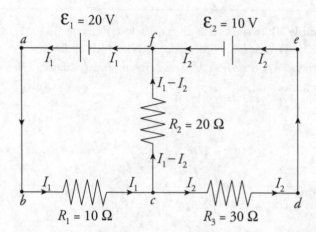

Now pick a loop; say, *abcfa*. Starting at *a*, we go to *b*, then across R_1 in the direction of the current, so the potential drops by $I_1 R_1$. Then we move to *c*, then up through R_2 in the direction of the current, so the potential drops by $(I_1 - I_2)R_2$. Then we reach *f*, turn left and travel through $\mathcal{E}_1$ from the negative to the positive terminal, so the potential increases by $\mathcal{E}_1$. We now find ourselves back at *a*. By the Loop Rule, the total change in potential around this closed loop must be zero, and

$$-I_1 R_1 - (I_1 - I_2)R_2 + \mathcal{E}_1 = 0 \qquad (1)$$

Since we have two unknowns (I_1 and I_2), we need two equations. Pick another loop; let's choose *cdefc*. From *c* to *d*, we travel across the resistor in the direction of the current, so the potential drops by $I_2 R_3$. From *e* to *f*, we travel through $\mathcal{E}_2$ from the positive to the negative terminal, so the potential *drops* by $\mathcal{E}_2$. Heading down from *f* to *c*, we travel across R_2 but in the direction opposite to the current, so the potential *increases* by $(I_1 - I_2)R_2$. At *c*, our loop is completed, so

$$-I_2 R_3 - \mathcal{E}_2 + (I_1 - I_2)R_2 = 0 \qquad (2)$$

Substituting in the given numerical values for R_1, R_2, R_3, $\mathcal{E}_1$, and $\mathcal{E}_2$, and simplifying, these two equations become

$$3I_1 - 2I_2 = 2 \qquad (1')$$
$$2I_1 - 5I_2 = 1 \qquad (2')$$

Solving this pair of simultaneous equations, we get

$$I_1 = \tfrac{8}{11} \text{ A} = 0.73 \text{ A} \quad \text{and} \quad I_2 = \tfrac{1}{11} \text{ A} = 0.09 \text{ A}$$

So the current through R_2 is $I_1 - I_2 = \tfrac{7}{11} \text{ A} = 0.64 \text{ A}$.

The choice of directions of the currents at the beginning of the solution was arbitrary. Don't worry about trying to guess the actual direction of the current in a particular branch. Just pick a direction, stick with it, and obey the Junction Rule. At the end, when you solve for the values of the branch current, a negative value will alert you that the direction of the current is actually opposite to the direction you originally chose for it in your diagram.

Example 10 In the circuit diagram below, Resistors ① and ③ have fixed resistances (R_1 and R_3, respectively), which are known. Resistor ② is a variable (or adjustable) resistor, and the resistance of Resistor ④ is unknown.

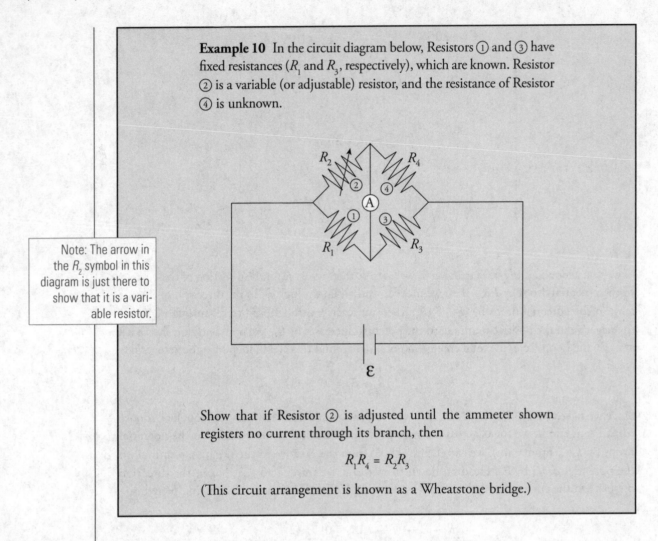

Note: The arrow in the R_2 symbol in this diagram is just there to show that it is a variable resistor.

Show that if Resistor ② is adjusted until the ammeter shown registers no current through its branch, then

$$R_1 R_4 = R_2 R_3$$

(This circuit arrangement is known as a Wheatstone bridge.)

Solution. If no current flows through the ammeter, then Resistors ① and ③ are in series, and so are Resistors ② and ④. Therefore, the current flowing through ① and ③ is equal to $I_{1\text{-}3} = \mathcal{E}/(R_1 + R_3)$. Similarly, the current flowing through ② and ④ is $I_{2\text{-}4} = \mathcal{E}/(R_2 + R_4)$. Now consider the three points marked a, b, and c:

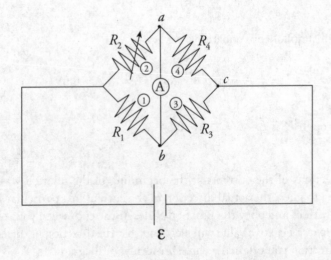

Since no current flows between points a and b, there must be no potential difference between a and b; they're at the same potential. So the voltage drop from a to c must equal the voltage drop from b to c. The voltage drop from a to c is

$$V_{ac} = I_{2\text{-}4}R_4 = \frac{\varepsilon}{R_2 + R_4}R_4$$

and the voltage drop from b to c is

$$V_{bc} = I_{1\text{-}3}R_3 = \frac{\varepsilon}{R_1 + R_3}R_3$$

Setting these equal to each other, we get

$$\frac{\varepsilon}{R_2 + R_4}R_4 = \frac{\varepsilon}{R_1 + R_3}R_3$$

$$\frac{R_4}{R_2 + R_4} = \frac{R_3}{R_1 + R_3}$$

$$R_4(R_1 + R_3) = R_3(R_2 + R_4)$$

$$R_1R_4 + R_3R_4 = R_2R_3 + R_3R_4$$

$$R_1R_4 = R_2R_3$$

By knowing R_1, R_2, and R_3, we can solve for the unknown resistance: $R_4 = R_2R_3/R_1$.

RESISTANCE–CAPACITANCE (RC) CIRCUITS

Capacitors are typically charged by batteries. Once the switch in the diagram on the left is closed, electrons are attracted to the positive terminal of the battery and leave the top plate of the capacitor, leaving a top plate that is now positively charged. Electrons accumulate on the bottom plate of the capacitor, so the bottom plate becomes negatively charged. This continues until the voltage across the capacitor plates matches the emf of the battery. The current then stops, and the capacitor is fully charged.

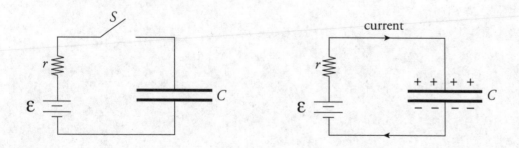

Charging or Discharging a Capacitor in an RC Circuit

When a battery is hooked up to an uncharged capacitor and current begins to flow, the current is not constant. So far, we have assumed that the currents in our circuits have been constant, but RC circuits are different.

Charging a Capacitor

Consider the RC circuit shown below, where the charge on the capacitor is zero:

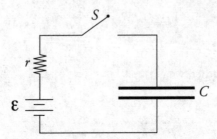

When the switch is closed (time $t = 0$), there is no charge on the capacitor, which means that there's no voltage across the capacitor, so Kirchhoff's Loop Rule gives us $\varepsilon - IR = 0$, and the initial current in the circuit is $I = \varepsilon/R$. But, as time passes, charge begins to accumulate on the capacitor, and a voltage is created that opposes the emf of the battery. At any time t after the switch is closed, Kirchhoff's Loop Rule gives us:

$$\varepsilon - I(t)R - V(t) = 0$$

where $I(t)$ is the current at time t and $V(t)$ is the voltage across the capacitor at time t. Since $V(t) = Q(t)/C$ and $I(t) = dQ(t)/dt$, the equation above becomes

$$\varepsilon - R\frac{dQ}{dt} - \frac{Q}{C} = 0 \quad \Rightarrow \quad \frac{dQ}{dt} = \frac{C\varepsilon - Q}{RC}$$

We will solve for $Q(t)$ using separation of variables.

$$\frac{dQ}{C\varepsilon - Q} = \frac{dt}{RC}$$

Now integrate both sides to get

$$\int_0^q \frac{dQ}{C\varepsilon - Q} = \int_0^t \frac{dt}{RC}$$

$$-\left[\ln(C\varepsilon - q) - \ln(C\varepsilon)\right] = \frac{t}{RC}$$

$$\left(\ln\left(\frac{C\varepsilon - q}{C\varepsilon}\right)\right) = -\frac{t}{RC}$$

$$\frac{C\varepsilon - q}{C\varepsilon} = e^{-\frac{t}{RC}}$$

$$C\varepsilon - q = C\varepsilon e^{-\frac{t}{RC}}$$

$$q(t) = C\varepsilon\left(1 - e^{-\frac{t}{RC}}\right)$$

This gives us an equation, $q(t)$, for how the charge on the capacitor changes over time.

The product $C\varepsilon$ is equal to the final charge on the capacitor (remember that $Q = CV$). The quantity RC that appears in the exponent is called the **time constant** for the circuit and is represented by τ:

$$\tau = RC$$

Therefore, letting Q_f replace the quantity $C\varepsilon$, the equation above becomes

$$Q(t) = Q_f(1 - e^{-t/\tau}) \quad \text{(Eq. 1)}$$

What does the time constant, τ, say about the circuit? It says that, after a time interval of τ has elapsed, the charge on the capacitor is

$$Q(\tau) = Q_f(1 - e^{-t/\tau}) = Q_f(1 - e^{-1}) \approx 0.63Q_f$$

Therefore, the greater the value of τ, the more slowly the capacitor charges. By differentiating Eq. 1 with respect to t,

$$\frac{dQ}{dt} = \frac{Q_{ff}}{\tau}e^{-t/\tau} = \frac{Q}{RC}e^{-t/\tau} = \frac{\varepsilon}{R}e^{-t/\tau}$$

we can get the current in the circuit as a function of time:

$$I(t) = \frac{\varepsilon}{R}e^{-t/\tau} \quad \text{(Eq. 2)}$$

The time constant $\tau = RC$ tells us the time required for the current to drop off by a factor of e during the charging process.

Therefore, the charge builds up gradually on the capacitor,

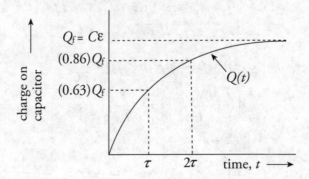

as the current in the circuit drops gradually to zero:

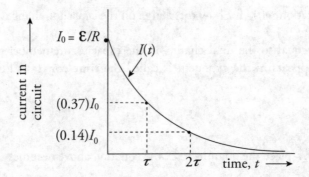

Discharging a Capacitor

Consider the RC circuit shown below, where the charge on the capacitor is Q_0:

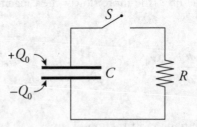

When the switch is closed (time $t = 0$), the voltage across the capacitor is $V_0 = Q_0/C$, so $V_0 - IR = 0$, and the initial current in the circuit is $I_0 = V_0/R$. But as time passes, charge leaks off the capacitor, and its voltage decreases, which causes the current to decrease.

The equation for $Q(t)$ can be derived by the method of separation of variables as in the charging example. It is

$$Q(t) = Q_0 e^{-t/RC} = Q_0 e^{-t/\tau}$$

The voltage across the plates as a function of time is

$$V(t) = \frac{Q(t)}{C} = \frac{Q_0}{C} e^{-t/\tau}$$

and, by using the equation $I(t) = -dQ/dt$, the current in the circuit is

$$I(t) = \frac{Q_0}{\tau} e^{-t/\tau} = \frac{Q_0}{RC} e^{-t/\tau} = \frac{V_0}{R} e^{-t/\tau}$$

Example 11 In the circuit below, $\mathcal{E} = 20$ V, $R = 1{,}000\ \Omega$, and $C = 2$ mF. If the capacitor is initially uncharged, how long will it take (after the switch S is closed) for the capacitor to be 99% charged?

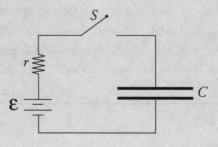

Solution. The equation $Q(t) = Q_f(1 - e^{-t/\tau})$, where $\tau = RC$, gives the charge on the capacitor as a function of time. If we want $Q(t)$ to equal $0.99Q_f$, then we have to wait until time t, when $e^{-t/\tau} = 0.01$. Solving this equation for t, we get

$$e^{-t/\tau} = 0.01$$
$$-\frac{t}{\tau} = \ln(0.01)$$
$$t = -\ln(0.01)\tau$$
$$= (4.6)\tau$$

Since the time constant is

$$\tau = RC = (1{,}000\ \Omega)(2\ \text{mF}) = 2\ \text{s}$$

the time required is

$$(4.6)\tau = (4.6)(2\ \text{s}) = 9.2\ \text{s}$$

Example 12

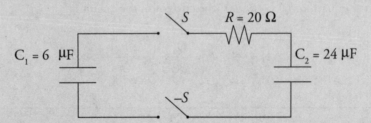

Capacitors 1 and 2, of capacitance $C_1 = 6\ \mu\text{F}$ and $C_2 = 24\ \mu\text{F}$, respectively, are connected in a circuit as shown above with a resistor of resistance $R = 20\ \Omega$ and two switches. Capacitor 1 is initially charged to a voltage $V_o = 30\ \text{V}$, and capacitor 2 is initially uncharged. Both of the switches S are then closed simultaneously.

(a) What is the initial current in the circuit?

(b) What are the final charges on each of the capacitors 1 and 2 after equilibrium has been reached?

Solution.

(a) Use Ohm's law to solve for the initial current. The voltage will decrease with time, but the initial voltage is 30 V.

$$V = IR \quad \Rightarrow \quad 30 = I(20) \quad \Rightarrow \quad I = 1.5\ \text{A}$$

(b) The initial charge stored on capacitor 1 is $Q = CV = (6)(30) = 180\ \mu\text{C}$. This charge will redistribute until equilibrium is reached and both capacitors have the same potential difference across them. Rearrange $Q = CV$ to solve for the potential difference across each capacitor: $V = \dfrac{Q}{C}$.

$$V = \frac{Q_1}{C_1} = \frac{Q_2}{C_2} \Rightarrow \frac{Q_1}{6} = \frac{180 - Q_1}{24}$$

$$Q_1 = 36\ \mu\text{C}_2 \text{ and } Q_2 = 144\ \mu\text{C}$$

Chapter 14 Drill

The answers and explanations can be found in Chapter 17.

Section I: Multiple Choice

1. A wire made of brass and a wire made of silver have the same length, but the diameter of the brass wire is 4 times the diameter of the silver wire. The resistivity of brass is 5 times greater than the resistivity of silver. If R_B denotes the resistance of the brass wire and R_S denotes the resistance of the silver wire, which of the following is true?

    (A) $R_B = \frac{5}{16}R_S$

    (B) $R_B = \frac{4}{5}R_S$

    (C) $R_B = \frac{5}{4}R_S$

    (D) $R_B = \frac{5}{2}R_S$

    (E) $R_B = \frac{16}{5}R_S$

2. For an ohmic conductor, doubling the voltage without changing the resistance will cause the current to

    (A) decrease by a factor of 4
    (B) decrease by a factor of 2
    (C) remain unchanged
    (D) increase by a factor of 2
    (E) increase by a factor of 4

3. A given circuit uses 125 watts of power. If the circuit has a total resistance of 5 Ω, at what voltage must it be operating?

    (A) 5 V
    (B) 15 V
    (C) 25 V
    (D) 45 V
    (E) 625 V

4. A standard light bulb in the United States is 60 W (watts). The standard wall outlet voltage in the United States is 120 V, but in Europe, the standard wall outlet voltage is 240 V. If this 60 W light bulb could be plugged into a socket in Europe, what would be true about how bright the bulb was?

    (A) The bulb would be four times as bright.
    (B) The bulb would be twice as bright.
    (C) The bulb would be the same brightness.
    (D) The bulb would be one-half as bright.
    (E) The bulb would be one-quarter as bright.

5. An ammeter is a device than can be arranged in a circuit to measure the current flowing through a particular part of the circuit. Which of the following sets of characteristics should it possess?

    (A) It should have high resistance and should be arranged in series with the part being measured.
    (B) It should have low resistance and should be arranged in series with the part being measured.
    (C) It should have high resistance and should be arranged in parallel with the part being measured.
    (D) It should have low resistance and should be arranged in parallel with the part being measured.
    (E) It should have the same resistance as the part being measured and should be arranged in series with the part being measured.

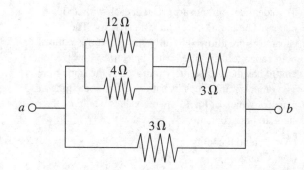

6. Determine the equivalent resistance between points a and b in the figure above.

    (A) 0.167 Ω
    (B) 0.25 Ω
    (C) 0.333 Ω
    (D) 1.5 Ω
    (E) 2 Ω

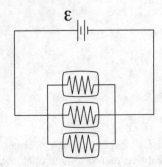

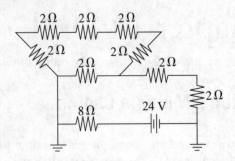

7. Three identical light bulbs are connected to a source of emf, as shown in the diagram above. What will happen if the middle bulb burns out?

(A) All the bulbs will go out.
(B) The light intensity of the other two bulbs will decrease (but they won't go out).
(C) The light intensity of the other two bulbs will increase.
(D) The light intensity of the other two bulbs will remain the same.
(E) More current will be supported by the source of emf.

8. Two identical circuits are created, each with an ideal battery. The circuits each have a capacitor. The capacitors are allowed to fully charge. At this point, both capacitors are fitted with dielectrics, however one circuit has the ideal battery removed at the same time. The other circuit keeps its battery. After some time has elapsed, which of the following values would be the same in both capacitors?

(A) Potential energy
(B) Charge
(C) Capacitance
(D) Voltage
(E) Electric field

9. What is the current through the 8-ohm resistor in the circuit shown above?

(A) 0.5 A
(B) 1.0 A
(C) 1.25 A
(D) 1.5 A
(E) 3.0 A

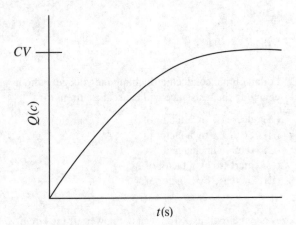

10. The graph above shows the charge vs. time for an RC circuit with voltage V and capacitance C. The slope of this graph represents

(A) the total charge on the capacitor plates
(B) the potential energy of the capacitor
(C) the resistance of the circuit
(D) the instantaneous voltage of the capacitor
(E) the instantaneous current of the circuit

11. Which of the following combinations of values for total resistance, R, and capacitance, C, would produce an RC circuit that reached its maximum charge (on the capacitor) most quickly?

(A) $R = 4\ \Omega$; $C = 20\ \mu F$
(B) $R = 6\ \Omega$; $C = 25\ \mu F$
(C) $R = 8\ \Omega$; $C = 30\ \mu F$
(D) $R = 4\ \Omega$; $C = 35\ \mu F$
(E) $R = 8\ \Omega$; $C = 40\ \mu F$

Section II: Free Response

1. Consider the following circuit:

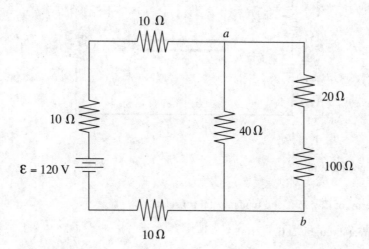

(a)   At what rate does the battery deliver energy to the circuit?

(b)   Find the current through the 20 Ω resistor.

(c)   (i)   Determine the potential difference between points *a* and *b*.

   (ii)   At which of these two points is the potential higher?

(d)   Find the energy dissipated by the 100 Ω resistor in 10 s.

(e)   Given that the 100 Ω resistor is a solid cylinder that's 4 cm long, composed of a material whose resistivity is 0.45 Ω·m, determine its radius.

2. The diagram below shows an uncharged capacitor, two resistors, and a battery whose emf is ε.

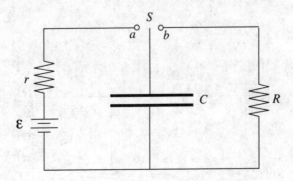

The switch S is turned to point a at time $t = 0$.

(Express all answers in terms of C, r, R, $ε$, and fundamental constants.)

(a) Determine the current through r at time $t = 0$.

(b) Compute the time required for the charge on the capacitor to reach one-half its final value.

(c) When the capacitor is fully charged, which plate is positively charged?

(d) Determine the electrical potential energy stored in the capacitor when the current through r is zero.

When the current through r is zero, the switch S is then moved to point b; for the following parts, consider this event time $t = 0$.

(e) Determine the current through R as a function of time.

(f) Find the power dissipated in R as a function of time.

(g) Determine the total amount of energy dissipated as heat by R.

Summary

General Information

○ Current is the flow rate of charge: $I = \dfrac{dQ}{dt}$. The units are amps (A).

○ The resistance of material depends on the resistivity, length, and cross-sectional area: $R = \dfrac{\rho L}{A}$. The units of resistance are ohms (Ω).

○ Many electrical devices obey Ohm's law:

$$\Delta V = IR$$

○ Power is the rate that electrical energy is transferred: $P = IV = I^2 R = \dfrac{V^2}{R}$.

Circuits with Resistors

○ The equivalent resistance of resistors added in series is the sum of the resistance of each device:

$$R_{eq} = R_1 + R_2 + \ldots = \sum_i R_i$$

○ The reciprocal of the equivalent resistance of resistors added in parallel is the sum of the reciprocals of each resistor:

$$\frac{1}{R_{eq}} = \frac{1}{R_1} + \frac{1}{R_2} + \ldots = \sum_i \frac{1}{R_i}$$

Kirchhoff's Rules:

1. **The Loop Rule.** The sum of the potential differences around a closed loop in a circuit must be zero: $\sum_{loop} V = 0$.

2. **The Junction (Node) Rule.** The total current that enters a junction must equal the total current that leaves the junction.

Circuits with Resistors and Capacitors (RC Circuits)

o When a battery is connected to a resistor and uncharged capacitors in a circuit, the charge builds up on the capacitor exponentially. This is due to the fact that, as more charge is added to the capacitor, it becomes more difficult for additional charge to be added. The equation for the charge buildup is $q(t) = C\varepsilon(1 - e^{-\frac{t}{RC}})$.

o The quantity RC is called the time constant: $\tau = RC$.

o During this time, the current is decreasing exponentially because the capacitor is acting like an opposing battery as it charges up. The equation for the current is $I(t) = \frac{\varepsilon}{R}(e^{-\frac{t}{RC}})$.

o When a charged capacitor that has an initial voltage, V_0, and initial charge, Q_0, is connected in a circuit to a resistor, both the charge on the plate and the current through the circuits decrease exponentially. The equations for the charge and the current are given by

$$Q(t) = Q_0(e^{-\frac{t}{RC}}) \text{ and } I(t) = \frac{V_0}{R}(e^{-\frac{t}{RC}})$$

Chapter 15
Magnetic Fields

INTRODUCTION

In Chapter 12, we learned that electric charges are the sources of electric fields and that other charges experience an electric force in those fields. The charges generating the field were assumed to be at rest because, if they weren't, then another force field would have been generated in addition to the electric field. Electric charges *that move* are the sources of **magnetic fields**, and other charges that move can experience a magnetic force in these fields. Not only do moving charges create magnetic fields, but certain materials that are called **ferromagnetic** can produce permanent magnets. Iron, nickel, and cobalt are three examples of ferromagnetic materials. When a bar magnet is suspended from a pivot, one end of the magnet will point north, and one end will point south. The region where the field is strongest in the magnet is called the **pole**, and thus a permanent magnet will have a **north pole** and a **south pole**. Opposite poles attract and like poles repel. When a compass (a small suspended magnet) is placed in a magnetic field, the north pole of the compass will thus point to the south pole of the magnet. We define the direction of the magnetic field to be parallel to the direction that the north pole of a compass needle is pointing.

One common misconception is to confuse magnetic effects and electrostatic effects. Since north poles repel each other, and positive charges repel each other, some students assume north poles repel stationary positive charges. This is incorrect because magnetic fields (and thus north and south poles) only exert forces on *moving* charges. And the force is perpendicular to both the velocity of the charged particle and the magnetic field.

THE MAGNETIC FORCE ON A MOVING CHARGE

If a particle with charge q moves with velocity $\mathbf{v}$ through a magnetic field $\mathbf{B}$, it will experience a magnetic force, $\mathbf{F}_{\mathrm{B}}$, with magnitude

$$F_{\mathrm{B}} = |q|vB\sin\theta \qquad (1)$$

where θ is the angle between $\mathbf{v}$ and $\mathbf{B}$. From this equation, we can see that if the charge is at rest, then $v = 0$, which immediately gives us $F_{\mathrm{B}} = 0$. This tells us that magnetic forces only act on moving charges. Also, if v is parallel (or antiparallel) to $\mathbf{B}$, then $F_{\mathrm{B}} = 0$ since, in either of these cases, $\sin\theta = 0$. So, only charges that cut across the magnetic field lines will experience a magnetic force. Furthermore, the magnetic force is maximized when v is perpendicular to $\mathbf{B}$, because if $\theta = 90°$, then $\sin\theta$ is equal to 1, its maximum value.

The direction of $\mathbf{F}_{\mathrm{B}}$ is always perpendicular to both $\mathbf{v}$ and $\mathbf{B}$ and depends on the sign of the charge q and the direction of $\mathbf{v} \times \mathbf{B}$ (which is given by the right-hand rule).

$$\text{direction of } \mathbf{F}_{\mathrm{B}} = \begin{cases} \text{same as the direction of } \mathbf{v} \times \mathbf{B} \text{ if } q \text{ is positive} \\ \text{opposite to the direction of } \mathbf{v} \times \mathbf{B} \text{ if } q \text{ is negative} \end{cases} \qquad (2)$$

Equations (1) and (2) can be summarized by a single equation:

$$\boxed{\mathbf{F}_{M} = q\mathbf{v} \times \mathbf{B}}$$

Note that there are fundamental differences between the electric force and magnetic force on a charge. First, a magnetic force acts on a charge only if the charge is moving; the electric force acts on a charge whether it moves or not. Second, the direction of the magnetic force is always perpendicular to the magnetic field, while the electric force is always parallel (or antiparallel) to the electric field.

The SI unit for the magnetic field is the **tesla** (abbreviated as **T**) which is one newton per ampere-meter [1 T = 1 N/(A·m)]. Another common unit for magnetic field strength is the **gauss** (abbreviated as **G**): 1 G = 10^{-4} T.

> The Earth's magnetic field depends on location, but typically varies between 25 µT and 65 µT. This is often strong enough to impact experiments involving magnetic fields. Thus, it can be a source of error if not included in calculations.

Example 1 A charge $+q = +6 \times 10^{-6}$ C moves with speed $v = 4 \times 10^5$ m/s through a magnetic field of strength $B = 0.4$ T, as shown in the figure below. What is the magnetic force experienced by q?

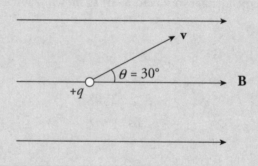

Solution. The magnitude of F_B is

$$F_B = qvB \sin \theta = (6 \times 10^{-6} \text{ C})(4 \times 10^5 \text{ m/s})(0.4 \text{ T}) \sin 30° = 0.48 \text{ N}$$

By the right-hand rule, the direction of $\mathbf{v} \times \mathbf{B}$ is into the plane of the page, which is symbolized by $\otimes$. Because q is a positive charge, the direction of F_B on q is also into the plane of the page.

Example 2 A particle of mass m and charge $+q$ is projected with velocity **v** (in the plane of the page) into a uniform magnetic field **B** that points into the page. How will the particle move?

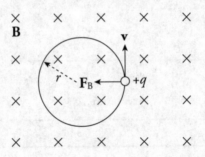

Solution. Since **v** is perpendicular to **B**, the particle will feel a magnetic force of strength qvB, which will be directed perpendicular to **v** (and to **B**) as shown below.

Since $\mathbf{F}_B$ is always perpendicular to **v**, the particle will undergo uniform circular motion; $\mathbf{F}_B$ will provide the centripetal force. Notice that, because $\mathbf{F}_B$ is always perpendicular to **v**, the magnitude of **v** will not change, just its direction. *Magnetic forces cannot change the speed of a charged particle, they can only change its direction of motion.* The radius of the particle's circular path is found from the equation $F_B = F_C$:

$$qvB = \frac{mv^2}{r} \;\Rightarrow\; r = \frac{mv}{qB}$$

Example 3 A particle of charge $-q$ is shot into a region that contains an electric field, **E**, crossed with a perpendicular magnetic field, **B**. If $E = 2 \times 10^4$ N/C and $B = 0.5$ T, what must be the speed of the particle if it is to cross this region without being deflected?

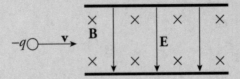

Solution. If the particle is to pass through undeflected, the electric force it feels has to be canceled by the magnetic force. In the diagram, the electric force on the particle is directed upward (since the charge is negative and **E** is downward), and the magnetic force is directed downward (since the charge is negative and **v** × **B** is upward). So $\mathbf{F}_E$ and $\mathbf{F}_B$ point in opposite directions. In order for their magnitudes to balance, qE must equal qvB, so v must equal E/B, which in this case gives

$$v = \frac{E}{B} = \frac{2 \times 10^4 \text{ N/C}}{0.5 \text{ T}} = 4 \times 10^4 \text{ m/s}$$

Example 4 A particle with charge $+q$, traveling with velocity **v**, enters a uniform magnetic field **B**, as shown below. Describe the particle's subsequent motion.

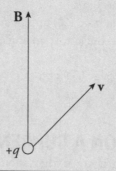

Solution. If the particle's velocity were parallel to **B**, then it would be unaffected by **B**. If **v** were perpendicular to **B**, then it would undergo uniform circular motion (as we saw in Example 2). In this case, **v** is neither purely parallel nor perpendicular to **B**. It has a component ($\mathbf{v}_1$) that's parallel to **B** and a component ($\mathbf{v}_2$) that's perpendicular to **B**.

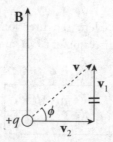

Component $\mathbf{v}_1$ will not be changed by **B**, so the particle will continue upward in the direction of **B**. However, the presence of $\mathbf{v}_2$ will create circular motion. The superposition of these two types of motion will cause the particle's trajectory to be a *helix*; it will spin in circular motion while traveling upward with the speed $v_1 = v \sin \phi$:

Magnetic Field Drill

Answers and explanations can be found in Chapter 17.

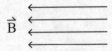

The magnetic field defined in the figure above has a magnitude of 5.0 T. For each particle described below, when the particle is placed in the magnetic field, find the force and direction exerted on the particle by the magnetic field, along with the acceleration of the particle.

1. A proton at rest
2. An electron moving down at 5 m/s
3. A proton moving right (→) at 5 m/s
4. An electron traveling into the page (⊗) at 5 m/s
5. A positron moving up and to the left, at an angle of 30 degrees with the horizontal, at 5 m/s
6. A proton moving out of the page (⊙) at 5 m/s

THE MAGNETIC FORCE ON A CURRENT-CARRYING WIRE

Since magnetic fields affect moving charges, they should also affect current-carrying wires. After all, a wire that contains a current contains charges that move.

Let a wire of length ℓ be immersed in magnetic field **B**. If the wire carries a current I, then the magnitude of the magnetic force it feels is

$$F_B = I\ell B \sin \theta$$

where θ is the angle between ℓ and **B**. Here, the direction of ℓ is the direction of the current, I. The direction of $\mathbf{F}_B$ is the same as the direction of $\ell \times \mathbf{B}$, and these properties can be summarized in a single equation:

$$d\mathbf{B} = \frac{\mu_0}{4\pi} \frac{Id\ell \times \mathbf{r}}{r^3}$$

The symbol μ_0 denotes a fundamental constant called the permeability of free space. Its value is

$$\mu_0 = 4\pi \times 10^{-7} \text{ N/A}^2 = 4\pi \times 10^{-7} \text{ T} \cdot \text{m/A}$$

Example 5 A U-shaped wire of mass m is lowered into a magnetic field **B** that points out of the plane of the page. How much current I must pass through the wire in order to cause the net force on the wire to be zero?

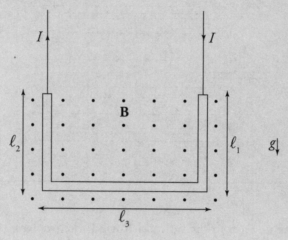

Solution. The total magnetic force on the wire is equal to the sum of the magnetic forces on each of the three sections of wire. The force on the first section (the right, vertical one), $\mathbf{F}_{B1}$, is directed to the left (applying the right-hand rule to $\ell_1 \times \mathbf{B}$; ℓ_1 points downward), and the force on the third piece (the left, vertical one), $\mathbf{F}_{B3}$, is directed to the right. Since these pieces are the same length, these two oppositely directed forces have the same magnitude, $I\ell_1 B = I\ell_2 B$, and they cancel. So the net magnetic force on the wire is the magnetic force on the middle piece. Since ℓ_2 points to the left and **B** is out of the page, $\ell_2 \times \mathbf{B}$ [and, therefore, $\mathbf{F}_{B2} = I(\ell_2 \times \mathbf{B})$] is directed upward.

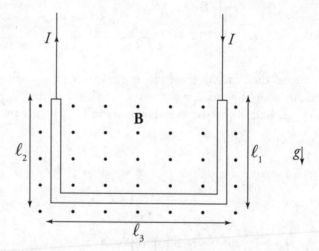

Since the magnetic force on the wire is $I\ell_2 B$, directed upward, the amount of current must create an upward magnetic force that exactly balances the downward gravitational force on the wire. Because the total mass of the wire is m, the resultant force (magnetic + gravitational) will be zero if

$$I\ell_2 B = mg \quad \Rightarrow \quad I = \frac{mg}{\ell_2 B}$$

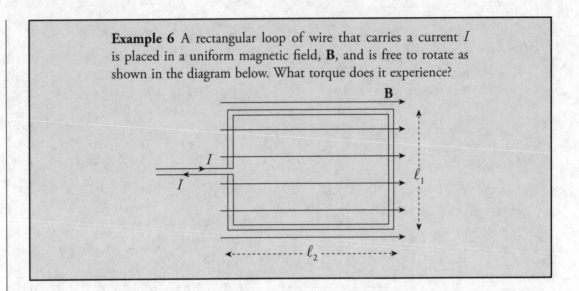

Example 6 A rectangular loop of wire that carries a current I is placed in a uniform magnetic field, **B**, and is free to rotate as shown in the diagram below. What torque does it experience?

Solution. Ignoring the tiny gap in the vertical left-hand wire, we have two wires of length ℓ_1 and two of length ℓ_2. There is no magnetic force on either of the sides of the loop of length ℓ_2 because the current on the top side is parallel to **B** and the current on the bottom side is antiparallel to **B**. The magnetic force on the right-hand side points out of the plane of the page, while the magnetic force on the left-hand side points into the plane of the page.

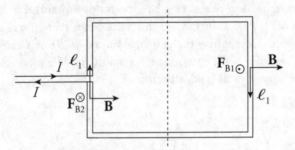

If the loop is free to rotate, then each of these two forces exerts a torque that tends to turn the loop in such a way that the right-hand side rises out of the plane of the page and the left-hand side rotates into the page. Relative to the axis shown above (which cuts the loop in half), the torque of $\mathbf{F}_{B1}$ is

$$\tau_1 = rF_{B1} \sin \theta = (\tfrac{1}{2}\ell_2)(I\ell_1 B) \sin 90° = \tfrac{1}{2}I\ell_1\ell_2 B$$

and the torque of $\mathbf{F}_{B2}$ is

$$\tau_2 = rF_{B2} \sin \theta = (\tfrac{1}{2}\ell_2)(I\ell_1 B) \sin 90° = \tfrac{1}{2}I\ell_1\ell_2 B$$

Since both these torques rotate the loop in the same direction, the net torque on the loop is

$$\tau_1 + \tau_2 = I\ell_1\ell_2 B$$

Example 7 The middle portion of the wire shown below is bent into the shape of a semicircle of radius r. The wire carries a current I. What's the total magnetic force that acts on the wire in the field, **B**?

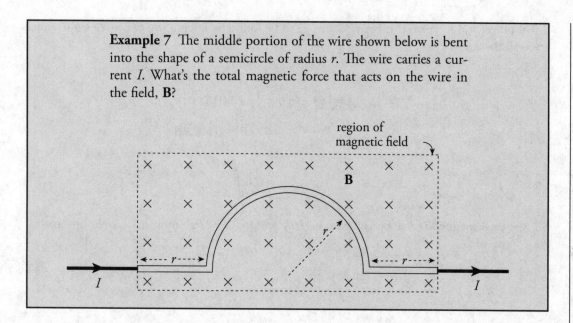

Solution. Let's first find the magnetic force that acts on the two straight sections. The magnitude of the magnetic force on each section is $I\ell B = IrB$, and since ℓ points to the right in both sections, the magnetic force on each of them points upward.

On the semicircle, the direction of ℓ is not constant, so we have to split this section into small pieces with lengths $d\ell = r\,d\theta$ and then use the equation $d\mathbf{F} = I(d\ell \times \mathbf{B})$. In the diagram below, the direction of $d\mathbf{F}$ is radially away from the center of the semicircle.

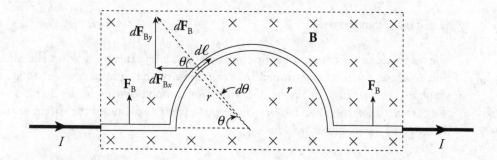

The total horizontal force on the semicircle, F_{Bx}, can be found by integrating $dF_{Bx} = dF\cos\theta$ from $\theta = 0$ to $\theta = \pi$:

$$F_{Bx} = \int dF_{Bx} = \int dF\cos\theta = \int I(r\,d\theta)B\cos\theta$$

$$= IrB\int_0^\pi \cos\theta\,d\theta$$

$$= IrB[\sin\theta]_0^\pi$$

$$= 0$$

The total vertical force on the semicircle, F_{By}, can be found by integrating $dF_{By} = dF \sin \theta$ from $\theta = 0$ to $\theta = \pi$:

$$F_{By} = \int dF_{By} = \int dF \sin \theta = \int I(r\,d\theta)B \sin \theta$$
$$= IrB \int_0^\pi \sin \theta \, d\theta$$
$$= IrB[-\cos \theta]_0^\pi$$
$$= 2IrB$$

So, the total magnetic force on the wire is $(IrB,\ \text{upward}) + (2IrB,\ \text{upward}) + (IrB,\ \text{upward}) = 4IrB$, upward.

MAGNETIC FIELDS CREATED BY CURRENT-CARRYING WIRES

As we said at the beginning of this chapter, the sources of magnetic fields are electric charges that move; they may spin, circulate, move through space, or flow down a wire. For example, consider a long, straight wire that carries a current I. The current generates a magnetic field in the surrounding space of magnitude

$$B = \frac{\mu_0}{2\pi}\frac{I}{r}$$

where r is the distance from the wire.

> You can use your pencil as a prop to represent the wire to help correctly identify the direction of the magnetic field.

The magnetic field lines are actually circles whose centers are on the wire. The direction of these circles is determined by a variation of the right-hand rule. Imagine grabbing the wire in your right hand with your thumb pointing in the direction of the current. Then the direction in which your fingers curl around the wire gives the direction of the magnetic field line.

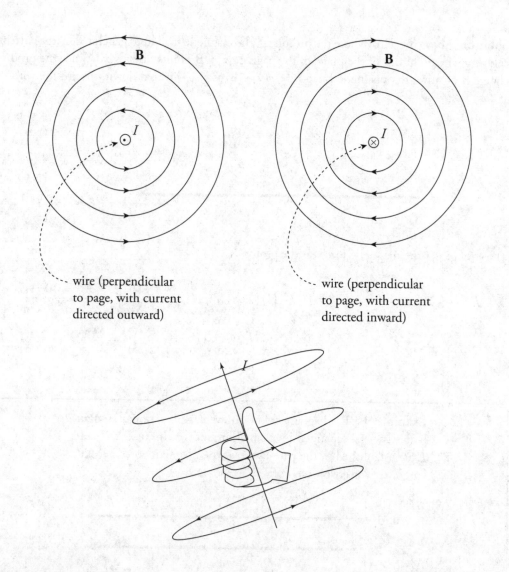

wire (perpendicular
to page, with current
directed outward)

wire (perpendicular
to page, with current
directed inward)

Example 8 The diagram below shows a proton moving with a speed of 2×10^5 m/s, initially parallel to, and 4 cm from, a long, straight wire. If the current in the wire is 20 A, what's the magnetic force on the proton?

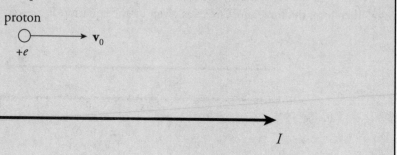

proton

+e

$\mathbf{v}_0$

I

Solution. Above the wire (where the proton is), the magnetic field lines generated by the current-carrying wire point out of the plane of the page, so $\mathbf{v}_0 \times \mathbf{B}$ points downward. Since the proton's charge is positive, the magnetic force $\mathbf{F}_B = q(\mathbf{v}_0 \times \mathbf{B})$ is also directed down, toward the wire.

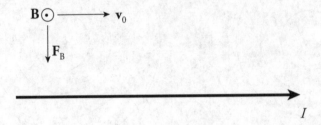

The strength of the magnetic force on the proton is

$$F_B = q v_0 B = e v_0 \frac{\mu_0}{2\pi} \frac{I}{r} = (1.6 \times 10^{-19} \text{ C})(2 \times 10^5 \text{ m/s}) \frac{4\pi \times 10^{-7} \text{ N/A}^2}{2\pi} \frac{20 \text{ A}}{0.04 \text{ m}}$$

$$= 3.2 \times 10^{-18} \text{ N}$$

Example 9 The diagram below shows a pair of long, straight, parallel wires, separated by a small distance, r. If currents I_1 and I_2 are established in the wires, what is the magnetic force per unit length they exert on each other?

Solution. To find the force on Wire 2, consider the current in Wire 1 as the source of the magnetic field. Below Wire 1, the magnetic field lines generated by Wire 1 point into the plane of the page. Therefore, the force on Wire 2, as given by the equation $\mathbf{F}_{B2} = I_2(\boldsymbol{\ell}_2 \times \mathbf{B}_1)$, points upward.

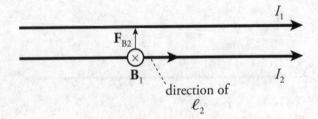

The magnitude of the magnetic force per unit length felt by Wire 2, due to the magnetic field generated by Wire 1, is found this way:

$$F_{B2} = I_2 \ell_2 B_1 = I_2 \ell_2 \frac{\mu_0}{2\pi}\frac{I_1}{r} \quad \Rightarrow \quad \frac{F_{B2}}{\ell_2} = \frac{\mu_0}{2\pi}\frac{I_1 I_2}{r}$$

By Newton's Third Law, this is the same force that Wire 1 feels due to the magnetic field generated by Wire 2. The force is attractive because the currents point in the same direction; if one of the currents were reversed, then the force between the wires would be repulsive. (And we don't mean "gross" repulsive, we mean physics-style repulsion, opposing in direction.)

The Magnetic Field of a Solenoid

Imagine taking a long piece of wire and winding it in a helix around the length of a hollow tube; this is a **solenoid**. Let N be the total number of turns and L the length of the solenoid; then the number of turns per unit length is N/L, which we'll call n. If a current I is established in the wire, then a magnetic field is generated in the space that surrounds the solenoid. If the solenoid is much longer than it is wide, then the magnetic field inside will be parallel to the central axis and nearly uniform. The strength of the field inside is given by the equation

$$B = \mu_0 n I$$

So, the magnetic field inside a solenoid can be increased either by winding the wire more tightly (to increase n) or by increasing the current.

Solenoids have low resistance, so when one is added to a circuit, it needs a resistor to be added in with it. Otherwise, the large current that would be produced without the presence of a resistor could damage other circuit elements connected to the solenoid.

> **Example 10** A tightly wound solenoid has a length of 30 cm, a diameter of 2 cm, and contains a total of 10,000 turns. If it carries a current of 5 A, what's the magnitude of the magnetic field inside the solenoid?

Solution. Since the solenoid is so much longer than it is wide, the equation $B = \mu_0 n I$ can be used to calculate the field inside:

$$B = \mu_0 n I = \mu_0 \frac{N}{L} I = (4\pi \times 10^{-7}\ \text{T·m/A})\ \frac{10,000}{0.30\ \text{m}}(5\ \text{A}) = 0.2\ \text{T}$$

Note that there is a difference between B created externally versus being self-generated. The forces from the self-generated field will cancel out and thus the field does not need to be considered.

THE BIOT–SAVART LAW

In this section, you'll learn one way to determine the magnetic field created by an electric current. Let's say we want to determine the magnitude and direction of the magnetic field at some point P that is created by the current I in the diagram below.

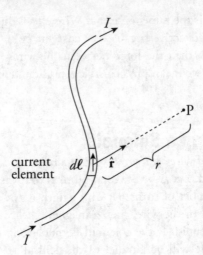

Consider a small section of the current, an infinitesimal section of length $d\ell$, and assign it the direction of the current so we obtain a **current element**, $d\ell$. Let **r** be the position vector from the current element to the point P, and form the unit vector $\hat{\mathbf{r}} = \mathbf{r}/r$. Then the magnitude and direction of the magnetic field at P, due to the current element is

$$d\mathbf{B} = \frac{\mu_0}{4\pi}\frac{I(d\boldsymbol{\ell}\times\hat{\mathbf{r}})}{r^2}$$

This is called the **Biot–Savart law**. By adding all the contributions to the magnetic field, we can get the total magnetic field due to the entire current. On the equation sheet for the free-response section, this information will be represented as follows:

$$d\mathbf{B} = \frac{\mu_0}{4\pi}\frac{Id\boldsymbol{\ell}\times\mathbf{r}}{r^3}$$

Example 11 Use the Biot–Savart law to show that the magnetic field due to an infinitely long straight wire carrying a current I is given by the equation $B = (\mu_0/2\pi)I/R$, where R is the distance from the wire. You should use the integration formula:

$$\int_{-\infty}^{\infty} \frac{dx}{(x^2 + a^2)^{3/2}} = \frac{2}{a^2}$$

Solution. Refer to the diagram below:

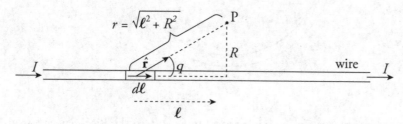

The direction of $d\boldsymbol{\ell} \times \hat{\mathbf{r}}$ is out of the plane of the page (because P is above the wire). The contribution from the current element to the total magnetic field at P is given by the Biot–Savart law:

$$dB = \frac{\mu_0}{4\pi} \frac{I |d\boldsymbol{\ell} \times \hat{\mathbf{r}}|}{r^2} = \frac{\mu_0}{4\pi} \frac{I \, d\ell \sin\theta}{r^2}$$

From the diagram, we see that $\sin\theta = R/r = R/\sqrt{\ell^2 + R^2}$, so

$$dB = \frac{\mu_0 I R}{4\pi} \frac{d\ell}{(\ell^2 + R^2)^{3/2}}$$

and, integrating along the entire wire, we get

$$B = \int dB = \int_{-\infty}^{\infty} \frac{\mu_0 I R}{4\pi} \frac{d\ell}{(\ell^2 + R^2)^{3/2}} = \frac{\mu_0 I R}{4\pi} \int_{-\infty}^{\infty} \frac{d\ell}{(\ell^2 + R^2)^{3/2}} = \frac{\mu_0 I R}{4\pi} \left(\frac{2}{R^2}\right) = \frac{\mu_0}{2\pi} \frac{I}{R}$$

Example 12 A circular loop of wire has radius R. If the loop carries a current I, what's the magnetic field created at the center of the loop?

Solution.

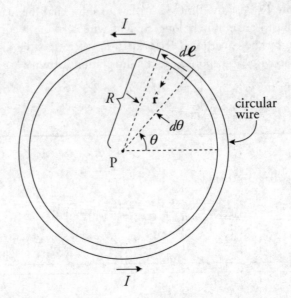

Every current element $d\ell$ is perpendicular to $\hat{\mathbf{r}}$, and $r = R$, so the Biot–Savart law gives us:

$$dB = \frac{\mu_0}{4\pi}\frac{I d\ell \sin\theta}{r^2} = \frac{\mu_0}{4\pi}\frac{I d\ell \sin 90°}{R^2} = \frac{\mu_0 I}{4\pi R^2}d\ell$$

Writing $d\ell = R\,d\theta$, we integrate around the circle to get

$$B = \int dB = \int_0^{2\pi} \frac{\mu_0 I}{4\pi R^2}R\,d\theta = \frac{\mu_0 I}{4\pi R}\int_0^{2\pi}d\theta = \frac{\mu_0 I}{2R}$$

If the current is counterclockwise (as shown), then $d\ell \times \hat{\mathbf{r}}$, and the magnetic field points out of the plane of the page, but if the current is clockwise, the magnetic field at the center points into the page.

AMPERE'S LAW

In Chapter 12, we learned about Gauss's law, which can be used to calculate the electric field due to a configuration of electric charges. Gauss's law says that the electric flux through a closed Gaussian surface (which is actually the integral of $\mathbf{E} \cdot d\mathbf{A}$ over the area of the surface) is equal to a constant $(1/\varepsilon_0)$ times the total electric charge enclosed by the surface.

There's an analogous law for calculating the magnetic field due to a configuration of electric currents. This law says that the integral of $\mathbf{B} \cdot d\mathbf{s}$ around a closed loop (called an **Amperian loop**) is equal to a constant (μ_0) times the total current that passes through the loop:

$$\oint_{\text{loop}} \mathbf{B} \cdot d\mathbf{s} = \mu_0 I_{\text{through loop}}$$

This is **Ampere's law,** and we'll use it to find the magnetic field at points on an Amperian loop surrounding the source current. (The circle on the integral sign in Ampere's law emphasizes that the integral must always be taken around a *closed* path.)

Example 13 Use Ampere's law to show that the magnetic field due to an infinitely long straight wire carrying a current I is given by the equation $B = (\mu_0/2\pi)I / R$, where R is the distance from the wire.

Solution. When we did this in Example 11 (using the Biot–Savart law), we ran into a messy integral. Using Ampere's law will be much easier.

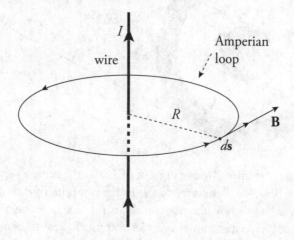

First, the magnetic field must have the same magnitude at every point on the Amperian circle, and the direction of the field must always be perpendicular to the circle (which means it points radially away from or toward the wire) or always tangent to the circle. We can rule out the first case in two ways: (1) Magnetic field lines *always* form closed curves that encircle the current that generates them; (2) If **B** were perpendicular to the Amperian circle, then $\mathbf{B} \cdot d\mathbf{s}$ would be zero everywhere along the circle, so Ampere's law would give $0 = \mu_0 I$, a contradiction. So **B** must be tangent to the Amperian circle.

Since **B** has constant magnitude on the circle and is always parallel to $d\mathbf{s}$ (which is a small section of the Amperian loop),

$$\mathbf{B} \cdot d\mathbf{s} = B \, ds \cos \theta = B \, ds$$

The current that passes through the loop is I, so Ampere's law gives us

$$\oint_{\text{loop}} B \, ds = \mu_0 I$$

$$B \oint_{\text{loop}} ds = \mu_0 I$$

$$B(2\pi R) = \mu_0 I$$

$$B = \frac{\mu_0 I}{2\pi R}$$

Example 14 The figure below is a cutaway view of a long coaxial cable that's composed of a solid cylindrical conductor (of radius R_1) surrounded by a thin conducting cylindrical shell (of radius R_2).

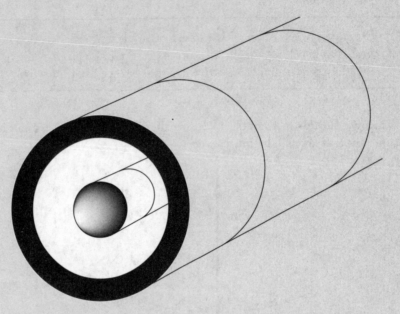

The inner cylinder carries a current of I_1 and the outer cylindrical shell carries a smaller current of I_2 in the opposite direction. Use Ampere's law to find the magnitude of the magnetic field (a) in the space between the inner cylinder and the shell, and (b) outside the outer shell.

Solution.

(a) Construct an Amperian loop of radius r in the space between the solid cylinder and the outer shell. The magnetic field must be of constant magnitude along (and tangent to) this loop.

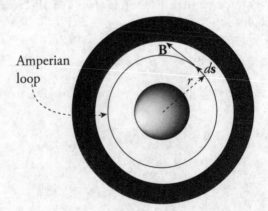

The only current that passes through the loop is I_1 (the current I_2 in the outer shell does not pierce the interior of the Amperian loop), so Ampere's law gives us:

$$\oint_{\text{loop}} B \, ds = \mu_0 I_1$$

$$B(2\pi r) = \mu_0 I_1$$

$$B = \frac{\mu_0}{2\pi} \frac{I_1}{r} \quad (R_1 < r < R_2)$$

(b) Now construct an Amperian loop that encircles the entire cable.

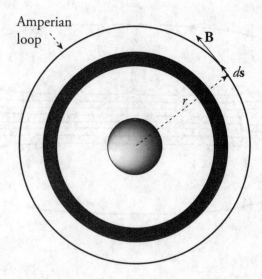

Unlike in the loop constructed in part (a), *both* currents pass through this loop. If they had the same direction, we'd add them to get the total current, but since the currents here have *opposite* directions, the net current through the loop is $I_1 - I_2$. Ampere's law then gives us

$$\oint_{\text{loop}} B \, ds = \mu_0 \left(I_1 - I_2 \right)$$

$$B(2\pi r) = \mu_0 \left(I_1 - I_2 \right)$$

$$B = \frac{\mu_0}{2\pi} \frac{\left(I_1 - I_2 \right)}{r} \quad (r > R_2)$$

Note that if $I_1 = I_2$, then the magnetic field outside the cable is zero.

Example 15 A tightly wound solenoid with n coils per length has a current I running through it. Use Ampere's law to show the magnetic field inside this ideal solenoid is given by the equation: $B = \mu_0 nI$.

Solution.

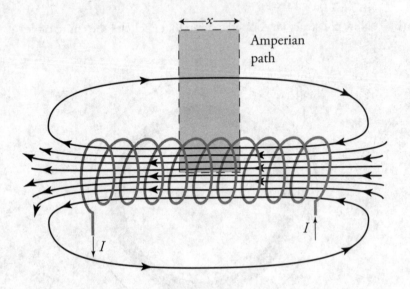

Consider the diagram above. The two vertical sections of the Amperian path do not matter because the path is perpendicular to the magnetic field (remember Ampere's law is a dot product for $\mathbf{B} \cdot d\mathbf{s}$). The top section of the path is approximated as zero for an ideal solenoid because the field is not weak outside a tightly wound solenoid. Therefore, the only part of the path we will use is the length x inside of the solenoid. We will assume N total coils pass through the shaded region of our Amperian path.

$$\oint_{path} B \, ds = \mu_0 I$$

$$Bx = \mu_0 NI$$

$$B = \frac{\mu_0 NI}{x}$$

$$B = \mu_0 n \, I$$

Example 16 The figure below is a cross section of a *toroidal solenoid* (a solenoid bent into a doughnut shape) of inner radius R_1 and outer radius R_2. It consists of N windings, and the wire carries a current I. What's the magnetic field in each of the following regions?

(a) $r < R_1$

(b) $R_1 < r < R_2$

(c) $r > R_2$

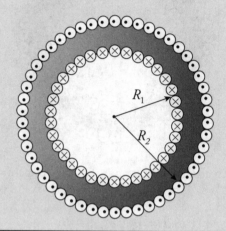

Solution.

(a) If we construct a circular Amperian loop of radius $r < R_1$, then no current will pass through it.

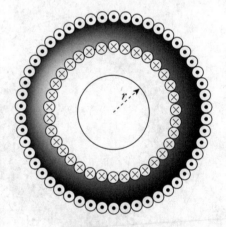

Since $\oint_{\text{loop}} \mathbf{B} \cdot d\mathbf{s} = \oint_{\text{loop}} B \, ds = B(2\pi r)$ and $I_{\text{through loop}} = 0$, Ampere's law immediately tells us that $B = 0$ in this region (the hole of the doughnut).

(b) Construct a circular Amperian loop of radius r such that $R_1 < r < R_2$. Each of the N windings that brings current into the plane of the page passes through the loop, but the N windings that bring current out of the plane of the page do not pass through the loop.

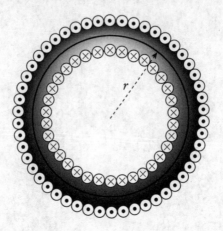

So, the net current through the loop is NI, and Ampere's law gives us:

$$\oint_{\text{loop}} B \, ds = \mu_0 I_1$$

$$B(2\pi r) = \mu_0 NI$$

$$B = \frac{\mu_0}{2\pi} \frac{NI}{r} \quad (R_1 < r < R_2)$$

(c) Construct a circular Amperian loop of radius $r > R_2$.

Note that the results of this example show that for an ideal toroidal solenoid, the magnetic field is contained within the space enclosed by the windings.

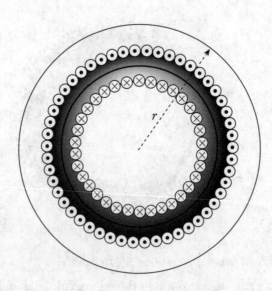

For every current contribution of I punching upward through the loop, there is an equal current punching downward through the loop, so the net current through the loop is zero, and $B = 0$ in this region.

Chapter 15 Drill

The answers and explanations can be found in Chapter 17.

Section I: Multiple Choice

1. Which of the following statements is/are true?

   I. Magnetic force can never do work on a charged particle.
   II. Magnetic force can never change the velocity of a charged particle.
   III. A charged particle will always experience a magnetic force if it moves through a magnetic field.

   (A) I only
   (B) I and II only
   (C) II and III only
   (D) III only
   (E) None of the above

2. The velocity of a particle of charge $+4.0 \times 10^{-9}$ C and mass 2×10^{-4} kg is perpendicular to a 0.1-tesla magnetic field. If the particle's speed is 3×10^4 m/s, what is the acceleration of this particle due to the magnetic force?

   (A) 0.0006 m/s²
   (B) 0.006 m/s²
   (C) 0.06 m/s²
   (D) 0.6 m/s²
   (E) None of the above

3. In the figure below, what is the direction of the magnetic force $\mathbf{F}_B$?

   (A) To the right
   (B) Downward, in the plane of the page
   (C) Upward, in the plane of the page
   (D) Out of the plane of the page
   (E) Into the plane of the page

4. In the figure below, what must be the direction of the particle's velocity, $\mathbf{v}$?

   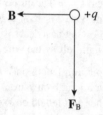

   (A) To the right
   (B) Downward, in the plane of the page
   (C) Upward, in the plane of the page
   (D) Out of the plane of the page
   (E) Into the plane of the page

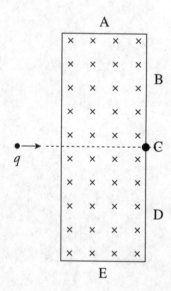

5. The picture above shows a positively charged particle, q, about to enter a rectangular area with a magnetic field directed into the plane of the page. Assume that the particle WILL escape the magnetic field. If it does so, it will exit

   (A) somewhere along length A
   (B) somewhere along length B
   (C) at point C
   (D) somewhere along length D
   (E) somewhere along length E

6. A straight wire of length 2 m carries a 10-amp current. How strong is the magnetic field at a distance of 2 cm from the wire?

   (A) 1×10^{-6} T
   (B) 1×10^{-5} T
   (C) 2×10^{-5} T
   (D) 1×10^{-4} T
   (E) 2×10^{-4} T

7. Two long, straight wires are hanging parallel to each other and are 1 cm apart. The current in Wire 1 is 5 A, and the current in Wire 2 is 10 A in the same direction. Which of the following best describes the magnetic force per unit length felt by the wires?

   (A) The force per unit length on Wire 1 is twice the force per unit length on Wire 2.
   (B) The force per unit length on Wire 2 is twice the force per unit length on Wire 1.
   (C) The force per unit length on Wire 1 is 0.0003 N/m, away from Wire 2.
   (D) The force per unit length on Wire 1 is 0.001 N/m, toward Wire 2.
   (E) The force per unit length on Wire 1 is 0.001 N/m, away from Wire 2.

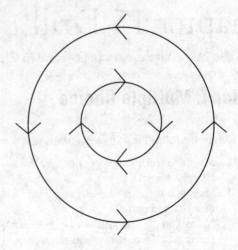

8. In the figure above, two concentric rings have current running through them in the directions shown. Assuming that both currents can be adjusted to any magnitude other than 0 (but always in the directions shown), where is it possible for the net magnetic field to be 0 ?

   I. Inside the inner ring
   II. Between the two rings
   III. Outside the outer ring

   (A) I only
   (B) I and III only
   (C) II only
   (D) II and III only
   (E) None of the above

9. How many windings must a solenoid of length 80 cm have in order to establish a magnetic field of strength 0.2 T inside the solenoid, if it carries a current of 20 amps?

   (A) 1,000
   (B) 6,400
   (C) 10,000
   (D) 32,000
   (E) 64,000

10. The value of $\oint \mathbf{B} \cdot ds$ along a closed path in a magnetic field $\mathbf{B}$ is 6.28×10^{-6} T • m. What is the total current that passes through this closed path?

    (A) 0.1 A
    (B) 0.5 A
    (C) 1 A
    (D) 4 A
    (E) 5 A

Section II: Free Response

1. The diagram below shows a simple mass spectrograph. It consists of a source of ions (charged atoms) that are accelerated (essentially from rest) by the voltage V and enter a region containing a uniform magnetic field, **B**. The polarity of V may be reversed so that both positively charged ions (cations) and negatively charged ions (anions) can be accelerated. Once the ions enter the magnetic field, they follow a semicircular path and strike the front wall of the spectrograph, on which photographic plates are constructed to record the impact.

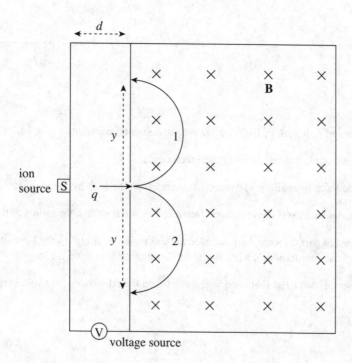

(a)   What is the acceleration of an ion of charge q just before it enters the magnetic field?

(b)   Find the speed with which an ion of charge q enters the magnetic field.

(c)   (i)   Which semicircular path, 1 or 2, would a cation follow?

(ii)   Which semicircular path, 1 or 2, would an anion follow?

(d)   Determine the mass of a cation entering the apparatus in terms of y, q, B, and V.

(e)   Once a cation of charge q enters the magnetic field, how long does it take to strike the photographic plate?

(f)   What is the work done by the magnetic force in the spectrograph on a cation of charge q?

2. A wire of diameter d and resistivity ρ is bent into a rectangular loop (of side lengths a and b) and fitted with a small battery that provides a voltage V. The loop is placed at a distance c from a very long, straight wire that carries a current I in the direction indicated in the diagram.

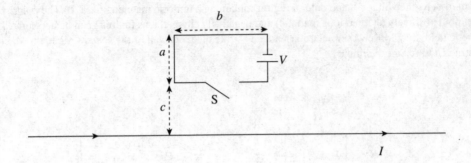

(Express all answers in terms of a, b, c, d, ρ, V, I, m, B, x, and fundamental constants.)

(a)   When the switch S is closed, find the current in the rectangular loop.

(b)   What is the magnetic force (magnitude and direction) exerted on the loop by the long, straight wire?

(c)   The wire of the rectangular loop is then reshaped into a circle. What will be the radius of the circular loop?

(d)   If the loop constructed in part (c) were then threaded around the long, straight wire (so that the straight wire passed through the center of the circular loop), what would be the magnetic force on the loop now?

(e)   In the following diagram, two fixed L-shaped wires, separated by a distance x, are connected by a wire that's free to slide vertically.

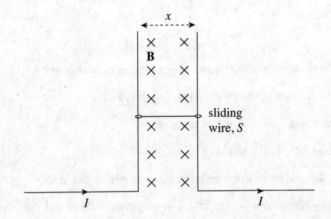

The mass of the sliding wire, S, is m. If the sliding wire S crosses a region that contains a uniform magnetic field B, how much current must be carried by the wire to keep S from sliding down (due to its weight)?

3. The figure below shows two long, straight wires connected by a circular arc of radius x that subtends a central angle ϕ. The current in the wire is I.

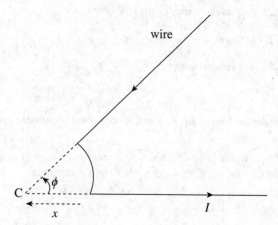

(a) Find the magnetic field (magnitude and direction) created at point C. Write your answer in terms of x, ϕ, I, and fundamental constants.

(b) A particle of charge $+q$ is placed at point C and released. Find the magnetic force on the particle.

(c) A second long, straight wire is set up perpendicular to the plane of the page through C, carrying the same current, I (directed out of the page), as the wire pictured in the diagram. Determine the magnetic force per unit length between the wires.

4. For a conducting rod that carries a current I, the current density is defined as the current per unit area: $J = I/A$.

Part 1. A homogeneous cylindrical rod of radius R carries a current whose current density, J, is uniform (constant); that is, J does not vary with the radial distance, r, from the center of the rod.

(a)   Determine the total current, I, in the rod.

(b)   Calculate the magnitude of the magnetic field for

(i) $r < R$

(ii) $r > R$, writing your answers in terms of r, R, I, and fundamental constants

Part 2. A nonhomogeneous cylindrical rod of radius R carries a current whose current density, J, varies with the radial distance, r, from the center of the rod according to the equation $J = \sigma r$, where σ is a constant.

(c)   What are the units of σ ?

(d)   Determine the total current, I, in the rod.

(e)   Calculate the magnitude of the magnetic field for

(i) $r < R$

(ii) $r > R$, writing your answers in terms of r, R, and I

Summary

Magnetic Force

o   When a charged particle moves through a magnetic field, it will experience a force if the velocity is not parallel with the magnetic field. The force is given by the following equation:

$$\mathbf{F}_B = q(\mathbf{v} \times \mathbf{B})$$

o   Notice that this is a cross product of the velocity and the magnetic field vectors, so the directions of the force can be obtained using the right-hand rule from Chapter 4.

o   Since the force will always be perpendicular to the velocity, the force will not change the speed of the particle; it will only change its direction. This force can be used to cause the particle to travel in a circle.

o   A current-carrying wire will also experience a force when it is in a magnetic field because the wire has a moving charge within it. The force is given by the following equation:

$$\mathbf{F}_B = I(\boldsymbol{\ell} \times \mathbf{B})$$

Magnetic Fields

o   One can calculate the magnetic field created by a current using the Biot–Savart law and Ampere's law. Ampere's law is easier to use, but the magnetic field created by the current needs to be more symmetric for it to be used.

o   The Biot–Savart law states

$$B = \frac{\mu_o}{4\pi} \int \frac{I d\boldsymbol{\ell} \times \hat{\mathbf{r}}}{r^2}$$

o   Ampere's law is $\oint_{loop} \mathbf{B} \cdot d\mathbf{s} = \mu_0 I_{\text{through loop}}$.

o   Use Ampere's law to find the magnetic field due to the following current-carrying devices: long wires, coaxial cables, solenoids, and toroids.

o   The magnitude of the magnetic field at a distance r from a long, straight wire is $B = \frac{\mu_0 I}{2\pi r}$.

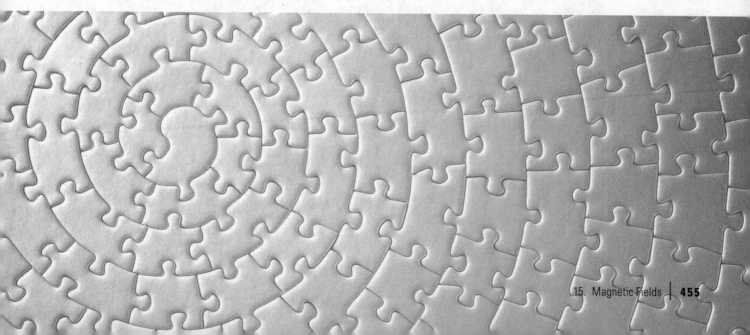

Chapter 16
Electromagnetism

INTRODUCTION

In Chapter 15, we learned that electric currents generate magnetic fields, and we will now see how magnetism can generate electric currents.

MOTIONAL EMF

The figure below shows a conducting wire of length ℓ moving with constant velocity $\mathbf{v}$ in the plane of the page through a uniform magnetic field $\mathbf{B}$ that's perpendicular to the page. The magnetic field exerts a force on the moving conduction electrons in the wire. With $\mathbf{B}$ pointing into the page, the direction of $\mathbf{v} \times \mathbf{B}$ is upward, so the magnetic force, $\mathbf{F}_B$, on these electrons (which are negatively charged) is downward.

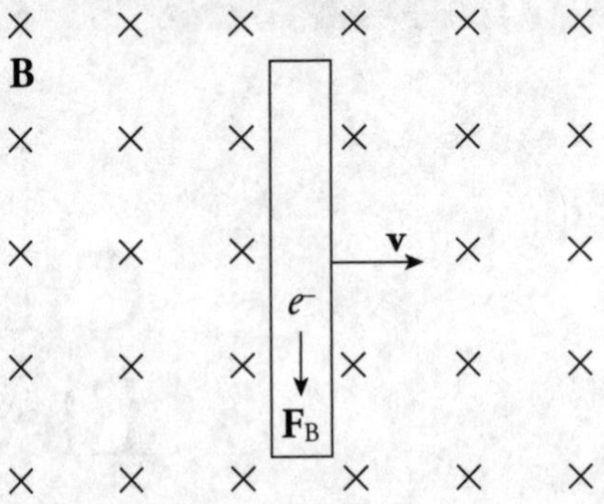

As a result, electrons will be pushed to the lower end of the wire, which will leave an excess of positive charge at its upper end. This separation of charge creates a uniform electric field, $\mathbf{E}$, within the wire (pointing downward).

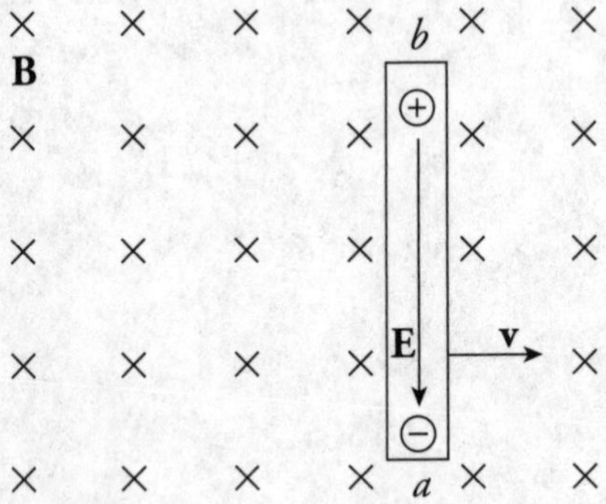

A charge q in the wire experiences two forces: an electric force, $\mathbf{F}_E = q\mathbf{E}$, and a magnetic force,

$$\mathbf{F}_B = q(\mathbf{v} \times \mathbf{B})$$

If q is negative, $\mathbf{F}_E$ is upward and $\mathbf{F}_B$ is downward; if q is positive, $\mathbf{F}_E$ is downward and $\mathbf{F}_B$ is upward. So, in both cases, the forces act in opposite directions. Once the magnitude of $\mathbf{F}_E$ equals the magnitude of $\mathbf{F}_B$, the charges in the wire are in electromagnetic equilibrium. This occurs when $qE = qvB$; that is, when $E = vB$.

The presence of the electric field creates a potential difference between the ends of the rod. Since negative charge accumulates at the lower end (which we'll call point *a*) and positive charge accumulates at the upper end (point *b*), point *b* is at a higher electric potential.

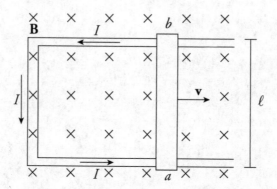

The potential difference V_{ba} is equal to $E\ell$ and, since $E = vB$, the potential difference can be written as $vB\ell$.

Now, imagine that the rod is sliding along a pair of conducting rails connected at the left by a stationary bar. The sliding rod now completes a rectangular circuit, and the potential difference, V_{ba}, causes current to flow.

The motion of the sliding rod through the magnetic field creates an electromotive force called **motional emf**:

$$\mathcal{E} = vB\ell$$

The existence of a current in the sliding rod causes the magnetic field to exert a force on it. Using the formula $\mathbf{F}_B = I(\ell \times \mathbf{B})$, the fact that ℓ points upward (in the direction of the current) and $\mathbf{B}$ is into the page tells us that the direction of $\mathbf{F}_B$ on the rod is to the left. An external agent must provide this same amount of force to the right to maintain the rod's constant velocity and keep the current flowing. The power that the external agent must supply is $P = Fv = I\ell Bv$, and the electrical power delivered to the circuit is $P = IV_{ba} = I\mathcal{E} = IvB\ell$. Notice that these two expressions are identical. The energy provided by the external agent is transformed first into electrical energy and then thermal energy as the conductors making up the circuit dissipate heat.

FARADAY'S LAW OF ELECTROMAGNETIC INDUCTION

Electromotive force can be created by the motion of a conductor through a magnetic field, but there's another way to create an emf from a magnetic field.

Magnetic Flux

Think back to Chapter 12 and electric flux. The electric flux through a surface of area A is equal to the product of A and the electric field that's perpendicular to it. That is, $\Phi_E = E_\perp A = \mathbf{E} \cdot \mathbf{A} = EA \cos \theta$. If $\mathbf{E}$ varies over the area A, then we write $\Phi_E = \int \mathbf{E} \cdot d\mathbf{A}$.

The idea of magnetic flux is exactly the same. The **magnetic flux**, Φ_B, through an area A is equal to the product of A and the magnetic field perpendicular to it: $\Phi_B = B_\perp A = \mathbf{B} \cdot \mathbf{A} = BA \cos \theta$. Again, if $\mathbf{B}$ varies over the area, then we must write:

$$\Phi_B = \int \mathbf{B} \cdot d\mathbf{A}$$

The SI unit for magnetic flux, the tesla·meter^2, is called a **weber** (abbreviated as **Wb**).

Example 1 The figure below shows two views of a circular loop of radius 3 cm placed within a uniform magnetic field, **B** (magnitude 0.2 T).

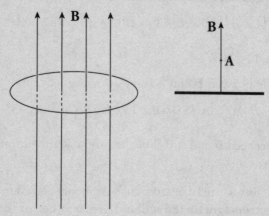

(a) What's the magnetic flux through the loop?

(b) What would be the magnetic flux through the loop if the loop were rotated 45°?

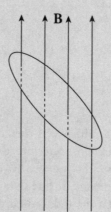
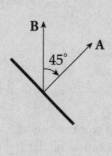

(c) What would be the magnetic flux through the loop if the loop were rotated 90°?

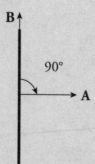

Solution.

(a) Since **B** is parallel to **A**, the magnetic flux is equal to BA:

$$\Phi_B = BA = B \cdot \pi r^2 = (0.2 \text{ T}) \cdot \pi(0.03 \text{ m})^2 = 5.7 \times 10^{-4} \text{ T·m}^2$$

So $\Phi_B = 5.7 \times 10^{-4}$ Wb.

(b) Since the angle between **B** and **A** is $45°$, the magnetic flux through the loop is

$$\Phi_B = BA \cos 45° = B \cdot \pi r^2 \cos 45° = (0.2 \text{ T}) \cdot \pi(0.03 \text{ m})^2 \cos 45° = 4.0 \times 10^{-4} \text{ Wb}$$

(c) If the angle between **B** and **A** is $90°$, the magnetic flux through the loop is zero because $\cos 90° = 0$.

The concept of magnetic flux is crucial because changes in magnetic flux induce emf. According to **Faraday's law of electromagnetic induction**, the magnitude of the emf induced in a circuit is equal to the rate of change of the magnetic flux through the circuit. This can be written mathematically in the form

$$\left| \mathcal{E}_{avg} \right| = \left| \frac{\Delta \Phi_B}{\Delta t} \right|$$

or, if we let $\Delta t \to 0$, we get

$$\left| \mathcal{E} \right| = \left| \frac{d \Phi_B}{dt} \right|$$

> The AP Physics C Table of Information uses the subscript M instead of B for this formula. In physics, these two are interchangeable when referring to magnetism.

This induced emf can produce a current, which will then create its own magnetic field. The direction of the induced current is determined by the polarity of the induced emf and is given by **Lenz's law**: The induced current will always flow in the direction that opposes the change in magnetic flux that produced it. If this were not so, then the magnetic flux created by the induced current would magnify the change that produced it, and energy would not be conserved. Lenz's law can be included mathematically with Faraday's law by the introduction of a minus sign; this leads to a single equation that expresses both results:

$$\mathcal{E}_{avg} = -\frac{\Delta \Phi_B}{\Delta t} \qquad \text{or} \qquad \boxed{\mathcal{E} = -\frac{d \Phi_M}{dt}}$$

Example 2 The circular loop of Example 1 rotates at a constant angular speed through 45° in 0.5 s.

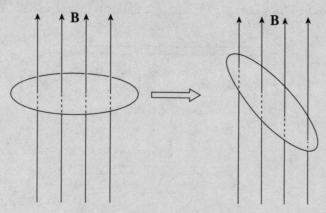

(a) What's the induced emf in the loop?

(b) In which direction will current be induced to flow?

Solution.

(a) As we found in Example 1, the magnetic flux through the loop changes when the loop rotates. Using the values we determined earlier, Faraday's law gives

$$\mathcal{E}_{avg} = -\frac{\Delta \Phi_B}{\Delta t} = -\frac{(4.0 \times 10^{-4} \text{ Wb}) - (5.7 \times 10^{-4} \text{ Wb})}{0.5 \text{ s}} = 3.4 \times 10^{-4} \text{ V}$$

(b) The original magnetic flux was 5.7×10^{-4} Wb upward and was decreased to 4.0×10^{-4} Wb. So the change in magnetic flux is -1.7×10^{-4} Wb upward or, equivalently, $\Delta \Phi_B = 1.7 \times 10^{-4}$ Wb downward. To oppose this change, we would need to create some magnetic flux upward. The current would be induced in the counterclockwise direction (looking down on the loop) because the right-hand rule tells us that the current would produce a magnetic field that would point up.

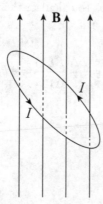

The current will flow only while the loop rotates because emf is induced only when magnetic flux is changing. If the loop rotates 45° and then stops, the current will disappear.

Example 3 Again, consider the conducting rod that's moving with constant velocity **v** along a pair of parallel conducting rails (separated by a distance ℓ), within a uniform magnetic field, **B**:

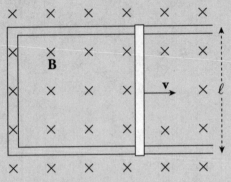

Find the induced emf and the direction of the induced current in the rectangular circuit.

Solution. The area of the rectangular loop is ℓx, where x is the distance from the left-hand bar to the moving rod:

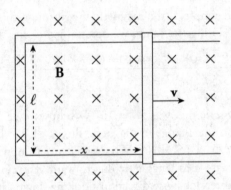

Because the area is changing, the magnetic flux through the loop is changing, which means that an emf will be induced in the loop. To calculate the induced emf, we first write $\Phi_B = BA = B\ell x$. Then since $\Delta x/\Delta t = v$, we get

$$\mathcal{E}_{avg} = -\frac{\Delta \Phi_B}{\Delta t} = -\frac{\Delta(B\ell x)}{\Delta t} = -B\ell\frac{\Delta x}{\Delta t} = -B\ell v$$

We can figure out the direction of the induced current from Lenz's law. As the rod slides to the right, the magnetic flux into the page increases. How do we oppose a flux that increases into the page? By producing out-of-the-page flux. In order for the induced current to generate a magnetic field that points out of the plane of the page, the current must be directed counter-clockwise (according to the right-hand rule).

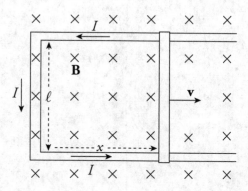

Note that the magnitude of the induced emf and the direction of the current agree with the results we derived earlier in the section on motional emf.

This example also shows how a violation of Lenz's law would lead directly to a violation of the Law of Conservation of Energy. The current in the sliding rod is directed upward, as given by Lenz's law, so the conduction electrons are drifting downward. The force on these drifting electrons—and thus, the rod itself—is directed to the left, opposing the force that's pulling the rod to the right. If the current were directed downward, in violation of Lenz's law, then the magnetic force on the rod would be to the right, causing the rod to accelerate to the right with ever-increasing speed and kinetic energy, without the input of an equal amount of energy from an external agent.

Example 4 A permanent magnet creates a magnetic field in the surrounding space. The end of the magnet at which the field lines emerge is designated the **north pole** (N), and the other end is the **south pole** (S):

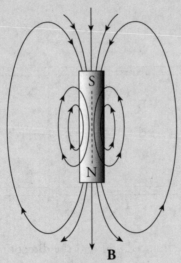

(a) The figure below shows a bar magnet moving down, through a circular loop of wire. What will be the direction of the induced current in the wire?

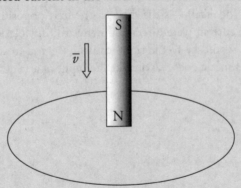

(b) What will be the direction of the induced current in the wire if the magnet is moved as shown in the following diagram?

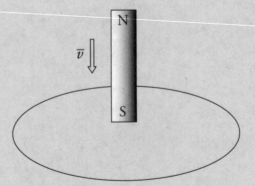

Solution.

(a) The magnetic flux down, through the loop, increases as the magnet is moved. By Lenz's law, the induced emf will generate a current that opposes this change. How do we oppose a change of *more flux downward*? By creating flux *upward*. So, according to the right-hand rule, the induced current must flow counterclockwise (because this current will generate an upward-pointing magnetic field, $B_{induced}$):

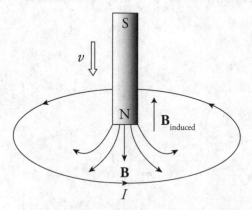

(b) In this case, the magnetic flux through the loop is upward. As the south pole moves closer to the loop, the magnetic field strength increases, so the magnetic flux through the loop increases upward. How do we oppose a change of *more flux upward*? By creating flux *downward*. Therefore, in accordance with the right-hand rule, the induced current will flow clockwise (because this current will generate a downward-pointing magnetic field):

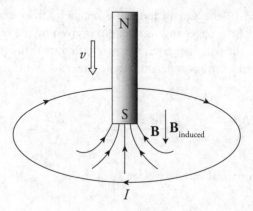

Example 5 A square loop of wire 2 cm on each side contains 5 tight turns and has a total resistance of 0.0002 Ω. It is placed 20 cm from a long, straight, current-carrying wire. If the current in the straight wire is increased at a steady rate from 20 A to 50 A in 2 s, determine the magnitude and direction of the current induced in the square loop. (Because the square loop is at such a great distance from the straight wire, assume that the magnetic field through the loop is uniform and equal to the magnetic field at its center.)

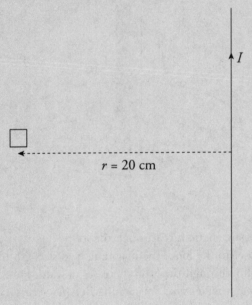

r = 20 cm

Solution. At the position of the square loop, the magnetic field due to the straight wire is directed out of the plane of the page, and its strength is given by the equation $B = (\mu_0/\pi 2)(I/r)$. As the current in the straight wire increases, the magnetic flux through the turns of the square loop changes, inducing an emf and current. There are $N = 5$ turns; Faraday's law becomes $\varepsilon_{avg} = -N(\Delta\Phi_B/\Delta t)$, and

$$\varepsilon_{avg} = -N\frac{\Delta\Phi_B}{\Delta t} = -N\frac{\Delta(BA)}{\Delta t} = -NA\frac{\Delta B}{\Delta t} = -NA\frac{\mu_0}{2\pi r}\frac{\Delta I}{\Delta t}$$

Substituting the given numerical values, we get

$$\varepsilon_{avg} = -NA\frac{\mu_0}{2\pi r}\frac{\Delta I}{\Delta t}$$

$$= -(5)(0.02\ \text{m})^2\frac{4\pi\times 10^{-7}\ \text{T}\cdot\text{m/A}}{2\pi(0.20\,\text{m})}\frac{(50\ \text{A} - 20\ \text{A})}{2\ \text{s}}$$

$$= 3\times 10^{-8}\ \text{V}$$

The magnetic flux through the loop is out of the page and increases as the current in the straight wire increases. To oppose an increasing out-of-the-page flux, the direction of the induced current should be clockwise, thereby generating an into-the-page magnetic field (and flux).

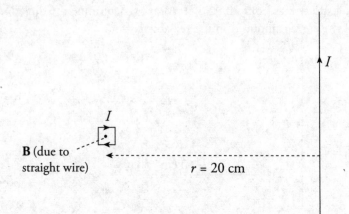

The value of the current in the loop will be

$$I = \frac{\varepsilon}{R} = \frac{3 \times 10^{-8} \text{ V}}{0.0002 \text{ } \Omega} = 1.5 \times 10^{-4} \text{ A}$$

Example 6 A rectangular loop of wire 10 cm by 4 cm has a total resistance of 0.005 Ω. It is placed 2 cm from a long, straight, current-carrying wire. If the current in the straight wire is increased at a steady rate from 20 A to 50 A in 2 s, determine the direction of the current induced in the rectangular loop.

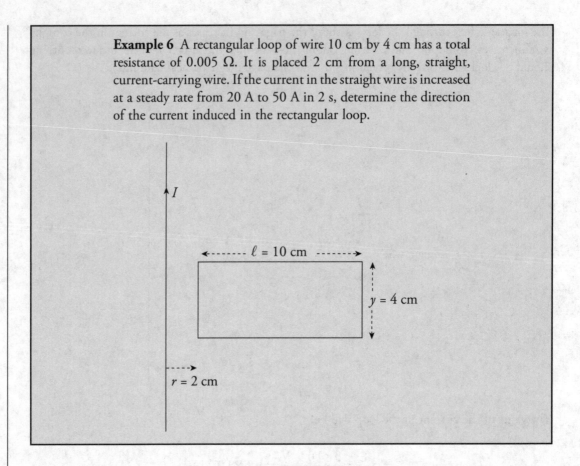

Solution. Unlike in the previous example, in this case, the magnetic field varies greatly over the interior of the rectangular loop, so we have to use integration to calculate the magnetic flux. Take a narrow strip of width dx and height y, whose area is $dA = y\,dx$:

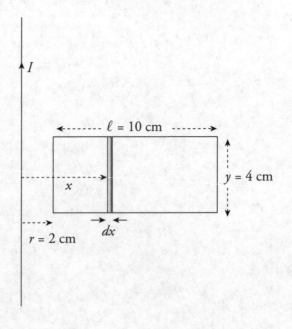

The magnetic field strength everywhere in this strip is $B = (\mu_0/2\pi)(I/x)$, so the magnetic flux through the strip is

$$d\Phi_B = B \cdot dA = B \cdot (y \ dx) = \frac{\mu_0}{2\pi} \frac{I}{x} \cdot y \ dx$$

Integrating from $x = r$ to $x = r + \ell$ gives the total magnetic flux through the loop.

$$\Phi_B = \int d\Phi_B = \int_{x=r}^{x=r+\ell} \frac{\mu_0}{2\pi} \frac{I}{x} \cdot y \ dx = \frac{\mu_0 I y}{2\pi} \int_r^{r+\ell} \frac{dx}{x} = \frac{\mu_0 I y}{2\pi} \left[\ln x\right]_r^{r+\ell} = \frac{\mu_0 I y}{2\pi} \ln \frac{r+\ell}{r}$$

Faraday's law then gives us

$$\varepsilon = -\frac{d\Phi_B}{dt} = -\frac{d}{dt}\left(\frac{\mu_0 I y}{2\pi} \ln \frac{r+\ell}{r}\right) = -\left(\frac{\mu_0 y}{2\pi} \ln \frac{r+\ell}{r}\right)\frac{dI}{dt}$$

Ignoring the minus sign (which only reminds us that Lenz's law has to be obeyed), the value of the current induced in the loop is

$$
\begin{aligned}
I = \frac{\varepsilon}{R} &= \left(\frac{\mu_0 y}{2\pi R} \ln \frac{r+\ell}{r}\right)\frac{dI}{dt} \\
&= \frac{(4\pi \times 10^{-7} \ \text{T} \cdot \text{m/A})(0.04 \ \text{m})}{2\pi(0.005 \ \Omega)} \cdot \ln \frac{2 \ \text{cm} + 10 \ \text{cm}}{2 \ \text{cm}} \cdot \frac{50 \ \text{A} - 20 \ \text{A}}{2 \ \text{s}} \\
&= 4.3 \times 10^{-5} \ \text{A}
\end{aligned}
$$

At the position of the rectangular loop, the magnetic field due to the straight wire is directed into the plane of the page. So, the magnetic flux through the loop is into the page and increases as the current in the straight wire increases. To oppose an increasing into-the-page flux, the direction of the induced current will be counterclockwise, generating an out-of-the-page magnetic field (and flux).

Induced Electric Fields

In Examples 4, 5, and 6, an electric current was induced in a stationary conducting loop by a changing magnetic flux through the loop, but what was the source of the force that pushed these charges around the loop? It was not the magnetic force, because the loop was stationary. It must therefore have been an electric force that was produced by an electric field. The changing magnetic field produced the electric field, which acted on the free charges in the conducting loop and caused the current.

Let's look at a specific situation. The figure below is a view down the central axis of an ideal solenoid (*ideal* means that the magnetic field is uniform and parallel to the axis and that no magnetic field exists outside). Surrounding the solenoid is a loop of wire.

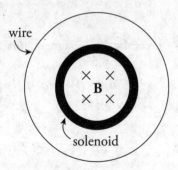

Now suppose that the current in the solenoid is increased, which increases the strength of its magnetic field. This changes the magnetic flux through the wire and induces a current (which is directed counterclockwise). The force that pushes the charges around this wire is $\mathbf{F}_E = q\mathbf{E}$; $\mathbf{E}$ is the induced electric field.

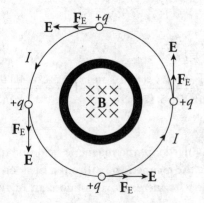

The work done on a charge q as it makes one revolution is F_E times the distance around the wire, $F_E \cdot 2\pi r = qE\cdot(2\pi r)$. So the work per unit charge, which is the definition of emf, is equal to $E(2\pi r)$. But Faraday's law says that the emf is equal to $-d\Phi_B/dt$.

$$E(2\pi r) = -d\Phi_B/dt$$

This equation can be generalized by saying that the work done per unit charge by the induced electric field around a closed path is equal to $\oint \mathbf{E} \cdot d\ell$. Therefore,

$$\oint \mathbf{E} \cdot d\ell = -\frac{d\Phi_B}{dt}$$

This is a restatement of Faraday's law that includes the electric field, **E**, induced by the changing magnetic flux. Notice that this electric field is different from the ones we studied in the previous chapters. Electric fields created by stationary source charges, called electrostatic fields, are conservative, meaning that the work done by them on charges moved along closed paths is always zero. However, as we've just seen in the example of the solenoid above, the electric field induced by a *changing magnetic flux* does not share this property. The work done by $\mathbf{E}_{induced}$ on a charge as it moves around a closed path is equal to $\oint q\mathbf{E} \cdot d\ell$, which is not zero because $d\Phi_B/dt$ is not zero. Because of this, the electric field induced by a changing magnetic flux is nonconservative.

> **Example 7** For the situation just described, assume that the solenoid has 15,000 turns per meter and a radius of $r = 2$ cm. The radius of the circular loop is $R = 4$ cm. If the current in the solenoid is increased at a rate of 10 A/s, what is the magnitude of the induced electric field at each position along the circular wire?

Solution. From Faraday's law, we have

$$\oint \mathbf{E} \cdot d\ell = -\frac{d\Phi_B}{dt}$$

$$E(2\pi R) = -\frac{d\Phi_B}{dt}$$

$$|E| = \frac{1}{2\pi R}\left|-\frac{d\Phi_B}{dt}\right|$$

Assuming that the magnetic field exists only inside the solenoid, the magnetic flux through the circular wire is $BA = B(\pi r^2)$, where $B = \mu_0 nI$:

$$\Phi_B = (\mu_0 nI) \cdot (\pi r^2)$$

Therefore,

$$\begin{aligned}
|E| &= \frac{1}{2\pi R}\left|-\frac{d\Phi_B}{dt}\right| \\
&= \frac{1}{2\pi R} \cdot \frac{d}{dt}(\mu_0 nI \cdot \pi r^2) \\
&= \frac{\mu_0 nr^2}{2R}\frac{dI}{dt} \\
&= \frac{(4\pi \times 10^{-7}\ \text{T} \cdot \text{m/A})(15{,}000/\text{m})(0.02\ \text{m})^2}{2(0.04\ \text{m})}(10\ \text{A/s}) \\
&= 9.4 \times 10^{-4}\ \text{V/m}
\end{aligned}$$

INDUCTANCE

Placing a capacitor in series with a resistor and battery in an electric circuit causes the current to drop exponentially from $\mathcal{E}/R$ (when the switch is closed) to zero as charge builds up on the capacitor and causes the voltage across the capacitor to oppose the emf of the battery. A similar thing happens when an inductor is placed in series with a resistor and a source of emf in an electric circuit.

An **inductor** is a circuit element that opposes changes in current. The prototypical inductor is a coil (a solenoid) that looks like this in diagrams:

How does this coil oppose changes in current in a circuit? Well, when current exists in the coil, a magnetic field is created, and magnetic flux passes through the loops. If the current changes, then the magnetic field changes proportionately, as does the magnetic flux through the loops. But, according to Faraday's law of induction, a changing magnetic flux induces an emf that opposes the change that produced it. In other words, the changing magnetic flux through the coil, due to the current in the coil itself, produces a **self-induced emf**. Since this self-induced emf opposes the change in the current that produced it, it's also called **back emf**.

Assume that the inductor contains N turns and that Φ_B is the magnetic flux through each turn. Then the total magnetic flux through the entire coil is $N\Phi_B$; this is proportional to the current, so $N\Phi_B = LI$ for some constant, L. The proportionality constant L is called the **inductance** (or **self-inductance**) of the coil:

$$L = \frac{N\Phi_B}{I}$$

The SI unit for inductance is the weber per ampere, which is renamed a **henry** (**H**).

Example 8 An ideal solenoid of cross-sectional area A and length ℓ contains n turns per unit length. What is its self-inductance?

Solution. The magnetic field inside an ideal solenoid is parallel to its central axis and has strength $B = \mu_0 nI$. Since the solenoid's length is ℓ, the total number of turns, N, is $n\ell$. Therefore,

$$L = \frac{N\Phi_B}{I} = \frac{N \cdot BA}{I} = \frac{n\ell \cdot BA}{I} = \frac{n\ell \cdot (\mu_0 nI)A}{I} = \mu_0 n^2 A \ell$$

Or, in terms of the total number of turns in the solenoid, $L = \mu_0 N^2 A / \ell$.

The self-induced emf in an inductor now follows directly from the definition of L and Faraday's law. Since $N\Phi_B = LI$, we have $N(d\Phi_B/dt) = L(dI/dt)$. But by Faraday's law, $\mathcal{E} = -N(d\Phi_B/dt)$. So,

$$\mathcal{E} = -L\frac{dI}{dt}$$

The self-induced emf is proportional to the rate of change of the current. The faster the current tries to change, the greater the self-induced (back) emf will be. But if the current is steady, then $\mathcal{E} = 0$.

RL Circuits

Consider the following circuit, which contains a battery, an inductor, and a resistor:

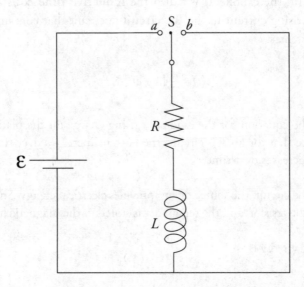

At time $t = 0$, the switch is moved to point a. Applying Kirchhoff's loop rule to the left-hand part of the circuit, we obtain

$$\varepsilon - iR - L\frac{di}{dt} = 0$$

where R is the resistance and L is the inductance. We apply the method of separation of variables to solve for the current in the circuit as a function of time. First, we rearrange the equation so that one variable is on the left side of the equation and the other variable is on the right side.

$$\frac{di}{\dfrac{\varepsilon}{R} - i} = \frac{R\,dt}{L}$$

We then integrate both sides of the equation:

$$\int_0^i \frac{di}{\frac{\varepsilon}{R}-i} = \int_0^t \frac{R\,dt}{L}$$

$$\ln\left(\frac{\varepsilon}{R}-i\right) - \ln\left(\frac{\varepsilon}{R}\right) = -\frac{R}{L}t$$

A little algebraic manipulation allows us to solve for the current as a function of time, $i(t)$:

$$i(t) = \frac{\varepsilon}{R}\left(1-e^{\frac{Rt}{L}}\right)$$

The ratio of L to R (in the exponent) is called the **inductive time constant,** $\tau_L = L/R$, and the equation for the rising current in the **RL circuit** (a circuit that contains a resistor and an inductor) can be written as

$$i(t) = \frac{\varepsilon}{R}\left(1-e^{-t/\tau_L}\right)$$

which is similar to the equation for the gradually rising charge on the plates of a capacitor in an RC circuit. Notice that for an RL circuit, the current increases over time, and for an RC circuit, the current decreases over time.

As current builds in the circuit, the source of emf provides electrical energy. Some of this energy is dissipated as heat by the resistor, and the remainder is stored in the magnetic field of the inductor.

The amount of stored energy is

$$\boxed{U_L = \frac{1}{2}LI^2}$$

where L is the inductance of the inductor carrying a current I.

Example 9 In a particular circuit, assume that $\varepsilon = 12$ V, $R = 40\ \Omega$, and $L = 5$ mH. How much energy is stored in the inductor's magnetic field when the current reaches its maximum steady-state value?

Solution. When the current in the circuit reaches its maximum steady-state value, $I = \mathcal{E}/R$. At this point, the energy stored in the magnetic field of the inductor is

$$U_L = \frac{1}{2}LI^2 = \frac{1}{2}L \cdot \left(\frac{\varepsilon}{R}\right)^2 = \frac{L\varepsilon^2}{2R^2} = \frac{\left(5 \times 10^{-3}\ \text{H}\right)\left(12\ \text{V}\right)^2}{2\left(40\ \Omega\right)^2} = 2.3 \times 10^{-4}\ \text{J}$$

Let's look back to the diagram from the beginning of this section. Once the current has reached its steady-state value of $\mathcal{E}/R$, imagine that you move the switch to point b. Without the inductor, the current would drop abruptly to zero (because the source of emf has been taken out of the circuit). But the inductor would oppose this abrupt decrease in the current by producing a self-induced emf. The presence of this emf would cause the current to die out gradually, according to the equation

$$i(t) = \frac{\varepsilon}{R} e^{-t/\tau_L}$$

Example 10

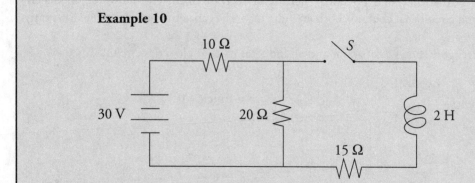

A circuit is connected as shown above.

(a) Determine the current through the 10 Ω resistor when the switch is open.

(b) Determine the current through the 15 Ω resistor when the switch is first closed.

(c) Determine the current through the 10 Ω resistor when the switch has been closed a long time.

Solution.

(a) When the switch is open, the circuit is a basic series circuit with a total resistance of 30 Ω. Use Ohm's law to solve for the current:

$$V = IR \quad \Rightarrow \quad 30 = I(30) \quad \Rightarrow \quad I = 1\ \text{A}$$

(b) When the switch is first closed, the inductor will not let the current go through that part of the branch because it opposes changes in current. Therefore, the current through the 15 Ω resistor is zero. (The current is 1 A in the rest of the circuit.)

(c) After the switch has been closed a long time, the inductor acts like a wire because the current is constant. Now we solve for the resistance of the parallel part of the circuit.

$$\frac{1}{R} = \frac{1}{15} + \frac{1}{20} \Rightarrow R = 8.57 \ \Omega$$

Then the total resistance is that value added to 10 Ω: R_{tot} = 18.57 Ω.

The current leaving the battery will be the current through the 10 Ω resistor. Solve for that current using Ohm's law.

$$V = IR \quad \Rightarrow \quad 30 = I(18.57) \quad \Rightarrow \quad I = 1.62 \ \text{A}$$

LC Circuits

The RC circuit we looked at in Chapter 14 and the RL circuit that we just studied are similar in that they both experience an exponential increase or decrease of current. However, an **LC circuit**, which is a circuit that contains both an inductor and a capacitor, behaves quite differently.

The following figure shows a capacitor and an inductor in a simple series circuit.

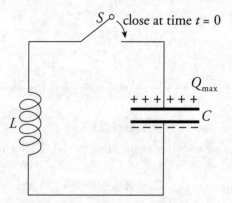

Let's assume that initially there is no current and that the capacitor is charged; at time $t = 0$, the switch S is closed and the circuit is complete. The presence of the inductor prevents the capacitor from discharging abruptly, so the current rises gradually, and the energy in the electric field of the capacitor is transferred to energy in the magnetic field of the inductor. Once the capacitor has discharged, all the energy is in the inductor's magnetic field. The current in the inductor (now at its maximum) delivers charge to the capacitor, but in the opposite direction from its original charge configuration. The current gradually returns to zero as the magnetic-field energy in the inductor is transformed back to electric-field energy in the capacitor.

The capacitor starts to discharge again, but this time, it sends current in the opposite direction. The current rises gradually, reaching a maximum value as the charge on the capacitor reaches zero. The inductor continues to deliver charge to the capacitor as the current gradually returns to zero, and it finds itself right back where it started. The circuit oscillates, and this defines one cycle of oscillation.

Applying Kirchhoff's loop equation to this circuit, we obtain

$$L\frac{dI}{dt} + \frac{q}{C} = 0$$

The current is the derivative as a function of time, so this equation can be rewritten as

$$-L\frac{d^2q}{dt^2} = \frac{q}{C}$$

This can be rewritten as

$$\frac{d^2q}{dt^2} = -\frac{1}{LC}q$$

This is a second-order linear differential equation of the same form we studied in Chapter 10, when we studied oscillations. In this case, the equation above is in the same form as that for a mass oscillating on the end of a spring,

$$\frac{d^2y}{dt^2} = -\omega^2 y$$

Thus, we can infer that, in the case of an inductor and capacitor in a circuit, $\omega = \sqrt{\dfrac{1}{LC}}$. The angular frequency, ω, is related to the frequency of oscillation according to $\omega = 2\pi f$.

The frequency of oscillation is related to inductance and capacitance by the equation

$$f = \frac{1}{2\pi\sqrt{LC}}$$

The charge on the capacitor, as a function of time (assuming that it possessed its maximum charge, Q_{max}, at time $t = 0$) is

$$Q(t) = Q_{max}\cos(\omega t)$$

where $\omega = 2\pi f$.

> **Example 11** Find an equation for the current in the LC circuit as a function of time. How long does it take for the circuit to complete a full oscillation?

Solution. By definition, the current is the time derivative of the charge. So,

$$I(t) = \frac{dQ}{dt} = \frac{d}{dt}\left(Q_{max} \cos(\omega t)\right) = -\omega Q_{max} \sin(\omega t)$$

The time required to complete a cycle (the period) would be the reciprocal of the frequency:

$$T = \frac{1}{f} = 2\pi\sqrt{LC}$$

MAXWELL'S EQUATIONS

We'll finish up this chapter with a set of four equations that embody the subject of electromagnetism.

1. Gauss's Law

This law, first studied in Chapter 12, gives us a method for calculating electric fields and is particularly useful when the system we're working with possesses symmetry.

$$\Phi_E = \oint E \cdot dA = \frac{Q_{enc}}{\varepsilon_0}$$

It can also be shown as this:

$$\int_{\substack{closed \\ surface}} E \cdot dA = \frac{Q_{enclosed}}{\varepsilon_0}$$

2. Gauss's Law for Magnetic Fields

We mentioned earlier that magnetic field lines always form closed loops that encircle the current that generates them. (Remember: This property is not shared by electrostatic fields, which radiate away from positive source charges and toward negative ones.) The fact that magnetic field lines always close upon themselves tells us that the magnetic flux through any closed surface must be zero; as much magnetic flux will enter the closed surface as will exit it. Mathematically, this says:

$$\int_{\substack{closed \\ surface}} B \cdot dA = 0$$

Another statement of this law is that there are no magnetic monopoles. If there were, a single isolated magnetic north pole would generate a magnetic field that radiates away, and an isolated magnetic south pole would generate a magnetic field that radiates inward, toward the pole. A closed surface surrounding such a magnetic charge (if one existed) would have a nonzero flux through it. No magnetic monopoles like this have ever been observed.

> If you tried to isolate the poles of a bar magnet by cutting it in half, you'd only end up creating two smaller bar magnets, each with a north and south pole.

3. Faraday's Law

As we discussed in this chapter, a changing magnetic flux induces an emf; this also means that a changing magnetic field produces an electric field. The equation is

$$\varepsilon = \oint \mathbf{E} \cdot d\ell = -\frac{d\Phi_B}{dt}$$

This equation is typically used to find the emf and the electric field induced by a changing magnetic flux.

4. The Ampere–Maxwell Law

The final equation in this set begins with Ampere's law, which provides us with a method for calculating magnetic fields and is particularly useful when systems are symmetrical. It reads

$$\oint B \cdot ds = \mu_0 I_{\text{enclosed}}$$

You may notice that the first two equations listed in this section possess a sort of symmetry; there's a Gauss's law for E-fields and one for B-fields. The third equation says that a changing B-field produces an E-field. One question we haven't asked yet, which is of interest here, is *Does a changing E-field produce a B-field?* The answer is "yes," and the equation can be amended in the following way [Maxwell added the missing piece, which is proportional to the rate of change of the electric flux, $\mu_0 \varepsilon_0 (d\Phi_E/dt)$]:

$$\oint \mathbf{B} \cdot ds = \mu_0 I + \mu_0 \varepsilon_0 \frac{d\Phi_E}{dt}$$

This shows that a changing electric field, which appears on the right-hand side of the equation [in the term $\mu_0 \varepsilon_0 (d\Phi_E/dt)$], will produce a magnetic field, which appears on the left-hand side of the equation.

The quantity $\varepsilon_0 (d\Phi_E/dt)$ has units of current and is called displacement current, I_D. It differs from the current I, which is conduction current (I_C), because while I_C is composed of moving electric charge, I_D is not. The Ampere–Maxwell equation can be written in terms of conduction current and displacement current as follows:

$$\oint \mathbf{B} \cdot ds = \mu_0 (I_C + I_D)$$

Chapter 16 Drill

The answers and explanations can be found in Chapter 17.

Section I: Multiple Choice

1. A metal rod of length L is pulled upward with constant velocity **v** through a uniform magnetic field **B** that points out of the plane of the page.

What is the potential difference between points a and b?

(A) 0

(B) $\frac{1}{2} v \, BL$, with point a at the higher potential

(C) $\frac{1}{2} v \, BL$, with point b at the higher potential

(D) vBL, with point a at the higher potential

(E) vBL, with point b at the higher potential

2. A circular disk of radius a is rotating at a constant angular speed ω in a uniform magnetic field, **B**, which is directed out of the plane of the page.

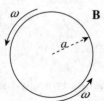

Determine the induced emf between the center of the disk and the rim.

(A) $\frac{1}{2} \omega \mathbf{B} a$

(B) $\frac{1}{2} \mathbf{B} a$

(C) $\frac{1}{2} \omega \mathbf{B} a^2$

(D) $\omega \mathbf{B} a^2$

(E) $2\pi \omega \mathbf{B} a^2$

3. A conducting rod of length 0.2 m and resistance 10 Ω between its endpoints slides without friction along a U-shaped conductor in a uniform magnetic field **B** of magnitude 0.5 T perpendicular to the plane of the conductor, as shown in the diagram below.

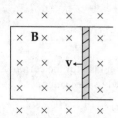

If the rod is moving with velocity **v** = 3 m/s to the left, what is the magnitude and direction of the current induced in the rod?

Current	Direction
(A) 0.03 A	down
(B) 0.03 A	up
(C) 0.3 A	down
(D) 0.3 A	up
(E) 3 A	down

4. In the figure below, a small, circular loop of wire (radius r) is placed on an insulating stand inside a hollow solenoid of radius R. The solenoid has n turns per unit length and carries a current I. If the current in the solenoid is decreased at a steady rate of a amps/s, determine the induced emf, $\mathcal{E}$, and the direction of the induced current in the loop.

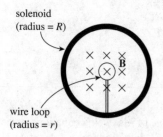

solenoid (radius = R)

wire loop (radius = r)

(A) $\varepsilon = \mu_0 \pi n r^2 a$; induced current is clockwise
(B) $\varepsilon = \mu_0 \pi n r^2 a$; induced current is counterclockwise
(C) $\varepsilon = \mu_0 \pi n R^2 a$; induced current is clockwise
(D) $\varepsilon = \mu_0 \pi n R^2 a$; induced current is counterclockwise
(E) $\varepsilon = \mu_0 \pi I n R^2 a$; induced current is counterclockwise

5. In the figure below, a permanent bar magnet is pulled upward with a constant velocity through a loop of wire.

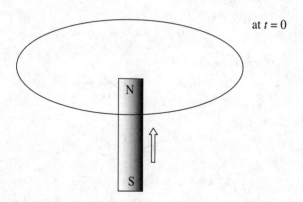

at $t = 0$

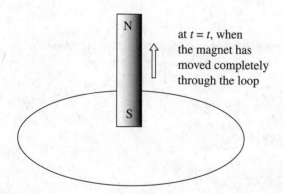

at $t = t$, when the magnet has moved completely through the loop

Which of the following best describes the direction(s) of the current induced in the loop (looking down on the loop from above)?

(A) Always clockwise
(B) Always counterclockwise
(C) First clockwise, then counterclockwise
(D) First counterclockwise, then clockwise
(E) No current will be induced in the loop.

6. A square loop of wire (side length = s) surrounds a long, straight wire such that the wire passes through the center of the square.

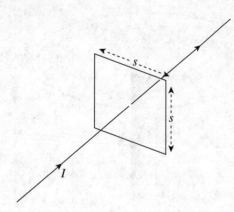

If the current in the wire is I, determine the current induced in the square loop.

(A) $\dfrac{2\mu_0 I s}{\pi\left(1+\sqrt{2}\right)}$

(B) $\dfrac{\mu_0 I s}{\pi\sqrt{2}}$

(C) $\dfrac{\mu_0 I s}{\pi}$

(D) $\dfrac{\mu_0 I s \sqrt{2}}{\pi}$

(E) 0

Questions 7–9

A circuit contains a solenoid of inductance L in series with a resistor of resistance R and a battery with terminal voltage ε. At time $t = 0$, a switch is closed and the circuit is completed.

7. How long does it take for the current to reach $\dfrac{3}{4}$ of its maximum (steady-state) value?

(A) $(\ln 4)(L/R)$

(B) $(\ln \dfrac{3}{4})(L/R)$

(C) $(\ln \dfrac{4}{3})(L/R)$

(D) $(\ln \dfrac{4}{3})(R/L)$

(E) $(\ln 4)(R/L)$

8. When the current reaches its maximum value, how much energy is stored in the magnetic field of the solenoid?

(A) $L^2\varepsilon^2/(4R^2)$
(B) $L^2\varepsilon^2/(2R^2)$
(C) $L\varepsilon^2/(4R^2)$
(D) $L\varepsilon^2/(2R^2)$
(E) 0

9. When the current reaches its maximum value, what is the total magnetic flux through the solenoid?

(A) $L\varepsilon$
(B) $L\varepsilon/R$
(C) $\varepsilon/(RL)$
(D) RL/ε
(E) 0

10. Which one of Maxwell's equations states that a changing electric field produces a magnetic field?

(A) Gauss's law
(B) Gauss's law for magnetism
(C) Biot–Savart law
(D) Ampere–Maxwell law
(E) Faraday's law

Section II: Free Response

1. The diagram below shows two views of a metal rod of length ℓ rotating with constant angular speed ω about an axis that is in the plane of the page. The rotation takes place in a uniform magnetic field **B** whose direction is parallel to the angular velocity ω.

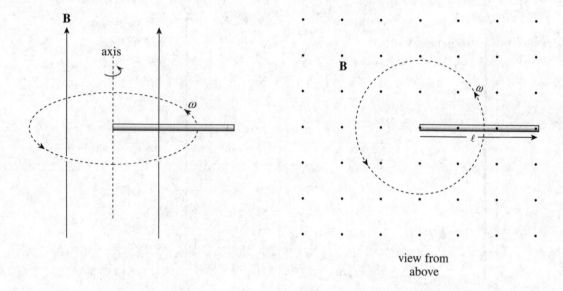

view from above

(a)   What is the emf induced between the ends of the rod?

(b)   What is the polarity (+ or −) of the rotating end?

In the following diagram, a metal rod of length ℓ moves with constant velocity **v** parallel to a long, straight wire carrying a steady current I. The lower end of the rod maintains a distance of a from the straight wire.

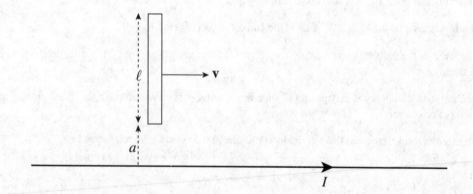

(c)   What is the emf induced between the ends of the rod?

(d)   What is the polarity (+ or −) of the end that is farther from the straight wire?

2. A rectangular loop of wire (side lengths a and b) rotates with constant angular speed ω in a uniform magnetic field **B**. At time $t = 0$, the plane of the loop is perpendicular to **B**, as shown in the figure on the left. The magnetic field **B** is directed to the right (in the $+x$ direction), and the rotation axis is the y-axis (with ω in the $+y$ direction), and the four corners of the loop are labeled 1, 2, 3, and 4. (Express your answers in terms of a, b, ω, B, and fundamental constants.)

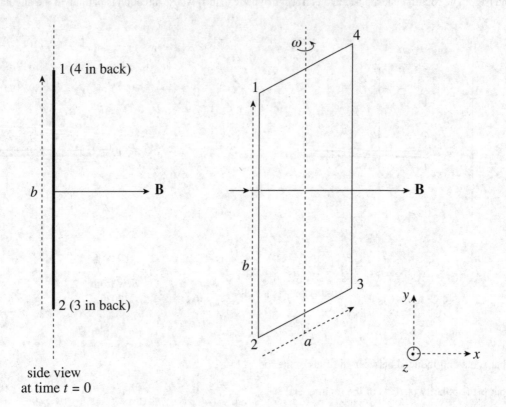

side view
at time $t = 0$

(a) Find a formula that gives the magnetic flux $\mathbf{\Phi B}$ through the loop as a function of time, t.

(b) Find a formula that gives the emf induced in the loop as a function of time, t.

(c) If the total resistance of the loop is R, what is the current induced in the loop?

(d) When $\omega t = \pi/2$, is the induced current in the loop directed from point 1 to point 2 ($-y$ direction) or from point 2 to point 1 ($+y$ direction)?

(e) Find the rate at which energy is dissipated (as joule heat) in the wires that comprise the loop, and the amount of energy dissipated per revolution.

(f) Find the external torque required to keep the loop rotating at the constant angular speed ω.

3. The figure below shows a toroidal solenoid of mean radius R and N total windings. The cross-sections of the toroid are circles of radius a (which is much smaller than R, so variations in the magnetic field strength within the space enclosed by the windings may be neglected).

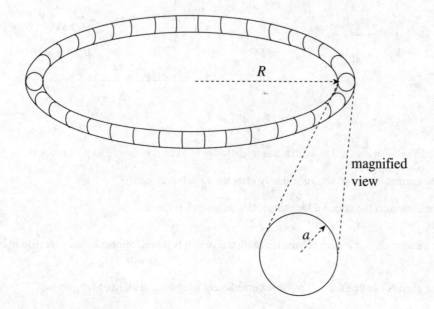

(a) Use Ampere's law to find the magnetic field strength within the toroid. Write your answer in terms of N, I, R, and fundamental constants.

A circular loop of wire of radius $2a$ is placed around the toroid as shown:

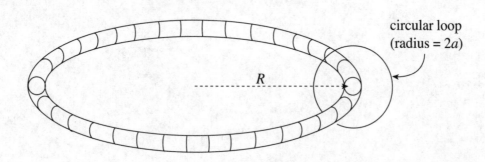

Assume that the current in the toroid is varied sinusoidally according to the equation $I(t) = I_0 \sin \omega t$, where I_0 and ω are fixed constants.

(b) Determine the emf induced in the circular wire loop.

(c) Determine the electric field induced at the position of the circular wire loop.

(d) What is the self-inductance of the toroidal solenoid?

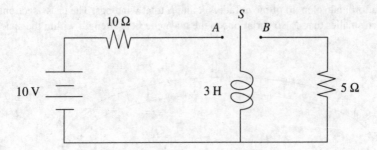

4. A circuit is connected as shown above. The switch *S* is initially open. Then it is moved to position *A*.

(a) Determine the current in the circuit immediately after the switch is closed.

(b) Determine the current in the circuit a long time after the switch is closed.

Some time after the steady-state situation has been reached, the switch is moved almost instantaneously from position *A* to position *B*.

(c) Determine the current through the 5 Ω resistor immediately after the switch has been moved.

(d) Determine the potential difference across the inductor immediately after the switch has been closed.

Summary

Motional Emf

○ A conducting bar with a component of its velocity perpendicular to the magnetic field will have an induced emf across its length. This is due to the fact that each charge in the bar will have a force on it and the charges will separate until the magnetic force and the electric force caused by the attraction of the separate charges are equal. The motional emf is given by the equation $\varepsilon = B\ell v$.

Magnetic Flux, Faraday's Law, and Lenz's Law

○ Magnetic flux is determined by the number of magnetic field lines that pass through a surface. It is given by the dot product of the magnetic field and area vector: $\Phi_B = \mathbf{B} \cdot \mathbf{A} = \int \mathbf{B} \cdot d\mathbf{A}$.

○ When the magnetic flux is changing, an emf is induced in a conductor. This is known as Faraday's Law of Electromagnetic Induction. The general form of Faraday's law is

$$\varepsilon = -\frac{d\Phi_B}{dt} = -\int \mathbf{E} \cdot d\ell$$

○ The first part of the equation is used to calculate the induced emf from a changing magnetic flux and the last part can be used to calculate the induced electric field created in the conductor. The last part of the equation comes from $\Delta V = -\int \mathbf{E} \cdot d\ell$, where the potential difference is replaced with the emf.

○ Lenz's law is used to determine the direction of the induced current and explains the negative sign in Faraday's law. The induced current will flow in the direction that opposes the change in the magnetic flux that produced it.

Inductance

o   An inductor is an electric device that has a self-induced emf that opposes the change in the current in the circuit. The inductance, L, is given by the equation $L = \dfrac{N\Phi_B}{I}$.

o   The emf of an inductor is $\varepsilon = -L\dfrac{dI}{dt}$, where the negative sign indicates the emf is opposing the change in the current through the circuit.

o   The energy stored in the magnetic field on an inductor is $U_L = \dfrac{1}{2}LI^2$.

Resistor–Inductor (RL) Circuits

o   When a switch is closed in a typical RL circuit, the current increases exponentially over time. This is due to the fact that the inductor opposes the change in the current so it will not let it increase from zero to a maximum instantaneously. The equation for the current is $I(t) = \dfrac{\varepsilon}{R}\left(1 - e^{-t/\left(\frac{L}{R}\right)}\right)$, where the time constant is now given by $\tau_L = \dfrac{L}{R}$.

Inductor–Capacitor (LC) Circuits

o   When a switch is closed in a typical LC circuit, the energy oscillates between being stored in the electric field of the capacitor and the magnetic field of the inductor. The frequency of the oscillation is $f = \dfrac{1}{2\pi\sqrt{LC}}$ and the charge of the plate is given by $Q(t) = Q_{max}\cos(\omega t)$.

Chapter 17
Chapter Drills:
Answers and
Explanations

CHAPTER 4 DRILL

Section I: Multiple Choice

1. **B** Find the components of vector **C** by subtracting the components of vector **B** from the components of vector **A**: **C** = (15 – 3, 8 – 3) = (12, 5). Notice that vector **C** forms a 5:12:13 right-angled triangle with the coordinate axes. Therefore, vector **C** has a magnitude of 13. The correct answer is (B).

2. **C** The components of vectors **a** and **b** must sum to the components of vector **c**. Looking at the x-components, $2 + x = 3$, which means $x = 1$. Eliminate (A) and (B), since they don't have the correct x-component. Looking at the y-components, $y + 3 = -10$, which means $y = -13$. Eliminate (D) and (E), since they don't have the correct y-component. The correct answer is (C).

3. **A** Use the dot product to find the angle between vectors **u** and **v**.

$$\mathbf{u} \cdot \mathbf{v} = uv \cos \theta$$

$$\theta = \cos^{-1} \frac{\mathbf{u} \cdot \mathbf{v}}{uv}$$

$$\theta = \cos^{-1} \frac{(-2)(4) + (2)(5)}{\sqrt{(8)}\sqrt{41}}$$

$$\theta = 83.7°$$

4. **E** The cross product will be perpendicular to both vectors **v** and **B**. Since the cross product is directed out of the page, both vectors **v** and **B** must be on the plane of the page. Therefore, B_z must be zero. Eliminate (B) and (C). Furthermore, the y-component of vector **B** must be positive, since the cross product is directed out of the page; a negative y-component of vector **B** would result in a cross product directed into the page. Eliminate (A). Since the cross product only has a z-component, find the 2×2 determinant of the x- and y-components of vectors **v** and **B** to ultimately find that $B_y = 6$. Eliminate (D). The correct answer is (E).

5. **B** First, work out the cross product **B** × **C**.

$$\begin{vmatrix} \hat{\mathbf{i}} & -\hat{\mathbf{j}} & \hat{\mathbf{k}} \\ 2 & 1 & -1 \\ 1 & -3 & 2 \end{vmatrix} = \hat{\mathbf{i}}(2-3) - \hat{\mathbf{j}}(4+1) + \hat{\mathbf{k}}(-6-1) = -\hat{\mathbf{i}} - 5\hat{\mathbf{j}} - 7\hat{\mathbf{k}}$$

Now work out the dot product of **A** with this cross product vector. $\mathbf{A} \cdot (\mathbf{B} \times \mathbf{C}) = (4)(-1) + (5)(-5) + (8)(-7) = -4 - 25 - 56 = -85$. The correct answer is (B).

Section II: Free Response

1. (a) (i) The question asks for $F_{parallel}$, the component of **F** that aligns with **d**. Since **d** points in the x-direction, the question ultimately asks for the x-component of **F**. $F_x = F\cos\theta = (10\ N)\cos(30°) = 5\sqrt{3}\ N$.

   (ii) Using the equation given in the question, $W = F_{parallel}d = (5\sqrt{3}\ N)(10\ m) = 85\ J$.

   (b) $\mathbf{F} \cdot \mathbf{d} = (5\sqrt{3}\ N)(10\ m) + (5\ N)(0\ m) = 85\ J$.

   (c) The answers in (a)(ii) and (b) are the same. This is an expected result since the dot product of any two vectors **A** and **B** is defined as $\mathbf{A} \cdot \mathbf{B} = AB\cos\theta = A_{parallel}B$.

   (d) The cross product of **F** and **d** would point perpendicular to both vectors. Since both **F** and **d** are on the plane of the page, their cross product would be perpendicular to the plane of the page. Using the right-hand rule, the index finger points with **F**, the middle finger points with **d**, and the thumb, which gives the direction of the cross product, points into the plane of the page.

CHAPTER 5 DRILL

Section I: Multiple Choice

1. **A** Traveling once around a circular path means that the final position is the same as the initial position. Therefore, the displacement is zero. The average speed, which is total *distance* traveled divided by elapsed time, cannot be zero. Since the velocity changed (because its direction changed), there was a nonzero acceleration. Therefore, only (I) is true.

2. **C** By definition $\bar{\mathbf{a}} = \Delta\mathbf{v} / \Delta t$. We determine $\Delta\mathbf{v} = \mathbf{v}_2 - \mathbf{v}_1 = \mathbf{v}_2 + (-\mathbf{v}_1)$ geometrically as follows:

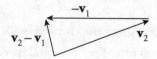

Since Δt is a positive scalar, the direction of $\bar{\mathbf{a}}$ is the same as the direction of $\Delta\mathbf{v}$, which is displayed above; (C) is best.

3. **C** Any object in free fall will experience an acceleration of g, downward, which is (C). The launch conditions do not have any impact.

4. **C** The baseball is still under the influence of Earth's gravity. Its acceleration throughout the *entire* flight is constant, equal to g downward.

5. **C** The velocity can be found by taking the derivative of position.

$$\mathbf{v}(t) = \frac{d}{dt}\mathbf{x}(t) = \frac{d}{dt}(-3t^3 + t^2 + 6t) = -9t^2 + 2t + 6$$

To find the maximum, set the derivative of this equation equal to 0 and use the resulting time.

$$\frac{d}{dt}(-9t^2 + 2t + 6) = 0 \rightarrow -18t + 2 = 0 \rightarrow t = \frac{1}{9}$$

$$\mathbf{v}_{max} = -9\left(\frac{1}{9}\right)^2 + 2\left(\frac{1}{9}\right) + 6 \approx 6.11 \text{ m/s}$$

6. **D** Use Big Five #5 with $v_0 = 0$ (calling *down* the positive direction):

$$v^2 = v_0^2 + 2a\Delta s = 2a\Delta s \quad \Rightarrow \quad \Delta s = \frac{v^2}{2a} = \frac{v^2}{2g} = \frac{(30 \text{ m/s})^2}{2(10 \text{ m/s}^2)} = 45 \text{ m}$$

7. **C** Apply Big Five #3 to the vertical motion, calling *down* the positive direction:

$$\Delta y = v_{0y}t + \tfrac{1}{2}a_y t^2 = \tfrac{1}{2}a_y t^2 = \tfrac{1}{2}g t^2 \quad \Rightarrow \quad t = \sqrt{\frac{2\Delta y}{g}} = \sqrt{\frac{2(80 \text{ m})}{10 \text{ m/s}^2}} = 4 \text{ s}$$

Note that the stone's initial horizontal speed ($v_{0x} = 10$ m/s) is irrelevant.

8. **B** First, we determine the time required for the ball to reach the top of its parabolic trajectory (which is the time required for the vertical velocity to drop to zero).

$$v_y \overset{\text{set}}{=} 0 \quad \Rightarrow \quad v_{0y} - gt = 0 \quad \Rightarrow \quad t = \frac{v_{0y}}{g}$$

The total flight time is equal to twice this value:

$$T = 2t = 2\frac{v_{0y}}{g} = 2\frac{v_0 \sin\theta_0}{g} = \frac{2(10 \text{ m/s})\sin 30°}{10 \text{ m/s}^2} = 1 \text{ s}$$

9. **C** After 4 seconds, the stone's vertical speed has changed by $\Delta v_y = a_y t = (10 \text{ m/s}^2)(4 \text{ s}) = 40$ m/s. Since $v_{0y} = 0$, the value of v_y at $t = 4$ is 40 m/s. The horizontal speed does not change. Therefore, when the rock hits the water, its velocity has a horizontal component of 30 m/s and a vertical component of 40 m/s.

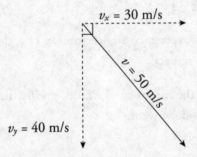

By the Pythagorean Theorem, the magnitude of the total velocity, v, is 50 m/s.

10. **E** Since the acceleration of the projectile is always downward (because it's gravitational acceleration), the vertical speed decreases as the projectile rises and increases as the projectile falls. Choices (A), (B), (C), and (D) are all false.

11. **A** In order to maximize the horizontal distance, you would want the minimum possible angle in this situation because the range will continually decrease as θ increases beyond 45°. This would result in the object reaching height h at the same moment it has traveled the horizontal distance d. Focusing on the vertical components first, we can use one of our Big 5 equations to say

$$d = v_0 t + \frac{1}{2} at^2 \rightarrow h = (v\sin\theta)t + \frac{1}{2}(-g)t^2$$

We can substitute t out of the equation by using our horizontal components.

$$v_x = \frac{d_x}{t} \rightarrow t = \frac{d_x}{v_x} = \frac{d}{v\cos\theta}$$

Putting this value into the first equation gives us

$$h = v\sin\theta\left(\frac{d}{v\cos\theta}\right) + \frac{1}{2}(-g)\left(\frac{d}{v\cos\theta}\right)^2 \rightarrow h = d\tan\theta - \frac{\left(\frac{1}{2}\right)gd^2}{v^2\cos^2\theta}$$

12. **D** The displacement of the object is the area between the velocity function and the t-axis. Divide the area up to three seconds into one rectangle created for the first second and one triangle for the next two seconds.

$$Area = (4)(1) + \frac{1}{2}(2)(4) = 8$$

Remember that the area is the displacement, not the position. Since the object started at $x = 3$ m, when $t = 0$ s, it is now 8 meters away from that point. Therefore, its position is 11 m.

Section II: Free Response

1. (a) At time $t = 1$ s, the car's velocity starts to decrease as the acceleration (which is the slope of the given v-vs.-t graph) changes from positive to negative.

    (b) The average velocity between $t = 0$ and $t = 1$ s is $\frac{1}{2}(v_{t=0} + v_{t=1}) = \frac{1}{2}(0 + 20 \text{ m/s}) = 10$ m/s, and the average velocity between $t = 1$ and $t = 5$ is $\frac{1}{2}(v_{t=1} + v_{t=5}) = \frac{1}{2}(20 \text{ m/s} + 0) = 10$ m/s. The two average velocities are the same.

(c) The displacement is equal to the area bounded by the graph and the t-axis, taking areas above the t-axis as positive and those below as negative. In this case, the displacement from $t = 0$ to $t = 5$ s is equal to the area of the triangular region whose base is the segment along the t-axis from $t = 0$ to $t = 5$ s:

$$\Delta s \ (t = 0 \text{ to } t = 5 \text{ s}) = \tfrac{1}{2} \times \text{base} \times \text{height} = \tfrac{1}{2} (5 \text{ s})(20 \text{ m/s}) = 50 \text{ m}$$

The displacement from $t = 5$ s to $t = 7$ s is equal to the negative of the area of the triangular region whose base is the segment along the t-axis from $t = 5$ s to $t = 7$ s:

$$\Delta s \ (t = 5 \text{ s to } t = 7 \text{ s}) = -\tfrac{1}{2} \times \text{base} \times \text{height} = -\tfrac{1}{2} (2 \text{ s})(10 \text{ m/s}) = -10 \text{ m}$$

Therefore, the displacement from $t = 0$ to $t = 7$ s is

$$\Delta s \ (t = 0 \text{ to } t = 5 \text{ s}) + \Delta s \ (t = 5 \text{ s to } t = 7 \text{ s}) = 50 \text{ m} + (-10 \text{ m}) = 40 \text{ m}$$

(d) The acceleration is the slope of the v-vs.-t graph. The segment of the graph from $t = 0$ to $t = 1$ s has a slope of $a = \Delta v / \Delta t = (20 \text{ m/s} - 0)/(1 \text{ s} - 0) = 20 \text{ m/s}^2$, and the segment of the graph from $t = 1$ s to $t = 7$ s has a slope of $a = \Delta v / \Delta t = (-10 \text{ m/s} - 20 \text{ s})/(7 \text{ s} - 1 \text{ s}) = -5 \text{ m/s}^2$. Therefore, the graph of a vs. t is

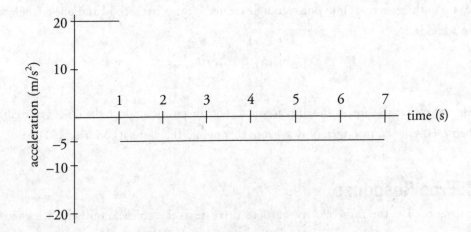

(e) One way to determine the displacement is to determine equations for $v(t)$ and integrate to get $s(t)$. The segment from $t = 0$ to $t = 1$ s connects the points $(0, 0)$ and $(1, 20)$; the slope is 20, and we get $v = 20t$. The segment from $t = 1$ to $t = 7$ connects the points $(1, 20)$ and $(7, -10)$; the slope is -5, and we get $v = -5t + 25$. In summary,

$$v(t) = \begin{cases} 20t & 0 \le t \le 1 \\ -5t + 25 & 1 \le t \le 7 \end{cases}$$

Therefore, since $s(t) = \int v(t)\,dt$, we find that

$$s(t) = \begin{cases} 10t^2 & 0 \le t \le 1 \\ -\frac{5}{2}t^2 + 25t - \frac{25}{2} & 1 \le t \le 7 \end{cases}$$

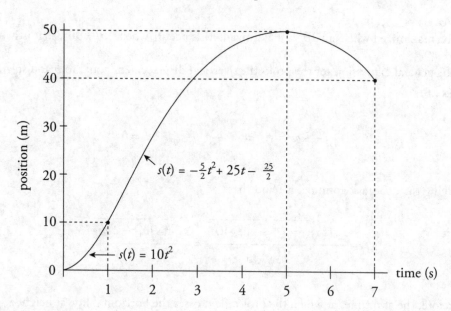

2. (a) The maximum height of the projectile occurs at the time at which its vertical velocity drops to zero:

$$v_y \overset{\text{set}}{=} 0 \quad \Rightarrow \quad v_{0y} - gt = 0 \quad \Rightarrow \quad t = \frac{v_{0y}}{g}$$

The vertical displacement of the projectile at this time is computed as follows:

$$\Delta y = v_{0y}t - \tfrac{1}{2}g t^2 \quad \Rightarrow \quad H = v_{0y}\frac{v_{0y}}{g} - \tfrac{1}{2}g\left(\frac{v_{0y}}{g}\right)^2 = \frac{v_{0y}^2}{2g} = \frac{v_0^2 \sin^2 \theta_0}{2g}$$

(b) The total flight time is equal to twice the time computed in part (a):

$$T = 2t = 2\frac{v_{0y}}{g}$$

The horizontal displacement at this time gives the projectile's range:

$$\Delta x = v_{0x}t \quad \Rightarrow \quad R = v_{0x}T = \frac{v_{0x}\cdot 2v_{0y}}{g} = \frac{2v_0^2 \sin\theta_0 \cos\theta_0}{g} \quad \text{or} \quad \frac{v_0^2 \sin 2\theta_0}{g}$$

(c) For any given value of v_0, the range,

$$\Delta x = v_{0x}t \implies R = v_{0x}T = \frac{v_{0x} \cdot 2v_{0y}}{g} = \frac{2v_0^2 \sin\theta_0 \cos\theta_0}{g} \text{ or } \frac{v_0^2 \sin 2\theta_0}{g}$$

will be maximized when $\sin 2\theta_0$ is maximized. This occurs when $2\theta_0 = 90°$, that is, when $\theta_0 = 45°$.

(d) Set the general expression for the projectile's vertical displacement equal to h and solve for the two values of t:

$$v_{0y}t - \tfrac{1}{2}gt^2 \overset{\text{set}}{=} h \implies \tfrac{1}{2}gt^2 - v_{0y}t + h = 0$$

Applying the quadratic formula, we find that

$$t = \frac{v_{0y} \pm \sqrt{(-v_{0y})^2 - 4(\tfrac{1}{2}g)(h)}}{2(\tfrac{1}{2}g)} = \frac{v_{0y} \pm \sqrt{v_{0y}^2 - 2gh}}{g}$$

Therefore, the two times at which the projectile crosses the horizontal line at height h are

$$t_1 = \frac{v_{0y} - \sqrt{v_{0y}^2 - 2gh}}{g} \quad \text{and} \quad t_2 = \frac{v_{0y} + \sqrt{v_{0y}^2 - 2gh}}{g}$$

so the amount of time that elapses between these events is

$$\Delta t = t_2 - t_1 = \frac{2\sqrt{v_{0y}^2 - 2gh}}{g}$$

3. (a) If we need to find how long the cannonball takes to reach the plane of the wall, we are dealing with the horizontal direction. The only equation we need is $x = v_x t$.

$$220 = (50 \cos 40°)t \implies t = 5.74 \text{ s}$$

(b) To determine whether the cannonball hits the wall, we need to know the vertical displacement of the ball when it reaches the plane of the wall.

$$\Delta y = v_{0y}t + \frac{1}{2}gt^2$$

$$\Delta y = (50\sin 40°)(5.74) + \frac{1}{2}(-9.8)(5.74)^2$$

$$\Delta y = 22.8 \text{ m}$$

Since the wall is 30 m tall, the cannonball strikes the wall 7.2 m below the top of the wall.

4. (a) Integrating $a(t)$ with respect to time gives the velocity, $v(t)$:

$$v(t) = \int a(t)\,dt = \int 6t\,dt = 3t^2 + v_0 \quad \Rightarrow \quad v(t) = 3t^2 + 2$$

Setting this equal to 14, we solve for t:

$$v(t) = 14 \quad \Rightarrow \quad 3t^2 + 2 = 14 \quad \Rightarrow \quad t = 2 \text{ s}$$

(Note that we discarded the solution $t = -2$.)

(b) Integrating $v(t)$ with respect to time gives the position, $x(t)$:

$$x(t) = \int v(t)\,dt = \int (3t^2 + 2)\,dt = t^3 + 2t + x_0 \quad \Rightarrow \quad x(t) = t^3 + 2t + 4$$

Therefore, the particle's position at $t = 3$ s is

$$x(3) = \left[t^3 + 2t + 4 \right]_{t=3} = 37 \text{ m}$$

CHAPTER 6 DRILL

Section I: Multiple Choice

1. **E** An action-reaction pair must satisfy two conditions: (1) equal in magnitude but opposite in direction, and (2) the forces must involve the same two objects. In (I), gravity is provided by the Earth and is affecting the box. The normal force is provided by the table and is also affecting the box. Therefore, (A), (B), and (D) can be eliminated. In (II), both friction provided by the table and the tension provided by the string affect the box. This allows us to eliminate (C). The two forces in (III) both involve only the table and the box, so (E) is correct.

2. **D** First, draw a free-body diagram.

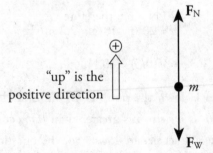

The person exerts a downward force on the scale, and the scale pushes up on the person with an equal (but opposite) force, F_N. Thus, the scale reading is F_N, the magnitude of the normal force. Since $F_N - F_w = ma$, we have $F_N = F_w + ma = (800 \text{ N}) + [800 \text{ N}/(10 \text{ m/s}^2)](5 \text{ m/s}^2) = 1{,}200 \text{ N}$.

3. **D**

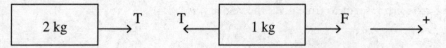

As shown in the figure above, draw free-body diagrams for each of the two blocks, and sum the forces for each block. Then solve the system of equations.

We have assumed that the force is pulling in the positive direction. So we get the following equations:

First block, 2 kg block: Second block, 1 kg block

$$\mathbf{F}_T - 0 = ma$$ $$\mathbf{F} - \mathbf{F}_T = ma$$

$$\mathbf{F}_T = 2a$$ $$20 \text{ N} - \mathbf{F}_T = 1a$$

If we add the 2 equations together, we get $20 = 3a$, so $a = 20/3$ m/s², or 6.7 m/s².

Since $\mathbf{F}_T = 2a$, $t = 2(20/3) = 40/3$ N = 13.3 N.

4. **C**

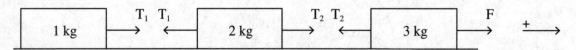

As shown in the figure above, draw free-body diagrams for each block, and sum the forces for each block. Then solve the system of equations. Again, positive is to the right.

For the 1.0 kg block: For the 2.0 kg block: For the 3.0 kg block:

$$\mathbf{F}_{T1} - 0 = ma$$ $$\mathbf{F}_{T2} - \mathbf{F}_{T1} = ma$$ $$\mathbf{F} - \mathbf{F}_{T2} = ma$$

$$\mathbf{F}_{T1} = (1)a$$ $$\mathbf{F}_{T2} - \mathbf{F}_{T1} = (2)a$$ $$20 \text{ N} - \mathbf{F}_{T2} = (3)a$$

$$\mathbf{F}_{T1} = a$$ $$\mathbf{F}_{T2} - a = 2a \text{ (substitute for } \mathbf{F}_{T1})$$

$$\mathbf{F}_{T2} = 3a \text{ (Use this to substitute into the third equation)}$$

$$20 \text{ N} - 3a = 3a$$

20 N = 6a, therefore
$a = 20/6 = 10/3 = 2.2$ m/s²

$\mathbf{F}_{T1} = a = 3.3$ N, then $\mathbf{F}_{T2} = 3a = 3(10/3) = 10$ N.

5. **A**   The force pulling the block down the ramp is $mg \sin \theta$, and the maximum force of static friction is $\mu_s F_N = \mu_s mg \cos \theta$. If $mg \sin \theta$ is greater than $\mu_s mg \cos \theta$, then there is a net force down the ramp, and the block will accelerate down. So, the question becomes, "Is $\sin \theta$ greater than $\mu_s \cos \theta$?" Since $\cos 30° \approx 0.87$ and $\mu_s = 0.5$, the answer is "yes."

6. **D**

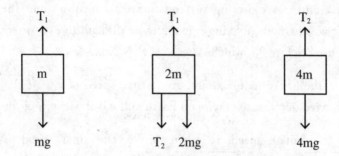

As shown in the figure above, draw free-body diagrams for each block and sum the forces for each block to develop a system of equations to solve for a. We will assume clockwise torque as positive. For the small block of mass m, $\mathbf{F}_{T1} - mg = ma$, so $\mathbf{F}_{T1} = mg + ma$, so $\mathbf{F}_{T1} = m(g + a)$. For the block of mass $2m$, $\mathbf{F}_{T2} + 2mg - \mathbf{F}_{T1} = 2ma$, so $\mathbf{F}_{T2} - \mathbf{F}_{T1} = -2mg + 2ma$, therefore $\mathbf{F}_{T2} - \mathbf{F}_{T1} = 2m(a - g)$. For the block of mass $4m$, $4mg - \mathbf{F}_{T2} = 4ma$, $\mathbf{F}_{T2} = -4ma + 4mg$, so $\mathbf{F}_{T2} = 4m(g - a)$.

Let's add the equations for the blocks of mass m and $2m$ to get, $\mathbf{F}_{T2} = m(g + a) + 2m(a - g)$. Set that equal to the $\mathbf{F}_{T2}$ from the $4m$ block equations to get $4m(g - a) = m(g + a) + 2m(a - g)$. Let's multiply $2m(a - g)$ by -1 to get our a and g terms in the same form.

$$4m(g - a) = m(g + a) - 2m(g - a)$$

$$6m(g - a) = m(g + a)$$

$$6g - 6a = g + a$$

$$5g = 7a, \text{ so } a = 5/7(g) \text{ m/s}^2$$

7. **E**

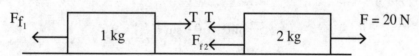

Like the figure above, draw free-body diagrams for each block, and sum the forces for each block. Then solve the system of equations. Again, positive is to the right.

For the 1.0 kg block:

$$\mathbf{F}_T - \mathbf{F}_{f1} = ma$$

$$\mathbf{F}_T - (0.2)(mg) = ma$$

$$\mathbf{F}_T - (0.2)(1)(10) = (1)a$$

$$\mathbf{F}_T - 2 = a$$

$$\mathbf{F}_T = a + 2$$

For the 2.0 kg block:

$$\mathbf{F} - \mathbf{F}_T - \mathbf{F}_{f2} = ma$$

Now substitute into the second equation for the 2 kg block:

$$20 - (a + 2) - (0.2)(2)(10) = (2)a$$

$20 - a - 2 - 4 = 2a$, so, collecting terms, $14 = 3a$, so $a = 14/3 = 4.7 \text{ m/s}^2$

Since $\mathbf{F}_T = a + 2$, $\mathbf{F}_T = 4.7 + 2 = 6.7 \text{ N}$

8. **B** In solving this question, be careful to avoid using any equations that involve time. The value of time will be different in the two situations, making it much more difficult to use correctly. If we want to avoid time, we must use Big Five #5, which states $v^2 = v_0^2 + 2a(x - x_0)$.

Because the initial speed is 0, the first situation would give a final speed of $v = \sqrt{(2a(x - x_0))} = \sqrt{(2g\sin\theta d)}$, where g is the acceleration due to gravity on Earth and d is the length of the ramp.

In the second situation, the initial speed is again 0, so the final speed would be $v = \sqrt{(2a(x - x_0))} = \sqrt{(2(2g\sin\theta)d)} = 2\sqrt{(g\sin\theta d)}$, which is $\sqrt{2}$ times the original, making (B) the correct choice.

9. **B** The people will be safe as long as there is some normal force from the track against the car. This will be true as long as the centripetal force is greater than gravity alone. Therefore, the critical point is when those two forces are exactly equal.

$$F_g = F_c$$
$$mg = \frac{mv^2}{r}$$
$$v = \sqrt{gr} = \sqrt{(10 \text{ m/s}^2)(25\,\text{m})} = 5\sqrt{10} \text{ m/s} \approx 15.8 \text{ m/s}$$

10. **B** The pulley system shown would allow a person to lift the block with a force equal to $\frac{1}{6}$ of the block's weight. Therefore, the force $F = \frac{mg}{6}$.

For a frictionless inclined plane of angle θ, the applied force must equal the parallel component of the force of gravity.

$$F_{app} = F_{g,\|} \rightarrow \frac{mg}{6} = mg\sin\theta \rightarrow \theta = \sin^{-1}(\frac{1}{6}) \approx 9.6°$$

11. **B** When the bucket is at the lowest point in its vertical circle, it feels a tension force $\mathbf{F}_T$ upward and the gravitational force $\mathbf{F}_w$ downward. The net force toward the center of the circle, which is the centripetal force, is $F_T - F_w$. Thus,

$$F_T - F_w = m\frac{v^2}{r} \implies v = \sqrt{\frac{r(F_T - mg)}{m}} = \sqrt{\frac{(0.60 \text{ m})[50 \text{ N} - (3 \text{ kg})(10 \text{ N/kg})]}{3 \text{ kg}}} = 2 \text{ m/s}$$

12. **C** When the bucket reaches the topmost point in its vertical circle, the forces acting on the bucket are its weight, $\mathbf{F}_w$, and the downward tension force, $\mathbf{F}_T$. The net force, $\mathbf{F}_w + \mathbf{F}_T$, provides the centripetal force. In order for the rope to avoid becoming slack, $\mathbf{F}_T$ must not vanish. Therefore, the cut-off speed for ensuring that the bucket makes it around the circle is the speed at which $\mathbf{F}_T$ just becomes zero; any greater speed would imply that the bucket would make it around. Thus,

$$F_w + F_T = m\frac{v^2}{r} \implies F_w + 0 = m\frac{v_{cut\text{-}off}^2}{r} \implies v_{cut\text{-}off} = \sqrt{\frac{rF_w}{m}} = \sqrt{gr}$$
$$= \sqrt{(10 \text{ m/s}^2)(0.60 \text{ m})}$$
$$= 2.4 \text{ m/s}$$

13.　**D**　Centripetal acceleration is given by the equation $a_c = v^2/r$. Since the object covers a distance of $2\pi r$ in 1 revolution each second,

$$a_c = \frac{v^2}{r} = \frac{(2\pi r)^2}{r} = 4\pi^2 r$$

Section II: Free Response

1.　(a)　The forces acting on the crate are $\mathbf{F}_T$ (the tension in the rope), $\mathbf{F}_w$ (the weight of the block), $\mathbf{F}_N$ (the normal force exerted by the floor), and $\mathbf{F}_f$ (the force of kinetic friction):

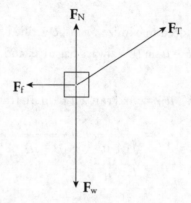

(b)　First, break $\mathbf{F}_T$ into its horizontal and vertical components:

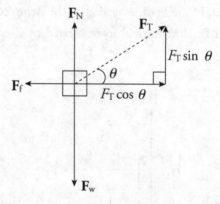

Since the net vertical force on the crate is zero, we have

$$F_N + F_T \sin \theta = F_w, \text{ so } F_N = F_w - F_T \sin \theta = mg - F_T \sin \theta$$

(c)　From part (b), we see that the net horizontal force acting on the crate is

$$F_T \cos \theta - F_f = F_T \cos \theta - \mu F_N = F_T \cos \theta - \mu(mg - F_T \sin \theta)$$

so the crate's horizontal acceleration across the floor is

$$a = \frac{F_{net}}{m} = \frac{F_T \cos \theta - \mu(mg - F_T \sin \theta)}{m}$$

(d) In order to maximize the crate's acceleration, we want to maximize the net horizontal force, $F_T \cos\theta - \mu(mg - F_T \sin\theta) = F_T(\cos\theta + \mu\sin\theta) - \mu mg$. Since F_T, μ, m, and g are all fixed, maximizing the net horizontal force depends on maximizing the expression $\cos\theta + \mu\sin\theta$.

Call this $f(\theta)$. To determine the value of θ at which $f(\theta) = \cos\theta + \mu\sin\theta$ attains an extreme value, we take the derivative of f and set it equal to zero.

$$\begin{aligned} f'(\theta) &= -\sin\theta + \mu\cos\theta = 0 \\ \mu\cos\theta &= \sin\theta \\ \mu &= \tan\theta \\ \theta &= \tan^{-1}\mu \end{aligned}$$

That this value of θ does indeed maximize f can be verified by noticing that for θ in the interval $0 \le \theta \le \frac{1}{2}\pi$, $f''(\theta) = -\cos\theta - \mu\sin\theta$ is always negative, and

$$\begin{aligned} f(\tan^{-1}\mu) &= \cos(\tan^{-1}\mu) + \mu\sin(\tan^{-1}\mu) \\ &= \frac{1}{\sqrt{1+\mu^2}} + \mu\frac{\mu}{\sqrt{1+\mu^2}} \\ &= \sqrt{1+\mu^2} \end{aligned}$$

is greater than $f(0) = 1$ or $f(\frac{1}{2}\pi) = \mu$, the values of f at the endpoints of the interval.

2. (a) The forces acting on Block #1 are $\mathbf{F}_T$ (the tension in the string connecting it to Block #2), $\mathbf{F}_{w1}$ (the weight of the block), and $\mathbf{F}_{N1}$ (the normal force exerted by the tabletop) as seen in the following figure:

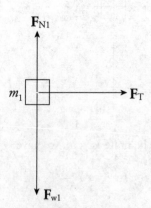

(b)   The forces acting on Block #2 are **F** (the pulling force), $\mathbf{F}_T$ (the tension in the string connecting it to Block #1), $\mathbf{F}_{w2}$ (the weight of the block), and $\mathbf{F}_{N2}$ (the normal force exerted by the tabletop) as seen in the following figure:

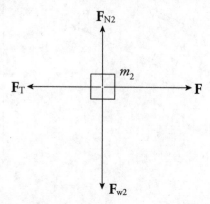

(c)   Newton's Second Law applied to Block #2 yields $F - F_T = m_2 a$, and applied to Block #1 yields $F_T = m_1 a$. Adding these equations, we find that $F = (m_1 + m_2)a$, so

$$a = \frac{F}{m_1 + m_2}$$

(d)   Substituting the result of part (c) into the equation $F_T = m_1 a$, we get

$$F_T = m_1 a = \frac{m_1 \times F}{m_1 + m_2}$$

(e)   (i)   Since the force **F** must accelerate all three masses—m_1, m, and m_2—the common acceleration of all parts of the system is

$$a = \frac{F}{m_1 + m + m_2}$$

(ii)   Let $\mathbf{F}_{T1}$ denote the tension force in the connecting string acting on Block #1, and let $\mathbf{F}_{T2}$ denote the tension force in the connecting string acting on Block #2. Then, Newton's Second Law applied to Block #1 yields $F_{T1} = m_1 a$ and applied to Block #2 yields $F - F_{T2} = m_2 a$. Therefore, using the value for a computed above, we get

$$\begin{aligned}
F_{T2} - F_{T1} &= (F - m_2 a) - m_1 a \\
&= F - (m_1 + m_2)a \\
&= F - (m_1 + m_2)\frac{F}{m_1 + m + m_2} \\
&= F\left(1 - \frac{m_1 + m_2}{m_1 + m + m_2}\right) \\
&= F\frac{m}{m_1 + m + m_2}
\end{aligned}$$

3. (a) First, draw free-body diagrams for the two boxes.

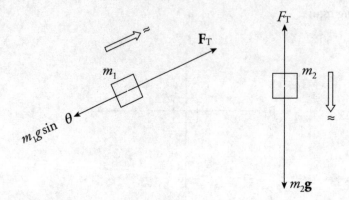

Applying Newton's Second Law to the boxes yields the following two equations:

$$F_T - m_1 g \sin \theta = m_1 a \qquad (1)$$

$$m_2 g - F_T = m_2 a \qquad (2)$$

Adding the equations allows us to solve for a:

$$m_2 g - m_1 g \sin \theta = (m_1 + m_2)a$$

$$a = \frac{m_2 - m_1 \sin \theta}{m_1 + m_2} g$$

(i) For a to be positive, we must have $m_2 - m_1 \sin \theta > 0$, which implies that $\sin \theta < m_2/m_1$, or, equivalently, $\theta < \sin^{-1}(m_2/m_1)$.

(ii) For a to be zero, we must have $m_2 - m_1 \sin \theta = 0$, which implies that $\sin \theta = m_2/m_1$, or, equivalently, $\theta = \sin^{-1}(m_2/m_1)$.

(b) Including the force of kinetic friction, the force diagram for m_1 is

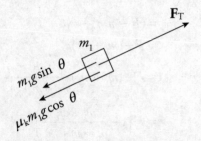

Since $F_f = \mu_k F_N = \mu_k m_1 g \cos \theta$, applying Newton's Second Law to the boxes yields the following two equations:

$$F_T - m_1 g \sin \theta - \mu_k mg \cos \theta = m_1 a \qquad (1)$$

$$m_2 g - F_T = m_2 a \qquad (2)$$

Adding the equations allows us to solve for a:

$$m_2 g - m_1 g \sin \theta - \mu_k mg \cos \theta = (m_1 + m_2) a$$

$$a = \frac{m_2 - m_1 (\sin \theta + \mu_k \cos \theta)}{m_1 + m_2} g$$

If we want a to be equal to zero (so that the box of mass m_1 slides up the ramp with constant velocity), then

$$m_2 - m_1 (\sin \theta + \mu_k \cos \theta) = 0$$

$$\sin \theta + \mu_k \cos \theta = \frac{m_2}{m_1}$$

4. (a) The forces acting on the skydiver are $\mathbf{F}_r$, the force of air resistance (upward), and $\mathbf{F}_w$, the weight of the skydiver (downward) as seen in the following diagram:

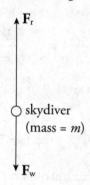

(b) Since $F_{net} = F_w - F_r = mg - kv = ma$, the skydiver's acceleration is

$$a = \frac{F_{net}}{m} = \frac{mg - kv}{m}$$

(c)   Terminal speed occurs when the skydiver's acceleration becomes zero, because then the descent velocity becomes constant. Setting the expression derived in part (b) equal to 0, we find the speed $v = v_t$ at which this occurs:

$$v = v_t \text{ when } a = 0 \quad \Rightarrow \quad \frac{mg - kv_t}{m} = 0 \quad \Rightarrow \quad v_t = \frac{mg}{k}$$

(d)   The skydiver's descent speed is initially v_0. However, once the parachute opens, the force of air resistance provides a large, speed-dependent upward acceleration, causing her descent velocity to decrease. The slope of the v-vs.-t graph (the acceleration) is not constant but instead decreases to zero as her descent speed decreases from v_0 to v_t. Therefore, the graph is not linear.

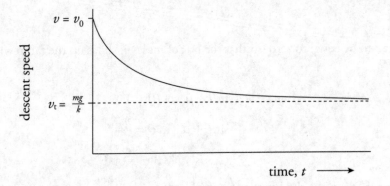

(e)   Since $a = dv/dt$ and, from part (b), $a = (mg - kv)/m$, we have

$$\frac{dv}{dt} = \frac{mg - kv}{m} \quad \Rightarrow \quad \frac{dv}{mg - kv} = \frac{dt}{m}$$

Integrating both sides of this equation gives

$$-\tfrac{1}{k}\ln|mg - kv| = \tfrac{1}{m}t + c$$

where c is a constant of integration. This equation can be rewritten in the form:

$$mg - kv = Ce^{(-k/m)t}$$
$$kv = mg - Ce^{(-k/m)t}$$

Since $v = v_0$ at $t = 0$, we can determine the constant C:

$$kv_0 = mg - C \quad \Rightarrow \quad C = mg - kv_0$$

Therefore, the equation for the skydiver's descent speed as a function of time is

$$v(t) = \tfrac{1}{k}\left(mg - (mg - kv_0)e^{(-k/m)t}\right)$$

5. (a) The forces acting on a person standing against the cylinder wall are gravity ($\mathbf{F}_w$, downward), the normal force from the wall ($\mathbf{F}_N$, directed toward the center of the cylinder), and the force of static friction ($\mathbf{F}_f$, directed upward), which are depicted in the following diagram:

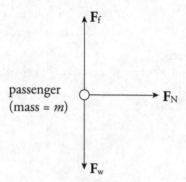

   (b) In order to keep the passenger from sliding down the wall, the maximum force of static friction must be at least as great as the passenger's weight: $F_{f\,(max)} \geq mg$. Since $F_{f\,(max)} = \mu_s F_N$, this condition becomes

$$\mu_s F_N \geq mg$$

Now, consider the circular motion of the passenger. Neither $\mathbf{F}_f$ nor $\mathbf{F}_w$ has a component toward the center of the path, so the centripetal force is provided entirely by the normal force

$$F_N = \frac{mv^2}{r}$$

Substituting this expression for F_N into the previous equation, we get

$$\mu_s \frac{mv^2}{r} \geq mg$$

$$\mu_s \geq \frac{gr}{v^2}$$

Therefore, the coefficient of static friction between the passenger and the wall of the cylinder must satisfy this condition in order to keep the passenger from sliding down.

   (c) Since the mass m canceled out in deriving the expression for μ_s, the conditions are independent of mass. Thus, the inequality $\mu_s \geq gr/v^2$ holds for both the adult passenger of mass m and the child of mass $m/2$.

6.   (a)   The forces acting on the car are gravity ($\mathbf{F}_w$, downward), the normal force from the road ($\mathbf{F}_N$, upward), and the force of static friction ($\mathbf{F}_f$, directed toward the center of curvature of the road), as seen in the diagram below:

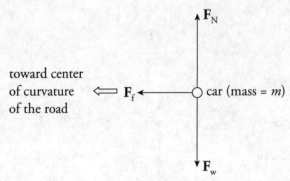

       (b)   The force of static friction [we assume static friction because we *don't* want the car to slide (that is, skid)] provides the necessary centripetal force.

$$F_f = \frac{mv^2}{r}$$

Therefore, to find the maximum speed at which static friction can continue to provide the necessary force, we write

$$F_{f\,(max)} = \frac{mv_{max}^2}{r}$$

$$\mu_s F_N = \frac{mv_{max}^2}{r}$$

$$\mu_s mg = \frac{mv_{max}^2}{r}$$

$$v_{max} = \sqrt{\mu_s g r}$$

       (c)   Ignoring friction, the forces acting on the car are gravity ($\mathbf{F}_w$, downward) and the normal force from the road ($\mathbf{F}_N$, which is now tilted toward the center of curvature of the road), which are shown in the following diagram:

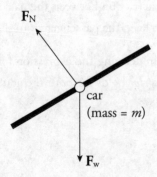

(d)   Because of the banking of the turn, the normal force is tilted toward the center of curvature of the road. The component of $\mathbf{F}_N$ toward the center can provide the centripetal force, making reliance on friction unnecessary.

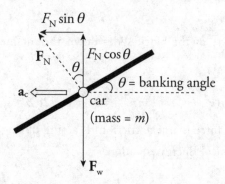

There's no vertical acceleration, so $F_N \cos \theta = F_w = mg = 0$, so $F_N = mg/\cos \theta$. The component of F_N toward the center of curvature of the turn, $F_N \sin \theta$, provides the centripetal force.

$$F_N \sin \theta = \frac{mv^2}{r}$$

$$\frac{mg}{\cos \theta} \sin \theta = \frac{mv^2}{r}$$

$$g \tan \theta = \frac{v^2}{r}$$

$$\theta = \tan^{-1} \frac{v^2}{gr}$$

CHAPTER 7 DRILL

Section I: Multiple Choice

1.   **A**   Since the force **F** is perpendicular to the displacement, the work it does is zero.

2.   **B**   By the work–energy theorem,

$$W = \Delta K = \tfrac{1}{2} m(v^2 - v_0^2) = \tfrac{1}{2}(4 \text{ kg})\left[(6 \text{ m/s})^2 - (3 \text{ m/s})^2\right] = 54 \text{ J}$$

3.   **D**   Because gravity is a conservative force, the amount of work required will be the same regardless of the chosen path.

4.   **E**   The motion of a box as it slides up a ramp will be parallel to the ramp. The normal force, by definition, must act perpendicular to the surface providing it. This means the normal force will be perpendicular to motion, resulting in 0 work done.

5. **A** Our basic equation for work is $W = Fd \cos \theta$. The normal force will affect the man the entire time he ascends, so F and d will be some nonzero values. Furthermore, the normal force will push up, and the man will move up, making the angle between the forces 0. Since $\cos(0) = 1$, the work done will be positive.

6. **D** The work done by gravity as the block slides down the inclined plane is equal to the potential energy at the top (mgh).

$$mgh = W = \Delta K = \tfrac{1}{2}m(v^2 - v_0^2) = \tfrac{1}{2}mv^2 \implies v = \sqrt{2gh} = \sqrt{2(10)(6.4)(\sin 30°)} = 8 \text{ m/s}$$

7. **D** Since a nonconservative force (namely, friction) is acting during the motion, we use the modified Conservation of Mechanical Energy equation.

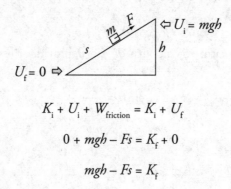

$$K_i + U_i + W_{\text{friction}} = K_i + U_f$$

$$0 + mgh - Fs = K_f + 0$$

$$mgh - Fs = K_f$$

8. **C** The force diagram would look like this:

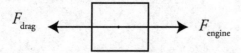

From here, simply apply Newton's Second Law, taking forward as the positive direction.

$$F_{\text{net}} = ma$$

$$F_{\text{engine}} - F_{\text{drag}} = ma$$

$$F_{\text{engine}} = ma + F_{\text{drag}}$$

$$= (1{,}000 \text{ kg})(4 \text{ m/s}^2) + (300 \text{ N})$$

$$= 4{,}300 \text{ N}$$

9. **E** Because the rock has lost half of its gravitational potential energy, its kinetic energy at the halfway point is half of its kinetic energy at impact. Since K is proportional to v^2, if $K_{\text{at halfway point}}$ is equal to $\tfrac{1}{2}K_{\text{at impact}}$, then the rock's speed at the halfway point is $\sqrt{1/2} = 1/\sqrt{2}$ its speed at impact.

10. **D** First, (B) and (C) can be eliminated because the units are not appropriate (due to the lack of t^3 in the denominator of the first term). Next, consider the energy that must be expended in moving the block to the appropriate position. First, the block will have kinetic energy because it will be in motion. Second, it will have the potential energy from its higher position. Last, some energy must be expended to overcome the work from friction.

$$E = KE + PE + W_f = \tfrac{1}{2}mv_f^2 + mgh + F_f d$$

If the motor applies a constant power, then the acceleration will be constant. This means the final speed of the block (v_f) must be double the average speed of the block (v_{avg}), which will simply be the distance divided by the time. Therefore, $v_f = \dfrac{2d}{t}$.

We can use simple trigonometry to show that the height of the ramp must be $h = d\sin\theta$.

The force of friction will be $F_f = \mu F_N$, where the normal force $F_N = mg\cos\theta$.

Substituting all these values into the first equation gives

$$E = \tfrac{1}{2}m\left(\frac{2d}{t}\right)^2 + mg(d\sin\theta) + (\mu mg\cos\theta)d$$

Finally, power is simple energy over time, so dividing everything by time gives

$$P = \frac{\dfrac{1}{2}m\left(\dfrac{2d}{t}\right)^2 + mg\left(d\sin\theta\right) + \left(\mu mg\cos\theta\right)d}{t}$$

which simplifies to

$$P = m\left(\frac{2d^2}{t^3} + \frac{gd\sin\theta}{t} + \frac{dg\mu\cos\theta}{t}\right)$$

Section II: Free Response

1. (a) Applying Conservation of Energy,

$$K_A + U_A = K_{at\,H/2} + U_{at\,H/2}$$
$$0 + mgH = \tfrac{1}{2}mv^2 + mg(\tfrac{1}{2}H)$$
$$\tfrac{1}{2}mgH = \tfrac{1}{2}mv^2$$
$$v = \sqrt{gH}$$

(b) Applying Conservation of Energy again,

$$K_A + U_A = K_B + U_B$$
$$0 + mgH = \tfrac{1}{2}mv_B^2 + 0$$
$$v_B = \sqrt{2gH}$$

(c) By the work–energy theorem, we want the work done by friction to be equal (but opposite) to the kinetic energy of the box at point B.

$$W = \Delta K = \tfrac{1}{2}m(v_C^2 - v_B^2) = -\tfrac{1}{2}m\,v_B^2 = -\tfrac{1}{2}m(\sqrt{2gH})^2 = -mgH$$

Therefore,

$$W = -mgH \;\Rightarrow\; -F_f x = -mgH \;\Rightarrow\; -\mu_k mgx = -mgH \;\Rightarrow\; \mu_k = H/x$$

(d) Apply Conservation of Energy (including the negative work done by friction as the box slides up the ramp from B to C).

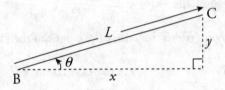

$$K_B + U_B + W_f = K_C + U_C$$
$$\tfrac{1}{2}m(\sqrt{2gH})^2 + 0 - F_f L = 0 + mgy$$
$$mgH + 0 - F_f L = 0 + mgy$$
$$mg(H-y) - (\mu_k mg\cos\theta)(L) = 0$$
$$\mu_k = \frac{H-y}{L\cos\theta} = \frac{H-y}{x}$$

2. (a) The centripetal acceleration of the car at point C is given by the equation $a = v_C^2/r$, where v_C is the speed of the car at C. To find v_C^2, we apply Conservation of Energy.

$$K_A + U_A = K_C + U_C$$
$$0 + mgH = \tfrac{1}{2}mv_C^2 + mgr$$
$$mg(H-r) = \tfrac{1}{2}mv_C^2$$
$$v_C^2 = 2g(H-r)$$

Therefore,

$$a_c = \frac{v_C^2}{r} = \frac{2g(H-r)}{r}$$

(b) When the car reaches point D, the forces acting on the car are its weight, $\mathbf{F}_w$, and the downward normal force, $\mathbf{F}_N$, from the track. Thus, the net force, $\mathbf{F}_w + \mathbf{F}_N$, provides the centripetal force. In order for the car to maintain contact with the track, $\mathbf{F}_N$ must not vanish. Therefore, the cut-off speed for ensuring that the car makes it safely around the track is the speed at which $\mathbf{F}_N$ just becomes zero; any greater speed would imply that the car would make it around. Thus,

$$F_w + F_N = m\frac{v^2}{r} \quad \Rightarrow \quad F_w + 0 = m\frac{v_{\text{cut-off}}^2}{r} \quad \Rightarrow \quad v_{\text{cut-off}} = \sqrt{\frac{rF_w}{m}} = \sqrt{gr}$$

(c) Using the cut-off speed calculated in part (b), we now apply Conservation of Mechanical Energy.

$$K_A + U_A = K_D + U_D$$
$$0 + mgH = \tfrac{1}{2}mv_{\text{cut-off}}^2 + mg(2r)$$
$$mgH = \tfrac{1}{2}m(gr) + mg(2r)$$
$$= \tfrac{5}{2}mgr$$
$$H = \tfrac{5}{2}r$$

(d) First, we calculate the car's kinetic energy at point B; then, we determine the distance x the car must travel from B to F for the work done by friction to eliminate this kinetic energy. So, applying Conservation of Mechanical Energy, we find

$$K_A + U_A = K_B + U_B$$
$$0 + mg(6r) = \tfrac{1}{2}mv_B^2 + 0$$
$$6mgr = \tfrac{1}{2}mv_B^2$$

Now, by the work–energy theorem,

$$W = \Delta K = \tfrac{1}{2}mv_F^2 - \tfrac{1}{2}mv_B^2 = -\tfrac{1}{2}mv_B^2 \quad \Rightarrow \quad -F_f x = -\tfrac{1}{2}mv_B^2$$
$$-\mu mg x = -6mgr$$
$$x = \frac{6r}{\mu} = \frac{6r}{0.5} = 12r$$

3. (a) The point where $x = x_1$ is a local minimum of the function $U(x)$, and the point where $x = x_3$ is a local maximum. Therefore, at each of these locations, the derivative of $U(x)$ must be equal to 0.

$$\frac{dU}{dx} = 0 \quad \Rightarrow \quad 3 - 3(x-3)^2 = 0 \quad \Rightarrow \quad (x-3)^2 = 1 \quad \Rightarrow \quad x = 2 \text{ or } 4$$

Therefore, $x_1 = 2$ m and $x_3 = 4$ m.

(b)   If the particle's total energy is $E_2 = \frac{1}{2}(E_1 + E_3) = \frac{1}{2}[U(x_1) + U(x_3)] = \frac{1}{2}[U(2) + U(4)] = \frac{1}{2}(4\text{ J} + 8\text{ J}) = 6\text{ J}$, then the particle can oscillate between the points marked a and b in the following figure:

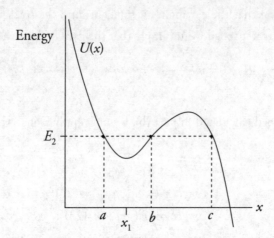

At points a and b, the object has no kinetic energy (since $U = E_2$ at these points), so these are the turning points at which the object momentarily comes to rest before being accelerated back toward $x = x_1$ (where the potential energy is minimized). The particle cannot be found at $x < a$ or within the interval $b < x < c$, since $U > E_2$ in these intervals (and this would imply a negative K, which is impossible). If $x > c$, then since U decreases, K increases: the particle would move off in the $+x$ direction with increasing speed.

(c)   Since $K + U = E$, we have $K = E - U$. Therefore,

$$
\begin{aligned}
K(x_1) &= E - U(x_1) \\
&= E - \left(3(x-1) - (x-3)^3\right)_{x=2} \\
&= 58\text{ J} - [(3 - (-1)]\text{ J} \\
&= 54\text{ J}
\end{aligned}
$$

Therefore,

$$K = \tfrac{1}{2}mv_1^2 \quad \Rightarrow \quad v_1 = \sqrt{\frac{2K}{m}} = \sqrt{\frac{2(54\text{ J})}{3\text{ kg}}} = 6\text{ m/s}$$

(d) Since $F(x) = -dU/dx = -[3 - 3(x-3)^2] = 3(x-3)^2 - 3$, dividing by m gives a.

$$a(x) = \frac{F(x)}{m} = \frac{[3(x-3)^2 - 3]\text{ J}}{3\text{ kg}} = (x-3)^2 - 1 \ \ (\text{in m/s}^2)$$

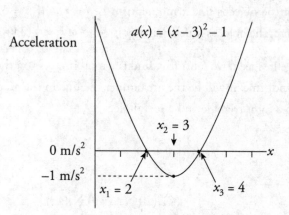

(e) At $x = \frac{1}{2}x_1$, the particle's energy is $E_3 = U(\frac{1}{2}x_1) = U(1) = \left[3(x-1) - (x-3)^3\right]_{x=1} = 8$ J.

This is the particle's total energy because it has no initial kinetic energy (it is released from rest). As the particle passes through $x = x_1$, its potential energy decreases to

$$U(x_1) = U(2) = \left[3(x-1) - (x-3)^3\right]_{x=2} = 4 \text{ J}$$

Therefore, since its total energy is 8 J, the particle's kinetic energy at $x = x_1$ must be $K = E - U = 8$ J $- 4$ J $= 4$ J. This implies that its speed is

$$K = \tfrac{1}{2}mv^2 \ \ \Rightarrow \ \ v = \sqrt{\frac{2K}{m}} = \sqrt{\frac{2(4\text{ J})}{3\text{ kg}}} = 1.6 \text{ m/s}$$

4. (a) Work equals the integral of the force and displacement.

$$W = \int F\, dx = \int_0^4 (3x + d)\, dx = \left(\frac{3}{2}x^2 + 5x\right)\Big|_0^4$$

$W = 44$ J

(b) The work–energy theorem states the work done on an object equals the change in kinetic energy of the object.

$$W = \Delta K$$
$$44 \text{ J} = \frac{1}{2}m\left(v_2{}^2 - v_1{}^2\right)$$
$$44 \text{ J} = \frac{1}{2}(6 \text{ kg})\left(v_2{}^2 - 2^2\right)$$
$$v_2 = 4.3 \text{ m/s}$$

CHAPTER 8 DRILL

Section I: Multiple Choice

1. **C** The magnitude of the object's linear momentum is $p = mv$. If $p = 6$ kg·m/s and $m = 2$ kg, then $v = 3$ m/s. Therefore, the object's kinetic energy is $K = \frac{1}{2} mv^2 = \frac{1}{2} (2$ kg$)(3$ m/s$)^2 = 9$ J.

2. **D** To find the change in speed, we must first find the total impulse on the object. Since impulse is the product of force and time, it will be the area under the line in the graph. We can do this by breaking the shape into a right triangle and a rectangle.

$$J = \left(\frac{1}{2}\right) bh + lw$$

$$= \left(\frac{1}{2}\right)(4 \text{ s})(20 \text{ N}) + (10 \text{ N})(4 \text{ s})$$

$$= 80 \text{ N·s}$$

Impulse is also equal to the change in momentum of an object.

$$80 \text{ N·s} = m\Delta v$$

$$\Delta v = \frac{80 \text{ N·s}}{8 \text{ kg}} = 10 \text{ m/s}$$

3. **E** The impulse delivered to the box, $J = \overline{F}\Delta t$, equals its change in momentum. Thus,

$$\overline{F}\Delta t = \Delta p = p_f - p_i = m(v_f - v_i) \quad \Rightarrow \quad \overline{F} = \frac{m(v_f - v_i)}{\Delta t} = \frac{(2 \text{ kg})(8 \text{ m/s} - 4 \text{ m/s})}{0.5 \text{ s}} = 16 \text{ N}$$

4. **D** The impulse delivered to the ball is equal to its change in momentum. The momentum of the ball was $m\mathbf{v}$ before hitting the wall and $m(-\mathbf{v})$ after. Therefore, the change in momentum is $m(-\mathbf{v}) - m\mathbf{v} = -2m\mathbf{v}$, so the magnitude of the momentum change (and the impulse) is $2m\mathbf{v}$.

5. **B** By definition of *perfectly inelastic*, the objects move off together with one common velocity, $\mathbf{v}'$, after the collision. By Conservation of Linear Momentum,

$$m_1\mathbf{v}_1 + m_2\mathbf{v}_2 = (m_1 + m_2)\mathbf{v}'$$

$$\mathbf{v}' = \frac{m_1 v_1 + m_2 v_2}{m_1 + m_2}$$

$$= \frac{(3 \text{ kg})(2 \text{ m/s}) + (5 \text{ kg})(-2 \text{ m/s})}{3 \text{ kg} + 5 \text{ kg}}$$

$$= -0.5 \text{ m/s}$$

Since the velocity after the collision is -0.5 m/s, the speed is 0.5 m/s.

6. **D** For this problem, we know the center of mass needs to be the 50 cm mark. Using the center of mass formula (and taking the left edge as our zero point), that gives us

$$x_{CM} = (m_1 x_1 + m_2 x_2 + m_3 x_3) / (m_1 + m_2 + m_3)$$

$$x_3 = [x_{CM}(m_1 + m_2 + m_3) - m_1 x_1 - m_2 x_2] / m_3$$

$$x_3 = [(0.5 \text{ m})(6 \text{ kg} + 2 \text{ kg} + 10 \text{ kg}) - (6 \text{ kg})(0 \text{ m}) - (2 \text{ kg})(1 \text{ m})]/(10 \text{ kg})$$

$$x_3 = 0.7 \text{ m} = 70 \text{ cm}$$

7. **C** Total linear momentum is conserved in a collision during which the net external force is zero. If kinetic energy is lost, then by definition, the collision is not elastic.

8. **B** First, replace each rod by concentrating its mass at its center-of-mass position:

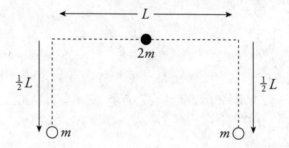

The center of mass of the two m's is at their midpoint, at a distance of $\frac{1}{2} L$ below the center of mass of the rod of mass $2m$:

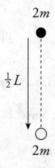

Now, applying the equation for locating the center of mass (letting $y = 0$ denote the position of the center of mass of the top horizontal rod), we find

$$y_{cm} = \frac{(2m)(0) + (2m)(\frac{1}{2} L)}{2m + 2m} = \frac{1}{4} L$$

9. **B** Because the collision is perfectly inelastic, the two vehicles will stick together after the collision. We can use Conservation of Momentum to determine the speed of the vehicles just after the crash. Taking the truck's direction as positive, we get

$$m_t v_t - m_c v_c = (m_t + m_c)v_f$$

$$v_f = \frac{(m_t v_t - m_c v_c)}{(m_t + m_c)}$$

$$v_f = [(2{,}000 \text{ kg})(21 \text{ m/s}) - (1{,}000 \text{ kg})(21 \text{ m/s})]/(2{,}000 \text{ kg} + 1{,}000 \text{ kg})$$

$$v_f = 7 \text{ m/s}$$

Now that we have the final speed of the combined vehicles, we can use it to determine the total kinetic energy of the system just after the collision.

$$K = \left(\frac{1}{2}\right)mv^2$$

$$K = \left(\frac{1}{2}\right)(2{,}000 \text{ kg} + 1{,}000 \text{ kg})(7 \text{ m/s})^2$$

$$K = 73{,}500 \text{ J}$$

Since work is equal to the change in kinetic energy, we must take the negative of this value. Another way of looking at it is that friction would have acted antiparallel to the direction of motion, so the θ in $W = Fd\cos\theta$ would be 180 degrees.

10. **D** First, find the collision location. The blocks are moving toward one another at a combined rate of 5 m/s, meaning 0.2 s will pass before the collision takes place. This places the collision location 60 cm away from the first block's starting location.

Next, calculate the velocity of the combined blocks post-collision using Conservation of Linear Momentum for a perfectly inelastic collision:

$$m_1\mathbf{v}_1 + m_2\mathbf{v}_2 = (m_1 + m_2)\mathbf{v}_f \rightarrow \mathbf{v}_f$$

$$= \frac{m_1\mathbf{v}_1 + m_2\mathbf{v}_2}{m_1 + m_2}$$

$$= \frac{(2\,\text{kg})(3\,\text{m/s}) + (5\,\text{kg})(-2\,\text{m/s})}{2\,\text{kg} + 5\,\text{kg}}$$

$$= \frac{-4}{7} \text{ m/s}$$

The final velocity came out to a negative, meaning the combined blocks will move back toward the starting position of the first block. This means they must travel 0.6 m before reaching one of the starting positions.

$$v = \frac{d}{t} \rightarrow t = \frac{d}{v} = \frac{0.6\,\text{m}}{\frac{4}{7}\,\text{m/s}} = 1.05\,\text{s}$$

Section II: Free Response

1. (a) First, draw a free-body diagram:

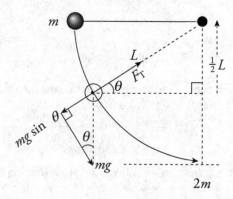

The net force toward the center of the steel ball's circular path provides the centripetal force. From the geometry of the diagram, we get Equation 1:

$$F_T - mg\sin\theta = \frac{mv^2}{L}$$

In order to determine the value of mv^2, we use Conservation of Mechanical Energy:

$$K_i + U_i = K_{ff} + U$$
$$0 + mgL = \tfrac{1}{2}mv^2 + mg(\tfrac{1}{2}L)$$
$$\tfrac{1}{2}mgL = \tfrac{1}{2}mv^2$$
$$mgL = mv^2$$

Substituting this result into Equation 1, we get

$$F_T - mg\sin\theta = \frac{mgL}{L}$$
$$F_T = mg(1 + \sin\theta)$$

Now, from the free-body diagram, we see that $\sin\theta = \tfrac{1}{2}L/L = \tfrac{1}{2}$, so

$$F_T = mg(1 + \tfrac{1}{2}) = \tfrac{3}{2}mg$$

(b)   Applying Conservation of Energy, we find the speed of the ball just before impact:

$$K_i + U_i = K_f + U_f$$
$$0 + mg\,L = \tfrac{1}{2}m\,v^2 + 0$$
$$v = \sqrt{2g\,L}$$

Because the collision is elastic, we know that kinetic energy will be conserved. Also, as is the case with all collisions, momentum will be conserved. Furthermore, we can neglect the second term on the left-hand side of each equation because the second object starts at rest. This gives us

$$\left(\frac{1}{2}\right)m_1 v_{1,0}{}^2 = \left(\frac{1}{2}\right)m_1 v_{1,f}{}^2 + \left(\frac{1}{2}\right)m_2 v_{2,f}{}^2$$

and

$$m_1 v_{1,0} = m_1 v_{1,f} + m_2 v_{2,f}$$

Solving the latter equation for $v_{1,f}$ and inserting that result into the former equation gives us

$$\left(\frac{1}{2}\right)m_1 v_{1,0}{}^2 = \left(\frac{1}{2}\right)m_1 \left[\frac{(m_1 v_{1,0} - m_2 v_{2,f})}{m_1}\right]^2 + \left(\frac{1}{2}\right)m_2 v_{2,f}{}^2$$

The algebra here is pretty intense, so we'll break it into a few steps.

First, drop the $\left(\dfrac{1}{2}\right)$ from each term and plug in the known values for m_1 (m), m_2 ($4m$), and $v_{1,0}\left(\sqrt{(2gL)}\right)$.

$$m\left[\sqrt{(2gL)}\right]^2 = m\left[\frac{m\sqrt{(2gL)} - 4mv_{2,f}}{m}\right]^2 + 4mv_{2,f}{}^2$$

Next, notice that all the m's cancel out.

$$\left[\sqrt{(2gL)}\right]^2 = \left[\sqrt{(2gL)} - 4v_{2,f}\right]^2 + 4v_{2,f}{}^2$$

Squaring the first two terms gives

$$2gL = 2gL - 8v_{2,f}\sqrt{(2gL)} + 16v_{2,f}{}^2 + 4v_{2,f}{}^2$$

Subtract $2gL$ from both sides.

$$0 = -8v_{2,f}\sqrt{(2gL)} + 16v_{2,f}{}^2 + 4v_{2,f}{}^2$$

Divide both sides by $4v_{2,f}$

$$0 = -2\sqrt{(2gL)} + 4v_{2,f} + v_{2,f}$$
$$2\sqrt{(2gL)} = 5v_{2,f}$$
$$v_{2,f} = \left(\frac{2}{5}\right)\sqrt{(2gL)}$$

(c) We can quote the result of Chapter 8, Example 8 to find the velocity of the ball immediately after the collision:

$$v_1' = \frac{m_1 - m_2}{m_1 + m_2} v_1 = \frac{m - 4m}{m + 4m} \sqrt{2gL} = -\tfrac{3}{5}\sqrt{2gL}$$

Now, applying Conservation of Mechanical Energy, we find

$$K_i + U_i = K_f + U_f$$
$$\tfrac{1}{2} m v_1'^2 + 0 = 0 + mgh$$
$$h = \frac{v_1'^2}{2g} = \frac{\left(\tfrac{3}{5}\sqrt{2gL}\right)^2}{2g} = \tfrac{9}{25}L$$

2. (a) By Conservation of Linear Momentum, $mv = (m + M)v'$, so $v' = \dfrac{mv}{m + M}$.

Now, by Conservation of Mechanical Energy,

$$K_i + U_i = K_f + U_f$$
$$\tfrac{1}{2}(m + M)v'^2 + 0 = 0 + (m + M)gy$$
$$\tfrac{1}{2}v'^2 = gy$$
$$\tfrac{1}{2}\left(\frac{mv}{m + M}\right)^2 = gy$$
$$v = \frac{m + M}{m}\sqrt{2gy}$$

(b) Use the result derived in part (a) to compute the kinetic energy of the block and bullet immediately after the collision:

$$K' = \tfrac{1}{2}(m + M)v'^2 = \tfrac{1}{2}(m + M)\left(\frac{mv}{m + M}\right)^2 = \tfrac{1}{2}\frac{m^2 v^2}{m + M}$$

Since $K = \tfrac{1}{2} mv^2$, the difference is

$$\Delta K = K' - K = \tfrac{1}{2}\frac{m^2 v^2}{m + M} - \tfrac{1}{2}mv^2$$
$$= \tfrac{1}{2}mv^2\left(\frac{m}{m + M} - 1\right)$$
$$= K\left(\frac{-M}{m + M}\right)$$

Therefore, the fraction of the bullet's original kinetic energy that was lost is $M/(m + M)$. This energy is manifested as heat (the bullet and block are warmer after the collision than before), and some was used to break the intermolecular bonds within the wooden block to allow the bullet to penetrate.

(c)   From the geometry of the diagram,

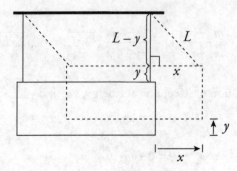

the Pythagorean Theorem implies that $(L-y)^2 + x^2 = L^2$. Therefore,

$$L^2 - 2Ly + y^2 + x^2 = L^2 \quad \Rightarrow \quad y = \frac{x^2}{2L}$$

(where we have used the fact that y^2 is small enough to be neglected). Substituting this into the result of part (a), we derive the following equation for the speed of the bullet in terms of x and L instead of y:

$$v = \frac{m+M}{m}\sqrt{2gy} = \frac{m+M}{m}\sqrt{2g\frac{x^2}{2L}} = \frac{m+M}{m}x\sqrt{\frac{g}{L}}$$

(d)   No; momentum is conserved only when the net external force on the system is zero (or at least negligible). In this case, the block and bullet feel a net nonzero force that causes it to slow down as it swings upward. Since its speed is decreasing as it swings upward, its linear momentum cannot remain constant.

3.   (a)   To apply Conservation of Linear Momentum to the collision, we recognize that momentum is a vector quantity and, therefore, linear momentum must be conserved separately in the horizontal (x) and vertical (y) directions. Before the collision, the linear momentum was horizontal only. Therefore, in the x direction:

$$mv = mv_1' \cos\theta_1 + mv_2' \cos\theta_2 \quad \Rightarrow \quad v = v_1' \cos\theta_1 + v_2' \cos\theta_2 \qquad (1)$$

and in the y direction:

$$0 = mv_1' \sin\theta_1 - mv_2' \sin\theta_2 \quad \Rightarrow \quad 0 = v_1' \sin\theta_1 - v_2' \sin\theta_2 \qquad (2)$$

Since the collision is elastic, kinetic energy is also conserved; thus,

$$K = K_1' + K_2'$$
$$\tfrac{1}{2}mv^2 = \tfrac{1}{2}mv_1'^2 + \tfrac{1}{2}mv_2'^2$$
$$v^2 = v_1'^2 + v_2'^2 \qquad (3)$$

In order to find $K_1' = \frac{1}{2}m{v_1'}^2$ in terms of K_1 and θ_1 only, we need to manipulate the equations to eliminate v_2 and θ_2. The following approach will do this. Rewrite Equations (1) and (2) in the following forms:

$$v_2' \cos\theta_2 = v - v_1' \cos\theta_1 \quad (1')$$
$$v_2' \sin\theta_2 = v_1' \sin\theta_1 \quad (2')$$

Square both equations and add, exploiting the trig identity $\cos^2\theta + \sin^2\theta = 1$:

$${v_2'}^2 \cos^2\theta_2 = v^2 - 2vv_1' \cos\theta_1 + {v_1'}^2 \cos^2\theta_1$$
$${v_2'}^2 \sin^2\theta_2 = {v_1'}^2 \sin^2\theta_1$$
$${v_2'}^2 = v^2 - 2vv_1' \cos\theta_1 + {v_1'}^2 \quad (4)$$

This eliminates θ_2. Now, to eliminate v_2, substitute the value of ${v_2}^2$ from Equation (3) into Equation (4):

$$v^2 - {v_1'}^2 = v^2 - 2vv_1' \cos\theta_1 + {v_1'}^2$$
$$-2{v_1'}^2 = -2vv_1' \cos\theta_1$$
$$v_1' = v\cos\theta_1$$

Therefore, we can write

$$K_1' = \left(\frac{1}{2}\right)m{v_1'}^2 = \left(\frac{1}{2}\right)m(v\cos\theta_1)^2 = K_1\cos\theta_1^{\,2}$$

(b) Since, by Equation (3), ${v_2'}^2 = v^2 - {v_1'}^2$, the result of part (a) gives

$$K_2' = \tfrac{1}{2}m{v_2'}^2 = \tfrac{1}{2}m(v^2 - {v_1'}^2) = \tfrac{1}{2}mv^2 - \tfrac{1}{2}m{v_1'}^2$$
$$= K_1 - K_1'$$
$$= K_1 - (\cos^2\theta_1)K_1$$
$$= (1 - \cos^2\theta_1)K_1$$
$$= (\sin^2\theta_1)K_1$$

(c) Square Equation (1) and Equation (2) from part (a) and add:

$$v^2 = {v_1'}^2 \cos^2\theta_1 + 2v_1'v_2' \cos\theta_1 \cos\theta_2 + {v_2'}^2 \cos^2\theta_2$$
$$0 = {v_1'}^2 \sin^2\theta_1 - 2v_1'v_2' \sin\theta_1 \sin\theta_2 + {v_2'}^2 \sin^2\theta_2$$

Adding gives

$$v^2 = {v_1'}^2 + 2v_1'v_2'(\cos\theta_1 \cos\theta_2 - \sin\theta_1 \sin\theta_2) + {v_2'}^2$$
$$v^2 = {v_1'}^2 + 2v_1'v_2' \cos(\theta_1 + \theta_2) + {v_2'}^2 \quad (5)$$

Now, since by Equation (3), $v^2 = v_1'^2 + v_2'^2$, combining this with equation (5) gives

$$2v_1'v_2' \cos(\theta_1 + \theta_2) = 0$$
$$\cos(\theta_1 + \theta_2) = 0$$
$$\theta_1 + \theta_2 = 90°$$

Thus, the objects' post-collision velocity vectors are perpendicular to each other.

CHAPTER 9 DRILL

Section I: Multiple Choice

1. **B** We use the equation $v = r\omega$:

$$v = r\omega = 0.06 \text{ m} \times \left(\frac{5 \text{ rev}}{\text{s}} \times \frac{2\pi \text{ rad}}{\text{rev}} \right) = (0.6)\pi \text{ m/s} = 1.9 \text{ m/s}$$

2. **E** Use the equation *distance = rate × time* with $v = r\omega$:

$$s = vt = r\omega t = 0.06 \text{ m} \times \left(\frac{5 \text{ rev}}{\text{s}} \times \frac{2\pi \text{ rad}}{\text{rev}} \right) \times \left(40 \text{ min} \times \frac{60 \text{ s}}{\text{min}} \right) = 4{,}500 \text{ m} = 4.5 \text{ km}$$

3. **D** In this question, time is neither given nor asked for. This means we should use the angular equivalent of Big Five #5.

$$\omega_f^2 = \omega_0^2 + 2\alpha\theta$$

$$\omega_f = \sqrt{(\omega_0^2 + 2\alpha\theta)}$$

$$\omega_f = \sqrt{(0 \text{ rad/s})^2 + 2(4 \text{ rad/s}^2)(50 \text{ rad})}$$

$$\omega_f = 20 \text{ rad/s}$$

4. **D** We know centripetal acceleration is ordinarily calculated as $a_c = \dfrac{v^2}{r}$. Instead, substitute rotational speed.

$$a_c = \frac{v^2}{r} = \frac{(\omega r)^2}{r} = \omega^2 r$$

Next, convert the given rotational speed into standard units.

$$120 \text{ rev/min} \times (1 \text{ min} / 60 \text{ s}) \times (2\pi \text{ rad} / 1 \text{ rev}) = 4\pi \text{ rad/s}$$

Therefore, the acceleration is

$$a_c = \omega^2 r = (4\pi \text{ rad/s})^2 (0.75 \text{ m}) = 12\pi^2 \text{ m/s}^2$$

5. **B** First, draw a quick sketch of the situation.

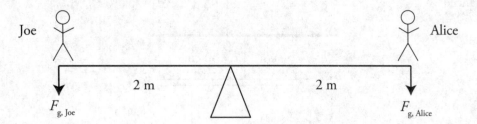

The gravity pulling down on each person will be the only source of torque in this situation. Furthermore, we can see that gravity will act perpendicular to the radius, allowing us to ignore the $\sin\theta$ term in torque. Because they're on opposite ends, the torques will act against one another rather than adding together. Because Joe is heavier and the two are equally distant from the fulcrum, the torque he provides will be greater. Let's make his torque positive. This gives us

$$\tau_{net} = \tau_{Joe} - \tau_{Alice}$$

$$= \left(F_{g, Joe}\right) - \left(F_{g, Alice}\right)$$

$$= \left(m_{Joe}g\right)\left(r_{Joe}\right) - \left(m_{Alice}g\right)\left(r_{Alice}\right)$$

$$= (60 \text{ kg})(10 \text{ m/s}^2)(2 \text{ m}) - (30 \text{ kg})(10 \text{ m/s}^2)(2 \text{ m})$$

$$= 600 \text{ N} \cdot \text{m}$$

6. **D** From the diagram,

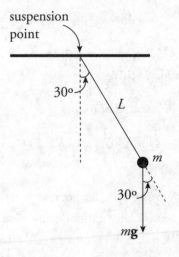

we calculate that

$$\tau = rF\sin\theta = Lmg\sin\theta$$

$$= (0.80 \text{ m})(0.50 \text{ kg})(10 \text{ N/kg})(\sin 30°)$$

$$= 2.0 \text{ N} \cdot \text{m}$$

7.   **B**   The stick will remain at rest in the horizontal position if the torques about the suspension point are balanced.

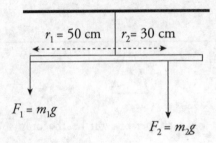

$$\tau_{CCW} = \tau_{CW}$$
$$r_1 F_1 = r_2 F_2$$
$$r_1 m_1 g = r_2 m_2 g$$
$$m_2 = \frac{r_1 m_1}{r_2} = \frac{(50 \text{ cm})(3 \text{ kg})}{30 \text{ cm}} = 5 \text{ kg}$$

8.   **A**   Each of the four masses is at distance L from the rotation axis, so

$$I = \sum m_i r_i^2 = m L^2 + m L^2 + m L^2 + m L^2 = 4m L^2$$

Note that the length of the rectangular frame, $\frac{8}{3} L$, is irrelevant here.

9.   **C**   An object sliding down a ramp converts potential energy to kinetic energy, resulting in a final speed of

$$PE = KE \rightarrow mgh = \tfrac{1}{2} mv^2 \rightarrow v = \sqrt{2gh}$$

In contrast, a cylinder rolling down a ramp will have a final speed of $v = \sqrt{\frac{4}{3} gh}$ (see page 251, Example 15 for a full explanation of this).

If we want the two speeds to be equal, that means

$$\sqrt{2gh} = \sqrt{\frac{4}{3} gH} \rightarrow h = \frac{2}{3} H$$

10.   **D**   Gravity cannot cause rotation of the cylinder because it acts at the center of mass, so $r = 0$. This eliminates (A) and (B). The normal force cannot cause rotation because it acts antiparallel to the radius vector, so $\sin \theta = \sin 180° = 0$. This eliminates (C). Finally, the friction that occurs cannot be kinetic because the question specifies that the cylinder rolls without slipping. This means the point of contact never slides along the surface, implying that the friction is static.

Section II: Free Response

1. (a) Apply Conservation of Mechanical Energy, remembering to take into account the translational kinetic energy of the falling block (m_2), the rising block (m_1), and the rotating pulley.

$$K_i + U_i = K_f + U_f$$

$$0 + m_2 g h = \left(\tfrac{1}{2} m_1 v^2 + \tfrac{1}{2} m_2 v^2 + \tfrac{1}{2} I \omega^2\right) + m_1 g h$$

$$= \left[\tfrac{1}{2} m_1 v^2 + \tfrac{1}{2} m_2 v^2 + \tfrac{1}{2}\left(\tfrac{1}{2} M R^2\right)\left(\frac{v}{R}\right)^2\right] + m_1 g h$$

$$(m_2 - m_1) g h = \tfrac{1}{2} v^2 \left(m_1 + m_2 + \tfrac{1}{2} M\right)$$

$$v = \sqrt{\frac{2(m_2 - m_1) g h}{m_1 + m_2 + \tfrac{1}{2} M}}$$

(b) Using the equation $v = R\omega$, we find

$$\omega = \frac{v}{R} = \frac{1}{R} \sqrt{\frac{2(m_2 - m_1) g h}{m_1 + m_2 + \tfrac{1}{2} M}}$$

(c) Since Block 2 (and Block 1) moved a distance h during the motion, a point on the rim of the pulley must also move through a distance h. Therefore, $\Delta\theta = h/R$.

(d) Use Big Five #1 (for translational motion):

$$\Delta t = \frac{\Delta s}{\bar{v}} = \frac{\Delta s}{\tfrac{1}{2} v} = \frac{2\Delta s}{v} = \frac{2h}{v} = 2h \sqrt{\frac{m_1 + m_2 + \tfrac{1}{2} M}{2(m_2 - m_1) g h}} = h \sqrt{\frac{2m_1 + 2m_2 + M}{(m_2 - m_1) g h}}$$

2. (a) Consider the following diagram of the disk rolling down the ramp:

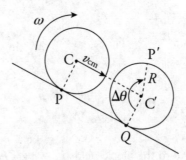

In a time interval Δt, the center of mass moves from C to C′, a distance equal to $v_{cm}\Delta t$. While this is occurring, point P on the rim moves to the new position P′. In order for there to be no slipping, the arc length P′Q, which is $R\Delta\theta$, must equal the straight line distance PQ. But PQ = CC′, so we have

$$v_{cm}\Delta t = R\Delta\theta \implies v_{cm}\Delta t = R\omega\Delta t \implies v_{cm} = R\omega$$

(b)   One way to derive the desired result is to notice that the velocity of T relative to P is equal to the velocity of T relative to C plus the velocity of C relative to P ($\mathbf{v}_{TP} = \mathbf{v}_{TC} + \mathbf{v}_{CP}$).

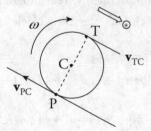

Since the velocity of C relative to P is the opposite of the velocity of P relative to C ($\mathbf{v}_{CP} = -\mathbf{v}_{PC}$), we have $\mathbf{v}_{TP} = \mathbf{v}_{TC} - \mathbf{v}_{PC}$. Relative to C, point T is moving forward with speed $R\omega$ and point P is moving backward with speed $R\omega$. Therefore, $\mathbf{v}_{TP} = (+R\omega) - (-R\omega) = +2R\omega = 2\mathbf{v}_{cm}$.

Another method is to think of the disk as executing pure rotation instantaneously about the contact point P. From this point of view (with P as the pivot), the distance to C is R, and the distance to T is $2R$. Therefore, the linear speed of C is $R\omega$ and that of T is $2R\omega$, which implies that the linear speed of point T is twice that of point C, as desired.

(c)   The speed with which the block rises, v_b, is equal to the speed at which the thread wraps around the cylinder, which is $2v_{cm}$, as shown in part (b). Therefore, differentiating the equation $v_b = 2v_{cm}$ with respect to time, we get $a_b = 2a_{cm}$. That is, the acceleration of the block is equal to twice that of the (center of mass of the) cylinder.

(d)   First, draw free-body diagrams for the block and the cylinder.

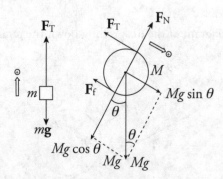

Newton's Second Law applied to the block yields $F_T - mg = ma_b$. Since $a_b = 2a_{cm}$, we write

$$F_T - mg = 2ma_{cm} \qquad (1)$$

Newton's Second Law applied to the forces on the cylinder gives us

$$Mg \sin\theta - F_f - F_T = Ma_{cm} \qquad (2)$$

The tension force and friction force each exert a torque about the center of mass of the cylinder. The torque of the friction force (clockwise) overwhelms the torque of the tension force (counterclockwise), which leads to clockwise acceleration. Therefore, $\tau_{net} = I\alpha$ becomes

$$RF_f - RF_T = \tfrac{1}{2}MR^2 \cdot \frac{a_{cm}}{R}$$
$$F_f - F_T = \tfrac{1}{2}Ma_{cm} \qquad (3)$$

Now for the algebra. Adding Equations (2) and (3) gives $Mg\sin\theta - 2F_T = \tfrac{3}{2}Ma_{cm}$. Multiplying both sides of Equation (1) by 2 gives $2F_T - 2mg = 4ma_{cm}$. Adding these last two equations allows us to find a_{cm}, the linear acceleration of the cylinder.

$$Mg\sin\theta - 2mg = \left(\tfrac{3}{2}M + 4m\right)a_{cm} \quad\Rightarrow\quad a_{cm} = \frac{M\sin\theta - 2m}{\tfrac{3}{2}M + 4m}g$$

(e) Since $a_b = 2a_{cm}$, the acceleration of the block is

$$a_b = \frac{2(M\sin\theta - 2m)}{\tfrac{3}{2}M + 4m}g$$

3. (a) Consider a rod of length $2L$ with the pivot at its center. Since the rod is uniform, its linear mass density is $M/(2L)$, so the mass of a segment of length dx is $M\,dx/(2L)$.

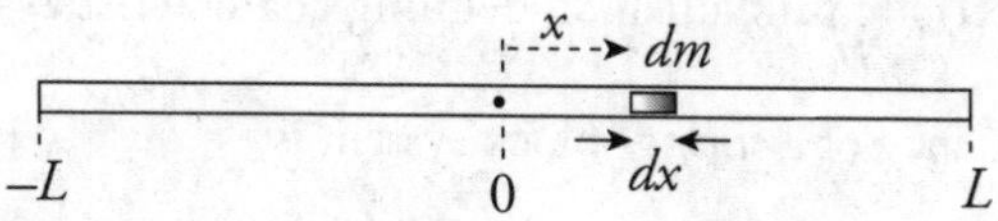

Applying the definition of I, we find

$$I = \int r^2\,dm = \int_{-L}^{L} x^2 \cdot \frac{M}{2L}\,dx = \frac{M}{2L}\left[\tfrac{1}{3}x^3\right]_{-L}^{L} = \frac{M}{2L}\left[\tfrac{2}{3}L^3\right] = \tfrac{1}{3}ML^2$$

(b) We apply Conservation of Angular Momentum to the collision of the bullet and assembly. Since the rods were at rest prior to the bullet's impact, only the bullet had angular momentum. With respect to the pivot point, its angular momentum was $Lm_b v_\perp$, where $v_\perp$ is the component of **v** that is perpendicular to the radius vector. From the definition of θ in the figure given with the question, $v_\perp = v\cos\theta$. After the collision, the angular momentum is equal to the angular velocity of the assembly times the rotational inertia of the four clay balls, the two rods, and the embedded bullet. We get

$$Lm_b v\cos\theta = \left[4mL^2 + 2\left(\tfrac{1}{3}ML^2\right) + m_b L^2\right]\omega_f$$
$$\omega_f = \frac{m_b v\cos\theta}{L\left(4m + \tfrac{2}{3}M + m_b\right)}$$

(c) Using the equation $v = R\omega$, we find that

$$v_f = L\omega_f = \frac{m_b v \cos\theta}{4m + \frac{2}{3}M + m_b}$$

(d) After the collision, the assembly has rotational kinetic energy. The ratio of this energy to the (translational) kinetic energy of the bullet before impact is

$$\frac{K'}{K} = \frac{\frac{1}{2}I\omega_f^2}{\frac{1}{2}m_b v^2} = \frac{\left(4mL^2 + \frac{2}{3}ML^2 + m_b L^2\right)\left[\dfrac{m_b v \cos\theta}{L\left(4m + \frac{2}{3}M + m_b\right)}\right]^2}{m_b v^2}$$

$$= \frac{m_b \cos^2\theta}{4m + \frac{2}{3}M + m_b}$$

CHAPTER 10 DRILL

Section I: Multiple Choice

1. **D** The acceleration of a simple harmonic oscillator is not constant, since the restoring force—and, consequently, the acceleration—depends on position. Therefore, (I) is false. However, both Statements (II) and (III) are fundamental, defining characteristics of simple harmonic motion.

2. **B** The equation for frequency of a spring–block system is

$$f_{spring} = \frac{1}{2\pi}\sqrt{(k/m)}$$

First, this means we can ignore amplitude since it has no effect on frequency. If we want the highest value for frequency, then we need to maximize k/m. Therefore, (B) is correct.

3. **B** Don't let the lack of numbers throw you. This is a Conservation of Energy problem.

$KE_{initial} + PE_{initial} = KE_{final} + PE_{final}$. At position 1 (initial), it is released from rest, so $KE_{initial} = 0$ and $PE_{initial} = mgh$. At the final position, at the equilibrium position of the pendulum, $KE_{final} = \frac{1}{2}mv^2$ and $PE_{final} = 0$. Therefore, the equation becomes

$$mgh = \frac{1}{2}mv^2$$

$$2gh = v^2$$

$$v = \sqrt{2gh}$$

4. **C** As we derived in Chapter 10, Example 2, the maximum speed of the block is given by the equation $v_{max} = A\sqrt{k/m}$. Therefore, v_{max} is inversely proportional to $\sqrt{m}$. If m is increased by a factor of 2, then v_{max} will decrease by a factor of $\sqrt{2}$.

5. **E** The spring will reach equilibrium when the restoring force and the gravitational force on the mass are equal in magnitude.

$$F_g = F_{rest}$$

$$mg = kx$$

$$x = mg/k$$

$$x = [(10 \text{ kg})(10 \text{ m/s}^2)] / (400 \text{ N/m}) = 0.25 \text{ m} = 25 \text{ cm}$$

It's important to remember the length being solved for here is how much the spring will stretch beyond its natural length, so we have to add this 25 cm to the natural length of 40 cm.

6. **E** In general, the effective spring constant of springs combined in this way is $k_{eff} = \dfrac{k_1 k_2}{k_1 + k_2}$. Since we want $k_{eff} = k_2$, that means

$$k_{eff} = k_2 = \frac{k_1 k_2}{k_1 + k_2} \Rightarrow 1 = \frac{k_1}{k_1 + k_2} \Rightarrow k_1 + k_2 = k_1 \Rightarrow k_2 = 0$$

Therefore, it is not possible to achieve this with both springs having nonzero values.

7. **A** First, we find the angular frequency:

$$\omega = 2\pi f = \sqrt{\frac{k}{m}} = \sqrt{\frac{400 \text{ N/m}}{4 \text{ kg}}} = 10 \text{ s}^{-1}$$

Therefore, the equation that gives the block's position, $x = A \sin(\omega t + \phi_0)$, becomes $x = 6 \sin(10t + \phi_0)$. Because the block is at $x = 6$ cm at time $t = 0$, the initial phase angle ϕ_0 must be $\frac{1}{2}\pi$.

8. **C** Begin by calculating the period of the spring.

$$T = 2\pi \sqrt{\frac{m}{k}} = 2\pi \sqrt{\frac{1 \text{ kg}}{\pi^2 \text{ N/m}}} = 2 \text{ s}$$

Next, calculate the time until the block returns to the tube opening. We can use one of our Big 5 equations for this:

$$\mathbf{v}_f = \mathbf{v}_0 + \mathbf{a}t \Rightarrow t = \frac{\mathbf{v}_f - \mathbf{v}_0}{\mathbf{a}} = \frac{(-10 \text{ m/s}) - (10 \text{ m/s})}{-10 \text{ m/s}^2} = 2 \text{ s}$$

Therefore, the block will return to the tube opening at the same time as the platform (having completed exactly one cycle). Thus, they will meet at position $x = 10$ cm.

9. **C** For small angular displacements, the period of a simple pendulum is essentially independent of amplitude.

10. **B** Think of the pendulum as undergoing circular motion. The centripetal force would be equal to the difference between these two forces.

$$F_c = F_T - F_G$$

$$mv^2/r = F_T - F_G$$

This shows that, as long as the pendulum is in motion (as long as $v > 0$), tension must be the larger of the two forces. However, when $v = 0$ (at the maximum displacement), the two will be equal. Therefore, (B) is the correct answer.

Section II: Free Response

1. (a) Since the spring is compressed to $\frac{3}{4}$ its natural length, the block's position relative to equilibrium is $x = -\frac{1}{4}L$. Therefore, from $F_S = -kx$, we find

$$a = \frac{F_S}{m} = \frac{-k(-\frac{1}{4}L)}{m} = \frac{kL}{4m}$$

(b) Let v_1 denote the velocity of Block 1 just before impact, and let v_1' and v_2' denote, respectively, the velocities of Block 1 and Block 2 immediately after impact. By Conservation of Linear Momentum, we write $mv_1 = mv_1' + mv_2'$, or

$$v_1 = v_1' + v_2' \qquad (1)$$

The initial kinetic energy of Block 1 is $\frac{1}{2}mv_1^2$. If half is lost to heat, then $\frac{1}{4}mv_1^2$ is left to be shared by Block 1 and Block 2 after impact: $\frac{1}{4}mv_1^2 = \frac{1}{2}mv_1'^2 + \frac{1}{2}mv_2'^2$, or

$$v_1^2 = 2v_1'^2 + 2v_2'^2 \qquad (2)$$

Square Equation (1) and multiply by 2 to give

$$2v_1^2 = 2v_1'^2 + 4v_1'v_2' + 2v_2'^2 \qquad (1')$$

then subtract Equation (2) from Equation (1'):

$$v_1^2 = 4v_1'v_2' \qquad (3)$$

Square Equation (1) again,

$$v_1^2 = v_1'^2 + 2v_1'v_2' + v_2'^2$$

and substitute into this the result of Equation (3).

$$4v_1'v_2' = v_1'^2 + 2v_1'v_2' + v_2'^2$$
$$0 = v_1'^2 - 2v_1'v_2' + v_2'^2$$
$$0 = (v_1' - v_2')^2$$
$$v_1' = v_2' \qquad (4)$$

Thus, combining Equations (1) and (4), we find that

$$v_1' = v_2' = \tfrac{1}{2}v_1$$

(c) When Block 1 reaches its new amplitude position, A', all of its kinetic energy is converted to elastic potential energy of the spring. That is,

$$K_1' \to U_S' \quad \Rightarrow \quad \tfrac{1}{2}mv_1'^2 = \tfrac{1}{2}kA'^2$$
$$A'^2 = \frac{m}{k}v_1'^2$$
$$A'^2 = \frac{m}{k}(\tfrac{1}{2}v_1)^2$$
$$A'^2 = \frac{mv_1^2}{4k} \qquad (1)$$

But the original potential energy of the spring, $U_S = \tfrac{1}{2}k(-\tfrac{1}{4}L)^2$, gave K_1 as

$$U_S \to K_1 \quad \Rightarrow \quad \tfrac{1}{2}k(-\tfrac{1}{4}L)^2 = \tfrac{1}{2}mv_1^2 \quad \Rightarrow \quad mv_1^2 = \tfrac{1}{16}kL^2 \qquad (2)$$

Substituting this result into Equation (1) gives

$$A'^2 = \frac{\tfrac{1}{16}kL^2}{4k} = \frac{L^2}{64} \quad \Rightarrow \quad A' = \tfrac{1}{8}L$$

(d) The period of a spring–block simple harmonic oscillator depends only on the spring constant k and the mass of the block. Since neither of these changes, the period will remain the same; that is, $T' = T_0$.

(e) As we showed in part (b), Block 2's velocity as it slides off the table is $\tfrac{1}{2}v_1$ (horizontally). The time required to drop the vertical distance H is found as follows (calling *down* the positive direction):

$$\Delta y = v_{0y} + \tfrac{1}{2}gt^2 \quad \Rightarrow \quad H = \tfrac{1}{2}gt^2 \quad \Rightarrow \quad t = \sqrt{\frac{2H}{g}}$$

Therefore,

$$R = (\tfrac{1}{2}v_1)t = \tfrac{1}{2}v_1\sqrt{\frac{2H}{g}}$$

Now, from Equation (2) of part (c), $v_1 = \sqrt{kL^2/16}$, so

$$R = \tfrac{1}{2}\sqrt{\frac{kL^2}{16m}}\sqrt{\frac{2H}{g}} = \frac{L}{8}\sqrt{\frac{2kH}{mg}}$$

2. (a) By Conservation of Linear Momentum,

$$mv = (m+M)v' \quad \Rightarrow \quad v' = \frac{mv}{m+M}$$

(b) When the block is at its amplitude position (maximum compression of spring), the kinetic energy it (and the embedded bullet) had just after impact will become the potential energy of the spring

$$K' \to U_S$$

$$\tfrac{1}{2}(m+M)\left(\frac{mv}{m+M}\right)^2 = \tfrac{1}{2}kA^2$$

$$A = \frac{mv}{\sqrt{k(m+M)}}$$

(c) Since the mass on the spring is $m + M$, $f = \dfrac{1}{2\pi}\sqrt{\dfrac{k}{m+M}}$.

(d) The position of the block is given by the equation $x = A\sin(\omega t + \phi_0)$, where $\omega = 2\pi f$ and A is the amplitude. Since $x = 0$ at time $t = 0$, the initial phase, ϕ_0, is 0. From the results of parts (b) and (c), we have

$$x = A\,\sin(2\pi ft) = \frac{mv}{\sqrt{k(m+M)}}\sin\left(t\sqrt{\frac{k}{m+M}}\right)$$

3. (a) By Conservation of Mechanical Energy, $K + U = E$, so

$$\tfrac{1}{2}Mv^2 + \tfrac{1}{2}k(\tfrac{1}{2}A)^2 = \tfrac{1}{2}kA^2$$

$$\tfrac{1}{2}Mv^2 = \tfrac{3}{8}kA^2$$

$$v = A\sqrt{\frac{3k}{4M}}$$

(b) Since the clay ball delivers no horizontal linear momentum to the block, horizontal linear momentum is conserved. Thus,

$$Mv = (M+m)v'$$

$$v' = \frac{Mv}{M+m} = \frac{MA}{M+m}\sqrt{\frac{3k}{4M}} = \frac{A}{M+m}\sqrt{\frac{3kM}{4}}$$

(c)   Applying the general equation for the period of a spring–block simple harmonic oscillator,

$$T = 2\pi\sqrt{\frac{M+m}{k}}$$

(d)   The total energy of the oscillator after the clay hits is $\frac{1}{2}kA'^2$, where A' is the new amplitude. Just after the clay hits the block, the total energy is

$$K' + U_S = \tfrac{1}{2}(M+m)v'^2 + \tfrac{1}{2}k(\tfrac{1}{2}A)^2$$

Substitute for v' from part (b), set the resulting sum equal to $\frac{1}{2}kA'^2$, and solve for A':

$$\tfrac{1}{2}(M+m)\left(\frac{A}{M+m}\sqrt{\frac{3kM}{4}}\right)^2 + \tfrac{1}{2}k(\tfrac{1}{2}A)^2 = \tfrac{1}{2}kA'^2$$

$$\frac{A^2 \cdot 3k\,M}{8(M+m)} + \tfrac{1}{8}kA^2 = \tfrac{1}{2}kA'^2$$

$$A^2\left(\frac{3M}{M+m}+1\right) = 4A'^2$$

$$A' = \tfrac{1}{2}A\sqrt{\frac{3M}{M+m}+1}$$

(e)   No, since the period depends only on the mass and the spring constant k.

(f)   Yes. For example, if the clay had landed when the block was at $x = A$, the speed of the block would have been zero immediately before the collision and immediately after. No change in the block's speed would have meant no change in K, so no change in E, so no change in $A = \sqrt{2E/k}$.

4. (a)   Since the weight provides a clockwise torque (which is negative), we have $\tau = -dMg \sin\theta$.

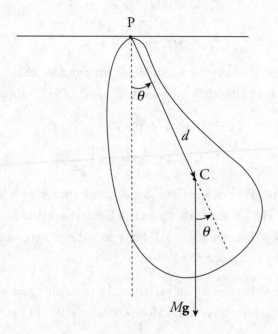

(b) If $\sin \theta \approx \theta$, then the equation in part (a) becomes

$$\tau = -dMg\theta \qquad (1)$$

(c) Since $\tau = I\alpha$, Equation (1) implies $I\alpha = -dMg\,\theta$. But by definition, $\alpha = d^2\theta/dt^2$, so we get

$$\frac{d^2\theta}{dt^2} = -\frac{dMg}{I}\theta \qquad (2)$$

Equation (2) is identical to the equation given in the statement of the question with $z = \theta$ and $b = dMg/I$. Therefore,

$$T = \frac{2\pi}{\sqrt{b}} = \frac{2\pi}{\sqrt{dMg/I}} = 2\pi\sqrt{\frac{I}{dMg}}$$

(d) Using the result of part (c) with $I = \frac{1}{3}ML^2$ and $d = \frac{1}{2}L$, we get

$$T = 2\pi\sqrt{\frac{\frac{1}{3}ML^2}{\frac{1}{2}L \cdot Mg}} = 2\pi\sqrt{\frac{2L}{3g}}$$

CHAPTER 11 DRILL

Section I: Multiple Choice

1. **E** The key word here is "altitude." An object's altitude is how high it is above the surface of the Earth, but this is not the relevant distance when dealing with gravity. We have to instead measure to the center of the Earth. So if an object has an altitude of $2r_E$, then the distance we use to find the initial force of gravity will be $3r_E$.

$$F_{g,\,0} = (GmM)/(3r_E)^2$$

$$F_{g,\,0} = (GmM)/(9r_E^{\,2})$$

To find the new force of gravity, we must double the altitude, making it $4r_E$. But we still have to add an r_E to convert from altitude, giving a new distance of $5r_E$ in the gravity formula.

$$F_{g,\,f} = (GmM)/(5r_E)^2$$

$$F_{g,\,f} = (GmM)/(25r_E^{\,2}) = \left(\frac{9}{25}\right)F_{g,\,0}$$

2. **E** Mass is an intrinsic property of an object that does not change with location. This eliminates (A) and (C). If an object's height above the surface of Earth is equal to $2R_E$, then its distance from the center of Earth is $3R_E$. Thus, the object's distance from Earth's center increases by a factor of 3, so its weight decreases by a factor of $3^2 = 9$.

3. **C** The gravitational force that the Moon exerts on the planet is equal in magnitude to the gravitational force that the planet exerts on the Moon (Newton's Third Law).

4. **C** The force of gravity we feel is inversely proportional to the square of the planet's radius. If gravity is multiplied by 4, then the radius will be multiplied by $\frac{1}{\sqrt{4}} = \frac{1}{2}$.

5. **E** Above the surface of a planet, the force of gravity will be $F_g = \frac{GMm}{r^2}$ where r is the distance from the center of the planet. Inside the surface of a planet, the force of gravity is $F_g = \frac{GMmr}{R^3}$ where r is again the distance from the center of the planet and R is the planet's full radius (see page 301 for a full explanation of these equations). Therefore, we must solve

$$\frac{GMm}{r^2} = \frac{GMm\left(\frac{1}{2}R_E\right)}{R_E^3} \Rightarrow r = \sqrt{2}R_E$$

However, this gives us the distance from the center of the planet. The question asks for the distance above the surface. Therefore, the correct answer is $\left(\sqrt{2} - 1\right)R_E$.

6. **E** The gravitational pull by Jupiter provides the centripetal force on its moon:

$$G\frac{Mm}{R^2} = \frac{mv^2}{R}$$

$$G\frac{M}{R} = v^2$$

$$= \left(\frac{2\pi R}{T}\right)^2$$

$$= \frac{4\pi^2 R^2}{T^2}$$

$$M = \frac{4\pi^2 R^3}{GT^2}$$

7. **E** Let the object's distance from Body A be x; then its distance from Body B is $R - x$. In order for the object to feel no net gravitational force, the gravitational pull by A must balance the gravitational pull by B. Therefore, if we let M denote the mass of the object, then

$$G\frac{m_A M}{x^2} = G\frac{m_B M}{(R-x)^2}$$

$$\frac{m}{x^2} = \frac{4m}{(R-x)^2}$$

$$\frac{(R-x)^2}{x^2} = \frac{4m}{m}$$

$$\left(\frac{R-x}{x}\right)^2 = 4$$

$$\left(\frac{R}{x} - 1\right)^2 = 4$$

$$\frac{R}{x} - 1 = 2$$

$$x = R/3$$

8. **B** Kepler's Third Law says that $T^2 \propto R^3$ for a planet with a circular orbit of radius R. Since $T \propto R^{3/2}$, if R increases by a factor of 9, then T increases by a factor of $9^{3/2} = (3^2)^{3/2} = 3^3 = 27$.

9. **D** The escape speed of a given planet of mass M and radius r is given by the equation $v = \sqrt{(2GM)/r}$. If the density remains constant, doubling the radius will result in the mass being multiplied by 8. Therefore,

$$v_{esc} = \sqrt{\frac{2G(8M)}{2r}} = 2\sqrt{\frac{2GM}{r}} = 2v$$

10. **E** The force of gravity is the centripetal force, so we can solve for the velocity of each satellite and compare them.

$$F_g = F_C$$
$$\frac{Gm_A m_p}{R_A{}^2} = \frac{m_A v_A{}^2}{R_A}$$
$$v_A = \sqrt{\frac{Gm_p}{R_A}} \ \& \ v_B = \sqrt{\frac{Gm_p}{R_B}}$$

Now since $R_B = 3R_A$,

$$v_A = \sqrt{\frac{Gm_p}{R_A}} \ \& \ v_B = \sqrt{\frac{Gm_p}{3R_A}} = \sqrt{\frac{1}{3}} \left(\sqrt{\frac{Gm_p}{R_A}} \right)$$
$$v_B = \sqrt{\frac{1}{3}} v_A = \frac{v_A}{\sqrt{3}}$$

Section II: Free Response

1. (a) Combining Newton's Second Law with the Law of Gravitation, we find

$$a_1 = \frac{F_{2\text{-on-}1}}{m_1} = \frac{G\frac{m_1 m_2}{(\frac{1}{2}R)^2}}{m_1} = \frac{4Gm_2}{R^2}$$

The direction of $\mathbf{a}_1$ is toward Sphere 2.

(b) Combining Newton's Second Law with the Law of Gravitation, we find

$$a_2 = \frac{F_{1\text{-on-}2}}{m_2} = \frac{G\frac{m_1 m_2}{(\frac{1}{2}R)^2}}{m_2} = \frac{4Gm_1}{R^2}$$

The direction of $\mathbf{a}_2$ is toward Sphere 1.

(c)   Since the two spheres start from rest, the total linear momentum of the system is clearly zero. Since no external forces act (we're in "deep space"), the total momentum must remain zero during their motion toward each other. Therefore, the magnitude of Sphere 1's linear momentum, $m_1 v_1$, must always equal the magnitude of Sphere 2's linear momentum, $m_2 v_2$. Thus, $v_2 = (m_1/m_2)v_1$. With this result, we can apply Conservation of Mechanical Energy:

$$K_i + U_i = K_f + U_f$$

$$0 - \frac{Gm_1 m_2}{R} = \tfrac{1}{2} m_1 v_1^2 + \tfrac{1}{2} m_2 v_2^2 - \frac{Gm_1 m_2}{R/2}$$

$$\frac{Gm_1 m_2}{R} = \tfrac{1}{2} m_1 v_1^2 + \tfrac{1}{2} m_2 \left(\tfrac{m_1}{m_2} v_1\right)^2$$

$$\frac{Gm_2}{R} = \tfrac{1}{2} v_1^2 \left(1 + \tfrac{m_1}{m_2}\right)$$

$$v_1 = \sqrt{\frac{2Gm_2}{R\left(1 + \tfrac{m_1}{m_2}\right)}}$$

$$= m_2 \sqrt{\frac{2G}{R(m_1 + m_2)}}$$

(d)   Using the result of part (c) and the relationship $v_2 = (m_1/m_2)v_1$, we get

$$v_2 = \tfrac{m_1}{m_2} v_1 = \tfrac{m_1}{m_2} \cdot m_2 \sqrt{\frac{2G}{R(m_1 + m_2)}} = m_1 \sqrt{\frac{2G}{R(m_1 + m_2)}}$$

(e)   The centripetal force on each sphere is provided by the gravitational pull by the other sphere.

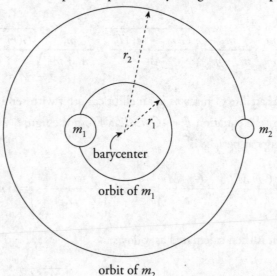

Therefore,

$$\frac{m_1 v_1^2}{r_1} = \frac{G m_1 m_2}{(r_1 + r_2)^2} = \frac{m_2 v_2^2}{r_2}$$

Since both spheres have the same orbit period, T, we get

$$\frac{m_1 v_1^2}{r_1} = \frac{m_2 v_2^2}{r_2} \quad \Rightarrow \quad \frac{m_1}{r_1}\left(\frac{2\pi r_1}{T}\right)^2 = \frac{m_2}{r_2}\left(\frac{2\pi r_2}{T}\right)^2 \quad \Rightarrow \quad m_1 r_1 = m_2 r_2 \qquad (1)$$

We can also derive (as we did in Chapter 11, Example 9) that

$$\frac{T^2}{(R_1 + R_2)^3} = \frac{4\pi^2}{G(m_1 + m_2)} \quad \Rightarrow \quad R_1 + R_2 = \sqrt[3]{\frac{G(m_1 + m_2)T^2}{4\pi^2}} \qquad (2)$$

Substituting the result of Equation (1), $r_2 = (m_1/m_2)r_1$, into Equation (2), we can solve for r_1:

$$r_1 + r_2 = \sqrt[3]{\frac{G(m_1 + m_2)T^2}{4\pi^2}}$$

$$r_1 + \frac{m_1}{m_2}r_1 =$$

$$r_1\left(1 + \frac{m_1}{m_2}\right) =$$

$$r_1 = \frac{m_2}{m_1 + m_2}\sqrt[3]{\frac{G(m_1 + m_2)T^2}{4\pi^2}}$$

and then for r_2:

$$r_2 = \frac{m_1}{m_2}r_1 = \frac{m_1}{m_2}\cdot\frac{m_2}{m_1 + m_2}\sqrt[3]{\frac{G(m_1 + m_2)T^2}{4\pi^2}} = \frac{m_1}{m_1 + m_2}\sqrt[3]{\frac{G(m_1 + m_2)T^2}{4\pi^2}}$$

2. (a) The total energy of a satellite of mass m in an elliptical orbit with semimajor axis a around a planet of mass M is given by the equation $E = -GmM/2a$. From the figure, we see that $2a = r_1 + r_2$, so the satellite's kinetic energy at perigee is

$$K_1 = E - U_1 = -\frac{GmM}{r_1 + r_2} - \left(-\frac{GmM}{r_1}\right) = GmM\left(\frac{1}{r_1} - \frac{1}{r_1 + r_2}\right) = GmM\frac{r_2}{r_1(r_1 + r_2)}$$

Its speed at this point is then calculated as follows:

$$\tfrac{1}{2}mv_1^2 = GmM\frac{r_2}{r_1(r_1 + r_2)} \quad \Rightarrow \quad v_1 = \sqrt{\frac{2GMr_2}{r_1(r_1 + r_2)}}$$

(b) Employing the same method as used in part (a), we find

$$K_2 = E - U_2 = -\frac{GmM}{r_1 + r_2} - \left(-\frac{GmM}{r_2}\right) = GmM\left(\frac{1}{r_2} - \frac{1}{r_1 + r_2}\right) = GmM\frac{r_1}{r_2(r_1 + r_2)}$$

so,

$$\tfrac{1}{2}mv_2^2 = GmM\frac{r_1}{r_2(r_1+r_2)} \implies v_2 = \sqrt{\frac{2GMr_1}{r_2(r_1+r_2)}}$$

(c) The ratio of v_1 to v_2 is

$$\frac{v_1}{v_2} = \frac{\sqrt{\dfrac{2GMr_2}{r_1(r_1+r_2)}}}{\sqrt{\dfrac{2GMr_1}{r_2(r_1+r_2)}}} = \frac{\sqrt{\dfrac{r_2}{r_1}}}{\sqrt{\dfrac{r_1}{r_2}}} = \sqrt{\frac{r_2}{r_1}} \cdot \sqrt{\frac{r_2}{r_1}} = \frac{r_2}{r_1}$$

(d) Since $\mathbf{v}_2$ is perpendicular to $\mathbf{r}_2$, the satellite's angular momentum is

$$L_2 = r_2 m v_2 = r_2 m\sqrt{\frac{2GMr_1}{r_2(r_1+r_2)}} = m\sqrt{\frac{2GMr_1r_2}{r_1+r_2}}$$

(e) The distance from X to the center of Earth is equal to the semimajor axis, $a = \tfrac{1}{2}(r_1 + r_2)$. (This is because the sum of the distances from *any* point on the ellipse to the two foci must equal $2a = r_1 + r_2$, and X is equidistant from the two foci, one of which is at the center of Earth.) Therefore, the kinetic energy of the satellite when it's at X is

$$K_X = E - U_X = -\frac{GmM}{r_1 + r_2} - \left[-\frac{GmM}{\tfrac{1}{2}(r_1+r_2)}\right] = \frac{GmM}{r_1+r_2}$$

so its speed at this point can be determined:

$$\tfrac{1}{2}mv_X^2 = \frac{GmM}{r_1+r_2} \implies v_X = \sqrt{\frac{2GM}{r_1+r_2}}$$

(f) [See Example 8(c)] Use Kepler's Third Law for elliptical orbits:

$$\frac{T^2}{a^3} = \frac{4\pi^2}{GM} \implies T = \sqrt{\frac{4\pi^2 a^3}{GM}} = \sqrt{\frac{4\pi^2[\tfrac{1}{2}(r_1+r_2)]^3}{GM}} = \sqrt{\frac{\pi^2(r_1+r_2)^3}{2GM}}$$

ELECTRIC FIELD DRILL

1. $\vec{F} = q\vec{E}$

    $F = (1.6 \times 10^{-19}\,\text{C})(5.0\,\text{N/C}) = 8.0 \times 10^{-19}\,\text{N}$ in the direction of the electric field (up)

    $F = ma$, so $a = F/m = (8.0 \times 10^{-19}\,\text{N})/(1.7 \times 10^{-27}\,\text{kg}) = 4.7 \times 10^{8}\,\text{m/s}^2$

2. $\vec{F} = q\vec{E} = (1.6 \times 10^{-19}\,\text{C})(5.0\,\text{N/C}) = 8.0 \times 10^{-19}\,\text{N}$ in the direction of the electric field (up)

    $a = F/m = (8.0 \times 10^{-19}\,\text{N})/(9.1 \times 10^{-31}\,\text{kg}) = 8.8 \times 10^{11}\,\text{m/s}^2$

3. $\vec{F} = q\vec{E} = (-1.6 \times 10^{-19}\,\text{C})(5.0\,\text{N/C}) = -8.0 \times 10^{-19}\,\text{N}$ in the direction opposite of the electric field (down)

    $a = F/m = (8.0 \times 10^{-19}\,\text{N})/(9.1 \times 10^{-31}\,\text{kg}) = 8.8 \times 10^{11}\,\text{m/s}^2$

4. Remember, velocity does not affect force exerted by an electric field.

    $F = (1.6 \times 10^{-19}\,\text{C})(5.0\,\text{N/C}) = 8.0 \times 10^{-19}\,\text{N}$ in the direction of the electric field (up)

    $a = F/m = (8.0 \times 10^{-19}\,\text{N})/(1.7 \times 10^{-27}\,\text{kg}) = 4.7 \times 10^{8}\,\text{m/s}^2$

5. Remember, velocity does not affect force exerted by an electric field.

6. $\vec{F} = q\vec{E} = (-1.6 \times 10^{-19}\,\text{C})(5.0\,\text{N/C}) = -8.0 \times 10^{-19}\,\text{N}$ in the direction opposite of the electric field (down)

    $a = F/m = (8.0 \times 10^{-19}\,\text{N})/(9.1 \times 10^{-31}\,\text{kg}) = 8.8 \times 10^{11}\,\text{m/s}^2$

7. Remember, velocity does not affect force exerted by an electric field.

    $\vec{F} = q\vec{E} = (1.6 \times 10^{-19}\,\text{C})(5.0\,\text{N/C}) = 8.0 \times 10^{-19}\,\text{N}$ in the direction of the electric field (up)

    $a = F/m = (8.0 \times 10^{-19}\,\text{N})/(9.1 \times 10^{-31}\,\text{kg}) = 8.8 \times 10^{11}\,\text{m/s}^2$

8. Remember, velocity does not affect force exerted by an electric field.

    $\vec{F} = q\vec{E} = (-1.6 \times 10^{-19}\,\text{C})(5.0\,\text{N/C}) = -8.0 \times 10^{-19}\,\text{N}$ in the direction opposite of the electric field (down)

    $a = F/m = (8.0 \times 10^{-19}\,\text{N})/(9.1 \times 10^{-31}\,\text{kg}) = 8.8 \times 10^{11}\,\text{m/s}^2$

CHAPTER 12 DRILL

Section I: Multiple Choice

1. **A** The magnitude of electric force will be

$$F_e = \left(\frac{1}{4\pi\varepsilon_0}\right)\left(\frac{q_1 q_2}{r^2}\right)$$

Multiplying each charge by 3 will result in the force being multiplied by 9. In contrast, doubling the distance between the charges will reduce the force by a factor of 4. Therefore, the force will be multiplied by $\frac{9}{4}$.

2. **C** First, take note of the particle's negative charge. This means it will move in the opposite direction of the electric field, thus escaping the field after traveling only 30 cm. Next, you can treat this as a kinematics problem since the constant electric field will result in a constant acceleration. If we define right as the positive direction, we get

$$F = ma = qE \rightarrow a = \frac{qE}{m}$$

$$d = v_0 t + \frac{1}{2}at^2 = 0 + \frac{1}{2}\left(\frac{qE}{m}\right)t^2 \rightarrow t = \sqrt{\frac{2md}{qE}} = \sqrt{\frac{2\left(10^{-10}\,\text{kg}\right)(0.3\,\text{m})}{\left(-5\times10^{-13}\,\text{C}\right)\left(-50\,\text{N/C}\right)}} = 1.55\,\text{s}$$

3. **D** First, use the fact that electric field lines point away from positive particles and toward negative particles. The original field, E, would point to the right (because the particle at A is positive). To amplify that field, the particle at B would have to be negative. This eliminates (A) and (B).

Next, use superposition to determine the strength of E that the new particle must create.

$$E_{net} = E_1 + E_2 + \dots$$

$$E_P = E_A + E_B$$

$$E_B = E_P - E_A$$

$$E_B = 3E - E = 2E$$

Finally, because electric field strength is proportional to the magnitude of the charge providing it, the charge at B must have double the magnitude of the charge at A.

4. **D** The acceleration of the small sphere is

$$a = \frac{F_E}{m} = \frac{1}{4\pi\varepsilon_0}\frac{Qq}{mr^2}$$

As r increases (that is, as the small sphere is pushed away), a decreases. However, since a is always positive, the small sphere's speed, v, is always increasing.

5. **A** At the beginning of the problem, there is no movement, so there is no kinetic energy. Once the object is released, the force of gravity and the electric force begin acting on the object in opposite directions (gravity pulling down, electric pushing up). At first, the gravitational force is stronger, so the object begins moving down. However, as the object gets closer to the particle, the electric force grows stronger until it eventually overcomes the gravitational force, but the object will continue falling at this point as only the acceleration (not the velocity) has become 0. Eventually, the electric force will force the object to a stop and begin repelling it. At the moment when it stops, the kinetic energy will again be 0, meaning it's the same as the original kinetic energy. Therefore, this will be the point at which 0 net work has been done, meaning the amount of work done by the electric force and the force of gravity will be equal (but in opposite directions). If we define the falling distance as x, then

$$W_E = W_g$$

$$\int_0^x F_E \, dx = \int_0^x F_g \, dx \rightarrow \int_0^x \frac{kq^2}{(1-x)^2} \, dx = \int_0^x mg \, dx$$

Evaluating the first integral, we find that

$$kq^2 \int_0^x \frac{1}{(1-x)^2} \, dx = kq^2 \left. \frac{1}{(1-x)} \right|_0^x = \frac{kq^2 x}{1-x}$$

Evaluating the gravity integral, we find that

$$mg \int_0^x dx = mgx$$

Therefore,

$$\frac{kq^2 x}{1-x} = mgx \Rightarrow x = 1 - \frac{kq^2}{mg} = 1 - \frac{\left(9 \cdot 10^9 \, \frac{\text{Nm}^2}{\text{C}^2}\right)\left(10^{-6} \, \text{C}\right)^2}{\left(10^{-3} \, \text{kg}\right)\left(10 \, \text{m/s}^2\right)} = 0.1 \, \text{m}$$

6. **B** The net electric flux through the Gaussian surface, Φ_E, is equal to $(1/\varepsilon_0)$ times the net charge *enclosed by the surface*, $Q_2 + Q_3$. However, the net electric field vector at P is equal to the sum of the electric field vectors due to each of the four charges individually.

7. **A** Since the charge is distributed uniformly throughout the sphere, the magnitude of its volume charge density is $\rho = Q/(\frac{4}{3}\pi R^3)$. Therefore, if we construct a spherical Gaussian surface of radius $r < R$, Gauss's law gives

$$\Phi_E = \frac{Q_{enclosed}}{\varepsilon_0} \quad \Rightarrow \quad E(4\pi r^2) = \frac{\rho \cdot \frac{4}{3}\pi r^3}{\varepsilon_0} \quad \Rightarrow \quad E = \rho\frac{r}{3\varepsilon_0}$$

$$E = \frac{Q}{\frac{4}{3}\pi R^3}\frac{r}{3\varepsilon_0}$$

$$E = \frac{1}{4\pi\varepsilon_0}\frac{Q}{R^3}r$$

Since the charge on the sphere is negative, the direction of **E** at any point is directed radially inward.

8. **A** By definition, an electric dipole consists of two equal but opposite charges. Therefore, a Gaussian surface enclosing both charges would enclose zero net charge. By Gauss's law, the electric flux is zero as well (since $\Phi_E = Q_{enclosed}/\varepsilon_0$).

9. **C** Points A and E are on the same equipotential line, so no work is done by the field to move the electron from A to E. Electric fields go in the direction of decreasing potential, so the field associated with these equipotential lines is generally going to the left. The field lines indicate the force on a positive test charge, so the force on an electron will be in the opposite direction. Therefore, the work done by the field on the electron is negative for the movement from E to C because the force on the electron is to the right and the movement is to the left.

10. **B** By definition, $W_E = -q\Delta V$, which gives

$$W_E = -q(V_B - V_A) = -(-0.05 \text{ C})(100 \text{ V} - 200 \text{ V}) = -5 \text{ J}$$

Note that neither the length of the segment AB nor that of the curved path from A to B is relevant.

11. **E** Anywhere along either line will be a location of 0 net electric potential. At any location on the lines, the two charges closest to that point will cancel each other's electric potential, and the two charges farthest from that point will cancel each other's electric potential.

12. **A** First, find the change in potential energy.

$$\Delta U = q\,\Delta V$$

$$= q(V_B - V_A)$$

Next, we know that the change in potential energy will be the negative change in kinetic energy.

$$\Delta K = -\Delta U$$

$$= q(V_A - V_B)$$

Because the particle started at rest ($K = 0$), the change in kinetic energy will be equal to its kinetic energy at position B.

$$\Delta K = K_B - K_A$$

$$= K_B - 0 = K_B = q(V_A - V_B)$$

From there, just use the formula for kinetic energy to solve for V.

$$K_B = q(V_A - V_B)$$

$$\left(\frac{1}{2}\right)mv_B^2 = q(V_A - V_B)$$

$$v_B = \sqrt{\frac{2q(V_A - V_B)}{m}}$$

Section II: Free Response

1. (a)   From the figure below, we have $F_{1\text{-}2} = F_1/\cos 45°$.

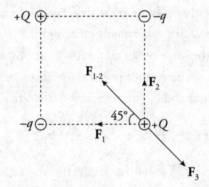

Since the net force on $+Q$ is zero, we want $F_{1\text{-}2} = F_3$. If s is the length of each side of the square, then:

$$F_{1\text{-}2} = F_3 \quad \Rightarrow \quad \frac{F_1}{\cos 45°} = F_3 \quad \Rightarrow \quad \frac{1}{\cos 45°} \frac{1}{4\pi\varepsilon_0} \frac{Qq}{s^2} = \frac{1}{4\pi\varepsilon_0} \frac{Q^2}{(s\sqrt{2})^2}$$

$$\sqrt{2} \cdot q = \frac{Q}{2}$$

$$q = \frac{Q}{2\sqrt{2}}$$

(b)   No. If $q = Q/2\sqrt{2}$, as found in part (a), then the net force on $-q$ is not zero.

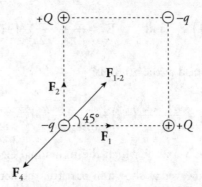

This is because $F_{1\text{-}2} \neq F_4$, as the following calculations show:

$$F_{1\text{-}2} = \frac{F_1}{\cos 45°} = \sqrt{2}\,\frac{1}{4\pi\varepsilon_0}\frac{Qq}{s^2} = \sqrt{2}\,\frac{1}{4\pi\varepsilon_0}\frac{Q\frac{Q}{2\sqrt{2}}}{s^2} = \frac{1}{8\pi\varepsilon_0}\frac{Q^2}{s^2}$$

but

$$F_4 = \frac{1}{4\pi\varepsilon_0}\frac{q^2}{(s\sqrt{2})^2} = \frac{1}{4\pi\varepsilon_0}\frac{\left(\frac{Q}{2\sqrt{2}}\right)^2}{(s\sqrt{2})^2} = \frac{1}{64\pi\varepsilon_0}\frac{Q^2}{s^2}$$

(c)   By symmetry, $E_1 = E_2$ and $E_3 = E_4$, so the net electric field at the center of the square is zero:

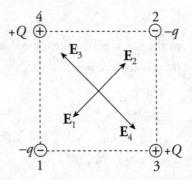

2.   (a)   The magnitude of the electric force on Charge 1 is

$$F_1 = \frac{1}{4\pi\varepsilon_0}\frac{(Q)(2Q)}{(a+2a)^2} = \frac{1}{18\pi\varepsilon_0}\frac{Q^2}{a^2}$$

The direction of $\mathbf{F}_1$ is directly away from Charge 2; that is, in the $+y$ direction, so

$$\mathbf{F}_1 = \frac{1}{18\pi\varepsilon_0}\frac{Q^2}{a^2}\hat{\mathbf{j}}$$

(b) The electric field vectors at the origin due to Charge 1 and due to Charge 2 are

$$\mathbf{E}_1 = \frac{1}{4\pi\varepsilon_0}\frac{Q}{a^2}\left(-\hat{\mathbf{j}}\right) \quad \text{and} \quad \mathbf{E}_2 = \frac{1}{4\pi\varepsilon_0}\frac{2Q}{(2a)^2}\left(+\hat{\mathbf{j}}\right) = \frac{1}{8\pi\varepsilon_0}\frac{Q}{a^2}\left(+\hat{\mathbf{j}}\right)$$

Therefore, the net electric field at the origin is

$$\mathbf{E} = \mathbf{E}_1 + \mathbf{E}_2 = \frac{1}{4\pi\varepsilon_0}\frac{Q}{a^2}\left(-\hat{\mathbf{j}}\right) + \frac{1}{8\pi\varepsilon_0}\frac{Q}{a^2}\left(+\hat{\mathbf{j}}\right) = \frac{1}{8\pi\varepsilon_0}\frac{Q}{a^2}\left(-\hat{\mathbf{j}}\right)$$

(c) No. The only point on the x-axis at which the individual electric field vectors point in opposite directions is the origin. However, as shown in part (b), the net electric field here is not zero. At all other points, the net electric field cannot be zero because the vectors are not in opposite directions.

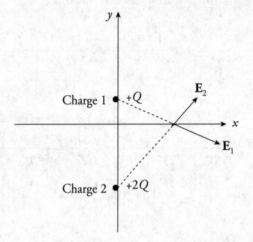

(d) Yes. There will be a point P on the y-axis between the two charges, where the electric fields due to the individual charges will cancel each other out.

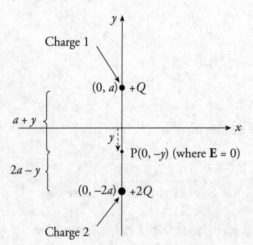

$$E_1 = E_2$$

$$\frac{1}{4\pi\varepsilon_0}\frac{Q}{(a+y)^2} = \frac{1}{4\pi\varepsilon_0}\frac{2Q}{(2a-y)^2}$$

$$\frac{1}{(a+y)^2} = \frac{2}{(2a-y)^2}$$

$$(2a-y)^2 = 2(a+y)^2$$

$$y^2 + 8ay - 2a^2 = 0$$

$$y = \frac{-8a \pm \sqrt{(8a)^2 - 4(-2a^2)}}{2}$$

$$= (-4 \pm 3\sqrt{2})a$$

Disregarding the value $y = (-4 - 3\sqrt{2})a$ (because it would place the point P below Charge 2 on the y-axis, where the electric field vectors do not point in opposite directions), we have that $\mathbf{E} = 0$ at the point P $= (0, -y) = (0, (4 - 3\sqrt{2})a)$.

(e)   Use the result of part (b) with Newton's Second Law:

$$\mathbf{a} = \frac{\mathbf{F}}{m} = \frac{-q\mathbf{E}}{m} = \frac{-q}{m}\left(\frac{1}{8\pi\varepsilon_0}\frac{Q}{a^2}(-\hat{\mathbf{j}})\right) = \frac{1}{8\pi\varepsilon_0}\frac{qQ}{ma^2}(+\hat{\mathbf{j}})$$

3.   (a)   Very close to the rod, the rod behaves as if it were "infinitely long." Construct a cylindrical Gaussian surface of radius x_1 and length L centered on the rod. By symmetry, the electric field at P_1 must point radially away from the rod, so the field at P_1 must point in the positive x direction.

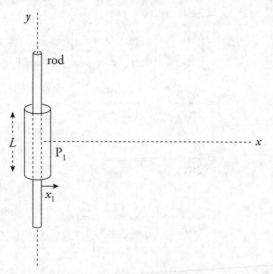

There is no electric flux through the ends of the cylinder (since $\mathbf{E}$ is parallel to the lids of the cylinder); the electric flux is perpendicular to the lateral surface area of the cylinder, so $\Phi_E = 0 + EA + 0 = E(2\pi x_1 \ell)$. Since the charge enclosed by the surface is $\lambda \ell$, Gauss's law gives

$$E(2\pi x_1 \ell) = \frac{\lambda \ell}{\varepsilon_0} \quad \Rightarrow \quad E = \frac{\lambda}{2\pi\varepsilon_0 x_1}$$

(b) The total charge on the rod is equal to its linear charge density times its total length: $Q = \lambda \ell$.

(c) If x_2 is not small compared to ℓ, then we must calculate the electric field at P_2 by integration. Consider two symmetrically located line elements along the rod, each of length dy, at a distance y above and below the origin.

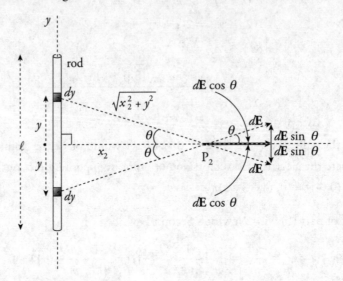

The vertical components of the electric fields at P_2 will cancel (by symmetry), leaving the net electric field at P_2 due to this pair of line elements equal to the sum of the horizontal components. Therefore,

$$dE_{total} = 2\,dE\cos\theta$$

$$E_{total} = \int_{y=0}^{y=\ell/2} 2\left(\frac{1}{4\pi\varepsilon_0}\frac{\lambda\,dy}{x_2^2+y^2}\right)\frac{x_2}{\sqrt{x_2^2+y^2}}$$

$$= \frac{\lambda x_2}{2\pi\varepsilon_0}\int_{y=0}^{y=\ell/2}\frac{dy}{(x_2^2+y^2)^{3/2}}$$

$$= \frac{\lambda x_2}{2\pi\varepsilon_0}\left(\frac{y}{x_2^2\sqrt{x_2^2+y^2}}\right)_{y=0}^{y=\ell/2}$$

$$= \frac{\lambda}{2\pi\varepsilon_0 x_2}\left(\frac{\ell/2}{\sqrt{x_2^2+(\ell/2)^2}}\right)$$

If $x_2 = \ell$, then

$$E_{total} = \frac{\lambda}{2\pi\varepsilon_0\ell}\left[\frac{\ell/2}{\sqrt{\ell^2+(\ell/2)^2}}\right] = \frac{\lambda}{4\pi\varepsilon_0}\frac{1}{\sqrt{5\ell^2/4}} = \frac{\lambda}{2\pi\varepsilon_0\ell\sqrt{5}} = \frac{Q/\ell}{2\pi\varepsilon_0\ell\sqrt{5}} = \frac{Q}{2\pi\varepsilon_0\ell^2\sqrt{5}}$$

4. (a) Volume charge density has units of charge per unit volume, that is, $[\rho] = C/m^3$. Since r/a has no units, ρ_s must also have units of C/m^3.

(b) Since ρ varies with the distance from the center, consider a thin spherical shell of radius R and thickness dR. Its volume is $dV = 4\pi R^2\, dR$, so its charge is

$$dQ = \rho\, dV = \rho_s \frac{R}{a} \cdot 4\pi R^2\, dR = \frac{4\pi \rho_s}{a} R^3\, dR$$

The charge on the entire sphere is found by integrating dQ from $R = 0$ to $R = a$:

$$Q = \int dQ = \int_{R=0}^{R=a} \frac{4\pi \rho_s}{a} R^3\, dR \qquad (1)$$

$$= \frac{4\pi \rho_s}{a} \frac{a^4}{4}$$

$$= \pi \rho_s a^3$$

(c) (i) Replace "$R = a$" with "$R = r$" in Equation (1) above to find the charge enclosed by a spherical Gaussian surface of radius $r < a$:

$$Q_{\text{within } r} = \int_{R=0}^{R=r} \frac{4\pi \rho_s}{a} R^3\, dR = \frac{4\pi \rho_s}{a} \frac{r^4}{4}$$

Using the result of part (b), we can write this expression for $Q_{\text{within } r}$ in terms of Q, the total charge on the full sphere:

$$Q_{\text{within } r} = \frac{4\pi \rho_s}{a} \frac{r^4}{4} = \frac{4\frac{Q}{a^3}}{a} \frac{r^4}{4} = \frac{Q}{a^4} r^4$$

So, by Gauss's law,

$$\Phi_E = \frac{1}{\varepsilon_0} \cdot Q_{\text{enclosed}}$$

$$E(4\pi r^2) = \frac{1}{\varepsilon_0} \cdot \frac{Q}{a^4} r^4$$

$$E = \frac{1}{4\pi \varepsilon_0} \frac{Q}{a^4} r^2 \quad (r < a)$$

(ii) Outside the sphere, $Q_{\text{enclosed}} = Q$, so

$$E = \frac{1}{4\pi \varepsilon_0} \frac{Q}{r^2} \quad (r \geq a)$$

(d)

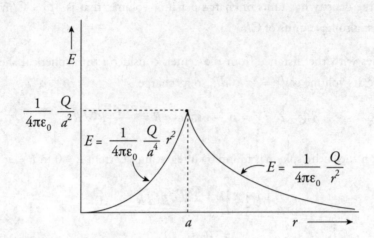

CHAPTER 13 DRILL

Section I: Multiple Choice

1. **E** If a dielectric is added, the electric field will be reduced (due to the electric field of the dielectric opposing the field of the capacitor's plates). Because $\Delta V = Ed$, voltage will decrease. If the battery were still attached, it would send more charge to restore the voltage, but that isn't the case here. The potential energy of a capacitor is given by the equation $U_{cap} = \dfrac{1}{2} Q\Delta V$. Charge hasn't changed, and voltage has decreased, which means potential energy will decrease as well. Finally, capacitance is a measure of $Q/\Delta V$. This quantity will increase, so the capacitance will be higher.

2. **C** By definition, $\Delta U_E = -W_E$, so if W_E is negative, then ΔU_E is positive. This implies that the potential energy, U_E, increases.

3. **D** You may be tempted to solve for the equivalent capacitance and then use that to determine the total charge stored on the capacitors, but we can already tell the potential difference across each capacitor is 9 V. The energy stored in a capacitor is given by the equation

$$U_c = \frac{1}{2}CV^2 = \frac{1}{2}(6 \times 10^{-6})(9)^2$$
$$U_c = 2.43 \times 10^{-4} \text{ J}$$

4. **C** Because **E** is uniform, the potential varies linearly with distance from either plate ($\Delta V = Ed$). Since points 2 and 4 are at the same distance from the plates, they lie on the same equipotential. (The equipotentials in this case are planes parallel to the capacitor plates.)

5. **D** Once the spheres are connected by a conducting wire, they quickly form a single equipotential surface. Since the potential on a sphere of radius r carrying charge q is given by the equation $(1/4\pi\varepsilon_0)(q/r)$, we have

$$V_{\text{Sphere \#1}} = V_{\text{Sphere \#2}} \implies \frac{1}{4\pi\varepsilon_0}\frac{q_1}{a} = \frac{1}{4\pi\varepsilon_0}\frac{q_2}{4a} \implies q_1 = \frac{q_2}{4}$$

Now, since $q_1 + q_2 = -Q + -Q = -2Q$, the fact that q_1 must equal $q_2/4$ means that $(q_2/4) + q_2 = -2Q$, which gives $q_2 = (-8/5)Q$ and $q_1 = (-2/5)Q$.

6. **A** Since Q cannot change and C is increased (because of the dielectric), $\Delta V = Q/C$ must decrease. Also, since $U_E = Q^2/(2C)$, an increase in C with no change in Q implies a decrease in U_E.

7. **D**

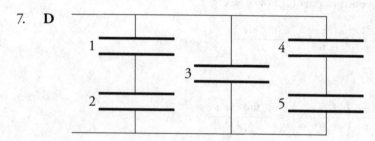

Capacitors 1 and 2 are in series, so their equivalent capacitance is $C_{1\text{-}2} = C/2$. (This is obtained from the equation $1/C_{1\text{-}2} = 1/C_1 + 1/C_2 = 1/C + 1/C = 2/C$.) Capacitors 4 and 5 are also in series, so their equivalent capacitance is $C_{4\text{-}5} = C/2$. The capacitances $C_{1\text{-}2}$, C_3, and $C_{4\text{-}5}$ are in parallel, so the overall equivalent capacitance is $(C/2) + C + (C/2) = 2C$.

Section II: Free Response

1. (a) The capacitance is $C = \varepsilon_0 A/d$. Since the plates are rectangular, the area A is equal to Lw, so $C = \varepsilon_0 Lw/d$.

   (b) Since the electron is attracted upward, the top plate must be the positive plate.

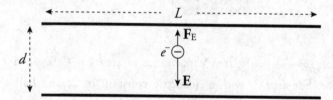

   (c) See explanation and illustration for part (b) above.

   (d) The acceleration of the electron is $a = F_E/m = qE/m = eE/m$, vertically upward. Therefore, applying Big Five #3 for vertical motion, $\Delta y = v_{0y}t + \frac{1}{2}a_y t^2$, we get

$$\Delta y = \frac{1}{2}\frac{eE}{m}t^2 \quad (1)$$

To find t, notice that $v_{0x} = v_0$ remains constant (because there is no horizontal acceleration). Therefore, the time necessary for the electron to travel the horizontal distance L is $t = L/v_0$. In this time, Δy is $d/2$, so Equation (1) becomes

$$\frac{d}{2} = \frac{1}{2}\frac{eE}{m}\left(\frac{L}{v_0}\right)^2 \quad \Rightarrow \quad E = \frac{dmv_0^2}{eL^2}$$

(e) Substituting the result of part (d) into the equation $\Delta V = Ed$ gives

$$\Delta V = \frac{d^2 m v_0^2}{eL^2}$$

Since $Q = C\Delta V$ (by definition), the result of part (a) now gives

$$Q = \frac{\varepsilon_0 L w}{d} \cdot \frac{d^2 m v_0^2}{eL^2} = \frac{\varepsilon_0 w d m v_0^2}{eL}$$

(f) Applying the equation $U_E = \frac{1}{2}C(\Delta V)^2$, we get

$$U_E = \frac{1}{2} \cdot \frac{\varepsilon_0 L w}{d} \cdot \left(\frac{d^2 m v_0^2}{eL^2}\right)^2 = \frac{\varepsilon_0 w d^3 m^2 v_0^4}{2e^2 L^3}$$

2. (a) If the initial capacitance of the plate is C_0, divide up the capacitor into 2 areas, area 1 and area 2.

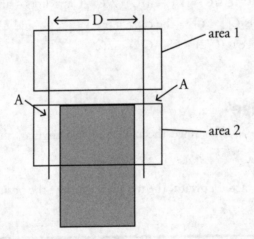

For area 1, since $C = \frac{\kappa \varepsilon_0 A}{d}$, the new capacitance of area 1 = $C_0/2$ since it is half the area of the original capacitor. For area 2, with a dielectric constant of 2, and half the area of the original capacitor, you wind up with the new capacitance of area 2 = $2 \times C_0/2$, the 2s cancel out, and the capacitance for area 2 = C_0. So if we add the capacitance for area 1 to the capacitance for area 2, we get $C_0/2 + C_0 = 3\,C_0/2$.

(b) If we assume the plates no longer have an area of A each, but are now infinitely long plates, with the positively charged plate on the left, and the negatively charged plate on the right, for both points S and T, the electric field will be constant and point from left to right $\rightarrow$. For point R, $E = 0$, since it will be pushed up as much as it will be pulled down, and they will cancel out.

3. (a) $Q = C\Delta V = (6\ \mu F)(12\ V) = 72$ micro Coulomb (μC) since capacitors in parallel have the same voltage across them.

(b) We have to take each of the parallel parts of the circuit and compute the capacitance for each parallel component, because we will then add the capacitance for each parallel equivalent capacitor.

We have 2 capacitors in series, so the equivalent capacitance for that leg of the circuit $1/C_{eq} = 1/6 + 1/3 = 3/6 = \frac{1}{2}$, so the equivalent capacitor of C_5 and C_6 is 2 μF.

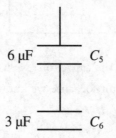

We find the equivalent capacitance for C_2 and C_3, which are both 4 μF, so $\frac{1}{4} + \frac{1}{4} = 2/4 = \frac{1}{2}$. Their equivalent capacitance is 2 μF.

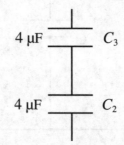

We now have C_4 in parallel with the equivalent capacitance from C_2 and C_3, so there are two capacitors in parallel. We add the capacitance, so we would have an equivalent capacitance of 4 μF.

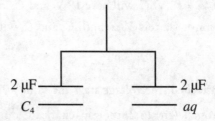

The figure below shows the single equivalent capacitance on each parallel leg of the circuit.

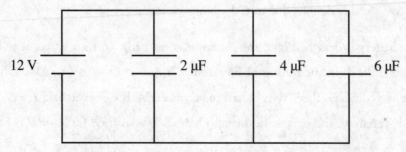

For capacitors in parallel, we add capacitance, so the equivalent capacitance for the entire circuit is 12 μF.

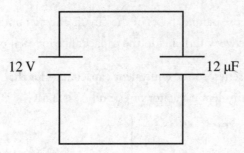

(c) To find the voltages across each of the capacitors, we use the equation $Q = C\Delta V$, so $\Delta V = Q/C$ for each of the parallel legs of the circuit, because we know each parallel leg has a 12 V voltage drop across each leg.

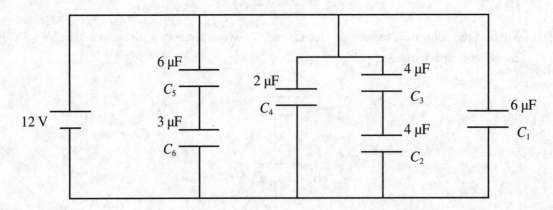

As shown in the figure, capacitors in series must have the same charge, Q, so they must have different voltages across C_5 and C_6 since they have different capacitances. So C_5 will be 4 V and C_6 will be 8 V, since $C_5 \times V_5 = C_6 \times V_6 = 12$ V. C_4 will have 12 V, and C_2 and C_3 must have the same charge since they are in series and must add to 4 V, so both C_2 and C_3 will be 6 V. For capacitor C_1, it will be 12 V.

4. (a) Since the battery remains attached, the voltage stays the same at 12 V. Since $C = \dfrac{\kappa \varepsilon_0 A}{d}$, with the dielectric constant going from 1 (air) to 2, the capacitance C doubles. Since $Q = C\Delta V$, C doubles and V is unchanged, Q doubles. Since $E = \Delta V/d$ and ΔV and d are unchanged, E stays the same. $U = \frac{1}{2}(C\Delta V^2)$, and U doubles due to C doubling.

(b) When the battery is detached (and the capacitor is isolated), Q, the charge, remains constant. C doubles due to the dielectric constant going from 1 (air) to 2. $Q = C\Delta V$, so $\Delta V = Q/C$, since Q stays the same and C doubles, then ΔV is halved to become 6 V. The electric field $E = \Delta V/d$, so since the voltage, ΔV, is halved, the electric field is also halved. U is also $Q^2/(2C)$, so U is halved since C is doubled.

(c) Since the battery remains attached, V is constant and stays the same. $C = \dfrac{\kappa \varepsilon_0 A}{d}$ and d changed to $2d$, then C halves. Since $Q = C\Delta V$, Q also halves since C halves and ΔV stays the same. For the electric field, $E = \Delta V/d$, since d doubles, the electric field halves. $U = \frac{1}{2}(C\Delta V^2)$, so U halves due to C being halved.

(d) Since the battery is detached (capacitor isolated), Q, the charge, stays the same. Since d changes to $2d$, $C = \dfrac{\kappa \varepsilon_0 A}{d}$, the capacitance, C halves with d changing to $2d$. So $\Delta V = Q/C$, since Q stays the same and C halves, then ΔV is doubled. For the electric field, $E = \Delta V/d$, since d doubles, and the voltage ΔV doubles, the electric field stays the same since $E = 2\Delta V/2d$. $U = Q^2/(2C)$ so U doubles due to C being halved.

CIRCUIT DRILL

1. We have 3 resistors in parallel, so $\dfrac{1}{R_{eq}} = \dfrac{1}{7} + \dfrac{1}{7} + \dfrac{1}{21} = \dfrac{7}{21} = \dfrac{1}{3}$, and $R_{eq} = 3\ \Omega$.

   $V = I \times R$, so $I = \dfrac{V}{R} = 9\text{ V}/3\ \Omega = 3\text{A}$ for the equivalent circuit. Since the 3 resistors are in parallel, they each have a 9 V voltage across the three resistors. For the 7 Ω resistors, 7 $\Omega \times I = 9$ V, so $I = 9/7$ A. For the 21 Ω resistor, 21 $\Omega \times I = 9$ V, so $I = 9/21 = 3/7$ A.

2. We have 2 resistors in parallel, the 3 Ω and 6 Ω, so $\dfrac{1}{R_{eq}} = \dfrac{1}{3} + \dfrac{1}{6} = \dfrac{3}{6} = \dfrac{1}{2}$, so $R_{eq} = 2\ \Omega$. The equivalent 2 Ω resistor is in series with a 4 Ω resistor, so the equivalent resistor for the entire circuit is a 6 Ω resistor. $V = I \times R$, so $I = \dfrac{V}{R} = 12\text{ V}/6\ \Omega = 2\text{A}$ for the equivalent circuit. For the 4 Ω resistor, $V = I \times R = 2$ A $\times$ 4 $\Omega = 8$ V. That means that the 3 Ω resistor and the 6 Ω resistor have a 4 V voltage across them (since they are in parallel). For the 3 Ω resistor, 3 $\Omega \times I = 4$ V, so $I = 4/3$ A. For the 6 Ω resistor, 6 $\Omega \times I = 4$ V, so $I = 4/6$ A $= 2/3$ A.

3. The 2 Ω and 4 Ω resistors are in series, so we have a 12 Ω and 6 Ω resistor in parallel, so $\dfrac{1}{R_{eq}} = \dfrac{1}{12} + \dfrac{1}{6} = \dfrac{3}{12} = \dfrac{1}{4}$ $R_{eq} = 4\ \Omega$ for the parallel equivalent resistor, which we add to the 3 Ω resistor since they are in series, for an equivalent resistance of 7 Ω for the circuit. $V = I \times R$, so $I = \dfrac{V}{R} = 14\text{ V}/7\ \Omega = 2$ A for the equivalent circuit. For the 3 Ω resistor, 3 $\Omega \times 2$ A $= V$, so $V = 6$ V. That means that the two parallel circuits have a voltage of 8 V across each of them. For the 12 Ω resistor, 12 $\Omega \times I = 8$ V, so $I = 8/12 = 2/3$ A. If 2/3 A is the current across the 12 Ω resistor, then the current across the 2 Ω resistor and the 4 Ω resistor is 2 A $-$ 2/3 A $= 4/3$ A. You can then figure out the voltage across the 2 Ω resistor and the 4 Ω resistor with $V = I \times R$. For the 2 Ω resistor,

$V = (4/3 \text{ A}) \times (2 \ \Omega) = 8/3 \text{ V}$, and for the 4 Ω resistor, $V = (4/3 \text{ A}) \times (4 \ \Omega) = 16/3 \text{ V}$. 8/3 + 16/3 = 24/3 = 8 V, which correctly adds to the 6 V across the 3 Ω resistor for a total of 14 V.

4. We have 2 sets of resistors in parallel. We have to find the equivalent resistance for each of the two pairs of resistors, and then add the two equivalent resistances since they are then in series. Let's find the equivalent resistance for the first pair, the 2 Ω resistor and the 3 Ω resistor. So $\frac{1}{R_{eq1}} = \frac{1}{2} + \frac{1}{3} = \frac{5}{6}$ $R_{eq1} = \frac{6}{5}$ Ω. Let's find the equivalent resistance for the second pair, the 4 Ω resistor and the 6 Ω resistor. So $\frac{1}{R_{eq2}} = \frac{1}{4} + \frac{1}{6} = \frac{5}{12}$ $R_{eq2} = \frac{12}{5}$ Ω. The total resistance for the two equivalent resistors in series is $\frac{6}{5}$ Ω $+ \frac{12}{5}$ Ω $= \frac{18}{5}$ Ω. $V = I \times R$, so $I = V/R = 9/(\frac{18}{5}) = \frac{5}{2}$ A. The current across the 2 Ω resistor is 3/2 A and the 3 Ω resistor is 1A, which both have a voltage of 3 V since they are in parallel. The current across the 4 Ω resistor is 3/2 A and the 6 Ω resistor is 1 A, and they both have a voltage of 6 V.

CHAPTER 14 DRILL

Section I: Multiple Choice

1. **A** Let ρ_S denote the resistivity of silver and let A_S denote the cross-sectional area of the silver wire. Then

$$R_B = \frac{\rho_B L}{A_B} = \frac{(5\rho_S)L}{4^2 A_S} = \frac{5}{16} \frac{\rho_S L}{A_S} = \frac{5}{16} R_S$$

2. **D** The equation $I = V/R$ implies that increasing V by a factor of 2 will cause I to increase by a factor of 2.

3. **C** Use the formula for power that contains resistance and voltage.

$$P = \frac{V^2}{R}$$
$$V = \sqrt{PR}$$
$$= \sqrt{(125 \text{ W})(5 \ \Omega)} = 25 \text{ V}$$

4. **A** Since Power $= V^2/R$, if you double the voltage, the power quadruples.

5. **B** As a rule, elements of a circuit arranged in series will have the same current flowing through them. This eliminates (C) and (D). Furthermore, as a measuring device, the ammeter should disturb the

circuit as little as possible. Introducing an ammeter with high resistance in series with the element being measured would cause the current to flow in a different way than it did before the ammeter was introduced. Keeping the ammeter's resistance as close to 0 as possible will produce the least change and is therefore the desirable outcome.

6. **E** The 12 Ω and 4 Ω resistors are in parallel and are equivalent to a single 3 Ω resistor, because 1/(12 Ω) + 1/(4 Ω) = 1/(3 Ω). This 3 Ω resistor is in series with the top 3 Ω resistor, giving an equivalent resistance in the top branch of 3 + 3 = 6 Ω. Finally, this 6 Ω resistor is in parallel with the bottom 3 Ω resistor, giving an overall equivalent resistance of 2 Ω, because 1/(6 Ω) + 1/(3 Ω) = 1/(2 Ω).

7. **D** If each of the identical bulbs has resistance R, then the current through each bulb is ε/R. This is unchanged if the middle branch is taken out of the parallel circuit. (What *will* change is the total amount of current provided by the battery.)

8. **C** The immediate effects will be the same in both. When a dielectric is inserted, the electric field is reduced, which in turn decreases voltage. This decrease in voltage causes a loss of potential energy. The difference between the two circuits is that the battery that is still attached can restore that lost voltage to its capacitor by sending additional charge. This additional charge restores not only the lost voltage, but also the electric field. The restored voltage combined with the increased charge means potential energy will be even greater than the original value. Capacitance, however, can be described by the equation $C = Q/\Delta V$, so the increases to charge and voltage cancel each other out in this regard, making capacitance the only value that stays the same for both capacitors.

9. **E** Since points a and b are grounded, they're at the same potential (call it zero).

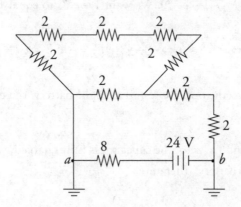

Traveling from b to a across the battery, the potential increases by 24 V, so it must decrease by 24 V across the 8 Ω resistor as we reach point a. Thus, $I = V/R = (24 \text{ V})/(8 \text{ } \Omega) = 3$ A.

10. **E** Slope is $\dfrac{\Delta y}{\Delta x}$, which in this case is $\dfrac{\Delta Q}{\Delta t}$, also known as current.

11. **A** Resistance times capacitance (RC) is known as the time constant. In an RC circuit, a lower time constant leads to a more rapid charging process. Of the combinations available here, (A) gives the smallest product.

Section II: Free Response

1. (a) The two parallel branches, the one containing the 40 Ω resistor and the other a total of 120 Ω, is equivalent to a single 30 Ω resistance. This 30 Ω resistance is in series with the three 10 Ω resistors, giving an overall equivalent circuit resistance of 10 + 10 + 30 + 10 = 60 Ω. Therefore, the current supplied by the battery is $I = V/R = (120 \text{ V})/(60 \text{ Ω}) = 2$ A, so it must supply energy at a rate of $P = IV = (2 \text{ A})(120 \text{ V}) = 240$ W.

   (b) Since three times as much current will flow through the 40 Ω resistor as through the branch containing 120 Ω of resistance, the current through the 20 Ω and 100 Ω resistors must be 0.5 A.

   (c) (i) $V_a - V_b = IR_{20} + IR_{100} = (0.5 \text{ A})(20 \text{ Ω}) + (0.5 \text{ A})(100 \text{ Ω}) = 60$ V.

   (ii) Point a is at the higher potential (current flows from high to low potential).

   (d) Because energy is equal to power multiplied by time, we get

   $$E = Pt = I^2Rt = (0.5 \text{ A})^2(100 \text{ Ω})(10 \text{ s}) = 250 \text{ J}$$

   (e) Using the equation $R = \rho L/A$, with $A = \pi r^2$, we find

   $$R = \frac{\rho L}{\pi r^2} \;\Rightarrow\; r = \sqrt{\frac{\rho L}{\pi R}} = \sqrt{\frac{(0.45 \text{ Ω·m})(0.04 \text{ m})}{\pi(100 \text{ Ω})}} = 0.0076 \text{ m} = 7.6 \text{ mm}$$

2. (a) The initial current, I_0, is ε/r.

   (b) Apply the equation $Q(t) = Q_f(1 - e^{-t/rC})$. We want $1 - e^{-t/rC}$ to equal 1/2, so

   $$1 - e^{-t/rC} = \tfrac{1}{2} \Rightarrow e^{-t/rC} = \tfrac{1}{2} \Rightarrow -\frac{t}{rC} = \ln\tfrac{1}{2} = -\ln 2 \Rightarrow t = (\ln 2)rC$$

   (c) Because it is connected to the positive terminal of the battery, the top plate will become positively charged.

   (d) When the current through r is zero, the capacitor is fully charged, with the voltage across its plates matching the emf of the battery. Therefore,

   $$U_E = \tfrac{1}{2}CV^2 = \tfrac{1}{2}C\mathcal{E}^2$$

   (e) The current established by the discharging capacitor decreases exponentially according to the equation $I(t) = I_0 e^{-t/\tau} = (\varepsilon/R)e^{-t/RC}$.

   (f) The power dissipated is given by the joule heating law, $P = I^2R$:

   $$P(t) = [I(t)]^2 R = \left(\frac{\mathcal{E}}{R} e^{-t/RC}\right)^2 R = \frac{\mathcal{E}^2}{R} e^{-2t/RC}$$

(g) We give two solutions. First, the total energy dissipated by the resistor will equal the integral of $P(t)$ from $t = 0$ to $t = \infty$:

$$E = \int_0^\infty P(t)\, dt$$

$$= \int_0^\infty \frac{\mathcal{E}^2}{R} e^{-2t/RC}\, dt$$

$$= \frac{\mathcal{E}^2}{R} \left[-\frac{RC}{2} e^{-2t/RC} \right]_0^\infty$$

$$= \frac{C\mathcal{E}^2}{2}$$

Alternatively, simply notice that all the energy stored in the capacitor will be dissipated as heat by the resistor R. But from part (d), we know that the initial energy stored in the capacitor (before discharging) was $\frac{1}{2} C\mathcal{E}^2$.

MAGNETIC FIELD DRILL

1. $\vec{F} = |q| v \vec{B} \sin \theta$, but since $v = 0$, no force or acceleration

2. $\vec{F} = |q| v \vec{B} \sin \theta$, $\theta = 90$ degrees, so $\sin(90) = 1$

   $F = (1.6 \times 10^{-19} \text{ C})(5 \text{ m/s})(5.0 \text{ T})(1) = 4.0 \times 10^{-18}$ N, out of the page ($\odot$)

   $a = F/m = (4.0 \times 10^{-18} \text{ N})/(9.1 \times 10^{-31} \text{ kg}) = 4.4 \times 10^{12}$ m/s^2

3. $\vec{F} = |q| v \vec{B} \sin \theta$, but since $\theta = 180$ degrees, $\sin(180) = 0$, no force or acceleration

4. $\vec{F} = |q| v \vec{B} \sin \theta$, $\theta = 90$ degrees, so $\sin(90) = 1$

   $F = (1.6 \times 10^{-19} \text{ C})(5 \text{ m/s})(5.0 \text{ T})(1) = 4.0 \times 10^{-18}$ N, down

   $a = F/m = (4.0 \times 10^{-18} \text{ N})/(9.1 \times 10^{-31} \text{ kg}) = 4.4 \times 10^{12}$ m/s^2

5. $\vec{F} = |q| v \vec{B} \sin \theta$, $\theta = 30$ degrees, so $\sin(30) = 0.5$

   $F = (1.6 \times 10^{-19} \text{ C})(5 \text{ m/s})(5.0 \text{ T})(0.5) = 2.0 \times 10^{-18}$ N, out of the page ($\odot$)

   $a = F/m = (4.0 \times 10^{-18} \text{ N})/(9.1 \times 10^{-31} \text{ kg}) = 4.4 \times 10^{12}$ m/s^2

6. $\vec{F} = |q| v \vec{B} \sin \theta$, $\theta = 90$ degrees, so $\sin(90) = 1$

   $F = (1.6 \times 10^{-19} \text{ C})(5 \text{ m/s})(5.0 \text{ T})(1) = 4.0 \times 10^{-18}$ N, down

   $a = F/m = (4.0 \times 10^{-18} \text{ N})/(1.7 \times 10^{-27} \text{ kg}) = 2.4 \times 10^9$ m/s^2

CHAPTER 15 DRILL

Section I: Multiple Choice

1. **A** Statement (I) is true because magnetic force always acts perpendicular to velocity, meaning that the θ in $W = Fd \cos \theta$ will always be 90°. Statement (II) is false because magnetic force can change the direction of a moving charged particle (just not the speed). Statement (III) is also false because the particle will feel no force if it moves parallel or antiparallel to the magnetic field.

2. **C** The magnitude of the magnetic force is $F_B = qvB$, so the acceleration of the particle has magnitude

$$a = \frac{F_B}{m} = \frac{qvB}{m} = \frac{(4.0 \times 10^{-9} \text{ C})(3 \times 10^4 \text{ m/s})(0.1 \text{ T})}{2 \times 10^{-4} \text{ kg}} = 0.06 \text{ m/s}^2$$

3. **D** By the right-hand rule, the direction of $\mathbf{v} \times \mathbf{B}$ is into the plane of the page. Since the particle carries a negative charge, the magnetic force it feels will be out of the page.

4. **D** Since $\mathbf{F}_B$ is always perpendicular to $\mathbf{v}$, $\mathbf{v}$ cannot be upward or downward in the plane of the page; this eliminates (B) and (C). The velocity vector also cannot be to the right, since then $\mathbf{v}$ would be antiparallel to $\mathbf{B}$, and $\mathbf{F}_B$ would be zero, so (A) can be eliminated. Because the charge is positive, the direction of $\mathbf{F}_B$ will be the same as the direction of $\mathbf{v} \times \mathbf{B}$. In order for $\mathbf{v} \times \mathbf{B}$ to be downward in the plane of the page, the right-hand rule implies that $\mathbf{v}$ must be out of the plane of the page.

5. **B** When a charged particle enters an external magnetic field, it will begin to move in a circular motion. If the magnetic field is large enough, the particle will finish the circle and remain in the field indefinitely. Using the right-hand rule, we can see that the particle in question would immediately experience a force pushing toward the top of the page. Continuing to apply the right-hand rule would make the particle move in a counterclockwise circle. However, if that result is being cut short, it would be because the rectangle is not wide enough, meaning it could not complete its initial turn before exiting the field somewhere along length B.

6. **D** The strength of the magnetic field at a distance r from a long, straight wire carrying a current I is given by the equation $B = (\mu_0/2\pi)(I/r)$. Therefore,

$$\frac{\mu_0}{2\pi} \frac{I}{r} = \frac{(4\pi \times 10^{-7} \text{ T} \cdot \text{m/A})}{2\pi} \frac{10 \text{ A}}{0.02 \text{ m}} = 1 \times 10^{-4} \text{ T}$$

7. **D** By Newton's Third Law, neither (A) nor (B) can be correct. Also, as we learned in Chapter 15, Example 9, if two parallel wires carry current in the same direction, the magnetic force between them is attractive; this eliminates (C) and (E). Therefore, the answer must be (D). The strength of the magnetic field at a distance r from a long, straight wire carrying a current I_1 is given by the

equation $B_1 = (\mu_0/2\pi)(I_1/r)$. The magnetic force on a wire of length ℓ carrying a current I through a magnetic field **B** is $I(\ell \times \mathbf{B})$, so the force on Wire #2 (F_{B2}) due to the magnetic field of Wire #1 (B_1) is

$$F_{B2} = I_2 \ell B_1 = I_2 \ell \frac{\mu_0}{2\pi} \frac{I_1}{r}$$

which implies

$$\frac{F_{B2}}{\ell} = \frac{\mu_0}{2\pi} \frac{I_1 I_2}{r} = \frac{(4\pi \times 10^{-7} \ \mathrm{N/A^2})}{2\pi} \frac{(5 \ \mathrm{A})(10 \ \mathrm{A})}{0.01 \ \mathrm{m}} = 0.001 \ \mathrm{N/m}$$

8.   **B**   Following the right-hand rules for magnetic fields produced by current in a wire, the resulting fields would look like this:

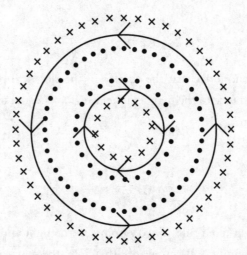

From this picture, we can see that the outer ring will produce a magnetic field into the page outside itself and a magnetic field out of the page inside itself. The inner ring will produce a magnetic field out of the page outside itself, and a field into the page inside itself. In order to have an area of 0 net magnetic field, fields in opposite directions will have to overlap. This can only happen outside the outer ring (into the page from the outer ring and out of the page from the inner ring) and inside the inner ring (into the page from the inner ring and out of the page from the outer ring).

9.   **B**   The strength of the magnetic field within a hollow, ideal solenoid is given by the equation $B = \mu_0 n I$, where n denotes the number of turns per unit length and I is the current in the solenoid. Therefore,

$$B = \mu_0 n I = \mu_0 \frac{N}{\ell} I \quad \Rightarrow \quad N = \frac{\ell B}{\mu_0 I}$$

This gives

$$N = \frac{\ell B}{\mu_0 I} = \frac{(0.80 \ \mathrm{m})(0.2 \ \mathrm{T})}{(4\pi \times 10^{-7} \ \mathrm{T \cdot m/A})(20 \ \mathrm{A})} = 6{,}400 \ \mathrm{turns}$$

10. **E** By Ampere's law, $\oint_{\text{loop}} \mathbf{B} \cdot d\mathbf{s} = \mu_0 I_{\text{through loop}}$. Therefore,

$$I_{\text{through loop}} = \frac{\oint_{\text{loop}} \mathbf{B} \cdot d\mathbf{s}}{\mu_0} = \frac{6.28 \times 10^{-6} \text{ T} \cdot \text{m}}{4\pi \times 10^{-7} \text{ T} \cdot \text{m}/\text{A}} = 5 \text{ A}$$

Section II: Free Response

1. (a) The acceleration of an ion of charge q is equal to F_E/m. The electric force is equal to qE, where $E = V/d$. Therefore, $a = qV/(dm)$.

   (b) Using $a = qV/(dm)$ and the equation $v^2 = v_0^2 + 2ad$, we get

   $$v^2 = 2\frac{qV}{dm}d \implies v = \sqrt{\frac{2qV}{m}}$$

   As an alternate solution, notice that the change in the electrical potential energy of the ion from the source S to the entrance to the magnetic-field region is equal to qV; this is equal to the gain in the particle's kinetic energy.

   Therefore,

   $$qV = \tfrac{1}{2}mv^2 \implies v = \sqrt{\frac{2qV}{m}}$$

   (c) (i) and (ii) Use the right-hand rule. Since **v** points to the right and **B** is into the plane of the page, the direction of $\mathbf{v} \times \mathbf{B}$ is upward. Therefore, the magnetic force on a positively charged particle (cation) will be upward, and the magnetic force on a negatively charged particle (anion) will be downward. The magnetic force provides the centripetal force that causes the ion to travel in a circular path. Therefore, a cation would follow Path 1 and an anion would follow Path 2.

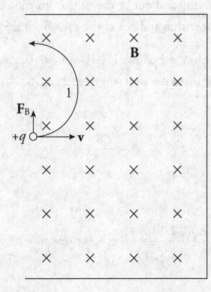 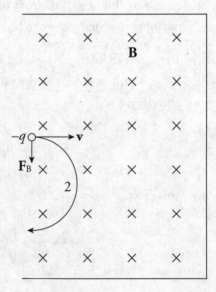

(d) Since the magnetic force on the ion provides the centripetal force,

$$qvB = \frac{mv^2}{r} \quad \Rightarrow \quad qvB = \frac{mv^2}{\frac{1}{2}y} \quad \Rightarrow \quad m = \frac{qBy}{2v}$$

Now, by the result of part (b),

$$m = \frac{qBy}{2\sqrt{\dfrac{2qV}{m}}} \quad \Rightarrow \quad m^2 = \frac{q^2B^2y^2}{\dfrac{8qV}{m}} \quad \Rightarrow \quad m^2 = \frac{mq^2B^2y^2}{8qV} \quad \Rightarrow \quad m = \frac{qB^2y^2}{8V}$$

(e) Since the magnetic force cannot change the speed of a charged particle, the time required for the ion to hit the photographic plate is equal to the distance traveled (the length of the semicircle) divided by the speed computed in part (b):

$$t = \frac{s}{v} = \frac{\pi \cdot \frac{1}{2}y}{\sqrt{\dfrac{2qV}{m}}} = \frac{1}{2}\pi y \sqrt{\frac{m}{2qV}}$$

(f) Since the magnetic force $\mathbf{F}_B$ is always perpendicular to a charged particle's velocity vector $\mathbf{v}$, it can do no work on the particle. Thus, the answer is zero.

2. (a) The current in the rectangular loop is equal to V divided by the resistance of the rectangular loop. Using the equation $R = \rho\ell/A$, we have

$$R = \frac{\rho\ell}{A} = \frac{\rho \cdot 2(a+b)}{\pi(\frac{1}{2}d)^2} = \frac{8\rho(a+b)}{\pi d^2}$$

Therefore, the current in the rectangular loop is

$$I_{loop} = \frac{V}{R} = \frac{\pi d^2 V}{8\rho(a+b)}$$

(b) We use the equation $\mathbf{F}_B = I(\ell \times \mathbf{B})$ to find the magnetic force on each side of the rectangular loop. By symmetry, the magnetic forces on the sides of length a have the same magnitude but opposite direction, so they cancel. The total magnetic force on the loop is equal to the sum of the magnetic forces on the sides of length b. Current in the rectangle is directed clockwise, so the current in the bottom wire of length b (the one closer to the wire) is directed to the left and the current in the top wire of length b is directed to the right. Currents that are parallel feel an attractive force, while currents that are antiparallel feel a repulsive force. Since the strength of the magnetic field at a distance

r from a long, straight wire carrying a current I is given by the equation $B = (\mu_0/2\pi)(I/r)$, we find that the magnetic force on the bottom wire is:

$$\mathbf{F}_{B1} = I_{\text{loop}}\, b \cdot \frac{\mu_0 I}{2\pi c}, \text{ upward (away from the long wire)},$$

and the magnetic force on the top wire is:

$$\mathbf{F}_{B2} = I_{\text{loop}}\, b \cdot \frac{\mu_0 I}{2\pi(c+a)}, \text{ downward (toward the long wire)}.$$

Since $\mathbf{F}_{B1}$ has a greater magnitude than $\mathbf{F}_{B2}$, and the forces point in opposite directions (with the direction of the net force equaling the direction of $\mathbf{F}_{B1}$), the magnitude of the total magnetic force on the loop is equal to the difference between the magnitudes of $\mathbf{F}_{B1}$ and $\mathbf{F}_{B2}$. Therefore,

$$\mathbf{F}_B = \mathbf{F}_{B1} + \mathbf{F}_{B2} = I_{\text{loop}}\, b \cdot \frac{\mu_0 I}{2\pi c} - I_{\text{loop}}\, b \cdot \frac{\mu_0 I}{2\pi(c+a)}, \text{ upward}$$

$$= \frac{\mu_0 I \cdot I_{\text{loop}}\, b}{2\pi}\left(\frac{1}{c} - \frac{1}{c+a}\right), \text{ upward}$$

$$= \frac{\mu_0 I \cdot I_{\text{loop}}\, b}{2\pi}\frac{a}{c(c+a)}, \text{ upward}$$

Substituting the result of part (a) for I_{loop} gives

$$\mathbf{F}_B = \frac{\mu_0 I \cdot \dfrac{\pi d^2 V}{8\rho(a+b)} b}{\pi}\frac{a}{c(c+a)}, \text{ upward}$$

$$= \frac{\mu_0 I \cdot d^2 V a\, b}{8\rho(a+b)(a+c)c}, \text{ upward (away from long wire)}$$

(c) The circumference of the circle will be equal to the perimeter of the rectangle; thus,

$$2\pi r = 2(a+b) \quad \Rightarrow \quad r = \frac{a+b}{\pi}$$

(d) If the long, straight wire passes through the center of the circular loop, then the magnetic field line of the straight wire would coincide with the current in the loop. Whether the current in the loop is parallel or antiparallel to the direction of the magnetic field line makes no difference; the magnetic force on the current-carrying loop would be zero (because $\ell \times \mathbf{B}$ would be 0).

(e) The current in the sliding wire is directed to the right, and the magnetic field is into the plane of the page, so the right-hand rule tells us that the direction of $\ell \times \mathbf{B}$ [and, therefore, of $\mathbf{F}_B = I(\ell \times \mathbf{B})$, the magnetic force on the sliding wire] will be upward. If this upward force is to balance the downward gravitational force, then, because $\ell = x$,

$$I x B = m g \quad \Rightarrow \quad I = \frac{mg}{xB}$$

3. (a) Neither of the straight portions of the wire contribute to the magnetic field at point C; this is because $\hat{\mathbf{r}}$ is parallel (or antiparallel) to ℓ, so the cross product $\ell \times r$, which appears in the Biot–Savart law, would be zero. Therefore, only the curved portion of the wire generates a magnetic field at C. Refer to the diagram below:

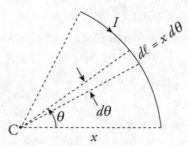

By the Biot–Savart law, the magnetic field at C due to a length $d\ell$ of the arc has magnitude

$$dB = \frac{\mu_0}{4\pi} \frac{I \cdot d\ell}{r^2} = \frac{\mu_0}{4\pi} \frac{I \cdot (x\,d\theta)}{x^2} = \frac{\mu_0 I}{4\pi x}\,d\theta$$

Integrating this from $\theta = 0$ to $\theta = \phi$ gives

$$B = \int dB = \frac{\mu_0 I}{4\pi x} \int_0^\phi d\theta = \frac{\mu_0 I \phi}{4\pi x}$$

By the right-hand rule, the direction of **B** at C is into the plane of the page. (Either curl the fingers of your right hand in the direction of the current in the arc and notice that your thumb points into the page, or apply the right-hand rule to $d\ell \times \hat{\mathbf{r}}$.)

(b) If the charged particle is placed at rest at point C, then it would feel no magnetic force. Only particles that move through magnetic fields feel a magnetic force.

(c) Since the current in the straight wire is antiparallel to the magnetic field at C, it will experience zero magnetic force ($\ell \times \mathbf{B} = 0$).

4. (a) The current in the rod is equal to the current density times the cross-sectional area of the rod: $I = JA = J(\pi R^2)$.

(b) (i) We apply Ampere's law to a circular loop of radius $r < R$. The magnetic field lines will be circles centered on the central axis of the rod (thus coinciding with the position of the Amperian loop drawn below):

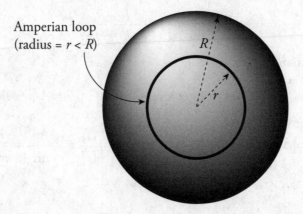

Amperian loop
(radius = $r < R$)

$$\oint_{\text{loop}} \mathbf{B} \cdot d\mathbf{s} = \mu_0 I_{\text{through loop}}$$

$$B(2\pi r) = \mu_0 \cdot JA_{\text{enclosed by loop}}$$

$$= \mu_0 \cdot J(\pi r^2)$$

$$B = \tfrac{1}{2}\mu_0 \cdot Jr$$

To write this result in terms of I, we simply note from part (a) that $J = I/(\pi R^2)$, so

$$B = \tfrac{1}{2}\mu_0 \cdot Jr = \tfrac{1}{2}\mu_0 \cdot \frac{I}{\pi R^2} \cdot r = \frac{\mu_0}{2\pi} \frac{I}{R^2} r$$

(ii)   Applying Ampere's law to an Amperian loop of radius $r > R$,

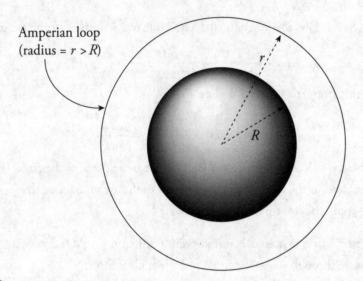

Amperian loop
(radius = $r > R$)

we get

$$\oint_{\text{loop}} \mathbf{B} \cdot d\mathbf{s} = \mu_0 I_{\text{through loop}}$$

$$B(2\pi r) = \mu_0 \cdot JA_{\text{enclosed by loop}}$$

$$= \mu_0 \cdot J(\pi R^2)$$

$$B = \frac{1}{2}\mu_0 J \frac{R^2}{r}$$

To write this result in terms of I, we once again use $J = I/(\pi R^2)$, giving

$$B = \tfrac{1}{2}\mu_0 J \frac{R^2}{r} = \tfrac{1}{2}\mu_0 \cdot \frac{I}{\pi R^2} \cdot \frac{R^2}{r} = \frac{\mu_0}{2\pi} \frac{I}{r}$$

which is certainly a familiar result!

(c)   Since current density is current per unit area, the units of J are A/m^2. Therefore,

$$J = \sigma\, r \quad \Rightarrow \quad [J] = [\sigma]\,[r] \quad \Rightarrow \quad [\sigma] = \frac{[J]}{[r]} = \frac{\text{A/m}^2}{\text{m}} = \frac{\text{A}}{\text{m}^3}$$

(d) Since the current density varies with the radial distance from the rod's center, construct a thin ring of width dr at radius r:

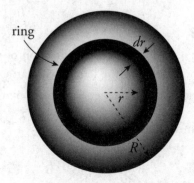

The area of this ring is $dA = (2\pi r)\, dr$, so the current through the ring is

$$dI = J\, dA = (\sigma r)(2\pi r\, dr) = 2\pi\sigma\, r^2\, dr$$

Therefore, the total current through the rod is

$$I = \int dI = \int_{r=0}^{r=R} (2\pi s \cdot r^2\, dr) = 2\pi\sigma\int_0^R r^2\, dr = \tfrac{2}{3}\pi\sigma R^3 \quad (1)$$

(e) (i) As in part (b), construct an Amperian loop of radius $r < R$. The current through such a loop is given by the result of Equation (1) above with r replacing R. Therefore,

$$\oint_{\text{loop}} \mathbf{B} \cdot d\mathbf{s} = \mu_0 I_{\text{through loop}}$$

$$B(2\pi r) = \mu_0 \cdot \left(\tfrac{2}{3}\pi\sigma r^3\right)$$

$$B = \tfrac{1}{3}\mu_0 \sigma r^2$$

To write this result in terms of I, we simply note from part (d) that $\sigma = I\big/\tfrac{2}{3}\pi R^3$, so

$$B = \tfrac{1}{3}\mu_0\sigma r^2 = \tfrac{1}{3}\mu_0 \cdot \frac{I}{\tfrac{2}{3}\pi R^3} \cdot r^2 = \frac{\mu_0}{2\pi}\frac{I}{R^3}r^2$$

(ii) We can either follow the same procedure as in (b) (ii)—and use the results of (d) and (e) (i)—or we can simply notice that outside the wire, the full current I would pass through our circular Amperian loop of radius $r > R$, so the magnetic field at such points must be given once again by the familiar formula

$$B = \frac{\mu_0}{2\pi}\frac{I}{r}$$

CHAPTER 16 DRILL

Section I: Multiple Choice

1. **E** Since **v** is upward and **B** is out of the page, the direction of **v** × **B** is to the right. Therefore, free electrons in the wire will be pushed to the left, leaving an excess of positive charge at the right. Therefore, the potential at point *b* will be higher than at point *a*, by $\mathcal{E} = vBL$ (motional emf).

2. **C** Consider a small radial segment of length *dr* as shown:

    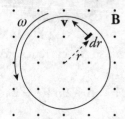

    Its velocity is $v = \omega r$, so the motional emf in this little piece is $d\mathcal{E} = (\omega r)B\,dr$. Integrating from $r = 0$ to $r = a$ gives the induced emf between the center and the rim:

    $$\mathcal{E} = \int d\mathcal{E} = \int_{r=0}^{r=a} \omega r B\,dr = \omega B \int_0^a r\,dr = \tfrac{1}{2}\omega B a^2$$

3. **A** The magnitude of the emf induced between the ends of the rod is $\mathcal{E} = BLv = (0.5\text{ T})(0.2\text{ m})(3\text{ m/s}) = 0.3$ V. Since the resistance is 10 Ω, the current induced will be $I = V/R = (0.3\text{ V})/(10\text{ }\Omega) = 0.03$ A. To determine the direction of the current, we can do the following: since positive charges in the rod are moving to the left and the magnetic field points into the plane of the page, the right-hand rule tells us that the magnetic force, $q\mathbf{v} \times \mathbf{B}$, points downward. The resulting force on the positive charges in the rod is downward, which means that so is the direction of the induced current.

4. **A** The magnetic field through the loop is $B = \mu_0 nI$. Since its area is $A = \pi r^2$, the magnetic flux through the loop is $\Phi_B = BA = (\mu_0 nI)(\pi r^2)$. If the current changes (with $\Delta I/\Delta t = -a$), then the magnetic flux through the loop changes, which, by Faraday's law, implies that an emf (and a current) will be induced. We get

    $$\mathcal{E} = -\frac{\Delta \Phi_B}{\Delta t} = -\frac{\Delta(\mu_0 nI \cdot \pi r^2)}{\Delta t} = -(\mu_0 n \pi r^2)\frac{\Delta I}{\Delta t} = -\mu_0 \pi n r^2(-a) = \mu_0 \pi n r^2 a$$

    Since the magnetic flux into the page is decreasing, the direction of the induced current will be clockwise (opposing a *decreasing into-the-page flux* means that the induced current will create more into-the-page flux).

5. **C** By definition, magnetic field lines emerge from the north pole and enter at the south pole. Therefore, as the north pole is moved upward through the loop, the upward magnetic flux increases. To oppose an increasing upward flux, the direction of the induced current will be clockwise (as seen from above) to generate some downward magnetic flux. Now, as the south pole moves away from the center of the loop, there is a decreasing upward magnetic flux, so the direction of the induced current will be counterclockwise.

6. **E** Since the current in the straight wire is steady, there is no change in the magnetic field, no change in magnetic flux, and, therefore, no induced emf or current.

7. **A** Use the equation $I(t) = I_{max}(1 - e^{-t/\tau_L})$. In order for $I(t)$ to equal $\frac{3}{4}I_{max}$, we must have $e^{-t/\tau_L} = \frac{1}{4}$. Therefore,

$$-t/\tau_L = \ln\frac{1}{4} = -\ln 4 \quad \Rightarrow \quad t = (\ln 4)\tau_L = (\ln 4)\frac{L}{R}$$

8. **D** The value of I_{max} is $\mathcal{E}/R$. Since the magnetic energy stored in an inductor is given by $U_B = \frac{1}{2}LI^2$, we have

$$U_B = \frac{1}{2}L\left(\frac{\mathcal{E}}{R}\right)^2 = \frac{L\mathcal{E}^2}{2R^2}$$

9. **B** By definition of self-inductance, $L = N\Phi_B/I$. The total magnetic flux through all the windings of the solenoid is $N\Phi_B$, which is equal to LI. Since $I = \mathcal{E}/R$, we have

$$\Phi_{B,total} = L(\mathcal{E}/R)$$

10. **D** Faraday's law shows how a changing B-field generates an E-field. The Ampere–Maxwell law shows the reverse: how a changing E-field generates a B-field.

Section II: Free Response

1. (a) Consider a small section of length dx of the rod at a distance x from the nonrotating end.

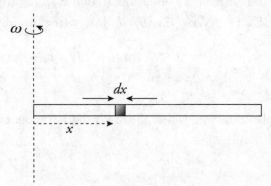

The velocity of this small piece is $v = \omega x$, so the motional emf induced between its ends is $d\mathcal{E} = vB\,dx = \omega xB\,dx$. Integrating this from $x = 0$ to $x = \ell$ gives

$$\mathcal{E} = \int d\mathcal{E} = \int_{x=0}^{x=\ell} \omega x\,B\,dx = \omega B\int_0^\ell x\,dx = \frac{1}{2}\omega B\ell^2$$

(b)   Refer to the view from above of the rod:

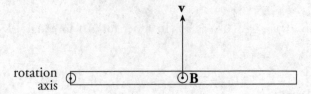

By the right-hand rule, the direction of **v** × **B** is to the right, so the magnetic force on free electrons will be to the left, leaving excess *positive* charge at the right (rotating) end.

(c)   Consider a small section of length *dy* at a distance *y* from the straight wire.

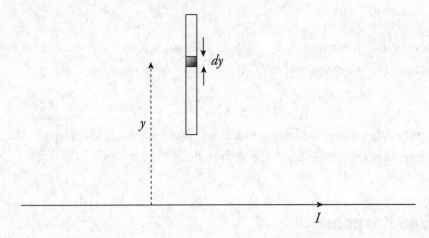

Since the magnetic field at the position of this section is $(\mu_0/2\pi)(I/y)$, the motional emf induced between its ends is $d\mathcal{E} = vB\,dy = v(\mu_0/2\pi)(I/y)\,dy$. Integrating this from $y = a$ to $y = a + \ell$ gives

$$\mathcal{E} = \int d\mathcal{E} = \int_{y=a}^{y=a+\ell} \frac{\mu_0}{2\pi}\frac{I}{y}v\,dy = \frac{\mu_0 Iv}{2\pi}\int_a^{a+\ell}\frac{dy}{y} = \frac{\mu_0 Iv}{2\pi}\ln\frac{a+\ell}{a}$$

(d)   The magnetic field above the wire (where the rod is sliding) is out of the plane of the page, so the direction of **v** × **B** is downward.

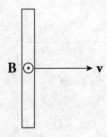

Therefore, free electrons in the rod will be pushed upward to the far end of the rod, causing it to become negatively charged.

2. (a) As the loop rotates, the angle between the normal to the loop's enclosed area and **B** changes according to the equation $\theta = \omega t$. Therefore, the magnetic flux through the loop is

$$\Phi_B(t) = BA\cos\theta = BA\cos\omega t = abB\cos\omega t$$

(b) According to Faraday's law, the induced emf is equal to the rate of change of magnetic flux through the loop (with a minus sign included to conform to Lenz's law). Therefore,

$$\mathcal{E} = -\frac{d\Phi_B}{dt} = -\frac{d}{dt}(abB\cos\omega t) = -[abB\omega(-\sin\omega t)] = ab\omega B\sin\omega t$$

(c) Using the result of part (b) with the equation $I = \mathcal{E}/R$, we find that the induced current is a function of time given by

$$I(t) = \frac{\mathcal{E}}{R} = \frac{ab\omega B\sin\omega t}{R}$$

(d) As the loop rotates from $\theta = 0$ to $\theta = \pi/2$, the magnetic flux (to the right, the $+x$ direction) through the loop decreases from $BA = abB$ to zero. To oppose this decreasing flux to the right, the current will be induced in a direction to generate a magnetic field with a component in the $+x$ direction. The current will be flowing from point 3 to 4 to 1 to 2 to 3.

(e) The rate at which energy is dissipated as heat in the loop is given by the equation $P = I^2R = (\mathcal{E}/R)^2R = \mathcal{E}^2/R$, so

$$P = \frac{\mathcal{E}^2}{R} = \frac{(ab\omega B\sin\omega t)^2}{R} = \frac{a^2b^2\omega^2B^2}{R}\sin^2\omega t$$

In one revolution, ωt increases from 0 to 2π, so the energy dissipated is

$$\begin{aligned}
E = \int P\,dt &= \int_{t=0}^{t=2\pi/w} \frac{a^2b^2\omega^2B^2}{R}\sin^2\omega t\,dt \\
&= \frac{a^2b^2\omega^2B^2}{R}\int_{t=0}^{t=2\pi/\omega}\sin^2\omega t\,dt \\
&= \frac{a^2b^2\omega^2B^2}{R}\int_{t=0}^{t=2\pi/\omega}\tfrac{1}{2}(1-\cos 2\omega t)\,dt \\
&= \frac{a^2b^2\omega^2B^2}{R}\left[\tfrac{1}{2}t - \tfrac{1}{4\omega}\sin 2\omega t\right]_{t=0}^{t=2\pi/\omega} \\
&= \frac{a^2b^2\omega^2B^2}{R}\cdot\frac{\pi}{\omega} \\
&= \frac{\pi a^2b^2\omega B^2}{R}
\end{aligned}$$

(f)   The work done by the externally provided torque is transformed into the heat lost in the loop. Equivalently, the rate at which the torque does work is equal to the rate at which heat energy is dissipated in the wires of the loop. Since the power associated with the external torque is $P = \omega\tau$, the result of part (e) gives

$$\tau\omega = \frac{a^2 b^2 \omega^2 B^2}{R}\sin^2\omega t \implies \tau = \frac{a^2 b^2 \omega B^2}{R}\sin^2\omega t$$

3.   (a)   Construct an Amperian loop of radius R whose center coincides with the center of the toroid. Then only the current in the inner set of N windings pierces through the area bounded by the loop, so Ampere's law becomes

$$\oint_{\text{loop}} B \cdot ds = \mu_0 I_{\text{through loop}}$$

$$B(2\pi R) = \mu_0 (NI)$$

$$B = \frac{\mu_0}{2\pi}\frac{NI}{R}$$

(b)   Because there is no magnetic field outside the toroid, the magnetic flux through the loop is equal to B times the cross-sectional area of the toroid (which is πa^2). Therefore, by Faraday's law,

$$\mathcal{E} = -\frac{d\Phi_B}{dt} = -\frac{d}{dt}(B \cdot \pi a^2)$$

$$= -\frac{d}{dt}\left(\frac{\mu_0 NI}{2\pi R} \cdot \pi a^2\right)$$

$$= -\frac{\mu_0 N a^2}{2R}\frac{dI}{dt}$$

$$= -\frac{\mu_0 N a^2}{2R} \cdot \frac{d}{dt}(I_0 \sin\omega t)$$

$$= -\frac{\mu_0 I_0 N a^2 \omega}{2R}\cos\omega t$$

(c)   Writing Faraday's law in the form

$$\oint_{\text{loop}} \mathbf{E} \cdot d\mathbf{s} = -\frac{d\Phi_B}{dt}$$

and applying it to the circular loop of radius $2a$, we find that

$$\oint_{\text{loop}} \mathbf{E} \cdot d\mathbf{s} = -\frac{d\Phi_B}{dt}$$

$$E(2\pi \cdot r_{\text{loop}}) = -\frac{d\Phi_B}{dt}$$

$$E(2\pi \cdot 2a) = -\frac{\mu_0 I_0 N a^2 \omega}{2R}\cos\omega t$$

$$E = -\frac{\mu_0 I_0 N a\omega}{8\pi R}\cos\omega t$$

(d) By definition, the self-inductance, L, is equal to $N\Phi_B/I$. Since

$$\Phi_B = B \cdot \pi a^2 = \frac{\mu_0 N I}{2\pi R} \cdot \pi a^2 = \frac{\mu_0 N I a^2}{2R}$$

as we calculated in the solution to part (b), we find that

$$L = \frac{N\Phi_B}{I} = \frac{N}{I} \cdot \frac{\mu_0 N I a^2}{2R} = \frac{\mu_0 N^2 a^2}{2R}$$

4. (a) When the switch is initially moved, the inductor prevents an instantaneous change in current. So the current in the inductor, and the rest of the circuit, is zero.

(b) After the steady-state condition is reached, the inductor has no emf (since $di/dt = 0$) so the current in the circuit can be found using Ohm's law.

$$V = IR$$

$$10 \text{ V} = I \cdot 10 \text{ W}$$

$$I = 1 \text{ A}$$

(c) The current in the inductor does not change immediately, so the current in the inductor is 1 A.

(d) The current through the resistor is also 1 A since it is in series with the inductor now. So the voltage drop across the resistor, V_R, can be found using Ohm's law, $V_R = (1 \text{ A})(5 \ \Omega) = 5$ V. The voltage across the inductor is the same (since the voltage drops around the loop must add up to zero). Thus,

$$\varepsilon_L = 5 \text{ V}$$

Chapter 18
Math for
AP Physics C

INTRODUCTION

Math is an integral part of doing physics. This chapter is intended to be a brief overview of the more common math tools that you'll want to be comfortable with when it comes to the AP Physics C exam. If you're taking AP Physics C, then you should have already taken or currently be taking a calculus course equivalent to AP Calculus AB. In other words, it's expected that you'll be familiar with topics in algebra, functions and relations, trigonometry, and the fundamentals of calculus. Here, we focus on the fundamentals of trigonometry and calculus that you'll need to know for the AP Physics C exam.

TRIGONOMETRY

Trigonometry finds its way into physics when it comes to vector quantities (see Chapter 4). A vector is a quantity that has magnitude and direction. So, a common way of representing a vector is with an arrow. The length of the arrow indicates the magnitude of the vector, and where the arrow points indicates the direction of the vector. When representing vectors in two or three dimensions, they can point in more than one coordinate direction. Therefore, vectors can be decomposed or broken down into components: a piece in each of the coordinate directions. Trigonometry can be used to find the various components in terms of the total vector magnitude and an angle between the vector and one of the coordinate axes. Consider the following vector, **A**.

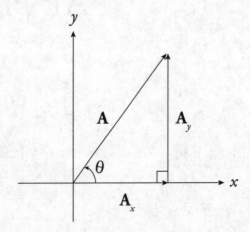

The components of vector **A** in the *x*- and *y*-directions are A_x and A_y, respectively. We can use "SOHCAHTOA" trigonometry to relate the components to the total vector magnitude and the labeled angle.

$$\sin\theta = \frac{opposite}{hypotenuse} = \frac{A_y}{A} \; ; \; \cos\theta = \frac{adjacent}{hypotenuse} = \frac{A_x}{A} \; ; \; \tan\theta = \frac{opposite}{adjacent} = \frac{A_y}{A_x}$$

The Pythagorean Theorem can also be used to relate the magnitudes of the components of a vector to the magnitude of the total vector: $A^2 = A_x^2 + A_y^2$. The Super Pythagorean Theorem can be used to relate the magnitudes of the components of a vector to the magnitude of the total vector for a vector in three dimensions: $A^2 = A_x^2 + A_y^2 + A_z^2$. These relationships, except for the Super Pythagorean Theorem, can be found on the exam equation sheet.

CALCULUS

The AP Physics C exam will require students to make use of a few fundamental techniques in calculus, particularly differentiation and integration. These techniques allow one to express one quantity in terms of precise, continuous changes in another quantity. For example, acceleration is the derivative of velocity with respect to time and displacement is the integral of velocity with respect to time. These techniques will be required when manipulating equations, deriving expressions, and interpreting graphs on the AP Physics C exam. Let's take a look at both of these techniques, starting with differentiation.

> **Looking for More Help with Your APs?**
> We now offer specialized AP tutoring and course packages that guarantee a 4 or 5 on the AP. To see which courses are offered and available, and to learn more about the guarantee, visit Princeton-Review.com/college/ap-test-prep

Differentiation

A derivative calculates the rate of change of a function. If we limit the changes to changes in time, then the derivative calculates instantaneous changes in time as opposed to changes over large periods of time. Looking over a larger period of time, the rate of change of a function, y, is often shown as $\Delta y / \Delta t$. Looking moment to moment, the rate of change is given by the derivative and is often shown as dy/dt. Alternatively, the derivate could be expressed as y' or, when the derivative is taken with respect to time, $\dot{y}$. If a second derivative is applied, the notation changes to $d^2 y / dt^2$, y'', or $\ddot{y}$. Functions on the AP Physics C exam will either be polynomial, exponential, logarithmic, or sinusoidal, so it will be important to know how to differentiate those functions. Here is a list of common derivatives to know for the AP Physics C exam, which can be found on the exam equation sheet.

$$\text{If } y(t) = t^n, \text{ then } y'(t) = nt^{n-1}$$

$$\text{If } y(t) = e^t, \text{ then } y'(t) = e^t$$

$$\text{If } y(t) = \ln(t), \text{ then } y'(t) = 1/t$$

$$\text{If } y(t) = \sin(t), \text{ then } y'(t) = \cos(t)$$

$$\text{If } y(t) = \cos(t), \text{ then } y'(t) = -\sin(t)$$

If the function, y, is the product of two other functions, then the Product Rule can be applied.

$$\text{If } y(t) = f(t) \cdot g(t), \text{ then } y'(t) = f(t) \cdot g'(t) + f'(t) \cdot g(t)$$

If the function, y, is the quotient of two other functions, then the Quotient Rule can be applied.

$$\text{If } y(t) = f(t)/g(t), \text{ then } y'(t) = [g(t) \cdot f'(t) - f(t) \cdot g'(t)]/[g(t)]^2$$

If the function, y, is defined as a function acting on another function, then the Chain Rule must be applied.

$$\text{If } y(t) = f(g(t)), \text{ then } y'(t) = f'(g(t)) \cdot g'(t)$$

Since the derivate determines the rate of change of a function, it can be interpreted as the slope of the tangent of the function. So, for example, if a velocity versus time graph is given on the exam and a question asks for the acceleration at a given time, since the acceleration is the derivative of velocity with respect to time, it can be found as the slope of the tangent to the function.

Integration

The antiderivative of a function is called the integral of the function. It calculates a precise, continuous sum of a quantity with respect to changes in another quantity. If we consider a function $y(t)$, this sum is shown as $\int y \, dt$. Functions on the AP Physics C exam will either be polynomial, exponential, logarithmic, or sinusoidal, so it will be important to know how to integrate those functions. Here is a list of common integrals to know for the AP Physics C exam, which can be found on the exam equation sheet.

$$\text{If } y(t) = t^n, \text{ then } \int y(t)\,dt = t^{n+1}/(n+1) + C$$

$$\text{If } y(t) = e^t, \text{ then } \int y(t)\,dt = e^t + C$$

$$\text{If } y(t) = 1/t, \text{ then } \int y(t)\,dt = \ln|t| + C$$

$$\text{If } y(t) = \sin(t), \text{ then } \int y(t)\,dt = -\cos(t)$$

$$\text{If } y(t) = \cos(t), \text{ then } \int y(t)\,dt = \sin(t)$$

Graphically, the integral can be interpreted as the area under the curve. So, for example, if a velocity versus time graph is given on the exam and a question asks for the displacement over a given time interval, since the displacement is the integral of velocity with respect to time, it can be found as the area under the curve.

Part VI
Practice Test 2

Practice Test 2

- Mechanics
- Electricity & Magnetism

AP® Physics C: Mechanics Exam

SECTION I: Multiple-Choice Questions

DO NOT OPEN THIS BOOKLET UNTIL YOU ARE TOLD TO DO SO.

At a Glance

Total Time
45 minutes
Number of Questions
35
Percent of Total Grade
50%
Writing Instrument
Pen required

Instructions

Section I of this examination contains 35 multiple-choice questions. Fill in only the ovals for numbers 1 through 35 on your answer sheet.

CALCULATORS MAY BE USED IN BOTH SECTIONS OF THE EXAMINATION.

Indicate all of your answers to the multiple-choice questions on the answer sheet. No credit will be given for anything written in this exam booklet, but you may use the booklet for notes or scratch work. After you have decided which of the suggested answers is best, completely fill in the corresponding oval on the answer sheet. Give only one answer to each question. If you change an answer, be sure that the previous mark is erased completely. Here is a sample question and answer.

Sample Question   Sample Answer

Chicago is a   Ⓐ ● Ⓒ Ⓓ Ⓔ
(A) state
(B) city
(C) country
(D) continent
(E) planet

Use your time effectively, working as quickly as you can without losing accuracy. Do not spend too much time on any one question. Go on to other questions and come back to the ones you have not answered if you have time. It is not expected that everyone will know the answers to all the multiple-choice questions.

About Guessing

Many candidates wonder whether or not to guess the answers to questions about which they are not certain. Multiple-choice scores are based on the number of questions answered correctly. Points are not deducted for incorrect answers, and no points are awarded for unanswered questions. Because points are not deducted for incorrect answers, you are encouraged to answer all multiple-choice questions. On any questions you do not know the answer to, you should eliminate as many choices as you can, and then select the best answer among the remaining choices.

GO ON TO THE NEXT PAGE.

ADVANCED PLACEMENT PHYSICS C TABLE OF INFORMATION

CONSTANTS AND CONVERSION FACTORS

Proton mass,	$m_p = 1.67 \times 10^{-27}$ kg	Electron charge magnitude,	$e = 1.60 \times 10^{-19}$ C
Neutron mass,	$m_n = 1.67 \times 10^{-27}$ kg	1 electron volt,	$1 \text{ eV} = 1.60 \times 10^{-19}$ J
Electron mass,	$m_e = 9.11 \times 10^{-31}$ kg	Speed of light,	$c = 3.00 \times 10^8$ m/s
Avogadro's number,	$N_A = 6.02 \times 10^{23}$ mol^{-1}	Universal gravitational constant,	$G = 6.67 \times 10^{-11}$ (N•m^2)/kg^2
Universal gas constant,	$R = 8.31$ J/(mol K)	Acceleration due to gravity	$g = 9.8$ m/s^2
Boltzmann's constant,	$k_B = 1.38 \times 10^{-23}$ J/K	at Earth's surface,	

1 unified atomic mass unit,	$1 \text{ u} = 1.66 \times 10^{-27} \text{ kg} = 931 \text{ MeV}/c^2$
Planck's constant,	$h = 6.63 \times 10^{-34} \text{ J•s} = 4.14 \times 10^{-15} \text{ eV•s}$
	$hc = 1.99 \times 10^{-25} \text{ J•m} = 1.24 \times 10^3 \text{ eV•nm}$
Vacuum permittivity,	$\varepsilon_0 = 8.85 \times 10^{-12} \text{ C}^2/(\text{N•m}^2)$
Coulomb's law constant,	$k = 1/(4\pi\varepsilon_0) = 9.0 \times 10^9 \text{ (N•m}^2)/\text{C}^2$
Vacuum permeability,	$\mu_0 = 4\pi \times 10^{-7} \text{ (T•m)/A}$
Magnetic constant,	$k' = \mu_0/(4\pi) = 1 \times 10^{-7} \text{ (T•m)/A}$
1 atmosphere pressure,	$1 \text{ atm} = 1.0 \times 10^5 \text{ N/m}^2 = 1.0 \times 10^5 \text{ Pa}$

UNIT SYMBOLS							
meter,	m	mole,	mol	watt,	W	farad,	F
kilogram,	kg	hertz,	Hz	coulomb,	C	tesla,	T
second,	s	newton,	N	volt,	V	degree Celsius,	°C
ampere,	A	pascal,	Pa	ohm,	Ω	electron volt,	eV
kelvin,	K	joule,	J	henry,	H		

VALUES OF TRIGONOMETRIC FUNCTIONS FOR COMMON ANGLES

θ	0°	30°	37°	45°	53°	60°	90°
$\sin \theta$	0	1/2	3/5	$\sqrt{2}/2$	4/5	$\sqrt{3}/2$	1
$\cos \theta$	1	$\sqrt{3}/2$	4/5	$\sqrt{2}/2$	3/5	1/2	0
$\tan \theta$	0	$\sqrt{3}/3$	3/4	1	4/3	$\sqrt{3}$	∞

PREFIXES

Factor	Prefix	Symbol
10^9	giga	G
10^6	mega	M
10^3	kilo	k
10^{-2}	centi	c
10^{-3}	milli	m
10^{-6}	micro	μ
10^{-9}	nano	n
10^{-12}	pico	p

The following assumptions are used in this exam.

I. The frame of reference of any problem is inertial unless otherwise stated.

II. The direction of current is the direction in which positive charges would drift.

III. The electric potential is zero at an infinite distance from an isolated point charge.

IV. All batteries and meters are ideal unless otherwise stated.

V. Edge effects for the electric field of a parallel-plate capacitor are negligible unless otherwise stated.

GO ON TO THE NEXT PAGE.

MECHANICS

$v_x = v_{x0} + a_x t$

$x = x_0 + v_{x0}t + \frac{1}{2}a_x t^2$

$v_x^2 = v_{x0}^2 + 2a_x(x - x_0)$

$\vec{a} = \dfrac{\Sigma \vec{F}}{m} = \dfrac{\vec{F}_{net}}{m}$

$\vec{F} = \dfrac{d\vec{p}}{dt}$

$\vec{J} = \int \vec{F}\,dt = \Delta \vec{p}$

$\vec{p} = m\vec{v}$

$\left|\vec{F}_f\right| \leq \mu \left|\vec{F}_N\right|$

$\Delta E = W = \int \vec{F}\cdot d\vec{r}$

$K = \frac{1}{2}mv^2$

$P = \dfrac{dE}{dt}$

$P = \vec{F}\cdot\vec{v}$

$\Delta U_g = mg\Delta h$

$a_c = \dfrac{v^2}{r} = \omega^2 r$

$\vec{\tau} = \vec{r}\times\vec{F}$

$\vec{\alpha} = \dfrac{\Sigma\vec{\tau}}{I} = \dfrac{\vec{\tau}_{net}}{I}$

$I = \int r^2 dm = \Sigma mr^2$

$x_{cm} = \dfrac{\Sigma m_i x_i}{\Sigma m_i}$

$v = r\omega$

$\vec{L} = \vec{r}\times\vec{p} = I\vec{\omega}$

$K = \frac{1}{2}I\omega^2$

$\omega = \omega_0 + \alpha t$

$\theta = \theta_0 + \omega_0 t + \frac{1}{2}\alpha t^2$

a	=	acceleration
E	=	energy
F	=	force
f	=	frequency
h	=	height
I	=	rotational inertia
J	=	impulse
K	=	kinetic energy
k	=	spring constant
ℓ	=	length
L	=	angular momentum
m	=	mass
P	=	power
p	=	momentum
r	=	radius or distance
T	=	period
t	=	time
U	=	potential energy
v	=	velocity or speed
W	=	work done on a system
x	=	position
μ	=	coefficient of friction
θ	=	angle
τ	=	torque
ω	=	angular speed
α	=	angular acceleration
ϕ	=	phase angle

$\vec{F}_s = -k\Delta\vec{x}$

$U_s = \frac{1}{2}k(\Delta x)^2$

$x = x_{max}\cos(\omega t + \phi)$

$T = \dfrac{2\pi}{\omega} = \dfrac{1}{f}$

$T_s = 2\pi\sqrt{\dfrac{m}{k}}$

$T_p = 2\pi\sqrt{\dfrac{\ell}{g}}$

$\left|\vec{F}_G\right| = \dfrac{Gm_1 m_2}{r^2}$

$U_G = -\dfrac{Gm_1 m_2}{r}$

ELECTRICITY & MAGNETISM

$\left|\vec{F}_E\right| = \dfrac{1}{4\pi\varepsilon_0}\left|\dfrac{q_1 q_2}{r^2}\right|$

$\vec{E} = \dfrac{\vec{F}_E}{q}$

$\oint \vec{E}\cdot d\vec{A} = \dfrac{Q}{\varepsilon_0}$

$E_x = -\dfrac{dV}{dx}$

$\Delta V = -\int \vec{E}\cdot d\vec{r}$

$V = \dfrac{1}{4\pi\varepsilon_0}\sum_i \dfrac{q_i}{r_i}$

$U_E = qV = \dfrac{1}{4\pi\varepsilon_0}\dfrac{q_1 q_2}{r}$

$\Delta V = \dfrac{Q}{C}$

$C = \dfrac{\kappa\varepsilon_0 A}{d}$

$C_p = \sum_i C_i$

$\dfrac{1}{C_s} = \sum_i \dfrac{1}{C_i}$

$I = \dfrac{dQ}{dt}$

$U_C = \frac{1}{2}Q\Delta V = \frac{1}{2}C(\Delta V)^2$

$R = \dfrac{\rho\ell}{A}$

$\vec{E} = \rho\vec{J}$

$I = Nev_d A$

$I = \dfrac{\Delta V}{R}$

$R_s = \sum_i R_i$

$\dfrac{1}{R_p} = \sum_i \dfrac{1}{R_i}$

$P = I\Delta V$

A	=	area
B	=	magnetic field
C	=	capacitance
d	=	distance
E	=	electric field
ε	=	emf
F	=	force
I	=	current
J	=	current density
L	=	inductance
ℓ	=	length
n	=	number of loops of wire per unit length
N	=	number of charge carriers per unit volume
P	=	power
Q	=	charge
q	=	point charge
R	=	resistance
r	=	radius or distance
t	=	time
U	=	potential or stored energy
V	=	electric potential
v	=	velocity or speed
ρ	=	resistivity
Φ	=	flux
κ	=	dielectric constant

$\vec{F}_M = q\vec{v}\times\vec{B}$

$\oint \vec{B}\cdot d\vec{\ell} = \mu_0 I$

$d\vec{B} = \dfrac{\mu_0}{4\pi}\dfrac{Id\vec{\ell}\times\vec{r}}{r^2}$

$\vec{F} = \int Id\vec{\ell}\times\vec{B}$

$B_s = \mu_0 n I$

$\Phi_B = \int \vec{B}\cdot d\vec{A}$

$\varepsilon = \oint \vec{E}\cdot d\vec{\ell} = -\dfrac{d\Phi_B}{dt}$

$\varepsilon = -L\dfrac{dI}{dt}$

$U_L = \frac{1}{2}LI^2$

GO ON TO THE NEXT PAGE.

GEOMETRY AND TRIGONOMETRY		**CALCULUS**

GEOMETRY AND TRIGONOMETRY

Rectangle

$A = bh$

Triangle

$A = \frac{1}{2}bh$

Circle

$A = \pi r^2$

$C = 2\pi r$

$s = r\theta$

Rectangular Solid

$V = \ell wh$

Cylinder

$V = \pi r^2 \ell$

$S = 2\pi r\ell + 2\pi r^2$

Sphere

$V = \frac{4}{3}\pi r^3$

$S = 4\pi r^2$

Right Triangle

$a^2 + b^2 = c^2$

$\sin\theta = \dfrac{a}{c}$

$\cos\theta = \dfrac{b}{c}$

$\tan\theta = \dfrac{a}{b}$

A = area
C = circumference
V = volume
S = surface area
b = base
h = height
ℓ = length
w = width
r = radius
s = arc length
θ = angle

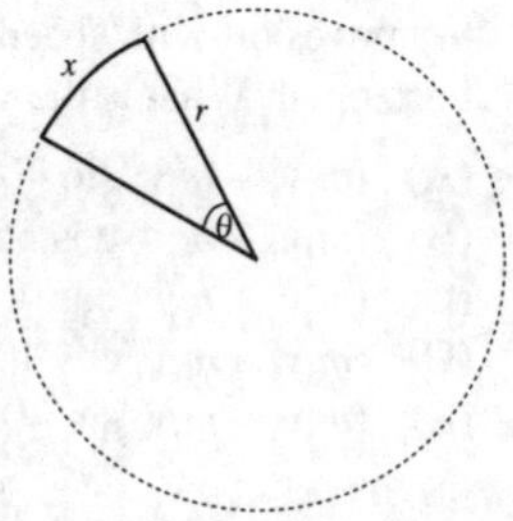

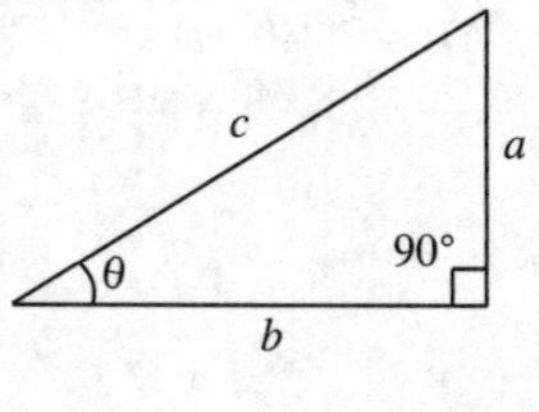

CALCULUS

$$\frac{df}{dx} = \frac{df}{du}\frac{du}{dx}$$

$$\frac{d}{dx}(x^n) = nx^{n-1}$$

$$\frac{d}{dx}(e^{ax}) = ae^{ax}$$

$$\frac{d}{dx}(\ln ax) = \frac{1}{x}$$

$$\frac{d}{dx}[\sin(ax)] = a\cos(ax)$$

$$\frac{d}{dx}[\cos(ax)] = -a\sin(ax)$$

$$\int x^n\,dx = \frac{1}{n+1}x^{n+1}, n \neq -1$$

$$\int e^{ax}\,dx = \frac{1}{a}e^{ax}$$

$$\int \frac{dx}{x+a} = \ln|x+a|$$

$$\int \cos(ax)\,dx = \frac{1}{a}\sin(ax)$$

$$\int \sin(ax)\,dx = -\frac{1}{a}\cos(ax)$$

VECTOR PRODUCTS

$$\vec{A} \cdot \vec{B} = AB\cos\theta$$

$$\left|\vec{A} \times \vec{B}\right| = AB\sin\theta$$

GO ON TO THE NEXT PAGE.

PHYSICS C: MECHANICS
SECTION I
Time—45 minutes
35 Questions

Directions: Each of the following questions or incomplete statements below is followed by five suggested answers or completions. Select the one that is best in each case and then mark it on your answer sheet.

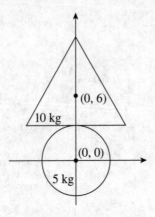

1. Find the location of the center of mass of the objects in the figure shown.

   (A) (0, 0)
   (B) (4, 0)
   (C) (0, 4)
   (D) (1, 5)
   (E) (5, 1)

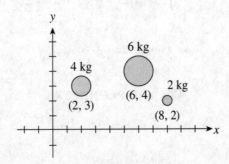

2. Find the center of mass of an object modeled as 3 separate masses on an x-y coordinate plane as shown in the figure. The first mass of 4 kg is at (2, 3). The second mass of 6 kg is at (6, 4). The third individual mass is 2 kg at (8, 2).

   (A) (4, 4.25)
   (B) (5, 3.33)
   (C) (5, 4)
   (D) (6, 4)
   (E) (6, 3.33)

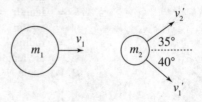

3. In the figure shown above, two balls with different masses, m_1 and m_2, collide. The first ball, of mass m_1, is moving to the right with speed v_1, and the second ball, of mass m_2, is at rest. After the collision, m_1 moves off with speed v_1' at a 40° angle below the horizontal, while m_2 moves off with speed v_2' at an angle of 35° above the horizontal. What is the value of v_2'?

   (A) $(m_1v_1 + m_1v_1'\sin 40°)/(m_2\cos 35°)$
   (B) $(m_1v_1 + m_1v_1'\cos 35°)/(m_2\cos 35°)$
   (C) $(m_1v_1 + m_1v_1'\cos 40°)/(m_2\cos 35°)$
   (D) $(m_1v_1 - m_1v_1'\cos 40°)/(m_2\cos 35°)$
   (E) $(m_1v_1 - m_1v_1'\sin 40°)/(m_2\cos 35°)$

4. Find the power delivered by the net force to a 20 kg mass at time = 3 seconds, given that the position of the mass is defined by the equation $x(t) = 5t^3 - 3t^2 + t$.

   (A) 5,734 watts
   (B) 12,355 watts
   (C) 21,233 watts
   (D) 47,040 watts
   (E) 198,240 watts

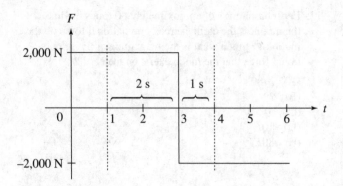

5. The graph shown in the figure above indicates the horizontal force on a pickup truck of mass 1,000 kg as a function of time. In the interval from 1 to 4 seconds, determine the change in the pickup truck's velocity.

(A) 0 m/s
(B) 1 m/s
(C) 1.5 m/s
(D) 2 m/s
(E) 3 m/s

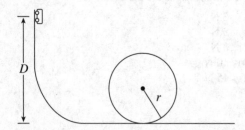

6. A roller coaster begins at a height D above the ground and completes a loop along its path with a radius of r as shown in the figure above. In order for the car to remain on the track throughout the loop, what is the minimum value for D in terms of the radius of the loop r? Assume the track is frictionless.

(A) r
(B) $3r/2$
(C) $2r$
(D) $5r/2$
(E) $3r$

7. The momentum of an object as a function of time is given by the equation $\mathbf{p}(t) = kt^3\mathbf{i}$, where k is a constant. What is the equation for the net force causing this motion?

(A) $\mathbf{F}(t) = kt^3\mathbf{i}$
(B) $\mathbf{F}(t) = kt^2\mathbf{i}$
(C) $\mathbf{F}(t) = 3kt^2\mathbf{i}$
(D) $\mathbf{F}(t) = 3kt\mathbf{i}$
(E) $\mathbf{F}(t) = (3/2)kt^2\mathbf{i}$

8. John pushes a crate across the floor of his warehouse (a horizontal surface) at a constant speed of 2 m/s. If the mass of the crate is 50 kg, find the power that John must supply given that $\mu_k = 0.2$.

(A) 50 watts
(B) 99 watts
(C) 153 watts
(D) 170 watts
(E) 196 watts

9. An 8,000 kg truck is accelerated uniformly from rest to a speed of 42 m/s in a time of 7 seconds. What is the average power required by the truck for the acceleration?

(A) 510 kW
(B) 1,008 kW
(C) 2,010 kW
(D) 3,500 kW
(E) 4,732 kW

10. What is the work done by an external force to compress a spring from its equilibrium position ($x = 0$) to some point x if the force required is given by $F(x) = 2kx^2$, where k is a constant?

    (A) $2kx^3/3$
    (B) $2kx^3$
    (C) $2kx^2$
    (D) kx^2
    (E) $kx^2/6$

11. A pickup truck with mass 1,000 kg is traveling east at 40 m/s. The driver sees a stop sign on the road 41 meters ahead. What force must the brakes exert in order to stop the truck in a distance of 40 meters, assuming constant acceleration?

    (A) 20,000 N east
    (B) 20,000 N west
    (C) 30,000 N east
    (D) 30,000 N west
    (E) 40,000 N west

12. An object experiences projectile motion when the following is true:

    (A) Gravity is the only force acting on it.
    (B) Both gravity and an initial force are acting on the object.
    (C) The object has an initial velocity straight up only.
    (D) The object experiences a non-constant acceleration.
    (E) The object has a horizontal velocity only.

13. Earth has a mass of approximately 80 times the mass of the moon. If the earth exerts a gravitational force on the moon of F, then what is the magnitude of the gravitational force that the moon exerts on the earth?

    (A) $F/80$
    (B) F
    (C) $9F$
    (D) $10F$
    (E) $80F$

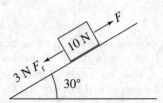

14. A crate weighing 10 N is on a ramp which is inclined at 30° to the horizontal. A 3 N force of friction F_f acts on the crate as it is pulled up the incline at a constant velocity with a force F which is parallel to the incline. Which of the following is the magnitude of F?

    (A) 3 N
    (B) 5 N
    (C) 8 N
    (D) 10 N
    (E) 15 N

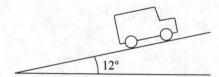

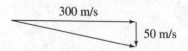

15. As shown in the figure above, a 3.0 x 10⁵ N truck is at rest on a hill that makes a 12° angle with the horizontal. Which of the following is equal to the component of the truck's weight that is parallel to the incline of the hill?

    (A) $(3.0 \times 10^5 \, \text{N})(\sin 12°)$
    (B) $(3.0 \times 10^5 \, \text{kg})(9.8 \, \text{m/s}^2)(\sin 12°)$
    (C) $(3.0 \times 10^5 \, \text{kg})(9.8 \, \text{m/s}^2)(\cos 12°)$
    (D) $(3.0 \times 10^5 \, \text{N})(\cos 12°)$
    (E) $(3.0 \times 10^5 \, \text{N})(\sin 78°)$

16. What is the benefit of using a multiple-pulley system to lift a piano up a flight of stairs instead of just carrying the piano up the stairs (assume ideal conditions and that the piano fits in the stairwell)?

    (A) The amount of time needed to move the piano is reduced.
    (B) The distance the piano must be moved is reduced.
    (C) The power that must be exerted will be reduced.
    (D) The amount of force needed to move the piano is reduced.
    (E) The amount of work needed to move the piano is reduced.

17. An airplane is flying at a speed of 300 m/s to the east with respect to the air. The air is traveling 50 m/s to the south with respect to the ground. What is the velocity of the plane with respect to the ground?

    (A) 350 m/s east
    (B) 300 m/s east
    (C) 250 m/s east
    (D) 304 m/s east-southeast
    (E) 350 m/s east-southeast

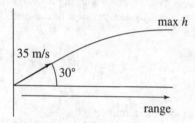

18. An airplane is launched at an angle of 30° with the horizontal at a speed of 35 m/s, but its engine fails immediately, and the pilot is able to glide the plane in a parabolic projectile and land safely. How many meters after launch does the plane land?

    (A) 25.9 m
    (B) 33.2 m
    (C) 47.6 m
    (D) 54.2 m
    (E) 69.3 m

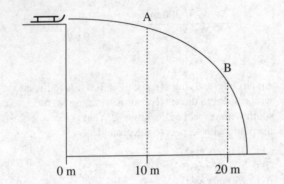

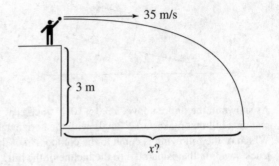

19. The diagram shown above represents a sled being driven off a cliff. Compared to the horizontal component of the sled's velocity at point A (ignoring friction and air resistance), the horizontal component of the sled's velocity at B will be

    (A) double
    (B) the same
    (C) half
    (D) one quarter
    (E) Cannot be determined

20. A 300 N girl exerts a force of 100 N on a 12,000 N motorboat as she pushes it away from the dock. What is the magnitude of the force that the boat exerts on the girl?

    (A) 100 N
    (B) 200 N
    (C) 300 N
    (D) 400 N
    (E) 12,000 N

21. Mary throws a baseball horizontally with an initial velocity of 35 m/s from a height of 3 m. How far will the ball travel before it reaches the ground?

    (A) 15 m
    (B) 21.2 m
    (C) 27.4 m
    (D) 31.2 m
    (E) 35 m

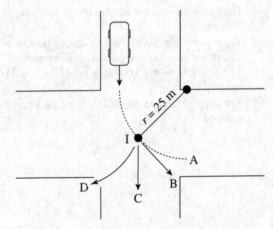

22. A 1,000 kg car travels at a constant speed of 10 m/s and turns at an intersection. The car follows a horizontal circular path with a radius of 20 meters to point I. At point I, the car hits an ice patch and loses all frictional force on its tires. Which path does the car then follow?

    (A) A
    (B) B
    (C) C
    (D) D
    (E) Cannot be determined

23. A 600 g toy train completes 10 laps of its circular track in 1 minute and 40 seconds. If the diameter of the track is 2 meters, find the magnitude of the centripetal force acting on the train.

    (A) 0.11 N
    (B) 0.18 N
    (C) 0.24 N
    (D) 0.28 N
    (E) 0.33 N

24. For the same toy train, what is the train's frequency of revolution?

    (A) 0.05 Hz
    (B) 0.1 Hz
    (C) 0.15 Hz
    (D) 0.2 Hz
    (E) 0.25 Hz

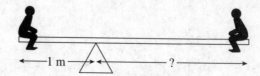

25. One child of mass 50 kg sits on a massless seesaw 1 meter from the fulcrum. How far from the fulcrum must a 20 kg child sit in order to maintain static equilibrium?

    (A) 1 m
    (B) 1.5 m
    (C) 2 m
    (D) 2.5 m
    (E) 3.5 m

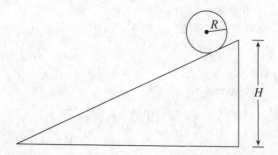

26. Find the speed of a disc of radius R which starts from rest and rolls down an incline of height H without slipping.

    (A) $(1/2)gH$
    (B) $(1/4)gH$
    (C) $(4/3)gH$
    (D) $\sqrt{\dfrac{4}{3}gH}$
    (E) $\sqrt{\dfrac{1}{3}gH}$

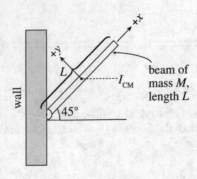

wall

$+x$

$+y$

L

I_{CM}

beam of
mass M,
length L

45°

27. A beam of mass M and length L has a moment of inertia about its center of $ML^2/12$. The beam is attached to a frictionless hinge at an angle of 45° to the horizontal and allowed to swing freely. Find the beam's initial angular acceleration.

(A) $3g\cos 45/2L$
(B) $3g\sin 45/2L$
(C) $3\cos 45/2$
(D) $3\cos 45/2L$
(E) $3g\cos 45/2$

28. As an object moves away from the surface of the earth, the graph of the gravitational potential energy as a function of distance, r, is

(A)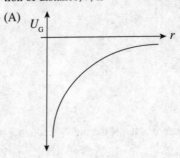

(B)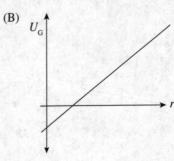

(C)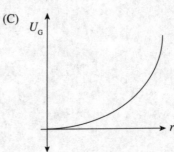

(D)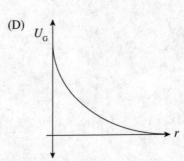

(E)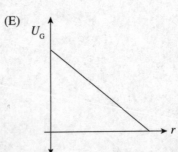

GO ON TO THE NEXT PAGE.

29. A mass-spring system of oscillation undergoes simple harmonic motion with amplitude B. At what location from the equilibrium position will the kinetic energy of the system equal its potential energy?

(A) $x = B$

(B) $x = B/2$

(C) $x = \dfrac{B}{\sqrt{2}}$

(D) $x = \dfrac{B}{\sqrt{3}}$

(E) $x = B/4$

30. For a simple pendulum, with no external forces acting on it, the total energy

(A) remains constant
(B) always equals zero
(C) oscillates between increasing and decreasing
(D) decreases as the object falls
(E) increases as the object falls

31. A planet has 5 times the mass of the earth and 2 times the radius. The acceleration due to gravity at the surface of the planet is closest to

(A) 6.25 m/s^2
(B) 12.5 m/s^2
(C) 25 m/s^2
(D) 10 m/s^2
(E) 49 m/s^2

32. Which of the following shows the graph for the elastic potential energy of a spring undergoing simple harmonic motion as a function of displacement?

(A)

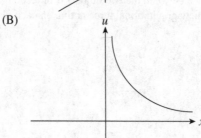

(B)

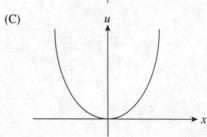

(C)

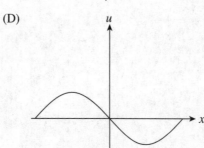

(D)

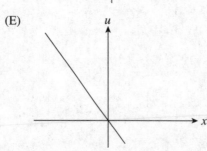

(E)

GO ON TO THE NEXT PAGE.

33. A pendulum with an adjustable length has a period of T when its length is L. What length is necessary for a period of $T/3$?

    (A) $L/27$
    (B) $L/9$
    (C) $L/3$
    (D) $3L$
    (E) $9L$

34. Two large objects of equal mass are separated by a distance R and exert a gravitational pull of magnitude F. If the distance between the two objects is reduced to $R/4$, what is the new gravitational force acting on each object?

    (A) $16F$
    (B) $4F$
    (C) $2F$
    (D) $F/4$
    (E) $F/16$

35. In a horizontal mass-spring system, the maximum displacement is A. When is the velocity the greatest?

    (A) $x = -A$
    (B) $x = A$
    (C) $x = -A/2$
    (D) $x = A/2$
    (E) $x = 0$

STOP

END OF SECTION I, MECHANICS

PHYSICS C: MECHANICS
SECTION II
Time—45 minutes
3 Questions

Directions: Answer all three questions. The suggested time is about 15 minutes per question for answering each of the questions, which are worth 15 points each. The parts within a question may not have equal weight.

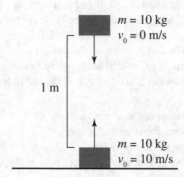

1. As shown above, two blocks of mass $m = 2$ kg are set to collide. The top block, Block 1, is 1 m above the bottom block, Block 2, and is released from rest. At that same moment, the bottom block is launched straight up with an initial speed of 10 m/s. When they collide, the collision will be perfectly inelastic.

   (a) How long after the blocks start moving will the collision occur?
   (b) How far above the ground will the collision occur?
   (c) What will be the speeds of the blocks just before collision?
   (d) After the collision, what is the highest point the blocks will reach?
   (e) What is the minimum speed with which the bottom block can be launched so that the blocks will reach a height of 1 m after the collision?

GO ON TO THE NEXT PAGE.

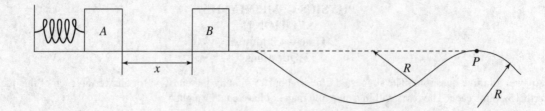

2. A massless spring with force constant k is attached at its left end to a wall, as shown above. Initially, block A and block B, each of mass M, are at rest on a frictionless, level surface, with block A in contact with the spring (but not compressing it) and block B a distance x from block A. Block A is then moved to the left, compressing the spring a distance of d, and held in place while block B remains at rest. First, block A is released, then as it passes the equilibrium position, it loses contact with the spring. After block A is released, it moves forward and has a perfectly inelastic collision with block B and then follows the frictionless, curved path shown above. The radius of the valley and the hill in the diagram are both R. Answer the following in terms of $M, k, d, x, g,$ and R.

(a) Determine the speed of block A just before it collides with block B.
(b) Determine the speed of block B just after the collision occurs.
(c) Determine the change in kinetic energy for the collision.
(d) Determine the normal force on the boxes when they are at position P, the top of the hill.

GO ON TO THE NEXT PAGE.

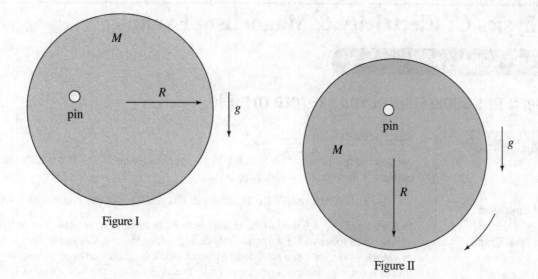

Figure I

Figure II

3. A disk of mass M and radius R is pinned half of the way along its radius, and held in a horizontal position, as shown in Figure I. The rotational inertia of the disk about its center is $\frac{1}{2}MR^2$. The disk is released at $t = 0$ s, falls to the vertical position shown in Figure II, and continues to rotate about the pin. Answer the following in terms of M, R, and g.

(a) Calculate the rotational inertia of the disk about the pin.
(b) Calculate the angular acceleration of the disk at $t = 0$ s.
(c) Calculate angular velocity of the disk when it is in the vertical position shown in Figure II.

Now the disk is stopped and brought to rest in the vertical position shown in Figure II. It is given a slight disturbance to an angle θ_0.

(d) Calculate the angular frequency of the oscillation.

STOP

END OF SECTION II, MECHANICS

AP® Physics C: Electricity & Magnetism Exam

SECTION I: Multiple-Choice Questions

DO NOT OPEN THIS BOOKLET UNTIL YOU ARE TOLD TO DO SO.

At a Glance

Total Time
45 minutes
Number of Questions
35
Percent of Total Grade
50%
Writing Instrument
Pen required

Instructions

Section I of this examination contains 35 multiple-choice questions. Fill in only the ovals for numbers 1 through 35 on your answer sheet.

CALCULATORS MAY BE USED IN BOTH SECTIONS OF THE EXAMINATION.

Indicate all of your answers to the multiple-choice questions on the answer sheet. No credit will be given for anything written in this exam booklet, but you may use the booklet for notes or scratch work. After you have decided which of the suggested answers is best, completely fill in the corresponding oval on the answer sheet. Give only one answer to each question. If you change an answer, be sure that the previous mark is erased completely. Here is a sample question and answer.

Sample Question   Sample Answer

Chicago is a   (A) ● (C) (D) (E)
(A) state
(B) city
(C) country
(D) continent
(E) planet

Use your time effectively, working as quickly as you can without losing accuracy. Do not spend too much time on any one question. Go on to other questions and come back to the ones you have not answered if you have time. It is not expected that everyone will know the answers to all the multiple-choice questions.

About Guessing

Many candidates wonder whether or not to guess the answers to questions about which they are not certain. Multiple-choice scores are based on the number of questions answered correctly. Points are not deducted for incorrect answers, and no points are awarded for unanswered questions. Because points are not deducted for incorrect answers, you are encouraged to answer all multiple-choice questions. On any questions you do not know the answer to, you should eliminate as many choices as you can, and then select the best answer among the remaining choices.

GO ON TO THE NEXT PAGE.

This page intentionally left blank.

GO ON TO THE NEXT PAGE.

PHYSICS C: ELECTRICITY & MAGNETISM
SECTION I
Time—45 minutes
35 Questions

Directions: Each of the following questions or incomplete statements below is followed by five suggested answers or completions. Select the one that is best in each case and then mark it on your answer sheet.

Questions 1–2

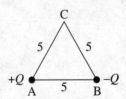

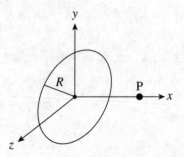

1. Two charges, one positive and the other negative, of equal magnitude Q, are placed at the corners of the base of an equilateral triangle with a side length 5, as shown above. What is the magnitude and direction of the net electric field at point C due to the two charges?

   (A) $kQ/4s^2$ in the positive x direction
   (B) $kQ/2s^2$ in the positive x direction
   (C) kQ/s^2 in the positive x direction
   (D) $kQ/2s^2$ in the positive y direction
   (E) $4kQ/s^2$ in the positive y direction

2. What is the net electric force on a positive test charge, q, placed at point C?

   (A) $kqQ/4s^2$ in the positive x direction
   (B) $kqQ/2s^2$ in the positive x direction
   (C) kqQ/s^2 in the positive x direction
   (D) $kqQ/2s^2$ in the positive y direction
   (E) kqQ/s^2 in the positive y direction

3. The electric field due to a ring of charge Q with radius R that lies in the y-z plane and is centered on the x-axis is $E = (Qx)/(4\pi\varepsilon_o(x^2 + R^2)^{3/2})$ in the positive x direction at point P. What is the distance, x, along the x-axis where the electric field is a maximum?

   (A) $R/8$
   (B) $R/4$
   (C) $R/2$
   (D) $\dfrac{R}{\sqrt{2}}$
   (E) R

GO ON TO THE NEXT PAGE.

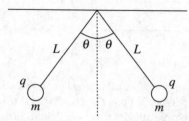

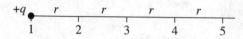

4. Two small spheres, each with mass m, are suspended at the ends of two strings of length L. Both spheres are charged with an equal amount of positive charge q. Each string makes an equal angle of θ with the vertical. Determine the magnitude of charge q on each sphere.

(A) $q = (2L\sin\theta)^2\sqrt{\dfrac{mg\tan\theta}{k}}$

(B) $q = (2L\sin\theta)\sqrt{\dfrac{mg\tan\theta}{k}}$

(C) $q = (L\sin\theta)\sqrt{\dfrac{mg\tan\theta}{k}}$

(D) $q = (2L\sin\theta)(mg\tan\theta/k)$

(E) $q = (2L\sin\theta)^2(mg\tan\theta/k)$

6. An electric field is created by a positive charge $+q$ located at point 1. Each point in the diagram above has the same distance, r, between them. If the electric field at point 2 is E_o, what is the electric field at point 5 in terms of E_o?

(A) $4E_o$
(B) $2E_o$
(C) $E_o/4$
(D) $E_o/9$
(E) $E_o/16$

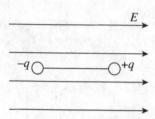

7. A dipole is placed in a uniform electric field as shown above. Which of the following statements is true about the net force on the dipole?

(A) The net force is zero.
(B) The net force causes the dipole to rotate in a clockwise direction.
(C) The net force causes the dipole to rotate in a counterclockwise direction.
(D) The net force is directed to the left.
(E) The net force is directed to the right.

8. A positively charged particle traveling with speed v enters a uniform electric field E that is perpendicular to the initial velocity. Which of the following paths describes the motion of the particle once it enters the field?

(A) Straight line parallel to the field
(B) Straight line perpendicular to the field
(C) Parabolic
(D) Circular
(E) Helical

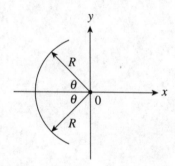

5. A negative charge is distributed uniformly around a wire arc of radius $R = 0.15$ m, linear charge density $\lambda = -3.0 \times 10^{-6}$ C/m, and $\theta = 45°$. Determine the total charge on the wire.

(A) -7.1×10^{-9} C
(B) -7.1×10^{-9} C/m
(C) -7.1×10^{-7} C
(D) -7.1×10^{-7} C/m
(E) -7.1×10^{-5} C/m

GO ON TO THE NEXT PAGE.

Questions 9–10

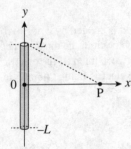

The above figure illustrates a line of charge q with length $2L$.

9. What is the value of the y-component of the electric field at point P?

(A) 0

(B) $(1/4\pi\varepsilon_0)\dfrac{q}{x^2}$

(C) $(1/4\pi\varepsilon_0)\dfrac{q}{\sqrt{x^2+L^2}}$

(D) $(1/4\pi\varepsilon_0)\dfrac{q}{x\sqrt{x^2+L^2}}$

(E) $(1/4\pi\varepsilon_0)\dfrac{q}{x\left(x^2+L^2\right)^{3/2}}$

10. What is the value of the x-component of the electric field at point P?

(A) 0

(B) $(1/4\pi\varepsilon_0)\dfrac{q}{x^2}$

(C) $(1/4\pi\varepsilon_0)\dfrac{q}{\sqrt{x^2+L^2}}$

(D) $(1/4\pi\varepsilon_0)\dfrac{q}{x\sqrt{x^2+L^2}}$

(E) $(1/4\pi\varepsilon_0)\dfrac{q}{x\left(x^2+L^2\right)^{3/2}}$

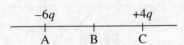

11. Two charges are located on the line shown above, one at point A and one at point C. The charge at point A is $-6q$ and the charge at point C is $+4q$. Point B is halfway between points A and C. Other than at infinity, the electric field strength is zero at a point on the line above in which of the following ranges?

(A) To the left of point A
(B) Between points A and B
(C) Between points B and C
(D) To the right of point C
(E) None. The electric field is zero only at infinity.

12. In a region of spherical symmetry, the electric potential as a function of distance x is given by the equation $V(x) = ax^3$. What is the value of the electric field at a point where $x = b$?

(A) $2ab^2$
(B) ab
(C) $3ab^2$
(D) $-3ab^2$
(E) $-2ab^3$

GO ON TO THE NEXT PAGE.

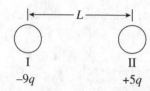

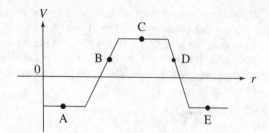

13. As shown in the diagram above, two identical conducting spheres are charged to $-9q$ and $+5q$, and separated by a distance L. The spheres are brought together, make contact with each other, and then separate again to a distance of L. What is the new charge on each sphere?

    (A) $-14q$
    (B) $-4q$
    (C) $-2q$
    (D) $+4q$
    (E) $+14q$

15. The graph above shows the electric potential V as a function of distance r. At which point is the electric field positive with respect to the positive value of r?

    (A) A
    (B) B
    (C) C
    (D) D
    (E) E

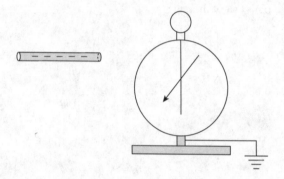

14. A negatively charged rod is brought near, but does not touch, a neutral, grounded electroscope. Holding the rod in place, the ground wire is removed from the electroscope. Which of the following is now true of the electroscope?

    (A) It is uncharged since it was originally neutral.
    (B) It is uncharged because the rod did not touch the electroscope.
    (C) It has a charge, but it's positive in certain places and negative in others.
    (D) It is negatively charged.
    (E) It is positively charged.

16. The diagram shows two spheres. Sphere II is neutral and grounded. A positively charged sphere I is brought close to but does not touch sphere II. Which of the following statements is true?

    (A) Negative charge flows from sphere II to the ground.
    (B) Negative charge flows from the ground to sphere II.
    (C) No charge flows since sphere II is neutral.
    (D) Positive charge flows from sphere II to the ground.
    (E) Positive charge flows from the ground to sphere II.

GO ON TO THE NEXT PAGE.

17. An isolated capacitor with no dielectric inserted between its plates has a potential difference V_o and a charge Q_o. After filling the space between the plates of the capacitor with oil, the difference in potential is V and the difference in charge is Q. Which of the following pairs of relationships is correct?

   (A) $Q = Q_o$ and $V > V_o$
   (B) $Q = Q_o$ and $V < V_o$
   (C) $Q > Q_o$ and $V = V_o$
   (D) $Q < Q_o$ and $V > V_o$
   (E) $Q > Q_o$ and $V > V_o$

18. A parallel-plate capacitor has a given capacitance of C_1. A second parallel plate capacitor has plates with twice the area and 4 times the distance between the plates of the first capacitor. The capacitance of the second capacitor in terms of C_1 is:

   (A) $4C_1$
   (B) $2C_1$
   (C) C_1
   (D) $(½)C_1$
   (E) $(¼)C_1$

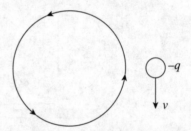

19. An electron enters a region where both a magnetic field B and electric field E act on the particle. The magnetic field B acts into the page, and the electric field acts between the two plates. What must be the direction of the electric field in order for the particle to move in a straight line?

   (A) Toward the top of the page
   (B) Toward the bottom of the page
   (C) Toward the left of the page
   (D) Toward the right of the page
   (E) Out of the page

20. A wire loop has a current I running through it counter-clockwise. An electron moving with velocity v downward is placed outside of the loop and in the plane of the loop. Which of the following describes the direction of the magnetic force on the electron from the magnetic field generated by the current in the wire loop?

   (A) Toward the top of the page
   (B) Toward the bottom of the page
   (C) Toward the left
   (D) Toward the right
   (E) Out of the page

GO ON TO THE NEXT PAGE.

Questions 21–22

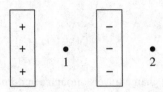

The above figure shows two oppositely charged parallel plates each producing a uniform electric field **E**.

21. What is the magnitude and direction of the electric field at point 1?

(A) $2E$ to the right
(B) E to the right
(C) 0 with no direction
(D) E to the left
(E) $2E$ to the left

22. What is the magnitude and the direction of the electric field at point 2?

(A) $2E$ to the right
(B) E to the right
(C) 0 with no direction
(D) E to the left
(E) $2E$ to the left

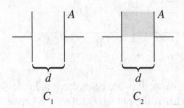

23. A capacitor has two parallel plates, each with area A separated by a distance d and a capacitance of C_1, as shown above. A second capacitor, also with plates with area A separated by distance d, is filled with a dielectric material with dielectric constant κ that fills in the space between the plates and covers half of the area of each plate. What is the capacitance of the second capacitor in terms of C_1?

(A) $\kappa C_1/4$
(B) $\kappa C_1/2$
(C) $(\kappa + 1)C_1/8$
(D) $(\kappa + 1)C_1/4$
(E) $(\kappa + 1)C_1/2$

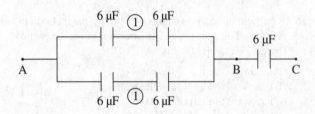

24. Five 6 μF capacitors are connected in a network as shown above. What is the equivalent capacitance of the circuit?

(A) 1 μF
(B) 3 μF
(C) 6 μF
(D) 9 μF
(E) 12 μF

GO ON TO THE NEXT PAGE.

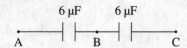

25. Two 6 μF capacitors are connected in the network shown above. What is the potential difference between points B and C if the circuit is connected to a 12 V battery?

    (A) 3 V
    (B) 4 V
    (C) 6 V
    (D) 9 V
    (E) 12 V

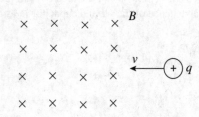

26. A particle of charge $+q$ enters a uniform magnetic field B directed into the page. The direction of the magnetic force F_B on the particle is

    (A) toward the top of the page
    (B) toward the bottom of the page
    (C) toward the left of the page
    (D) toward the right of the page
    (E) out of the page

27. A positive charge travels through a uniform magnetic field. The diagram above represents the direction of the particle's velocity (v) and magnetic force (F_B) on the charge at a certain point. What is the direction of the magnetic field?

    (A) Toward the top of the page
    (B) Toward the bottom of the page
    (C) Toward the left of the page
    (D) Into the page
    (E) Out of the page

28. When two identical parallel-plate capacitors are connected in series, which of the following statements is correct of the equivalent capacitance?

    (A) It is smaller than the capacitance of each capacitor.
    (B) It is bigger than the capacitance of each capacitor.
    (C) It is the same as the capacitance of each capacitor.
    (D) It depends on the charge stored on each capacitor.
    (E) It depends on the potential difference across both capacitors.

29. A sheet of glass is inserted between the plates of an isolated, charged parallel-plate capacitor. Which of the following statements is true?

    (A) The capacitance decreases.
    (B) The energy of the capacitor remains constant.
    (C) The charge on the plates decreases.
    (D) The electric field between the plates increases.
    (E) The potential difference across the capacitor decreases.

GO ON TO THE NEXT PAGE.

Questions 30–32

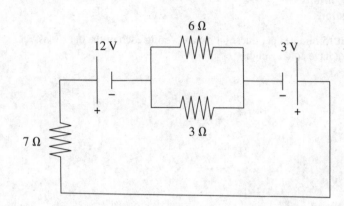

Questions 33–35

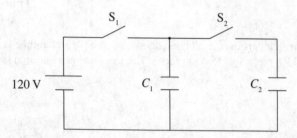

The above electric circuit includes one battery, two capacitors, and two switches. The voltage produced by the battery is 120 V. The capacitors have capacitance $C_1 = 9$ μF and $C_2 = 18$ μF. Initially, all the switches are open, and the capacitors are uncharged.

30. Calculate the current through the 7 Ω resistor in the diagram of the circuit above.

  (A) 1 A clockwise
  (B) 1 A counterclockwise
  (C) 2 A clockwise
  (D) 2 A counterclockwise
  (E) 3 A counterclockwise

31. Calculate the voltage drop across the 6 Ω and 3 Ω parallel resistors.

  (A) 1 V
  (B) 2 V
  (C) 3 V
  (D) 6 V
  (E) 9 V

32. Calculate the current through the 3 Ω resistor.

  (A) 0 A
  (B) 1/3 A
  (C) 2/3 A
  (D) 1 A
  (E) 2 A

33. Determine the charge on capacitor C_1 after switch S_1 is closed for a long time.

  (A) 1.08×10^{-6} C
  (B) 1.08×10^{-4} C
  (C) 1.08×1^{-3} C
  (D) 2.16×10^{-3} C
  (E) 2.16×10^{-6} C

34. Switch S_1 is opened and, afterward, switch S_2 is closed. Determine the charge on Capacitor C_1 after a long time.

  (A) 1.08×10^{-6} C
  (B) 2.16×10^{-6} C
  (C) 2.16×10^{-4} C
  (D) 3.6×10^{-4} C
  (E) 3.6×10^{-6} C

35. When S_1 is opened and switch S_2 is closed, determine the potential difference across capacitor C_1 when equilibrium is reached.

  (A) 120 V
  (B) 60 V
  (C) 40 V
  (D) 30 V
  (E) 20 V

STOP

END OF SECTION I, ELECTRICITY & MAGNETISM

PHYSICS C: ELECTRICITY & MAGNETISM
SECTION II
Time—45 minutes
3 Questions

Directions: Answer all three questions. The suggested time is about 15 minutes per question for answering each of the questions, which are worth 15 points each. The parts within a question may not have equal weight.

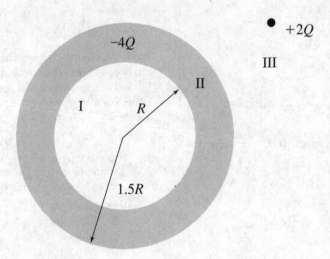

1. A spherical, metal shell of inner radius R and outer radius $1.5R$ has a charge of $-4Q$. A point charge of $+2Q$ is initially located outside the shell as shown above. Express all answers in terms of fundamental constants and given values.

   (a)

       i. Determine the charge on each surface of the spherical shell.

       ii. Sketch the electric field in regions I, II, and III.

GO ON TO THE NEXT PAGE.

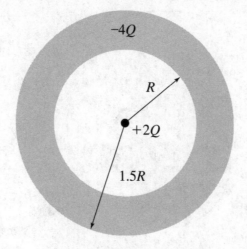

Now the $+2Q$ point charge is moved to the center of the spherical shell as shown above.

(b) Determine the electric field strength for the following radii.

    i. $r < R$

    ii. $R < r < 1.5R$

    iii. $r > 1.5R$

(c) Determine the potential difference between infinity and the outside surface of the spherical shell.

GO ON TO THE NEXT PAGE.

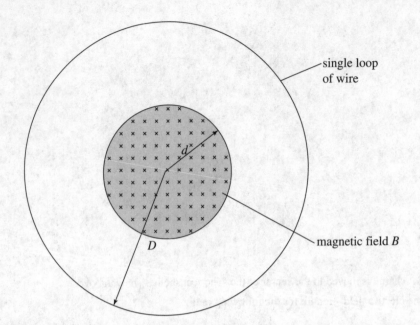

single loop
of wire

magnetic field B

d

D

2. A uniform magnetic field B is directed into the page, and exists in a circular region of radius d. A single loop of wire of radius D is placed concentrically around the magnetic field region in the plane of the page. The initial magnetic field strength is B_0. Calculate the following in terms of given values and fundamental constants.

    (a) Determine the initial flux through the loop of wire.

    At time $t = 0$ s, the magnetic field strength as a function of time t is given by the equation $B(t) = B_0 t^2$, where B_0 is a positive constant.

    (b) Determine the magnitude of the induced emf in the single loop.
    (c) Determine the direction of the induced current in the loop.

    The loop of wire has a resistance R.

    (d) Determine the energy dissipated in the loop up until a given time t_1.

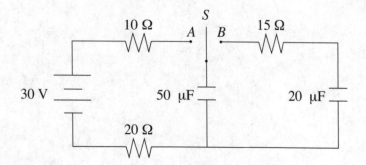

3. In the circuit shown above, the switch S is initially in the open position and both capacitors are initially uncharged. Then the switch is moved to position A.

(a) Determine the current through the 20 Ω resistor immediately after the switch is moved to position A.

(b) Sketch a graph of voltage vs. time for the voltage across the 10 Ω resistor.

After a long time, the switch is moved to position B.

(c) Determine the current through the 15 Ω resistor immediately after the switch is moved to position B.

(d) Determine the amount of charge stored on the upper plate of the 20 μF capacitor after a long time.

STOP

END OF EXAM

Practice Test 2:
Answers and
Explanations

MECHANICS

PRACTICE TEST 2 ANSWER KEY

1.	C		19.	B
2.	B		20.	A
3.	D		21.	C
4.	E		22.	B
5.	D		23.	C
6.	D		24.	B
7.	C		25.	D
8.	E		26.	D
9.	B		27.	A
10.	A		28.	A
11.	B		29.	C
12.	A		30.	A
13.	B		31.	B
14.	C		32.	C
15.	A		33.	B
16.	D		34.	A
17.	D		35.	E
18.	D			

PRACTICE TEST 2: MECHANICS ANSWERS AND EXPLANATIONS

Section I—Multiple-Choice

1. **C** Let x_{cm} denote the center of mass for the x-axis and let y_{cm} denote the center of mass for the y-axis. In addition to the equation for center of mass, we can determine that the center of mass for the x-axis must be zero because each of the figures has their center at 0 on the x-axis and is evenly distributed on either side of the x-axis. Therefore:

$$x_{cm} = \frac{m_1 x_1 + m_2 x_2}{m_1 + m_2} = \frac{(5\,\text{kg})(0) + (10\,\text{kg})(0)}{5\,\text{kg} + 10\,\text{kg}} = 0$$

$$y_{cm} = \frac{m_1 y_1 + m_2 y_2}{m_1 + m_2} = \frac{(5\,\text{kg})(0) + (10\,\text{kg})(6)}{5\,\text{kg} + 10\,\text{kg}} = \frac{60}{15} = 4$$

The center of mass is at (0, 4), answer (C).

2. **B** Let x_{cm} denote the center of mass for the x-axis and let y_{cm} denote the center of mass for the y-axis.

$$x_{cm} = \frac{m_1 x_1 + m_2 x_2 + m_3 x_3}{m_1 + m_2 + m_3}$$

$$x_{cm} = \frac{(4\text{kg})(2) + (6\text{kg})(6) + (2\text{kg})(8)}{4\text{kg} + 6\text{kg} + 2\text{kg}}$$

$$x_{cm} = \frac{8 + 36 + 16}{12} = \frac{60}{12} = 5$$

$$y_{cm} = \frac{m_1 y_1 + m_2 y_2 + m_3 y_3}{m_1 + m_2 + m_3}$$

$$y_{cm} = \frac{(4\text{kg})(3) + (6\text{kg})(4) + (2\text{kg})(2)}{4\text{kg} + 6\text{kg} + 2\text{kg}}$$

$$y_{cm} = \frac{12 + 24 + 4}{12} = \frac{40}{12} = 3.3\overline{3}$$

The center of mass of the object is at (5, 3.33), answer (B).

3. **D** We solve for the velocity of the second ball after the collision using the equation for Conservation of Momentum, where the initial velocity for the second ball, v_2, is 0. Therefore: $m_1v_1 + m_2v_2 = m_1v_1' \cos 40° + m_2v_2' \cos 35°$. Since $v_2 = 0$, then:

$$m_1v_1 = m_1v_1' \cos 40° + m_2v_2' \cos 35°$$

$$m_1v_1 - m_1v_1' \cos 40° = m_2v_2' \cos 35°$$

$$\frac{m_1v_1 - m_1v_1' \cos 40°}{m_2 \cos 35°} = v_2'$$

The answer is (D).

4. **E** To find the power delivered by the net force to a 20 kg mass at time $t = 3$ seconds, we will use the equation Power = (net force) * velocity. We will use the first derivative of $x(t)$ to get the velocity, and the second derivative of $x(t)$ to find the acceleration. Once we have the acceleration, we will find the net force from $F = ma$.

$$x(t) = 5t^3 - 3t^2 + t$$

$$v(t) = \frac{dx}{dt} = 15t^2 - 6t + 1$$

$$a(t) = \frac{dv}{dt} = \frac{d^2x}{dt^2} = 30t - 6$$

$$F_{net} = ma = (20 \text{ kg})(30t - 6)$$

$$F_{net} = 600t - 120$$

Power = Fv since the force is parallel to the velocity.

Power = $(600t - 120)(15t^2 - 6t + 1)$ at $t = 3$ seconds

Power = $(600(3) - 120)(15(3)^2 - 6(3) + 1)$

Power = $(1,800 - 120)(15(9) - 18 + 1)$

Power = $(1,680)(118)$

Power = 198,240 watts

The power = 198,240 watts, answer (E).

5. **D** The Impulse–Momentum Theorem states that $\mathbf{J} = \Delta\mathbf{p}$. By definition, $\mathbf{J}$ is the integral of force with respect to time, which can be found by calculating the area under the F-vs-t graph.

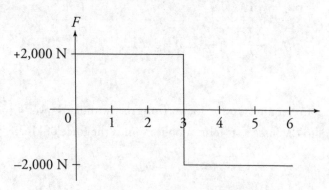

From 1–3 seconds, $F\Delta t = (2{,}000\ \text{N})(2\ \text{sec})$. Therefore, $J = 4{,}000\ \text{N·s}$.

From 3–4 seconds, $F\Delta t = (-2{,}000\ \text{N})(1\ \text{sec})$. Therefore, $J = -2{,}000\ \text{N·s}$.

For the whole interval 1–4 seconds, $J = 4{,}000\ \text{N·s} - 2{,}000\ \text{N·s} = 2{,}000\ \text{N·s}$, which means that Δp must also equal 2,000 N·s. Since $\Delta p = m\Delta v$, $2{,}000\ \text{N·s} = (1{,}000\ \text{kg})\ \Delta v$.

The change in the pickup truck's velocity, Δv, is 2 m/s, (D).

6. **D** In order for the car to remain on the track throughout the loop, centripetal acceleration, a_c, must always be equal to or exceed the acceleration due to gravity, $a_c = v^2/r \geq g$. At the highest point of the loop, the height = $2r$. Due to the nonconstant acceleration, Conservation of Mechanical Energy must be used to solve this problem for the minimum value of D.

$$KE_i + U_i = KE_f + U_f$$

$$mgD = \frac{1}{2}mv_f^2 + mg(2r)$$

$$gD = \frac{1}{2}v_f^2 + g(2r)$$

$$2gD = v_f^2 + 4gr$$

$$2gD - 4gr = v_f^2$$

Substitute $v_f^2 = gr$ since $\dfrac{v^2}{r} \geq g$

$$2gD - 4gr = gr$$

$$2D - 4r = r$$

$$2D = 5r$$

$$D = \frac{5r}{2}$$

In terms of r, the radius of the loop, D, must be at least $\dfrac{5r}{2}$. The correct answer is (D).

7. **C** The force is the derivative of $\mathbf{p}(t)$ with respect to time.

$$\mathbf{p}(t) = kt^3\mathbf{i}$$

$$\mathbf{F}(t) = \frac{d\mathbf{p}(t)}{dt} = \frac{d}{dt}(kt^3\mathbf{i}) = 3kt^2\mathbf{i}$$

The answer is (C).

8. **E** The power that John requires can be found from the formula: Power = (net force) × velocity. The net force is equal to the force that John supplies minus the force of kinetic friction.

$$F_{net} = F_{John} - F_{friction} = ma_x$$

Since this is at a constant speed, $a_x = 0$, then:

$$F_{John} - F_{friction} = 0$$

$$F_{John} = F_{friction} = \mu_k F_N = \mu_k mg = (0.2)(50 \text{ kg})(9.8\text{m/s}^2)$$

$$F_{John} = 98 \text{ N}$$

$$\text{Power} = \text{force} \times \text{velocity} = (98 \text{ N})(2 \text{ m/s}^2)$$

$$\text{Power} = 196 \text{ watts}$$

The power that John requires is 196 watts. The correct answer is (E).

9. **B** To find the average power required by the truck for acceleration, use the formula: average Power = net force × average velocity.

$$\text{Power}_{avg} = F_{net} \times v_{avg}$$

$$F_{net} = ma$$

$$a = \frac{\Delta v}{\Delta t} = \frac{42 \text{ m/s} - 0 \text{ m/s}}{7 \text{ sec}} = 6 \text{ m/s}^2$$

$$v_{avg} = \frac{v_f + v_i}{2} = \frac{42 + 0}{2} = 21 \text{ m/s}$$

$$\text{average Power} = (8,000)(6 \text{ m/s}^2)(21 \text{ m/s})$$

$$\text{average Power} = 1,008,000 \text{ W} = 1,008 \text{ kW}$$

The power required is 1,008,000 watts, or 1,008 kW. The correct answer is (B).

Note: Average Power could also be calculated by $P = W/t$, where W is the work done on the truck by the force. Work can be found using the Work–Kinetic Energy Theorem: $W_{net} = \Delta KE$.

10. **A** To find the work done to compress the spring, we will integrate the equation for $F(x)$ from 0 to position x.

$$F(x) = 2kx^2$$

$$\text{Work} = \int_0^x F(x)\,dx$$

$$\text{Work} = \int_0^x 2kx^2\,dx = 2k\int_0^x x^2\,dx$$

$$\text{Work} = (2k)\frac{x^3}{3}\bigg|_0^x$$

$$\text{Work} = 2k(\frac{x^3}{3} - 0)$$

$$\text{Work} = \frac{2kx^3}{3}$$

The work done is equal to answer (A), $\frac{2kx^3}{3}$.

11. **B** The force the brakes must exert in order to stop the truck in a distance of 40 meters, assuming constant acceleration, can be found from finding the net work. Since the net work = force × distance, the force can be found by dividing the net work by the stopping distance.

$$\text{Work} = \Delta KE = KE_{\text{final}} - KE_{\text{initial}}$$

$$\text{Work} = 0 - \frac{1}{2}mv_i^2$$

$$\text{Work} = -\frac{1}{2}(1{,}000 \text{ kg})(40 \text{ m/s})^2$$

$$\text{Work} = -800{,}000 \text{ J}$$

$$\text{Work} = \text{force} \times \text{distance}$$

$$\text{Force} = \frac{\text{work}}{\text{distance}}$$

$$\text{Force} = \frac{-800{,}000 \text{ J}}{40 \text{ m}}$$

$$\text{Force} = -20{,}000 \text{ N}$$

Taking east as the positive direction, the brakes must exert a 20,000 N stopping force west. The correct answer is (B).

12. **A** An object will experience projectile motion when it is not launched straight upward and experiences only the effect of acceleration due to gravity. Since $F = ma$, if the only acceleration is due to gravity, then the net force will also be the force due to the acceleration of gravity. The correct answer is (A).

13.   **B**   The gravitational force that the moon exerts on the earth is part of the action-reaction pair between the two objects, the earth and the moon. The two forces are equal in magnitude and opposite in direction. Since the earth exerts a gravitational force on the moon of F, then the magnitude of the gravitational force that the moon exerts on the earth must also be F. The correct answer is (B).

14.   **C**   Since the crate is pulled up the incline at a constant velocity, the acceleration is zero. When acceleration is zero, the net force must also equal zero. Since the net force is zero, we are going to set the sum of the forces in the x direction equal to zero and solve for F.

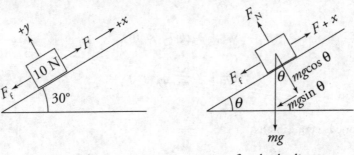

original figure   free-body diagram

$$F_{net_x} = F - F_f - mg\sin\theta = 0$$

$$F = F_f + mg\sin\theta$$

$$F = 3\text{ N} + 10(\sin 30°)$$

$$F = 3\text{ N} + 10\text{ N}(\frac{1}{2}) = 8\text{ N}$$

The correct answer is (C).

15.   **A**   As seen by the free-body diagram shown below, the weight of the truck has an x-component and a y-component, where the x-direction is parallel to the plane and the y-direction is perpendicular to the plane. The component of the truck's weight that is parallel to the incline of the hill will be the x-component of the weight of the truck.

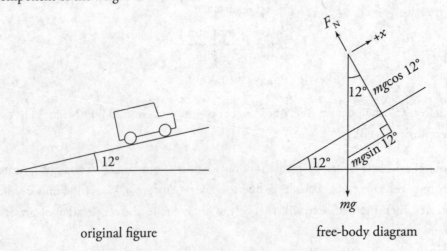

original figure   free-body diagram

The x-component of the truck's weight is $(3 \times 10^5 \text{ N})(\sin 12°)$, answer (A).

16. **D** The benefit of using a multiple-pulley system to lift a piano up a flight of stairs instead of just carrying the piano up the stairs is that the pulleys provide a mechanical advantage to a person. If you picked up the piano, you would need to lift all of its weight. With a double pulley, for example, a person pulling with a force F can lift an object with a weight of $2F$. A pulley system allows a person to lift an object with a reduced force. The correct answer is (D).

17. **D** The velocity of the plane with respect to the ground ($\mathbf{v}_{pg}$) is equal to the velocity of the plane relative to the air ($\mathbf{v}_{pa}$) plus the velocity of the air relative to the ground ($\mathbf{v}_{ag}$). Use the notation of north up, south down, west to the left, and east to the right. Setting up the velocity vectors as shown below, we use the Pythagorean Theorem to solve for the magnitude of the resulting hypotenuse.

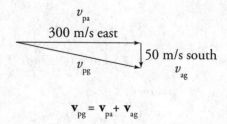

$$\mathbf{v}_{pg} = \mathbf{v}_{pa} + \mathbf{v}_{ag}$$

Since $\mathbf{v}_{pa}$ and $\mathbf{v}_{ag}$ are perpendicular, we can use the Pythagorean Theorem to find the magnitude of $\mathbf{v}_{pg}$.

$$v_{pg}^2 = v_{pa}^2 + v_{ag}^2$$

$$v_{pg}^2 = (300 \text{ m/s})^2 + (50 \text{ m/s})^2$$

$$v_{pg} = \sqrt{300^2 + 50^2} = 304.14 \text{ m/s}$$

The velocity of the plane relative to the ground is 304.14 m/s, answer (D).

18. **D** Since this is projectile motion, we analyze the horizontal and vertical motion in the problem separately. We use the vertical motion to solve for the time required to reach the maximum height of the projectile motion. This is the time the pilot has to land the plane since the engine has failed. We use the horizontal motion to determine the distance the plane went before the pilot landed it. Defining down as the negative y direction, we have the following:

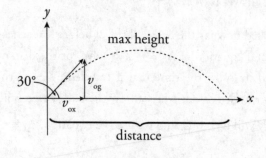

Horizontal

$$v_{ox} = (35 \text{ m/s})(\cos 30°)$$

$$v_{ox} = 30.3 \text{ m/s}$$

<u>Vertical</u>

$$v_{oy} = (35 \text{ m/s})(\sin 30°)$$

$$v_{oy} = 17.5 \text{ m/s}$$

At the maximum height, $v_{fy} = 0$.

$$v_{oy} = 17.5 \text{ m/s}$$

$$v_{fy} = 0$$

$$a = -g = -9.8 \text{ m/s}^2$$

Next, solve for time, t.

$$v = v_{oy} + at$$

$$0 = 17.5 \text{ m/s} + (-9.8 \text{ m/s}^2)t$$

$$9.8t = 17.5$$

$$t = 1.79 \text{ sec}$$

The horizontal distance is found using the formula $v_{0x} = $, so $\Delta x = (30.3 \text{ m/s})(1.79 \text{ sec})$. The horizontal distance = 54.24 m. Therefore, the pilot lands the plan 54.2 m from where it took off, answer (D).

19. **B** This problem asks about the horizontal component of the sled's projectile motion at point A compared to point B. We are ignoring friction and air resistance, so the answer relies on the knowledge that the horizontal component of velocity for projectile motion is a constant and does not change. The value will be the same at points A and B and all points along the horizontal axis. Therefore, the correct answer is (B).

20. **A** From Newton's Third Law, the force that the boat exerts on the girl is equal in magnitude and opposite in direction to the force that the girl exerts on the boat. These forces represent an action-reaction pair. The boat must exert a 100 N force on the girl since the girl exerts a 100 N force on the boat. The correct answer is (A).

21. **C** Since the question asked how far the ball will travel before it reaches the ground, we know that we will need the horizontal equation of $\Delta x = v_{0x}t$ to solve for the distance it travels. However, to find the distance, we need to know the time that it traveled. We use the vertical motion to solve for the time the baseball traveled. Since Mary threw the ball with a horizontal velocity of 35 m/s, the initial vertical velocity will be 0 m/s, and we use the equation, $\Delta y = v_{0y}t + \frac{1}{2}at^2$. Calling downward positive, $a = g$.

$$\Delta y = v_{0y}t + \frac{1}{2}gt^2$$

$$3 = \frac{1}{2}(9.8)t^2$$

$$\sqrt{\frac{2(3)}{9.8}} = t$$

$$\Delta x = v_{0x}t$$
$$\Delta x = (35 \text{ m/s})(.782 \text{ sec})$$
$$\Delta x = 27.37 \text{ m} = 27.4 \text{ m (rounded)}$$

The baseball travels 27.4 m before it hits the ground, (C).

22. **B** The centripetal force that was causing the car to turn the corner on a horizontal circular path was provided by the force of static friction between the bottom of each tire and the road. If the car hits an ice patch at point I, it will lose the frictional force on its tires causing the turn, and the car will then travel in the direction of its velocity, tangent to its path. This corresponds to path B, answer (B).

23. **C** To find the magnitude of the centripetal force acting on the train, you need the centripetal acceleration to use the equation $F_c = ma_c$. To find the centripetal acceleration, you use the equation $a_c = v^2/r$, where v is the speed. You find the speed from the equation v = distance/time. The distance that the toy train traveled is 10 laps around the track, where the distance around the track is the circumference = $2\pi r = \pi d$.

$$v = \frac{distance}{time} = \frac{10(\pi)(2 \text{ m})}{100 \text{ sec}} = 0.628 \text{ m/s}$$

$$a_c = \frac{v^2}{r} = \frac{(.628 \text{ m/s})^2}{1 \text{ m}} = 0.394 \text{ m/s}^2$$

$$F_c = ma_c = (0.6 \text{ kg})(0.394 \text{ m/s}^2)$$

$$F_c = 0.237 \text{ N} = 0.24 \text{ N (rounded)}$$

The train's centripetal force is 0.24 N, answer (C).

24. **B** To find the toy train's frequency of revolution, use the equation that frequency = 1/period. The period is the time it takes the train to do one lap.

$$\text{Frequency} = \frac{1}{T} \qquad (T = \text{period})$$

$$T = \frac{100 \text{ sec}}{10 \text{ laps}} = 10 \text{ sec.}$$

$$\text{Frequency} = \frac{1}{T} = \frac{1}{10 \text{ sec}} = 0.1 \text{ sec}^{-1}$$

The frequency = 0.1 Hz, (B).

25. **D** To achieve static equilibrium, the net torque acting on the seesaw must be equal to 0. Therefore, the torque due to the 20 kg child must be equal in magnitude to the torque due to the 50 kg child. Torque is equal to Force × lever arm.

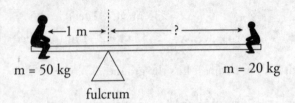

$$\tau_{net} = 0, \text{ so } \tau_{50kg} = \tau_{20kg}$$

$$mgl = mgl$$

$$(50 \text{ kg})(g)(1 \text{ m}) = (20 \text{ kg})(g)(l)$$

$$l = \frac{5}{2} = 2.5 \text{ m}$$

The 20 kg child must sit 2.5 m from the fulcrum to maintain static equilibrium. The correct answer is (D).

26. **D** Using Conservation of Mechanical Energy, $KE_i + U_i = KE_f + U_f + \frac{1}{2}I\omega^2$ where the last term represents the rotational kinetic energy. For the disc, $I_{cm} = \frac{1}{2}MR^2$, assuming its mass = M. Since the disc starts at rest, $KE_i = 0$ and $U_f = 0$.

$$K_i + U_i = K_f + U_f + \frac{1}{2}I\omega^2$$

$$MgH = \frac{1}{2}Mv_{cm}^2 + \frac{1}{2}\left(\frac{1}{2}MR^2\right)\omega^2$$

$$gH = \frac{1}{2}v_{cm}^2 + \frac{1}{4}R^2\omega^2$$

$$gH = \frac{1}{2}v_{cm}^2 + \frac{1}{4}(R\omega)^2 \qquad\qquad v = \omega R$$

$$gH = \frac{1}{2}v_{cm}^2 + \frac{1}{4}v_{cm}^2$$

$$gH = \frac{3}{4}v_{cm}^2$$

$$\frac{4}{3}gH = v_{cm}^2$$

$$\sqrt{\frac{4}{3}gH} = v_{cm}$$

The velocity at the center of mass of the disc = $\sqrt{\frac{4}{3}gH}$. The correct answer is (D).

27. **A** The question asks for the angular acceleration of a beam with mass M, length L, and moment of inertia about its center of $ML^2/12$. Since the beam is swinging about its end, and not the middle, you have to find the moment of inertia of the beam about its end, and then set the net torque = $I \times$ angular acceleration.

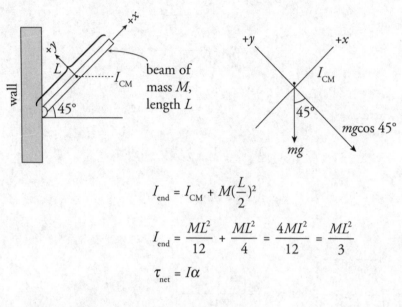

beam of mass M, length L

$mg\cos 45°$

$$I_{end} = I_{CM} + M\left(\frac{L}{2}\right)^2$$

$$I_{end} = \frac{ML^2}{12} + \frac{ML^2}{4} = \frac{4ML^2}{12} = \frac{ML^2}{3}$$

$$\tau_{net} = I\alpha$$

$$-Mg\cos 45 \times (L/2) = (ML^2/3)\alpha$$

$$\alpha = \frac{-Mg\cos 45\left(\dfrac{L}{2}\right)}{\dfrac{ML^2}{3}}$$

$$\alpha = \frac{-3g\cos 45}{2L}$$

The magnitude of the angular acceleration is $\dfrac{3g\cos 45}{2L}$, (A).

28. **A** The formula for gravitational potential energy is: $U = -GMm/r$, where M is the mass of the earth and m is the mass of the object. Potential energy is therefore always negative and increases (by becoming less negative) the farther an object travels from the earth. Choice (A) shows this.

29. **C** To find where on a mass-spring system with amplitude B the kinetic energy will equal the potential energy, we start with the Conservation of Energy equation for a mass-spring system: $E = KE + U$. On a mass-spring system, at amplitude B, potential energy is maximized, and kinetic energy is 0, so $E = 0 + (1/2)kB^2$, where k is the spring constant.

Therefore, $E = (1/2)kB^2$. At any position x, we have:

$$KE + U = E$$

$$\frac{1}{2}mv^2 + \frac{1}{2}kx^2 = E$$

$$\text{So, } v = \sqrt{\frac{E - \frac{1}{2}kx^2}{\frac{1}{2}m}}$$

To solve where $U = KE$:

$$\frac{1}{2}kx^2 = \frac{1}{2}mv^2$$

$$kx^2 = mv^2$$

$$x^2 = \frac{m}{k}v^2$$

$$\text{Since } v = \sqrt{\frac{E - \frac{1}{2}kx^2}{\frac{1}{2}m}} \text{ , then } v^2 = \frac{m}{k}\left(\frac{E - \frac{1}{2}kx^2}{\frac{1}{2}m}\right)$$

$$E = \frac{1}{2}kB^2$$

$$x^2 = \frac{m}{k}\left(\frac{\frac{1}{2}kB^2 - \frac{1}{2}kx^2}{\frac{1}{2}m}\right)$$

$$x^2 = \frac{m}{k}\left(\frac{kB^2 - kx^2}{m}\right)$$

$$x^2 = B^2 - x^2$$

$$2x^2 = B^2$$

$$x^2 = B^2/2$$

$$x = \frac{B}{\sqrt{2}}$$

The location on a mass-spring system where the kinetic energy equals the potential energy is $x = \frac{B}{\sqrt{2}}$, (C).

30. **A** For a simple pendulum with no external forces acting on it, the total energy remains constant and the potential energy is maximized at the endpoints of the oscillation where the speed of the pendulum is zero and the kinetic energy is zero, and the kinetic energy is maximized at the equilibrium position where the potential energy is zero. The correct answer is (A).

31. **B** Calculate the force due to gravity for Earth and the planet, then compare the two calculations to see how the acceleration can be found from the equation $F = mg$. On the surface of a planet with mass M and radius R, an object will feel a gravitational force equal to:

$$Fg = \frac{GMm}{R^2}$$

Therefore, the acceleration due to gravity is:

$$g = F_g/m = GM/R^2$$

On the Earth:

$$g = \frac{GM_E}{R_E^2}$$

$$g_{planet} = \frac{5GM_E}{4R_E^2} = \frac{5}{4}g_E$$

$$g_{planet} = \frac{5}{4}\left(10 \text{ m/s}^2\right) = 12.5 \text{ m/s}^2$$

So the acceleration due to gravity on the planet is (5/4)(acceleration due to gravity on Earth) = 12.5 m/s², (B).

32. **C** The formula for elastic potential energy stored in a spring is $U = kx^2/2$, where k is the spring constant and x is the displacement from equilibrium. The x^2 indicates a parabolic shape with vertex at $(0, 0)$.

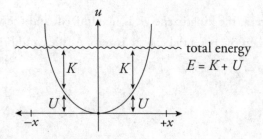

The graph in (C) properly represents this relationship.

33. **B** Use the equation on the exam equation sheet for the period of a pendulum, inserting T and L as given. Remember that 2π and g are constants, so solve for the new length, l, needed for $(\frac{1}{3})T$.

$$T_{new} = T/3 = 2\pi\sqrt{\frac{l}{g}}$$

$$T = 2\pi\sqrt{\frac{L}{g}}$$

To find $\dfrac{T}{3}$, rewrite the equation to read $\dfrac{1}{3}T = \dfrac{1}{3}\left(2\pi\sqrt{\dfrac{L}{g}}\right)$

$$\frac{1}{3}T = \frac{1}{3}\left(\frac{2\pi}{\sqrt{g}} \times \sqrt{L}\right)$$

$$\frac{1}{3}T = \left(\frac{2\pi}{\sqrt{g}}\right)\left(\frac{1}{3}\sqrt{L}\right)$$

$$\frac{1}{3}T = \left(\frac{2\pi}{\sqrt{g}}\right)\left(\sqrt{\frac{L}{9}}\right) = 2\pi\sqrt{\frac{l}{g}}$$

So $l = L/9$ will lead to $T/3$, answer (B).

34. **A** Use the equation for the magnitude of the force due to gravity and compare the values of the force calculated at a distance of R and at a distance of $R/4$.

$$F_R = F = \frac{Gm_1m_2}{R^2}$$

$$F_{R/4} = \frac{Gm_1m_2}{\left(\dfrac{R}{4}\right)^2} = \frac{16Gm_1m_2}{R^2} = 16F$$

The force at a distance of $R/4$ will be 16 times bigger than the force at R. (A) is the correct answer.

35. **E** In a mass-spring system, the kinetic energy is maximized, and the velocity is at its greatest at the equilibrium point, when the potential energy is zero. At the equilibrium point, $x = 0$, (E).

Section II—Free-Response

1. (a) The blocks will collide when the total distance traveled by the blocks is 1 m.

$$d_1 + d_2 = 1 \text{ m} \rightarrow (\tfrac{1}{2} gt^2) + (v_{2,0}t - \tfrac{1}{2} gt^2) = 1 \text{ m} \rightarrow v_{2,0}t = 1 \text{ m} \rightarrow t = \frac{1 \text{ m}}{v_{2,0}} = \frac{1 \text{ m}}{10 \text{ m/s}} = 0.1 \text{ s}$$

(b) The height of the collision will be equal to d_2.

$$d_2 = v_{2,0}t - \tfrac{1}{2} gt^2 = (10 \text{ m/s})(0.1 \text{ s}) - (\tfrac{1}{2})(10 \text{ m/s}^2)(0.1)^2 = 0.95 \text{ m}$$

(c) We can use one of our Big 5 equations to calculate the collision speeds of the blocks, $v_{1,c}$ and $v_{2,c}$.

$$v_{1,c} = v_{1,0} + at = (0 \text{ m/s}) + (10 \text{ m/s}^2)(0.1 \text{ s}) = 1 \text{ m/s}$$

$$v_{2,c} = v_{2,0} - at = (10 \text{ m/s}) - (10 \text{ m/s}^2)(0.1 \text{ s}) = 9 \text{ m/s}$$

Note the changing sign for the acceleration terms above. This is because gravity increases the speed of the falling block but decreases the speed of the launched block.

(d) First, solve for the speed of the blocks after the collision by using the equation for Conservation of Momentum in a perfectly inelastic collision.

$$m_1 \mathbf{v}_{1,c} + m_2 \mathbf{v}_{2,c} = (m_1 + m_2)\mathbf{v}_f \rightarrow \mathbf{v}_f$$

$$= \frac{m_1 \mathbf{v}_{1,c} + m_2 \mathbf{v}_{2,c}}{m_1 + m_2}$$

$$= \frac{(10 \text{ kg})(-1 \text{ m/s}) + (10 \text{ kg})(9 \text{ m/s})}{(10 \text{ kg} + 10 \text{ kg})} = 4 \text{ m/s}$$

Now use one of the Big 5 to solve for maximum height, using 4 m/s as the initial velocity and 0 m/s (the velocity at the peak) as the final.

$$v_f^2 = v_0^2 + 2ad \rightarrow d = \frac{(v_f^2 - v_0^2)}{2a} = \frac{(0 \text{ m/s})^2 - (4 \text{ m/s})^2}{2(-10 \text{ m/s}^2)} = 0.8 \text{ m}$$

Finally, don't forget to add this to the height at which the collision occurred (0.95 m) for a final height of 1.75 m.

(e) First, solve for the launch speed in terms of the time of collision. From our work in (a), we know

$$d_1 + d_2 = (\tfrac{1}{2} gt^2) + (v_{2,0}t - \tfrac{1}{2} gt^2) = v_{2,0}t = 1 \text{ m} \rightarrow v_{2,0} = \frac{1 \text{ m}}{t} \text{ or, alternatively, } t = \frac{1 \text{ m}}{v_{2,0}}$$

Next, solve for the collision speeds as done in (c).

$$v_{1,c} = v_{1,0} + at = v_{1,0} + g\left(\frac{1}{v_{2,0}}\right) = \frac{10 \text{ m/s}^2}{v_{2,0}}$$

$$v_{2,c} = v_{2,0} - at = v_{2,0} - g\left(\frac{1}{v_{2,0}}\right) = v_{2,0} - \frac{10 \text{ m/s}^2}{v_{2,0}}$$

Using these, calculate the post-collision speed as done in (d).

$$m_1\mathbf{v}_{1,c} + m_2\mathbf{v}_{2,c} = (m_1 + m_2)\mathbf{v}_f \rightarrow \mathbf{v}_f$$

$$= \frac{m_1\mathbf{v}_{1,c} + m_2\mathbf{v}_{2,c}}{m_1 + m_2}$$

$$= \frac{-(10 \text{ kg})\left(\dfrac{10 \text{ m/s}^2}{v_{2,0}} + 10 \text{ kg}\left(v_{2,0} - \dfrac{10 \text{ m/s}^2}{v_{2,0}}\right)\right)}{10 \text{ kg} + 10 \text{ kg}}$$

$$= \frac{v_{2,0} - \dfrac{20 \text{ m/s}^2}{v_{2,0}}}{2 \text{ kg}}$$

After the collision, the blocks will have to travel a distance equal to d_1 in order to reach 1 m height. The final step is to set up a kinematics equation using the last result as v_0, d_1 as the distance to travel, 0 m/s as v_f (since we want it to peak at this point), and gravity as the acceleration.

$$v_f^2 = v_0^2 + 2ad \rightarrow (0 \text{ m/s})^2$$

$$= \left(\frac{v_{2,0} - \dfrac{20 \text{ m/s}^2}{v_{2,0}}}{2 \text{ kg}}\right)^2 + 2\left(-10 \text{ m/s}^2\right)\left(\frac{\dfrac{1}{2}\left(10 \text{ m/s}^2\right)1 \text{ m}}{\left(v_{2,0}\right)^2}\right) \rightarrow v_{2,0}$$

$$= 2\sqrt{10} \text{ m/s}$$

2. (a) Use Conservation of Energy since all of the energy stored in the spring will be transferred into kinetic energy of block A. Realize the compression of the spring is given as d.

$$E_1 = E_2$$
$$\frac{1}{2}kd^2 = \frac{1}{2}Mv^2$$
$$\frac{kd^2}{M} = v^2$$

$$v = d\sqrt{\frac{k}{M}}$$

(b) Use Conservation of Linear Momentum for the collision. During a perfectly inelastic collision, the two blocks stick together and continue at the same velocity.

$$p_1 = p_2$$
$$Md\sqrt{\frac{k}{M}} = 2Mv_2$$
$$v_2 = \frac{d}{2}\sqrt{\frac{k}{M}}$$

(c)   The change in the kinetic energy during the collision is given by $\Delta K = K_2 - K_1$.

$$\Delta K = K_2 - K_1$$

$$\Delta K = \frac{1}{2}(2M)v_2^{\,2} - \frac{1}{2}(M)v_1^{\,2}$$

$$\Delta K = \frac{1}{2}(2M)\left(\frac{d}{2}\sqrt{\frac{k}{M}}\right)^2 - \frac{1}{2}(M)\left(d\sqrt{\frac{k}{M}}\right)^2$$

$$\Delta K = \frac{d^2 k}{4} - \frac{d^2 k}{2}$$

$$\Delta K = -\frac{d^2 k}{4}$$

(d)   (See the diagram below for more details.) The blocks are traveling along a curved path and, at the top of the hill, they are traveling with the same velocity as in part (b), since they are at the same elevation. They are also traveling in circular motion at the top of the hill. The gravitational force is pointing toward the center of the circle, so it will be positive and the normal force is pointing away from the center of the circle, so it will be negative.

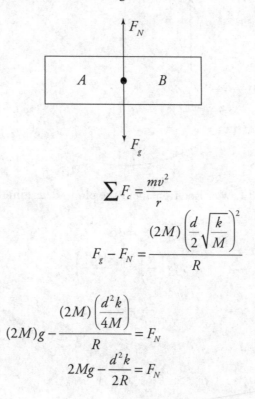

$$\sum F_c = \frac{mv^2}{r}$$

$$F_g - F_N = \frac{(2M)\left(\dfrac{d}{2}\sqrt{\dfrac{k}{M}}\right)^2}{R}$$

$$(2M)g - \frac{(2M)\left(\dfrac{d^2 k}{4M}\right)}{R} = F_N$$

$$2Mg - \frac{d^2 k}{2R} = F_N$$

3.   (a)   Use the parallel–axis theorem to determine the rotational inertia of the object about the pin.

$$I = I_{cm} + md^2$$
$$I = \frac{1}{2}MR^2 + M\left(\frac{R}{2}\right)^2$$
$$I = \frac{3}{4}MR^2$$

(b)   When the object is first released, gravity provides a torque about the pin causing the disk to rotate about the pin. A free-body diagram for the disk is shown here:

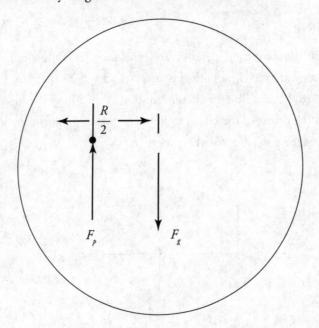

Use Newton's Second Law of Rotation about the pin to determine the angular acceleration.

$$\sum \tau_{pin} = I\alpha$$
$$Mg\left(\frac{R}{2}\right) = \left(\frac{3MR^2}{4}\right)\alpha$$
$$\frac{2g}{3R} = \alpha$$

(c) Use the Law of Conservation of Mechanical Energy to determine the angular velocity at the vertical position. All the initial gravitational potential energy is converted into rotational kinetic energy. The center of mass falls at half the distance of the radius.

$$E_1 = E_2$$

$$Mg\left(\frac{R}{2}\right) = \frac{1}{2}I\omega^2$$

$$\frac{MgR}{2} = \frac{1}{2}\left(\frac{3MR^2}{4}\right)\omega^2$$

$$\frac{4g}{3R} = \omega^2$$

$$\sqrt{\frac{4g}{3R}} = \omega$$

(d) Now the disk is undergoing simple harmonic motion because it is undergoing small angle oscillations. Derive a differential equation of the form $\frac{d^2x}{dt^2} = -\omega^2 x$ to determine ω. Start with Newton's Second Law of Rotation to end up with a differential equation because $\alpha = \frac{d^2\theta}{dt^2}$.

$$\sum \tau_{pin} = I\alpha$$

$$-Mg\sin\theta\left(\frac{R}{2}\right) = \left(\frac{3MR^2}{4}\right)\alpha$$

Using $\sin\theta \approx \theta$, for small θ,

$$-\frac{2g}{3R}\theta = \frac{d^2\theta}{dt^2}.$$

Therefore, $\omega = \sqrt{\frac{2g}{3R}}$.

ELECTRICITY & MAGNETISM
PRACTICE TEST 2 ANSWER KEY

1.	C		19.	B
2.	C		20.	C
3.	D		21.	A
4.	B		22.	C
5.	C		23.	E
6.	E		24.	B
7.	A		25.	C
8.	C		26.	B
9.	A		27.	E
10.	D		28.	A
11.	D		29.	E
12.	D		30.	B
13.	C		31.	B
14.	E		32.	C
15.	D		33.	C
16.	B		34.	D
17.	B		35.	C
18.	D			

PRACTICE TEST 2: ELECTRICITY & MAGNETISM ANSWERS AND EXPLANATIONS

Section I—Multiple-Choice

1. **C** Since the charge at A is positive, and the charge at B is negative, the electric field vector from point A, $\mathbf{E_A}$, will point away from point A, and the electric field vector from point B, $\mathbf{E_B}$, will point towards point B.

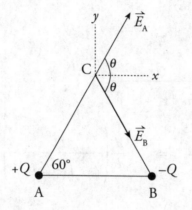 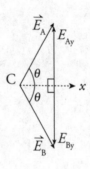

In the x direction:

$$E_{Ax} + E_{Bx} = E_{total}$$

$$\frac{k|Q|}{s^2}\cos 60° + \frac{k|-Q|}{s^2}\cos 60° = E_{total}$$

$$\frac{kQ}{s^2}\left(\frac{1}{2}\right) + \frac{kQ}{s^2}\left(\frac{1}{2}\right) = E_{total}$$

$$\frac{kQ}{s^2} = E_{total}$$

The y-components of $\mathbf{E_A}$ and $\mathbf{E_B}$, E_{Ay} and E_{By} will cancel out since they are equal length, but point in opposite directions. The x-components of $\mathbf{E_A}$ and $\mathbf{E_B}$, E_{Ax} and E_{Bx}, are equal in value and point in the same direction, and add together for the total electric field of kQ/s^2, where k is Coulomb's constant, pointing in the +x direction, (C).

2. **C** From question 1, the net electric field at point C is kQ/s^2, where k is Coulomb's constant, pointing in the +x direction. The net electric force at point C is calculated from the equation $\mathbf{F} = q\mathbf{E}$. If q is positive, $\mathbf{F}$ will point in the same direction as $\mathbf{E}$, in the +x direction. $\mathbf{F} = q(kQ/s^2) = kqQ/s^2$, in the +$x$ direction, (C).

3.   **D**   In this problem, you are given a ring of charge in the *yz*-plane that is centered on the *x*-axis = $(Qx)/$ $(4\pi\varepsilon_o(x^2 + R^2)^{3/2})$ in the positive *x*-direction at point P. To find the distance, *x*, along the *x*-axis where the electric field is a maximum, you take the derivative of the equation for the electric field, set it equal to zero, and solve for *x*.

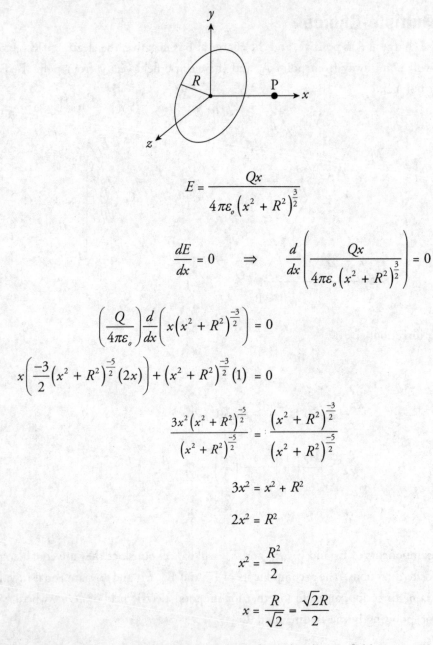

$$E = \frac{Qx}{4\pi\varepsilon_o \left(x^2 + R^2\right)^{\frac{3}{2}}}$$

$$\frac{dE}{dx} = 0 \quad \Rightarrow \quad \frac{d}{dx}\left(\frac{Qx}{4\pi\varepsilon_o \left(x^2 + R^2\right)^{\frac{3}{2}}}\right) = 0$$

$$\left(\frac{Q}{4\pi\varepsilon_o}\right)\frac{d}{dx}\left(x\left(x^2 + R^2\right)^{\frac{-3}{2}}\right) = 0$$

$$x\left(\frac{-3}{2}\left(x^2 + R^2\right)^{\frac{-5}{2}}(2x)\right) + \left(x^2 + R^2\right)^{\frac{-3}{2}}(1) = 0$$

$$\frac{3x^2\left(x^2 + R^2\right)^{\frac{-5}{2}}}{\left(x^2 + R^2\right)^{\frac{-5}{2}}} = \frac{\left(x^2 + R^2\right)^{\frac{-3}{2}}}{\left(x^2 + R^2\right)^{\frac{-5}{2}}}$$

$$3x^2 = x^2 + R^2$$

$$2x^2 = R^2$$

$$x^2 = \frac{R^2}{2}$$

$$x = \frac{R}{\sqrt{2}} = \frac{\sqrt{2}R}{2}$$

The solution for the distance, *x*, along the *x*-axis where the electric field is a maximum is when $x = \frac{R}{\sqrt{2}}$, answer (D).

4. **B** Each sphere has three forces acting on it: T, the tension in the string it is suspended from, F_g, its weight due to gravity, and F_E, the force from the electric field due to the other sphere, as shown in the free-body diagrams.

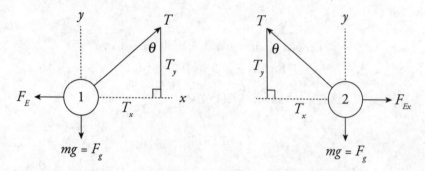

The tension force, T, can be resolved into x- and y-components. Since the spheres are not moving, the net forces in the x- and y-directions must be equal to zero, so we set up equations for the net forces in the x- and y-directions equal to zero for each sphere. Since the spheres are identical, we will use sphere 2 since $\mathbf{F}_E$ is pointing in the positive x direction. Start with a free-body diagram of the forces on each sphere.

$$\sum F_x = F_E - T\sin\theta = 0 \qquad\qquad T\sin\theta = F_E$$

$$\sum F_y = T\cos\theta - mg = 0 \qquad\qquad T\cos\theta = mg$$

To eliminate T, divide the F_x equation by the F_y equation:

$$\frac{T\sin\theta = F_E}{T\cos\theta = mg} \qquad\qquad \tan\theta = \frac{F_E}{mg}$$

$$\tan\theta = \frac{F_E}{mg}$$

From Coulomb's law, $F_E = \dfrac{kqq}{r^2}$. In this problem, $r = 2L\sin\theta$.

$$F_E = \frac{kqq}{\left(2L\sin\theta\right)^2}$$

Then, $\tan\theta = \dfrac{\dfrac{kqq}{\left(2L\sin\theta\right)^2}}{mg}$

$$\frac{mg\tan\theta}{k} = \frac{q^2}{\left(2L\sin\theta\right)^2}$$

$$q^2 = \frac{\left(2L\sin\theta\right)^2\left(mg\tan\theta\right)}{k}$$

$$q = 2L\sin\theta\sqrt{\frac{mg\tan\theta}{k}}$$

The answer is (B).

5. **C** The wire, at a distance $R = 0.15$ m from the origin, has a linear charge density, $\lambda = -3.0 \times 10^{-6}$ C/m, and $\theta = 45°$. To determine the total charge on the wire, you know that charge density $\lambda = Q/l$. You can find the length of the wire by determining the length of the arc the wire creates, and then you can solve for Q.

$$\text{Length of arc} = (\text{radius})(\text{angle in radians})$$
$$\text{Radius} = 0.15 \text{ m}$$
$$\text{Angle in radians} = 2(45°) = 90° = \frac{\pi}{2} \text{ radians}$$

$$l = (0.15 \text{ m})(\frac{\pi}{2} \text{ radians}) = 0.236 \text{ m}$$

$$Q = \lambda l = \left(-3.0 \times 10^{-6} \text{ C/m}\right)(0.236 \text{ m})$$

$$Q = -7.1 \times 10^{-7} \text{ C}$$

The answer is $Q = -7.1 \times 10^{-7}$ C, answer (C).

6. **E** The electric field can be found from the formula $E = kq/r^2$. If at point 2, which is a distance r from the point charge, the electric field is E_0, then the electric field at point 5, which is a distance of $4r$ from the charge, will be $E_0/16$, answer (E).

7. **A** The formula for the force from an electric field is $\mathbf{F}_E = q\mathbf{E}$. The positive charge feels a force to the right and the negative charge feels an equal but opposite force to the left.

The net force is zero, and the answer is (A).

8. **C** This is similar to a parabolic free fall mechanics question. There is no net force in the x direction (the direction of the initial velocity of the particle), so the x-component of velocity stays constant. The electric field exerts a force on the charge in the y direction, since $\mathbf{F}_E = q\mathbf{E}$. This produces an acceleration in the y direction, therefore the motion of the particle will be parabolic, answer (C).

9. **A** Since the charge is evenly distributed for length L on each side of the x-axis, the y-components with cancel each other out for a net y-component of zero, answer (A).

10. **D** This problem describes a uniformly charged rod on the y-axis, and asks for the x-component of the field at point P. The formula for the electric field for a uniformly charged rod is: $E = \dfrac{q}{4\pi\varepsilon_o a\sqrt{a+l}}$, where:

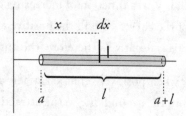

In this question, $a = x$, and $a + l = c$. To get c in terms of x and L, use the Pythagorean Theorem for triangle I:

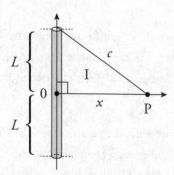

$$x^2 + L^2 = c^2$$

$$c = \sqrt{x^2 + L^2}$$

$$\text{So, } E = \frac{q}{4\pi\varepsilon_o x\sqrt{x^2 + L^2}}$$

Therefore, the answer is (D).

11. **D** Since field lines emanate away from a positive charge and toward a negative charge, the fields produced by each charge will be in opposite directions in the region to the left of point A and in the region to the right of point C. To make the magnitudes equal, point B must be located closer to the smaller charge (since $E = kq/r^2$). This can only happen to the right of point C. The correct answer is (D).

12. **D** To find the electric field, use the equation $E = -dV/dr$. Since $V(x) = ax^3$, E must equal $-3ax^2$. Plug in $x = b$, so the answer is (D).

13. **C** When the two conducting spheres are brought into contact with each other, they will balance the charge between the two spheres. So $-9q + 5q = -4q$, which will be distributed evenly over the two spheres, so each will have a charge of $-2q$. The correct answer is (C).

14. **E** While grounded, positive charge flows from the ground into the electroscope due to the nearness of the negatively charged rod. If the ground is removed from the electroscope, then it will have a positive charge to it. The answer is (E).

15. **D** The graph is showing potential, V, as a function of increasing distance r. $E = -dV/dr$, which is equal to the negative of the slope of the curve. To find the positive electric field, find the point where V is decreasing (i.e., when the slope is negative). Therefore, the answer is (D).

16. **B** Since sphere II is grounded, when positively charged sphere I is brought close to sphere II, it will have negative charge flow from the ground due to the attraction of the positive charge on sphere I, answer (B).

17. **B** Since the capacitor is isolated, it is not connected to a battery. So, when a dielectric is inserted, the charge Q_0 will remain constant, the capacitance must increase since the dielectric constant of the oil will be larger than air or a vacuum (dielectric constant is equal to 1). Since $Q = CV$, the voltage will have to drop. The answer is (B).

18. **D** To answer to this question, start with the equation for the capacitance of a parallel-plate capacitor: $C = \kappa \varepsilon_0 A/d$. With no dielectrics between the plates, κ, the dielectric constant, is equal to 1. If A is doubled and d is multiplied by four, C will now be $2/4 = \frac{1}{2}$ times its previous value, answer (D).

19. **B** Using the right-hand rule, the magnetic force exerted on the negative charge points downward. In order for the particle to continue to move in a straight line, the force due to the electric field $\mathbf{E}$ must be in the opposite direction as $\mathbf{F}_B$, so $\mathbf{F}_E$ must point up. Since the force due to an electric field on a negative charge will point in the opposite direction as the electric field (from the equation $\mathbf{F}_E = q\mathbf{E}$), the electric field $\mathbf{E}$ must point down toward the bottom of the page. The answer is (B).

20. **C** In order to answer this question, you need to use both versions of the right-hand rule—one to find the direction of the magnetic field produced by the current at the location of the moving charge, and one to find the direction of the force this magnetic field exerts on the charge. First, orient your thumb along the wire in the counterclockwise direction and curl your fingers. This shows that the magnetic field points into the page outside the wire loop and points out of the page inside the wire loop. The electron, therefore, experiences a magnetic field into the page. For the second right-hand rule, point your thumb down toward the bottom of the page in the direction of the velocity, and point your fingers into the plane of the page in the direction of the magnetic field. The palm of the hand (for a positive particle) points right, so for a negative particle, the back of the hand points to the left. This tells us that the force from the magnetic field, F_B, points to the left. Therefore, the correct answer is (C).

21. **A** Each plate produces an electric field E. The electric field produced by the positively charged plate points to the right at point 1, since the electric field points away from a positive charge. The electric field produced by the negatively charged plate points to the right at point 1, since the electric field points toward a negative charge. Since each plate has an electric field E, at point 1, the magnitude of the total electric field is $2E$ pointing towards the right, answer (A).

22. **C** Each plate produces an electric field E. The electric field produced by the positively charged plate points to the right at point 2, since the electric field points away from a positive charge. The electric field produced by the negatively charged plate points to the left at point 2, since the electric field points toward a negative charge. Since each plate has an electric field E, at point 2, the magnitude of the total electric field is 0, answer (C).

23. **E** Divide capacitor C_2 in half and calculate the capacitance for the top half of capacitor C_2 and the bottom half of capacitor C_2.

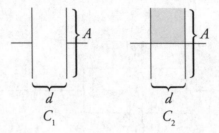

$$C_2 = (\kappa + 1)C_1/2$$

The correct answer is (E).

24. **B** To find the capacitance of the circuit, you must combine in four steps, as shown in the diagram below. First, find the equivalent capacitance of the two 6 μF capacitors in the top of the parallel branch of the circuit. Second, find the equivalent capacitance of the two 6 μF capacitors in the bottom of the parallel branch. Once you have completed the first two steps, you can find the equivalent capacitance for the parallel part of the circuit from A to B, and then, finally, do one last equivalent capacitance for the two remaining capacitors, one from A to B, and one from B to C. These steps are shown below.

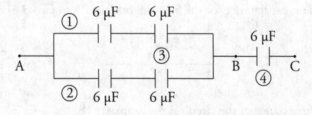

Step 1

Step 1 shows two capacitors in series on the top branch of the parallel circuit.

$$\frac{1}{C_{eq_{top}}} = \frac{1}{6} + \frac{1}{6} = \frac{2}{6} = \frac{1}{3} \quad \text{so } C_{eq_{top}} = 3 \text{ μF}$$

Step 2

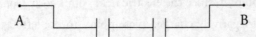

Step 2 shows two capacitors in series on the bottom branch of the parallel circuit.

$$\frac{1}{C_{eq_{bottom}}} = \frac{1}{6} + \frac{1}{6} = \frac{2}{6} = \frac{1}{3} \quad \text{so } C_{eq_{bottom}} = 3 \ \mu F$$

Step 3

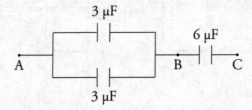

We now take the equivalent capacitance of the two 3 μF capacitors in parallel, and add capacitances.

$$C_{eq_{parallel}} = 3 \ \mu F + 3 \ \mu F = 6 \ \mu F$$

Step 4

Again, $\frac{1}{C_{eq_{total}}} = \frac{1}{6 \ \mu F} + \frac{1}{6} = \frac{1}{3 \ \mu F}$.

The total equivalent capacitance for the 5 capacitors from A to C is 3 μF.

The equivalent capacitance of the circuit is 3 μF, answer (B).

25. **C**
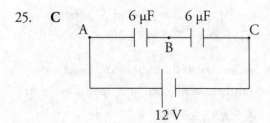

Since both capacitors have the same capacitance of 6 µF, the potential difference across each capacitor will be half of the total voltage of the battery, or 6 V. The answer is (C).

26. **B** Using the right-hand rule, the thumb points to the left in the direction of the velocity of the positive charge, and the fingers of the right hand point into the page in the direction of the magnetic field, **B**. The palm of the hand will point in the direction of the force from the magnetic field, which, at this instant, is toward the bottom of the page. As the particle continues in the magnetic field, it will experience uniform circular motion, but at the instant described, the force from the magnetic field $\mathbf{F}_B$ will be towards the bottom of the page, answer (B).

27. **E** Using the right-hand rule, point your thumb in the direction of the velocity vector and orient the palm of your hand in the direction of the force from the magnetic field, $\mathbf{F}_B$. Your hand will end up with your fingers pointing out of the page towards you, answer (E).

28. **A** Two capacitors in series have the equivalent capacitance equal to the sum of the inverses of the capacitance of each capacitor, which makes the equivalent capacitance smaller than either individual capacitor's capacitance, answer (A).

29. **E** The capacitance of a parallel-plate capacitor is given by $C = \kappa\varepsilon_0 A/d$, where κ is the dielectric constant. Inserting a dielectric like glass increases κ, and therefore increases C. Since the capacitor is isolated, it is not connected to a battery. So, when a dielectric is inserted, the charge Q will remain constant. Since $Q = CV$, we can see that the voltage or potential difference must decrease. The correct answer is (E).

30. **B** First, find the equivalent resistance for the two parallel resistors so that there is only one loop in the circuit to analyze: $1/R_{parallel} = 1/6 + 1/3 = 1/2$, so $R_{parallel} = 2\ \Omega$. Adding this to the 7 Ω resistor in series with it, we get that the equivalent resistance is 9 Ω. A battery supports current flow from its positive terminal to its negative terminal. The 12 V battery supports counterclockwise current flow, while the 3 V battery supports clockwise current flow. Since 12 V > 3 V, the current will travel counterclockwise, and the voltage available to the resistors is 12 V – 3 V = 9 V. (This can be shown using Kirchhoff's Loop Rule.) The total current is therefore 9 V / 9 Ω = 1 A. This current travels through the 7 Ω resistor, answer (B).

31. **B** The voltage drop across the 6 Ω resistor and the 3 Ω resistor must be the same because the resistors are parallel resistors. You can use $R_{parallel} = 2\ \Omega$ from the previous question. From $V = IR$, the voltage drop across those two resistors will be equal to 1 A × 2 Ω = 2 V, answer (B).

32. **C** To calculate the current across the 3 Ω resistor, from question 31, you know that the 3 Ω resistor has a voltage drop of 2 V. From $V = IR$, $I = V/R$. I across the 3 Ω resistor = 2 V/3 Ω = 2/3 A, answer (C).

33. **C** With switch S_1 closed and switch S_2 open, the circuit is one loop with the battery and capacitor C_1 as part of the loop. The current will flow in this circuit until the difference in potential between the battery and capacitor C_1 is zero. At that time, $V = 120$ V for capacitor C_1. Since $Q = C_1V$, $Q = (9 \times 10^{-6}$ F$)(120$ V$) = 1.08 \times 10^{-3}$ C = 1.08 mC, answer (C).

34. **D** With switch S_1 open and switch S_2 closed, the battery is no longer part of the circuit used to solve this question. Current will flow from C_1 to C_2 until the potential difference between the capacitors is zero. Due to the Law of Conservation of Charge, the final charge on each capacitor must add up to the original charge on capacitor $Q = 1.08 \times 10^{-3}$ C, from question 33.

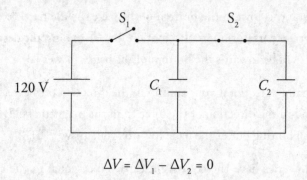

$$\Delta V = \Delta V_1 - \Delta V_2 = 0$$

$\Delta V_1 = \Delta V_2$ and $Q = Q_1 + Q_2$ where $Q = 1.08$ mC from question 33. Therefore, $Q_2 = Q - Q_1$

$$Q = \Delta V$$

$$\frac{Q_1}{C_1} = \frac{Q_2}{C_2}$$

$$\frac{Q_1}{C_1} = \frac{Q - Q_1}{C_2} = \frac{Q}{C_2} - \frac{Q_1}{C_2}$$

$$\frac{Q_1}{C_1} + \frac{Q_1}{C_2} = \frac{Q}{C_2}$$

$$\frac{Q_1(C_1 + C_2)}{C_1 C_2} = \frac{Q}{C_2}$$

$$Q_1 = \frac{QC_1}{C_1 + C_2} = \frac{\left(1.08 \times 10^{-3}\ \text{C}\right)\left(9.6 \times 10^{-6}\ \text{F}\right)}{\left(\left(9.6 \times 10^{-6}\ \text{F}\right)\right)\left(18 \times 10^{-6}\ \text{F}\right)}$$

$$Q_1 = 3.6 \times 10^{-4}\ \text{C}$$

The answer is (D), 3.6×10^{-4} C.

35. **C** With switch S_1 open and switch S_2 closed, just as in question 34, the potential difference across C_1 can be found from the formula $V = Q/C = (3.6 \times 10^{-4}$ C$)/(9 \times 10^{-6}$ F$) = 0.4 \times 10^2$ V = 40 V, answer (C).

Section II—Free Response

1. (a) (i) Consider the Gaussian sphere as shown below. Since the electric field is zero inside a conductor, the charge enclosed must be zero. Therefore, the charge on the inner surface is zero. This leaves the $-4Q$ charge on the outer surface, as shown below.

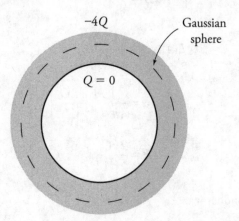

(ii) The electric field inside a conductor at equilibrium is always zero, so the electric field in region II is zero. Field lines originate on positive charges and end on negative charges. Since there is no excess charge on the inner surface of the sphere or within region I, the electric field in region I is also zero. The electric field in region III is shown below. The electric field lines will originate at the $+2Q$ charge and terminate at the outer surface of the spherical shell because the outer surface will have a charge of $-4Q$. Some of the $-4Q$ charge will be closer to the $+2Q$ charge, but that still leaves an excess of negative charge around the rest of the outer surface of the shell, so the field lines should be ending on the outer surface. All field lines need to be perpendicular to the surface of the shell.

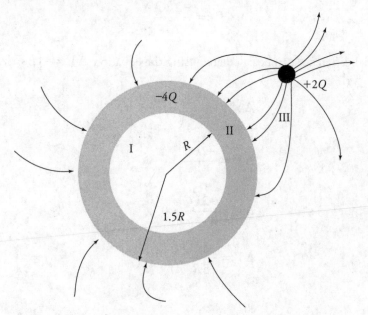

(b)   Use Gauss's law for each of the radii. The surface area of the Gaussian sphere is $4\pi r^2$, so $\oint dA = 4\pi r^2$. Also substitute $\dfrac{1}{4\pi\varepsilon_o} = k$ to simplify the expression.

(i)

$$\oint E \bullet dA = \frac{q_{in}}{\varepsilon_0}$$

$$E(4\pi r^2) = \frac{(+2Q)}{\varepsilon_0}$$

$$E = \frac{1}{4\pi r^2}\left(\frac{+2Q}{\varepsilon_0}\right)$$

$$E = \frac{2kQ}{r^2}$$

(ii) $E = 0$, within a conductor.

(iii)

$$\oint E \bullet dA = \frac{q_{in}}{\varepsilon_0}$$

$$E(4\pi r^2) = \frac{(+2Q - 4Q)}{\varepsilon_0}$$

$$E = \frac{1}{4\pi r^2}\left(\frac{-2Q}{\varepsilon_0}\right)$$

$$E = \frac{-2kQ}{r^2}$$

(c)   The potential difference can be calculated using the equation $\Delta V = -\int E \bullet dr$ as shown below.

$$\Delta V = -\int_{\infty}^{1.5R}\left(\frac{-2kQ}{r^2}\right)dr$$

$$\Delta V = \frac{-2kQ}{r}\Bigg|_{\infty}^{1.5R}$$

$$\Delta V = \frac{-2kQ}{1.5R} - 0$$

$$\Delta V = \frac{-4kQ}{3R}$$

2. (a) The magnetic flux is the dot product of the magnetic field and area vector that the magnetic field goes through. For this problem, the area goes out to a radius d because the field does not go out to the radius of the loop.

$$\Phi_B = B \cdot A$$
$$\Phi_B = B_o \cdot \pi d^2$$

(b) Use Faraday's law to determine the induced emf. We will ignore the negative sign in the final answer because we only need the magnitude of the emf.

$$\varepsilon = -\frac{d\phi_B}{dt}$$

$$\varepsilon = -\pi d^2 \frac{d}{dt}(B_o t^2)$$

$$\varepsilon = -\pi d^2 (2B_o t)$$

$$\varepsilon = -2B_o \pi d^2 t$$

$$|\varepsilon| = 2B_o \pi d^2 t$$

(c) Lenz's law states that the induced emf will produce a magnetic field to oppose the change in the external magnetic field. The external magnetic field going into the page is increasing with time, so to oppose the change in the external magnetic field, the induced current in the loop will create a magnetic field out of the page. To do this, the induced current must be counterclockwise.

(d) Power is the rate at which energy is dissipated. Power for a circuit is given by the equations $P = IV = \frac{V^2}{R} = I^2 R$. We will use $\frac{V^2}{R}$ for power since we know the voltage and the resistance.

$$P = \frac{dE}{dt}$$

$$\frac{V^2}{R} dt = dE$$

$$\int_0^{t_1} \frac{V^2}{R} dt = E$$

$$\int_0^{t_1} \frac{(2B_o \pi d^2 t)^2}{R} dt = E$$

$$\frac{4B_o^2 \pi^2 d^4 t_1^3}{3R} = E$$

3. (a) When the switch is first moved to position *A*, the 50 μF capacitor is acting like a wire because it does not oppose the current when it is uncharged. The current through the 20 Ω resistor is the same as the entire left side of the circuit because it is a series circuit. Use Ohm's law to solve for the current.

$$V = IR$$
$$30V = I(10\,\Omega + 20\,\Omega)$$
$$I = 1.0\text{ A}$$

(b) The initial voltage across the 10 Ω resistor can be calculated using Ohm's law.

$$V = IR$$
$$V = (1\text{ A})\,(10\,\Omega)$$
$$V = 10\text{ V}$$

For an RC circuit, the voltage across the resistor will decrease exponentially because the current decreases exponentially as the capacitor is charging. The sketch of the voltage is shown below.

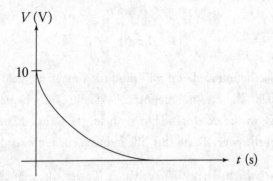

(c) When the switch is in position *A*, the capacitor will continue being charged until the voltage across the 50 μF capacitor becomes 30 V, which will happen after a long time. When the switch is moved to position *B*, the 20 μF capacitor acts like a wire because it is uncharged. Use Ohm's law to solve for the current through the 15 Ω resistor immediately after the switch is moved to position *B*.

$$V = IR$$
$$30\text{ V} = I(15\,\Omega)$$
$$I = 2\text{ A}$$

(d) The total charge stored on the 50 μF capacitor when the switch is at position *A* can be calculated using the equation

$$C = \frac{Q}{V}$$
$$50\ \mu F = \frac{Q}{30\text{ V}}$$
$$Q = 1,500\ \mu C$$

When the switch is moved to position B, this charge will distribute between the two capacitors until the system reaches equilibrium. The voltage across each capacitor will be the same when the system is in equilibrium. Use this information and the fact that $\Delta V = \dfrac{Q}{C}$ for a capacitor. Let q be the charge on the 20 μF capacitor, which means $1{,}500 - q$ is the charge on the 50 μF capacitor.

$$\frac{1{,}500 - q}{50} = \frac{q}{20}$$

$$30{,}000 - 20q = 50q$$

$$30{,}000 = 70q$$

$$q = 429\ \mu C$$

Completely darken bubbles with a No. 2 pencil. If you make a mistake, be sure to erase mark completely. Erase all stray marks.

1.

YOUR NAME: _____

SIGNATURE: _____ DATE: _____

HOME ADDRESS: _____

_____

PHONE NO.: _____

5. YOUR NAME

First 4 letters of last name				FIRST INIT	MID INIT

A A A A A A A
B B B B B B B
C C C C C C C
D D D D D D D
E E E E E E E
F F F F F F F
G G G G G G G
H H H H H H H
I I I I I I I
J J J J J J J
K K K K K K K
L L L L L L L
M M M M M M M
N N N N N N N
O O O O O O O
P P P P P P P
Q Q Q Q Q Q Q
R R R R R R R
S S S S S S S
T T T T T T T
U U U U U U U
V V V V V V V
W W W W W W W
X X X X X X X
Y Y Y Y Y Y Y
Z Z Z Z Z Z Z

IMPORTANT: Please fill in these boxes exactly as shown on the back cover of your test book.

2. TEST FORM

_____

3. TEST CODE **4. REGISTRATION NUMBER**

0	A	J	0	0	0	0	0	0	0	0
1	B	K	1	1	1	1	1	1	1	1
2	C	L	2	2	2	2	2	2	2	2
3	D	M	3	3	3	3	3	3	3	3
4	E	N	4	4	4	4	4	4	4	4
5	F	O	5	5	5	5	5	5	5	5
6	G	P	6	6	6	6	6	6	6	6
7	H	Q	7	7	7	7	7	7	7	7
8	I	R	8	8	8	8	8	8	8	8
9			9	9	9	9	9	9	9	9

6. DATE OF BIRTH

Month	Day		Year	
◯ JAN				
◯ FEB	0	0	0	0
◯ MAR	1	1	1	1
◯ APR	2	2	2	2
◯ MAY	3	3	3	3
◯ JUN		4	4	4
◯ JUL		5	5	5
◯ AUG		6	6	6
◯ SEP		7	7	7
◯ OCT		8	8	8
◯ NOV		9	9	9
◯ DEC				

7. GENDER
◯ MALE
◯ FEMALE

Mechanics

1. A B C D E
2. A B C D E
3. A B C D E
4. A B C D E
5. A B C D E
6. A B C D E
7. A B C D E
8. A B C D E
9. A B C D E
10. A B C D E
11. A B C D E
12. A B C D E
13. A B C D E
14. A B C D E
15. A B C D E
16. A B C D E
17. A B C D E
18. A B C D E
19. A B C D E
20. A B C D E
21. A B C D E
22. A B C D E
23. A B C D E
24. A B C D E
25. A B C D E
26. A B C D E
27. A B C D E
28. A B C D E
29. A B C D E
30. A B C D E
31. A B C D E
32. A B C D E
33. A B C D E
34. A B C D E
35. A B C D E

Electricity & Magnetism

1. A B C D E
2. A B C D E
3. A B C D E
4. A B C D E
5. A B C D E
6. A B C D E
7. A B C D E
8. A B C D E
9. A B C D E
10. A B C D E
11. A B C D E
12. A B C D E
13. A B C D E
14. A B C D E
15. A B C D E
16. A B C D E
17. A B C D E
18. A B C D E
19. A B C D E
20. A B C D E
21. A B C D E
22. A B C D E
23. A B C D E
24. A B C D E
25. A B C D E
26. A B C D E
27. A B C D E
28. A B C D E
29. A B C D E
30. A B C D E
31. A B C D E
32. A B C D E
33. A B C D E
34. A B C D E
35. A B C D E

The **Princeton Review®**

Completely darken bubbles with a No. 2 pencil. If you make a mistake, be sure to erase mark completely. Erase all stray marks.

1.

YOUR NAME: _____

SIGNATURE: _____ DATE: _____

HOME ADDRESS: _____

_____

PHONE NO.: _____

IMPORTANT: Please fill in these boxes exactly as shown on the back cover of your test book.

2. TEST FORM

_____

6. DATE OF BIRTH

Month	Day		Year	
○ JAN				
○ FEB	⓪	⓪	⓪	⓪
○ MAR	①	①	①	①
○ APR	②	②	②	②
○ MAY	③	③	③	③
○ JUN		④	④	④
○ JUL		⑤	⑤	⑤
○ AUG		⑥	⑥	⑥
○ SEP		⑦	⑦	⑦
○ OCT		⑧	⑧	⑧
○ NOV		⑨	⑨	⑨
○ DEC				

3. TEST CODE   **4. REGISTRATION NUMBER**

⓪ Ⓐ Ⓙ ⓪ ⓪ ⓪ ⓪ ⓪ ⓪ ⓪ ⓪
① Ⓑ Ⓚ ① ① ① ① ① ① ① ①
② Ⓒ Ⓛ ② ② ② ② ② ② ② ②
③ Ⓓ Ⓜ ③ ③ ③ ③ ③ ③ ③ ③
④ Ⓔ Ⓝ ④ ④ ④ ④ ④ ④ ④ ④
⑤ Ⓕ Ⓞ ⑤ ⑤ ⑤ ⑤ ⑤ ⑤ ⑤ ⑤
⑥ Ⓖ Ⓟ ⑥ ⑥ ⑥ ⑥ ⑥ ⑥ ⑥ ⑥
⑦ Ⓗ Ⓠ ⑦ ⑦ ⑦ ⑦ ⑦ ⑦ ⑦ ⑦
⑧ Ⓘ Ⓡ ⑧ ⑧ ⑧ ⑧ ⑧ ⑧ ⑧ ⑧
⑨ ⑨ ⑨ ⑨ ⑨ ⑨ ⑨ ⑨ ⑨

7. GENDER
○ MALE
○ FEMALE

The **Princeton Review®**

5. YOUR NAME

First 4 letters of last name					FIRST INIT	MID INIT
Ⓐ	Ⓐ	Ⓐ	Ⓐ	Ⓐ	Ⓐ	Ⓐ
Ⓑ	Ⓑ	Ⓑ	Ⓑ	Ⓑ	Ⓑ	Ⓑ
Ⓒ	Ⓒ	Ⓒ	Ⓒ	Ⓒ	Ⓒ	Ⓒ
Ⓓ	Ⓓ	Ⓓ	Ⓓ	Ⓓ	Ⓓ	Ⓓ
Ⓔ	Ⓔ	Ⓔ	Ⓔ	Ⓔ	Ⓔ	Ⓔ
Ⓕ	Ⓕ	Ⓕ	Ⓕ	Ⓕ	Ⓕ	Ⓕ
Ⓖ	Ⓖ	Ⓖ	Ⓖ	Ⓖ	Ⓖ	Ⓖ
Ⓗ	Ⓗ	Ⓗ	Ⓗ	Ⓗ	Ⓗ	Ⓗ
Ⓘ	Ⓘ	Ⓘ	Ⓘ	Ⓘ	Ⓘ	Ⓘ
Ⓙ	Ⓙ	Ⓙ	Ⓙ	Ⓙ	Ⓙ	Ⓙ
Ⓚ	Ⓚ	Ⓚ	Ⓚ	Ⓚ	Ⓚ	Ⓚ
Ⓛ	Ⓛ	Ⓛ	Ⓛ	Ⓛ	Ⓛ	Ⓛ
Ⓜ	Ⓜ	Ⓜ	Ⓜ	Ⓜ	Ⓜ	Ⓜ
Ⓝ	Ⓝ	Ⓝ	Ⓝ	Ⓝ	Ⓝ	Ⓝ
Ⓞ	Ⓞ	Ⓞ	Ⓞ	Ⓞ	Ⓞ	Ⓞ
Ⓟ	Ⓟ	Ⓟ	Ⓟ	Ⓟ	Ⓟ	Ⓟ
Ⓠ	Ⓠ	Ⓠ	Ⓠ	Ⓠ	Ⓠ	Ⓠ
Ⓡ	Ⓡ	Ⓡ	Ⓡ	Ⓡ	Ⓡ	Ⓡ
Ⓢ	Ⓢ	Ⓢ	Ⓢ	Ⓢ	Ⓢ	Ⓢ
Ⓣ	Ⓣ	Ⓣ	Ⓣ	Ⓣ	Ⓣ	Ⓣ
Ⓤ	Ⓤ	Ⓤ	Ⓤ	Ⓤ	Ⓤ	Ⓤ
Ⓥ	Ⓥ	Ⓥ	Ⓥ	Ⓥ	Ⓥ	Ⓥ
Ⓦ	Ⓦ	Ⓦ	Ⓦ	Ⓦ	Ⓦ	Ⓦ
Ⓧ	Ⓧ	Ⓧ	Ⓧ	Ⓧ	Ⓧ	Ⓧ
Ⓨ	Ⓨ	Ⓨ	Ⓨ	Ⓨ	Ⓨ	Ⓨ
Ⓩ	Ⓩ	Ⓩ	Ⓩ	Ⓩ	Ⓩ	Ⓩ

Mechanics

1. Ⓐ Ⓑ Ⓒ Ⓓ Ⓔ
2. Ⓐ Ⓑ Ⓒ Ⓓ Ⓔ
3. Ⓐ Ⓑ Ⓒ Ⓓ Ⓔ
4. Ⓐ Ⓑ Ⓒ Ⓓ Ⓔ
5. Ⓐ Ⓑ Ⓒ Ⓓ Ⓔ
6. Ⓐ Ⓑ Ⓒ Ⓓ Ⓔ
7. Ⓐ Ⓑ Ⓒ Ⓓ Ⓔ
8. Ⓐ Ⓑ Ⓒ Ⓓ Ⓔ
9. Ⓐ Ⓑ Ⓒ Ⓓ Ⓔ
10. Ⓐ Ⓑ Ⓒ Ⓓ Ⓔ
11. Ⓐ Ⓑ Ⓒ Ⓓ Ⓔ
12. Ⓐ Ⓑ Ⓒ Ⓓ Ⓔ

13. Ⓐ Ⓑ Ⓒ Ⓓ Ⓔ
14. Ⓐ Ⓑ Ⓒ Ⓓ Ⓔ
15. Ⓐ Ⓑ Ⓒ Ⓓ Ⓔ
16. Ⓐ Ⓑ Ⓒ Ⓓ Ⓔ
17. Ⓐ Ⓑ Ⓒ Ⓓ Ⓔ
18. Ⓐ Ⓑ Ⓒ Ⓓ Ⓔ
19. Ⓐ Ⓑ Ⓒ Ⓓ Ⓔ
20. Ⓐ Ⓑ Ⓒ Ⓓ Ⓔ
21. Ⓐ Ⓑ Ⓒ Ⓓ Ⓔ
22. Ⓐ Ⓑ Ⓒ Ⓓ Ⓔ
23. Ⓐ Ⓑ Ⓒ Ⓓ Ⓔ
24. Ⓐ Ⓑ Ⓒ Ⓓ Ⓔ

25. Ⓐ Ⓑ Ⓒ Ⓓ Ⓔ
26. Ⓐ Ⓑ Ⓒ Ⓓ Ⓔ
27. Ⓐ Ⓑ Ⓒ Ⓓ Ⓔ
28. Ⓐ Ⓑ Ⓒ Ⓓ Ⓔ
29. Ⓐ Ⓑ Ⓒ Ⓓ Ⓔ
30. Ⓐ Ⓑ Ⓒ Ⓓ Ⓔ
31. Ⓐ Ⓑ Ⓒ Ⓓ Ⓔ
32. Ⓐ Ⓑ Ⓒ Ⓓ Ⓔ
33. Ⓐ Ⓑ Ⓒ Ⓓ Ⓔ
34. Ⓐ Ⓑ Ⓒ Ⓓ Ⓔ
35. Ⓐ Ⓑ Ⓒ Ⓓ Ⓔ

Electricity & Magnetism

1. Ⓐ Ⓑ Ⓒ Ⓓ Ⓔ
2. Ⓐ Ⓑ Ⓒ Ⓓ Ⓔ
3. Ⓐ Ⓑ Ⓒ Ⓓ Ⓔ
4. Ⓐ Ⓑ Ⓒ Ⓓ Ⓔ
5. Ⓐ Ⓑ Ⓒ Ⓓ Ⓔ
6. Ⓐ Ⓑ Ⓒ Ⓓ Ⓔ
7. Ⓐ Ⓑ Ⓒ Ⓓ Ⓔ
8. Ⓐ Ⓑ Ⓒ Ⓓ Ⓔ
9. Ⓐ Ⓑ Ⓒ Ⓓ Ⓔ
10. Ⓐ Ⓑ Ⓒ Ⓓ Ⓔ
11. Ⓐ Ⓑ Ⓒ Ⓓ Ⓔ
12. Ⓐ Ⓑ Ⓒ Ⓓ Ⓔ

13. Ⓐ Ⓑ Ⓒ Ⓓ Ⓔ
14. Ⓐ Ⓑ Ⓒ Ⓓ Ⓔ
15. Ⓐ Ⓑ Ⓒ Ⓓ Ⓔ
16. Ⓐ Ⓑ Ⓒ Ⓓ Ⓔ
17. Ⓐ Ⓑ Ⓒ Ⓓ Ⓔ
18. Ⓐ Ⓑ Ⓒ Ⓓ Ⓔ
19. Ⓐ Ⓑ Ⓒ Ⓓ Ⓔ
20. Ⓐ Ⓑ Ⓒ Ⓓ Ⓔ
21. Ⓐ Ⓑ Ⓒ Ⓓ Ⓔ
22. Ⓐ Ⓑ Ⓒ Ⓓ Ⓔ
23. Ⓐ Ⓑ Ⓒ Ⓓ Ⓔ
24. Ⓐ Ⓑ Ⓒ Ⓓ Ⓔ

25. Ⓐ Ⓑ Ⓒ Ⓓ Ⓔ
26. Ⓐ Ⓑ Ⓒ Ⓓ Ⓔ
27. Ⓐ Ⓑ Ⓒ Ⓓ Ⓔ
28. Ⓐ Ⓑ Ⓒ Ⓓ Ⓔ
29. Ⓐ Ⓑ Ⓒ Ⓓ Ⓔ
30. Ⓐ Ⓑ Ⓒ Ⓓ Ⓔ
31. Ⓐ Ⓑ Ⓒ Ⓓ Ⓔ
32. Ⓐ Ⓑ Ⓒ Ⓓ Ⓔ
33. Ⓐ Ⓑ Ⓒ Ⓓ Ⓔ
34. Ⓐ Ⓑ Ⓒ Ⓓ Ⓔ
35. Ⓐ Ⓑ Ⓒ Ⓓ Ⓔ

NOTES